EFFICIENT ENERGY MANAGEMENT

EFFICIENT ENERGY MANAGEMENT

Methods for Improved Commercial and Industrial Productivity

HAROLD P. MAHON

University of Massachusetts, Boston

MIKLOS G. KISS

Electrowatt Engineering Services Ltd., Zurich

HANS J. LEIMER

Sulzer Energy Consulting, Inc., Winterthur

PRENTICE-HALL, INC., Englewood Cliffs, New Jersey 07632

Library of Congress Cataloging in Publication Data

Mahon, Harold P.
Efficient energy management.

Includes bibliographical references and index.
1. Energy conservation. 2. Energy policy. I. Kiss, Miklos G. II. Leimer, Hans Jörg. III. Title.
TJ163.3.M33 1983 658.2′6 82-18037
ISBN 0-13-247023-3

TO OUR PARENTS FOR OUR HERITAGE
AND TO OUR FAMILIES FOR THEIR SUPPORT

Editorial/production supervision by Ian List
and Zita de Schauensee
Interior design by Ian List
Cover design by Miriam Recio
Manufacturing buyer: Anthony Caruso

© 1983 by Prentice-Hall, Inc., Englewood Cliffs, N.J. 07632

All rights reserved. No part of this book
may be reproduced in any form or
by any means without permission in writing
from the publisher.

Printed in the United States of America

10 9 8 7 6 5 4 3 2 1

ISBN 0-13-247023-3

Prentice-Hall International, Inc., *London*
Prentice-Hall of Australia Pty. Limited, *Sydney*
Editora Prentice-Hall do Brasil, Ltda., *Rio de Janeiro*
Prentice-Hall Canada Inc. *Toronto*
Prentice-Hall of India Private Limited, *New Delhi*
Prentice-Hall of Japan, Inc., *Tokyo*
Prentice-Hall of Southeast Asia Pte. Ltd., *Singapore*
Whitehall Books Limited, *Wellington, New Zealand*

Contents

Preface

This book is for architects, designers, engineers, building operators, business owners and managers, and students. It is for people who are concerned with improving energy productivity and reducing costs.

Energy costs have increased sharply since the early 1970s due to dwindling resources and increased capital for production and utilization. This inexorable course will continue into the foreseeable future, at rates greater than average inflation. The profitability, if not the actual survival, of organizations supplying vital services and products in our culture, will be threatened by escalating operating costs.

History has repeatedly shown the danger to living standards of once powerful cultures from overdependence on dwindling supplies. Fortunately, today there is increasing awareness of the need for productivity improvements to save energy and conserve scarce resources. This will require concerted and continuous improvement in energy utilization.

1. We believe that organizations which are willing to commit themselves can deliberately improve their productivity.
2. We believe that within each organization there is a natural success system that wants to be activated.
3. We know that with full knowledge of how energy is used there is vast potential for improved energy utilization.
4. We know that the theory and practical measures explained in this book can help achieve this potential as they have helped others to achieve it.
5. We also know that it takes determination, persistence, and enthusiasm to maintain and continue making improvements in productivity.

6. We believe that there are many people and suppliers ready and able to provide help in achieving your productivity goals.
7. We trust that this book will provide support and stimulation to help you surpass them.

The practical reason for improving productivity is to save money. The potential impetus, therefore, arises from escalation in energy prices. European energy costs have traditionally been higher than those in North America due to scarcer resources and greater dependence on imports. Investments for the avoidance of energy costs have consequently been more rewarding. North America, on the other hand, has been fertile ground for the application of new design technologies. Some new buildings and industrial processes are among the most efficient found anywhere. This book represents experience from European and American economies that should provide useful perspective for a future of increased energy costs.

This book is the product of the authors' European and North American experience. Harold Mahon is a physicist at the University of Massachusetts in Boston; Miklos Kiss is an engineer with Electrowatt Engineering Services Ltd. in Zurich; and Hans Leimer is an engineer with Sulzer Energy Consulting, Inc. in Winterthur, Switzerland. This book also represents some of the insight which our colleagues have attempted to impart to us. These coworkers represent a broad spectrum of companies, government, and nonprofit organizations. Perhaps their most outstanding common thread has been their commitment to improving energy productivity. One of us, H.M., would especially like to thank Tom McNeil of UMB, Bill Jones of MIT, John Foss of HTEC, and the late Thola Theilhaber of Raytheon for their assistance and inspiration.

We thank the editorial staff at Prentice-Hall for their assistance. We express special appreciation to Zita de Schauensee for her attention to detail and editorial support.

Auburndale, Massachusetts Harold P. Mahon
Zurich, Switzerland Miklos G. Kiss
Winterthur, Switzerland Hans J. Leimer

1

Introduction

1.1 ADAPTING TO MORE COSTLY ENERGY

Within our lifetime conventional sources of energy will become so expensive that we will have to learn entirely new methods for obtaining and using energy. Nothing we can do will return us to the former times of plentiful, cheap fuels. There may be some confusing and misleading signals with the opening of new fields in Alaska, Mexico, and China, or because of temporary decreases in prices. Oil may appear to be plentiful and the public may not believe in the forthcoming shortage. Nevertheless, the oil and gas situation will become progressively tighter, with steep price increases for electricity as well. The price and availability of energy in the future will forever be fundamentally different from what it has been in the past.

The potential for saving energy with cost-effective measures is more than sufficient to enable many nations such as the United States to become energy independent. Cost-effective measures are those which return the investment with interest within a few years, from savings from avoided fuel costs. Energy-saving measures usually last longer than measures to increase fuel supplies and have a less adverse effect on the environment. Furthermore, and contrary to much popular notion, the potential savings are greatest in the areas of efficiency improvements rather than through belt-tightening, where the practical potential is small.

It is what people *believe,* rather than what may be true, that determines our approach to the energy problem. **The energy "crisis" is referred to by the media as a shortage of energy rather than a surplus of inefficiency.** The preoccupation with producing more oil is like trading a nickel for a dime because the nickel looks bigger. Although some planners understand the potential for cutting imports with efficiency improvements, many measures which would have promoted important savings have been rejected in favor of emphasis on production. The public and many legislators need to be convinced of the economic benefits of using more expensive fuels more efficiently.

Efficiency Improvements Cut Costs

Virtually all of our present commercial and industrial facilities were designed to run on low-cost and abundant energy that we could afford to use inefficiently. We will have to make the transition to methods that make more effective use of energy. Saving energy now simply makes good economic sense. There are other benefits. The threat to our health and quality of life; the difficulty of producing sufficient energy supplies; the depletion of local, economically producible fuels; the growing dependence on insecure sources of imported oil; shortages of investment capital—all these concerns diminish as the efficiency of energy use is improved.

In this book we will provide information which can help improve efficiency and cut operating costs for public facilities and businesses. We will also provide hard facts on the cost of saving energy which will be of interest both to business people and to those interested in future energy policy. We will discuss some of the methods for saving which enable the industrial countries of Europe to enjoy a high standard of living while using much less energy per person than is used in the United States (although Europeans, too, use much more energy than is necessary to satisfy their requirements). Some new buildings have been designed in North America which are among the most efficient anywhere. The spread of this information will truly represent a cross fertilization between economical design practices in Europe and some of the innovative ideas now brewing in the United States.

Using examples from North America and Europe, we will show how industrial engineers and architects, store owners, public officials, and others have improved the efficiency of their energy use in commercial buildings and industry. Considerable saving can be achieved with no-cost (operational) improvements in manufacturing processes and in the way equipment is used. Additional reductions with capital improvements will be advantageous in achieving a cost-effective balance with rising energy costs. The examples tabulated show both the profitability and the productivity of many capital investments.

1.2 ENERGY MANAGEMENT

Saving energy means using energy more efficiently to produce the goods and services we need. The authors do not believe that much saving can be achieved by asking people to freeze in the dark. The potential for saving is much greater with the large number of cost-effective measures that can improve productivity while assuring a pleasant and profitable working environment. Besides avoiding unnecessary waste of useful energy,

energy costs can be reduced by identifying new methods for performing a specific task with a smaller energy budget. Strategies for saving energy substitute improved management skills, capital, and technology for energy. This is a change from the earlier trend to substitute increasingly energy-intensive processes for capital.

Good energy management is founded upon 3 points:

1. Understanding of how energy is used
2. Knowledge of how to make improvements
3. Specific goals for achievement

A comprehensive efficiency improvement program (EIP) would consider occupational and environmental safety and assurance of a future energy supply in addition to improved economics. EIP organization is explained in Chapter 4, methodology is discussed in Chapter 5, examples are given in Chapters 6 and 7, and detailed calculations appear in Chapter 8.

Results

The following examples illustrate that better energy management can get results.

1. A report issued by the United States Department of Energy (DOE) on the effectiveness of the Federal Energy Management Program shows that the U.S. government's energy use is being reduced by approximately 2 percent per year. (1)* The government itself is the largest single user in the country (2.2 percent of the total). Its improved management has saved $400 million and energy use is expected to continue to decrease.
2. An electronics firm in Newton, Massachusetts, with a 160,000-square-foot (15,500 square meters) plant and 550 employees, has cut its electricity use by 54 percent since 1972 and its fuel oil use by 46.5 percent since 1973. The Honeywell sister facilities in Waltham, Massachusetts, covering 320,000 square feet (31,000 square meters), decreased electricity use by 60 percent and fuel oil consumption by 46 percent over a similar period. Savings in these plants were achieved not with the use of expensive computer control systems, but rather with improved energy accounting and such basic techniques as removing unnecessary lighting and shutting off lights, air conditioning, and fans when not in use. (2)
3. Business at Raytheon has grown 75 percent since 1973 to $2.8 billion annually, but its use of electricity and fuels has been reduced by about 25 percent. (3)
4. The telephone operations of American Telephone and Telegraph have grown by 44 percent during the five years following the oil embargo, but the company's energy use has dropped by 10 percent. Articles in local company newspapers give tips and keep employees informed on plans for energy-saving retrofits to their existing buildings and the dollar savings achieved. (4)

Obviously, energy saving does not mean putting business on hold.

The oil embargo and the sudden increases in the OPEC prices for oil at the end of 1973 and in 1974, stimulated intense public interest in saving energy. In the United States' industrial sector, typical reductions of 10 to 15 percent were achieved simply by

*References appear at the end of each chapter.

improved scheduling of product flows, reduction of heat losses in faulty equipment, and minor process improvements. (5) During this period, average energy use was reduced by 5 percent for non-communist, industrialized countries as a whole. (6) By 1975, Canadian energy requirements were 11.5 percent below their 1968 to 1973 trend. (7) To cite some European examples, the Ciba-Geigy Chemical Plant in Basel, Switzerland, reduced its energy use in 1974 by 10 percent over the pervious year. (8) The energy use in 31 buildings operated by the Swiss Federal Administration was reduced during the 1973–74 winter by 20 percent compared to the previous winter. Sweden had reduced the energy required per dollar of Gross National Product (GNP) in 1975 by 7.4 percent compared to 1973 requirements, and Denmark and Norway had reduced theirs by 8.7 percent. (7)

Continuing Savings Requires Leadership

Permanent and continuing savings depend to an important extent on the motivation of those who use energy. It is difficult to sustain a high level of concern for conservation over a protracted period of time if no actual shortage exists. If business management and the Federal administration want to reduce energy costs and foreign imports, it is their job to provide leadership and to stimulate action with a continuing program which repeatedly stresses the important reasons for saving energy. In preparing for a future with a more limited availability of energy, it is imperative to understand that this situation is authentic and long-term in nature; otherwise the urgency for adopting new policies will be under-estimated and our actions will be inappropriate. Saving is not a one-time activity. The rise of energy costs and the evolution of new knowledge will promote a continual re-evaluation of methods to improve efficiency.

1.3 POTENTIAL FOR IMPROVED EFFICIENCY

The demand for energy can be satisfied either by efficiency improvements so that existing supplies can satisfy future needs, or by finding new sources to meet present but avoidable waste. Although some new supplies will be needed in the future, efficiency improvements would relieve the pressure to develop hard-to-get and expensive sources rapidly and give more time for research to help lower their costs. Similar to repairing leaks in the city water system rather than getting a bigger pipe to keep it filled, efficiency improvements stretch our resources and they help to hold prices down. As Benjamin Franklin said, "A penny saved is a penny earned."

Many studies have shown that there is an enormous potential for saving. The United States Department of Energy examined methods for improving efficiency in the residential, commercial, transportation, and industrial sectors for Project Independence. (9) Cost-effective methods were identified which could save 15 million barrels of oil per day of equivalent energy (BPDE) by 1985. This potential, which was 230 percent greater than the level of imports of the time of the study, represented the largest single factor which could have resulted in achieving energy independence, had conservation options been promoted.

Had these measures been already in place in 1973, fuel consumption in the United States would have been less than 60 percent of its actual level. The United States could have been completely energy independent and it would have been more difficult for OPEC to have increased the price of oil by a factor of 4 or to have used it as a political weapon. By the end of the century, the potential savings could amount to 33.4 million BPDE. Other investigations have confirmed the large potential for savings. (See references 10–31.) Our energy needs in the 1990s could be 20 to 40 percent below what was previously expected. (26) Furthermore, the capital investment to achieve this improvement in efficiency would be less than the money required to increase energy production by a similar amount. (9)

Products can and will be made with less energy. Future expansion in the heavy industry sector will favor integrated energy facilities with heat cascading and cogeneration. The potential for savings in the steel, aluminum, paper, petroleum refining, and cement industries is given in a report out of MIT and the Thermo Electron Corporation. (27) These five energy-intensive industries could have reduced their fuel needs by an average of 32 percent by using cost-effective technology available at the time of the study. Subsequent price increases in energy have raised the level of improvement which would be cost-effective today if they could only obtain sufficient captial. (The ultimate, theoretical energy required to produce these products is an average of 700 percent lower than presently used in the United States.)

The comparison shown in Figure 1.1 indicates that West Germany is already using

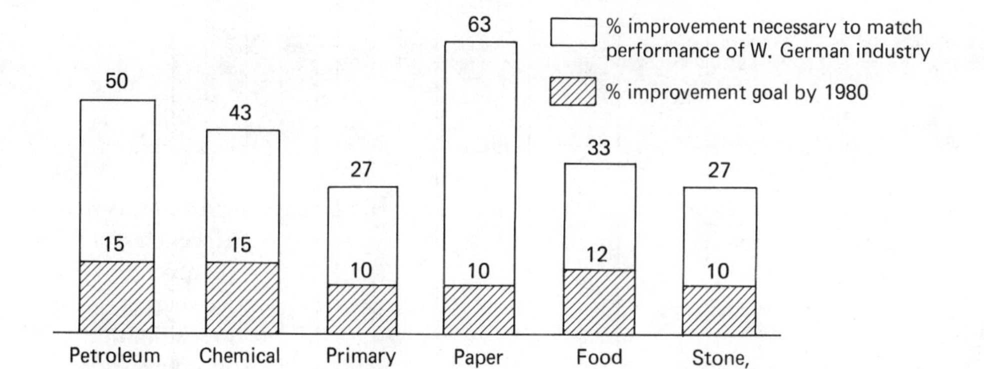

Figure 1.1. Comparison of the energy used by the United States, per unit of product, with the energy used by West Germany. (28) The United States uses more energy to make a ton of paper than to make a ton of steel. (27) The comparison with West Germany shows that paper can be made with 63 percent less energy,* which has nothing to do with comfort or life style. [DOE (28)]

*The theoretical minimum energy is some 400 times less, indicating that both the United States and West Germany have significant room for improvement. (27)

more efficient technology. (28) Policies to encourage replacement of some of the more inefficient energy processes in United States industry, as for example in the steel industry, would also improve its competitive export position. It will be imperative to adopt more efficient production techniques in order to remain competitive with other countries. This makes more sense than depleting the capital supply to find more energy to fuel outmoded and wasteful equipment.

It is estimated that by 1990, industry in North America could meet almost half of its

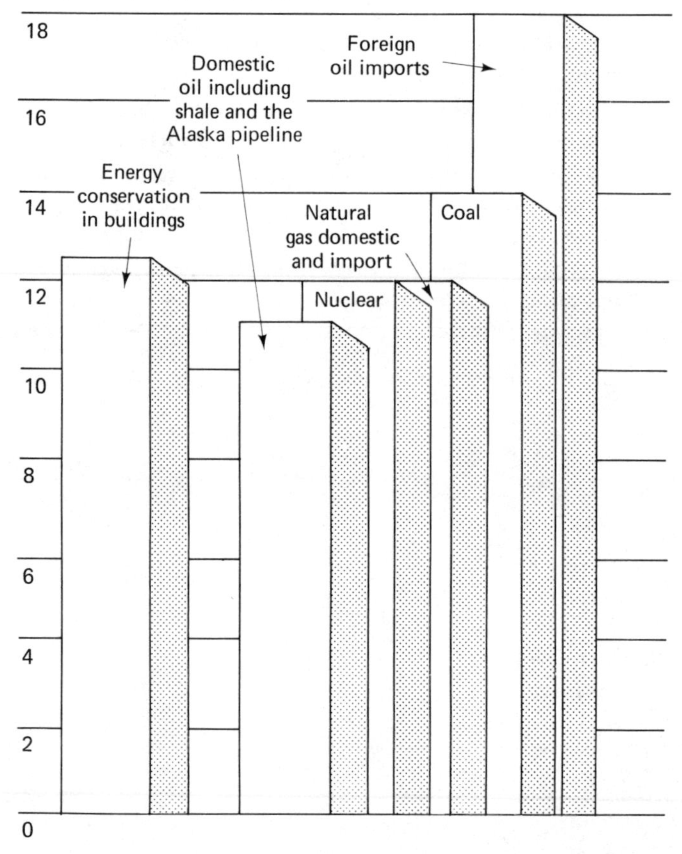

Figure 1.2. Comparison of the potential for saving in buildings with options to increase supply. This figure, from the 1974 AIA study done *before* the cost of some of these supply options had increased by many hundreds percent, considered those building investments with life cycle costs less than the cost of supply options. This study shows that the investment in making new and existing buildings more energy-efficient will return savings equivalent to the cost of over 12 million barrels of oil per day, an amount equal to the capacity of some supply options considered for Project Independence. [American Institute of Architects. (32)]

electricity needs by cogeneration with process steam. Western Germany already supplies almost 30 percent of its needs from cogeneration, compared to about 5 percent in the United States. (29) In addition, consider that the return of the capital investment in such a system occurs in half the time required for the payback when it is invested in new central station power plants.

A study by Arthur D. Little, Inc., indicates that existing technology, if used in office buildings, could save 50 percent of current energy requirements and 80 percent or more in new houses, as compared with past technology. (30) Such savings are also indicated by the economic impact assessment of the Building Energy Performance Standards (BEPS) proposed by the DOE. (31) Figure 1.2 illustrates the significance of the potential saving just in buildings, compared with several supply options. (32) With much of this, the capital cost is negative! The heating and cooling equipment can be reduced in size enough to pay for the other energy-saving modifications. (30) In new projects, sometimes very important items are sacrificed for the sake of the budget, items that may be paid for many times over during the life of the investment. Savings even more spectacular than those possible under the current industry standard could be achieved if focus were shifted from minimizing first costs to minimizing *life cycle costs*.

Following the 1973/74 oil price shock, owners concerned with operating costs asked designers to produce more efficient structures for operation in cold climates typical of Boston, Calgary, Manchester, Minneapolis, and Toronto. Many new buildings greatly exceeded ASHRAE 90-75 standards and required only one third to one fourth the 180,000 Btu/ft^2 (570 kWh/m^2) average annual energy use for existing U.S. buildings. The General Services Administration established the design standard of 55,000 Btu/ft^2 (175 kWh/m^2) for its new buildings.

As energy prices continued to rise, owners interested in future profitability insisted on even more efficient designs. With increasing frequency, new buildings are being announced with annual energy budgets below 40,000 Btu/ft^2 (125 kWh/m^2)—even operating as low as 25,000 Btu/ft^2 (80 kWh/m^2).

Such savings are not limited to new buildings, nor are they found only in North America. In one 70,000 ft^2 (6340 m^2) modern building in Zurich, Switzerland, which we will discuss later, savings of 60 percent in fuel oil and 85 percent in city water were demonstrated. (34)

One way to summarize the potential for saving is to note that it would allow today's economic output and living standard to be achieved with the energy use of the late 1950s.

Resistance to Saving?

The potential for increasing efficiency with existing technology is an enormous energy resource and it has a minimum of undesirable side effects. Since saving energy has its own reward—that of lowering operating costs—why is progress slow? We believe that it is because the users who stand to benefit receive inadequate, confusing, or even false information. We can cite four areas:

1. Traditions built on ample supplies at declining prices are slow to change. For example, news coverage of the United States natural gas shortage of the winter 1976–77 focused on the need

for more gas. The surplus of wasteful uses of energy, without which there would have been no shortage, was not mentioned.

2. Some people seem to believe that the only way to save is at the expense of production or occupant comfort, or that the uncertainty of increasing energy prices provides too much risk.
3. Energy producers frequently have much better access to capital than energy users.
4. Managers are rightfully wary of offers for products and services which are unnecessarily expensive for the energy saved, or ineffective, or even outright fraudulent.

This book will explain how much of this resistance can be avoided.

1.4 THE ECONOMICS OF SAVING ENERGY

The energy problem is that we do not have a benign, economical replacement for oil. One symptom of the energy problem has been the large price increase since the early 1970's. Another is the growing requirement for capital for new energy technology. Energy prices will increase by a much greater factor over the long term than already experienced during the last ten years. In addition to rising world demand, energy prices will increase because of depletion and super-inflation of energy production technologies.

Depletion of cheap and easy-to-use fuels forces producers to harsher environments and transportation over greater distances. New technology is required at costs that have been rising exponentially much faster (super-inflation) than in the general economy. The cost of energy conversion technology (electricity from coal and uranium, synthetic fuels from coal, oil shale, etc.) has also been increasing with super-inflation. Super-inflation of the cost of energy production technologies has defied worldwide efforts to reverse its trend, and led to postponement or abandonment of numerous nuclear plants and synthetic fuel projects.

The effect of super-inflation on energy prices has not yet made its full impact. Most fuel and electricity are produced by plants built during the former cheap-energy era, at capital costs 10 to 50 times less than for current technology. Future capital charges included in energy prices will become greater as the proportion of current, expensive technology in our supply system increases. As new production facilities are built to meet increased demand, or to replace existing plants as they become obsolete, the cost of energy they produce will make today's prices seem like a bargain.

Each example of former low prices paid by users only a few years ago should remind us of the opportunity we have today to mitigate the effect of future price increases.

The economics of saving energy *now* are favorably enhanced by the prospect of higher future energy prices. Capital investments will be repaid with savings that increase in value due to resource depletion and super-inflation. Furthermore, capital availability will become increasingly tighter forcing interest rates higher, as the pressure to switch from an oil-based economy increases. Fortunately, the cost of saving energy has increased much more slowly than the costs for energy production and conversion technology. Nevertheless, the threat of future energy shortages causes public confusion between producing more energy to feed present waste and the much less costly alternative of freeing existing capacity by saving. Rapid development of power and synthetic fuel plants could drive interest rates to greater levels than records previously set, which would raise the cost of energy saving. Efficient energy management *now* will divert dollars from needless burning to profits and vitally needed capital formation.

The shortage of capital will bring pressure for better management of capital resources. It is the efficiency of capital expenditures—the cost for a given benefit—which interests the energy manager, the owner, and the energy planner. We use *Specific Capital Productivity* (SCP) to enable comparison between different energy-saving measures. SCP is used both for (a) the investment per yearly average rate of fuel consumption saved, as well as for (b) the investment per unit of electric power saved. SCP is also used for comparison between the cost of saving and the cost of production. SCP is a measure of the cost of increasing energy availability.*

Throughout this book, efficiency improvements are summarized in tables with their costs, annual energy and dollar savings, and SCP. Tables listing SCP provide a useful comparison of the economic advantages of different measures. They also show, in contrast to lower expenditures to reduce energy needs, that usually the cost of adding energy production facilities is substantially more. In other words, one dollar spent on saving usually makes more energy available than the same dollar spent on production.

For example, the DOE study discussed above, on page 4, shows it would be cost effective to invest $470 billion over ten years to reduce energy demand by 15 million barrels of oil per day equivalent (BPDE).(9) By the year 2000, energy use could be reduced by 33.4 million BPDE with a cumulative cost of $966 billion. The overall SCP is $442/kW for the ten-year period, and $340/kW for the period to the year 2000. To put saving in perspective, consider that a SCP of $442/kW corresponds to spending $2 to save one gallon of oil per year. A 10% return on investment (in *constant value dollars*) would be considered an attractive commercial venture. This would correspond to spending 20¢ per gallon in 1973 dollars (or 35¢ per gallon in 1980 dollars) for the saved oil! Furthermore, the energy saved can be sold and used elsewhere.

Energy Saving Need Not Be at the Expense of Economic Growth

Energy saving is good for business, jobs, and regional prosperity. (35,36) A policy of continuous efficiency improvements can actually reverse energy growth while allowing for continuing economic activity.

Consider a building program which adds 5 percent to our commercial space each year. According to former practice, this would result in energy growth by at least 5 percent per year to provide for the heating and electrical needs of the new buildings if they were built according to former, cheap-energy practice. However, if the buildings are designed according to ASHRAE 90-75 standards or the DOE's building energy performance standards (BEPS), they will use only half the energy used by the average existing buildings, and some, as we have seen, will use less than one-fourth. Furthermore, assume that at today's energy prices and with today's technology there exists a 30 percent saving potential in existing buildings, but that due to imperfect owner motivation, institutional barriers, and other problems, this potential is realized over a period of ten years.

With conservation, a 5 percent increase in commercial floor space each year causes

*One $/kW is a measure of the cost to save (or produce) 8,760 kWh of fuel per year; that is, to save 30 million Btu (MBtu) or approximately 215 gallons of oil per year, or reduce connected electrical load by 1 kW. To convert to $ per barrels of oil per day equivalent energy ($/BPDE), multiply SCP in $/kW by 71.

only a 2.5 percent addition to commercial energy use, while saving in existing buildings cuts back their use by 3 percent each year, so that the result is negative energy growth. Similar processes could occur for the other energy sectors if energy saving would be given full government promotion.

The measures possible are so far-reaching, that one cannot be discouraged; nor can one conclude that energy saving does not work if all of the measures are not immediately put in place. Nor can it be concluded that this scenario represents a ten-year correction with 5 percent growth continuing thereafter. On into the future, it will be possible to make buildings more efficient than they have been in the past. The DOE analysis assumes little more than the comprehensive application of 1970s technology under the economics of higher energy prices.

1.5 PROFITABILITY WITH HIGHER PRICES

Over the long run, the only way to remain competitive with rising energy costs is to improve energy efficiency. Despite price increases which have already occurred, energy costs are still only a small part of total production or operating costs. However, if these are broken down into fixed and variable categories, energy costs form a significant proportion of the variable category. The after-tax profit will be extremely sensitive to small changes in the cost for energy.

Conservation in existing buildings could very well mean the difference between a profitable building and a non-profitable one that will suffer premature obsolescence. Escalation clauses cannot be a salvation forever. The competitor who is energy conscious and reduces operating costs significantly will attract tenants through lower rents. (37)

The following examples illustrate some of the savings which have been made in the face of rising energy costs:

1. Between 1973 and 1977 the price of natural gas used at One Shell Plaza in Houston, Texas, increased by 495 percent. The operation of their dual-duct HVAC system was modified to recycle building heat, domestic water supply temperature was lowered 20°F (11°C), and the use of the boiler for heating was limited to periods when the outside temperature dropped below 48°F (9°C) for more than 24 hours. As a result of these and other operational measures, the use of natural gas was cut from 40.5 million cubic feet per year (1.15 million cubic meters) to 12 million cubic feet per year (0.34×10^6 m^3). Continued attention to saving energy cut total use by an additional 10 percent in 1978. By 1979, the total energy use for this 1.6-million-square-foot building (150,000 square meters) had been lowered to an outstanding 46,687 Btu/ft^2 (147 kWh/m^2). (38)
2. In anticipation that by 1985 energy costs will rise from 4 percent to 15 percent of the production costs at TRW Inc., a vigorous energy management program has already been instituted. Although sales have nearly doubled to $3.3 billion since 1972, energy use has been reduced by 19 percent. Workers are given financial rewards for energy-saving ideas and production managers include "energy use forecasts" with quarterly sales, profits, and capital-spending projections. (3)

Figure 1.3. One Shell Plaza, Houston, Texas. Comfortable conditions have been maintained in this luxury office building while the use of natural gas was reduced by 70 percent. (37)

3. The price of electricity is up 99 percent since 1973 and the price of other fuels up 79 percent for New England Telephone. Despite a 10 percent increase in floor space, energy use has been decreased by 16 percent. (38)
4. As the University of Massachusetts moved from older rented quarters heated with district steam, to a new and larger all-electric campus, their energy costs jumped by a factor of 13. In a period of six years, energy costs were reduced over $5 million at a cost of $350,000, including the salary of a full-time manager to oversee a comprehensive energy-saving program that we will discuss later. Since the initiation of this program, its effect has been to reduce the cost of education by $100 per student per year.

SUMMARY

In summary, all future energy sources which will be available in significant quantities will cost appreciably more than the energy we are using today. The danger that confronts us now is that the warnings of future energy price increases will go unheeded, or that action will be delayed on the basis of what is easiest in the present rather than what is best for the future.

There could be some hard-felt shortages in the future, but they are not all inevitable. Avoiding them will present a variety of extraordinary challenges during the forthcoming years.

We can reduce the strain on our economy and environment, stretch the lifetime of the earth's petroleum reserves, and avoid any crisis if waste is eliminated and needed efficiency improvements are made now. Yet energy saving does not need to be justified on abstract arguments of national security or the threat of future shortages. **The most compelling argument for saving—for the individual, for business, and for government—is economic: the opportunity cost of not saving.**

REFERENCES

(1) James Gerstenzang, "Federal energy use drops, despite '77 rise," *Oregonian* (August 17, 1978), p. A4.

(2) "Savings Promised but Energy Budget Funds in Peril," *The News Tribune* (May 18, 1978), pp. 1 and 8.

(3) "Reaching for Fuel-Saving Ideas," *Time* (August 21, 1978), p. 74.

(4) "Bell System Technologists are Always Looking Around for Ways to Save Energy," *The New England Topics* 4, no. 6 (May 19, 1978), p. 11.

(5) David C. White, "Conserving, Finding and Directing Energy Resources," presented at the panel discussion of the Financial Executive Institute's 43rd annual conference at Honolulu, Hawaii, October 6–9, 1974 (Cambridge, Mass.: M.I.T. Energy Laboratory Working Paper No. 12, December 1974).

(6) Hilliard Roderick, "Energy and the Environment: A Conflict of Interest or Two Aspects of a Single Policy?" *OECD Observer,* no. 70 (June 1974).

(7) OECD, "Conservation of Energy, How are Countries Performing?" *OECD Observer,* no. 83 (September/October, 1976), pp. 4–9.

(8) G. Eigenmann, "Energiesparprogramm in der Chemischen Industrie" (Ruschlikon, Zurich: Gottlieb Duttweiler Institut Seminar; Wie spare Ich Energie? January 8–10, 1976).

(9) Marquis R. Seidel, "Economic Benefits from Energy Conservation," in *Forum Proceedings, Energy Conservation: A National Forum,* ed. T.N. Veziroglu, (Coral Gables, Florida: Clean Energy Research Institute, University of Miami, December 1975), pp. 553–84.

(10) E. P. Gyftopoulos, J.B. Dunlay, and S. Nydick, "A study of improved fuel effectiveness in the iron and steel and paper and pulp industries" (Waltham, Mass.: NSF Report by the Thermo Electron Corporation, March 1976).

(11) Marc Ross and Robert Williams, "The Potential for Fuel Conservation," *Tech. Rev.* (February 1977), pp. 49–57.

(12) Harry Perry, "Conservation of Energy," prepared by the Congressional Research Service, Library of Congress, for the U.S. Senate Committee on Interior and Insular Affairs (Washington, D.C.: U.S. Government Printing Office, 1972).

(13) NPC, "Potential for Energy Conservation in the United States 1974–1978," (Washington, D.C.: The National Petroleum Council, 1974).

(14) Robert Kupperman, "The Potential for Energy Conservation," a staff study by the Office of Emergency Preparedness, Executive Office of the United States (Washington, D.C.: U.S. Government Printing Office, 1972).

(15) Barry Hyman, "Initiatives in Energy Conservation," a report for the U.S. Senate Committee on Commerce (Washington, D.C.: U.S. Government Printing Office, 1973).

(16) Eric Hirst and John C. Moyers, "Potential for Energy Conservation Through Increased Efficiency of Use," *J. Environmental Systems,* 3 no. 2, (Summer, 1973), pp. 153–69.

(17) Keith Doig, "The National Energy Problem: Potential Energy Savings," (Houston Tex.: Shell Oil Company, October 1973).

(18) DOE, "Comparison of Energy Consumption Between West Germany and the U.S." (Washington, D.C.: U.S. Department of Energy, 1975).

(19) Lee Schipper and A. Lichtenberg, "Efficient Energy Use and Well-Being: the Swedish Example," Lawrence Berkeley Laboratory (Berkeley, Cal.: University of California, 1976).

(20) APS, *Physics and the Energy Problem,* ed. M.D. Fiske and W. Havens, Jr. (New York: American Physical Society, 1974).

(21) S. David Freeman, et al., *A Time to Choose—America's Energy Future,* a report of the Ford Foundation's Energy Policy Project (Cambridge, Mass.: Ballinger, 1974).

(22) G. Hatsopoulos, T. Widmer, E. P. Gyftopoulos, and R. Sant, "National Policy for Industrial Energy Conservation," (Waltham, Mass.: Thermo Electron Corporation, April 1977).

(23) W. Carnahan, K. Ford, et al., "Efficient Use of Energy; A Physics Perspective," in *Efficient Use of Energy,* (New York: American Physical Society, 1975).

(24) "A Study of Inplant Electric Power Generation in the Chemical, Petroleum Refining and Paper and Pulp Industries," a report for the DOE (Waltham, Mass.: Thermo Electron Corporation, 1976).

(25) T. Widmer and G. Hatsopoulos, "Summary Assessment of Electricity Cogeneration in Industry," (Waltham, Mass.: Thermo Electron Corporation, 1977).

(26) T. F. Widmer and E. P. Gyftopoulos, "Energy Conservation and a Healthy Economy," *Tech. Rev.* (June, 1977), pp. 31–40.

(27) E. P. Gyftopoulos and L. J. Lazaridis, *Potential Fuel Effectiveness in Industry* (Cambridge, Mass.: Ballinger, 1974), p. 8.

(28) Roger Sant, "Energy Conservation," presented by the Assistant Administrator for Conservation, Federal Energy Administration, at the Symposium for Public Awareness on Energy, Knoxville, Tennessee (February 27, 1976).

(29) Albert F. Plant, "Conservation Is the First Step," Chemical and Engineering News (November 8, 1975) p. 4.

(30) ADL, "Energy Conservation in New Building Design: An Impact Assessment of ASHRAE Standard 90-75," (Washington, D.C.: U.S. Department of Energy, 1976), 257 pp.

(31) DOE, Building Energy Performance Standards and associated documentation, (Washington, D.C.: U.S. Department of Energy, 1980).

(32) AIA, "Energy and the Built Environment " (Washington, D.C.: American Institute of Architects, 1974).

(33) Private communication with Fred Dubin, Princeton, New Jersey, June 21, 1980.

(34) "Energiesparen-Kostensenken," *Neue Zurcher Zeitung* (August 18, 1976), p. 29.

(35) Wilson Clark, "Conservation as an Energy Resource " (Sacramento, Cal.: Governor's Office for Issues and Planning, 1978).

(36) Patrick Forrester, et al., "New England Energy Policy Alternatives Study," A report done for the Department of Energy and the New England States (Boston, Mass.: The Massachusetts Energy Office, Office of the Governor, October, 1978).

(37) Private communication with C. J. Butler, Property Manager, Gerald D. Hines Interests, Houston, Texas, April 9, 1979.

(38) "New England Telephone consumed 5.3% less energy in 1977 than in 1976, but paid 5.7% more for it." *New England Telephone Management News* (Boston, March 8, 1978), p. 1.

2

Fundamentals of Energy Saving

Energy savings can be achieved by individuals in many different ways. An essential question is what would be the net saving in raw fuel resources? As an example, if one were to replace oil heating with electric heating, there could be an increase in the *local* efficiency of heat production. However, from a larger viewpoint, the improvement in the user's efficiency is offset by the use of three units of energy to produce one unit of electricity, one-third of which is produced from oil and gas.* (1)

The purpose of this chapter is to lay the groundwork for effective saving and to help you obtain the maximum benefit from energy with the least economic and environmental cost. In Section 2.1, some basic definitions of energy are explained. The economics of energy saving is discussed in 2.2. In Section 2.3, real and negative savings are described. The perspective of the energy saver, and the mental attitudes of the participants in energy saving, will affect its success. Some attitudes which can interfere with the success of saving energy will also be examined in this chapter.

2.1 GENERAL PRINCIPLES

The Choice of System Boundaries

Energy saving is not an end in itself to justify any means. Too often people try to improve the efficiencies of components without considering the function of the overall

*In 1977, 3.51 units of raw source energy were required for each unit of electricity sold in the U.S. (28.5 percent overall efficiency). The raw source energy used was 41 percent coal, 12 percent gas, 22 percent oil, 13 percent hydroelectricity, and 12 percent uranium.

system or its purpose. The importance of technical goals is elevated without recognizing that they are subordinate to quality of life and human productivity. The chosen boundaries of the system to be optimized should be large enough to include all of the consequences and yet small enough to be manageable.

Where Can Energy Be Saved?

Keeping the above in mind, lower operating costs and reduced environmental and social costs can be achieved at three different levels:

1. Improvements in supply
2. Minimizing the energy cost of products and services
3. Thrift in the rates of energy and product use

1. Under supply would come any improvement in extracting a natural resource and in converting it to a form more suitable for our needs as, for example, the conversion of crude oil to gasoline. Also included at this level is the use of alternative, less scarce, less harmful, and renewable energy resources.
2. The specific energy of products and services can be reduced through technical improvements in manufacturing processes and construction details (e.g., insulation), and improvements in operating methods. The specific energy is the energy used to provide each unit of a product or service as, for example, energy per passenger distance traveled, heat per degree-day, or kWh per ton of aluminum produced.
3. The rate at which a product or a service is used is called *demand*. Thrift could mean reducing demand, for example by choosing longer-lasting products, or by reducing demand for services (e.g., by heating and lighting only occupied spaces). Thrift could also mean reducing the intensity of demand, for example by heating and lighting at lower levels, by traveling in lighter vehicles, or by choosing a less-energy-intensive material than aluminum for cladding a building. The interrelationship of different ways for saving energy are summarized in Figure 2.1.

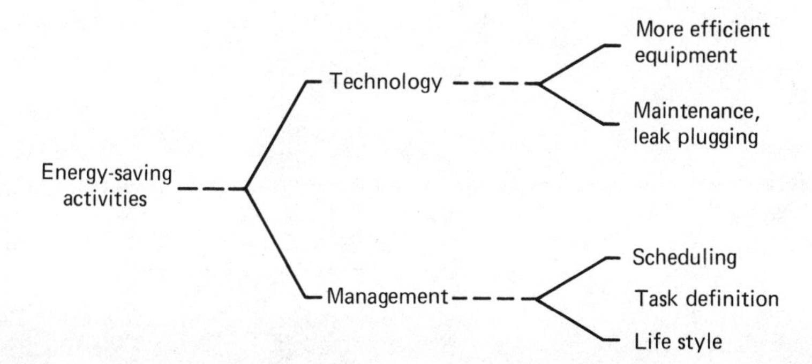

Figure 2.1. Factors into which the question of "How to save energy" can be analyzed. The largest improvements will come from measures that reduce losses and improve efficiency, rather than from austerity.

Energy Forms

The energy units used in this book refer to the energy content of the fuel or other energy form being discussed. Unless otherwise specified, the numbers given do not include losses from previous or subsequent conversion from one energy form to another (See "Energy Concepts," below). We use kWh as well as Btu as units of energy. One kWh of steam means the energy content of the steam and not that of the oil used to produce this steam.

Different energy forms have different impacts on resources and the environment. The number obtained by summing Btu's of steam and electricity, etc., used within a building, carries no information about its economic and environmental cost nor of the resource consumed. Therefore, a sum of different energy forms is ambiguous in terms of the value of saving it or the consequences of its use. A partial remedy would be to refer all energy to its primary form before summing, as is illustrated in Figure 2.2. (2) In this book we do not adhere to this convention primarily because of the unavailability of adequate data.

Energy forms differ with respect to their quality. Chemical or nuclear energy in the form of fuels is of potentially high quality. Mechanical work and electricity are the highest quality forms of energy. Matching the quality of the energy supply with the quality required by the task can improve the second law efficiency and reduce fuel costs.

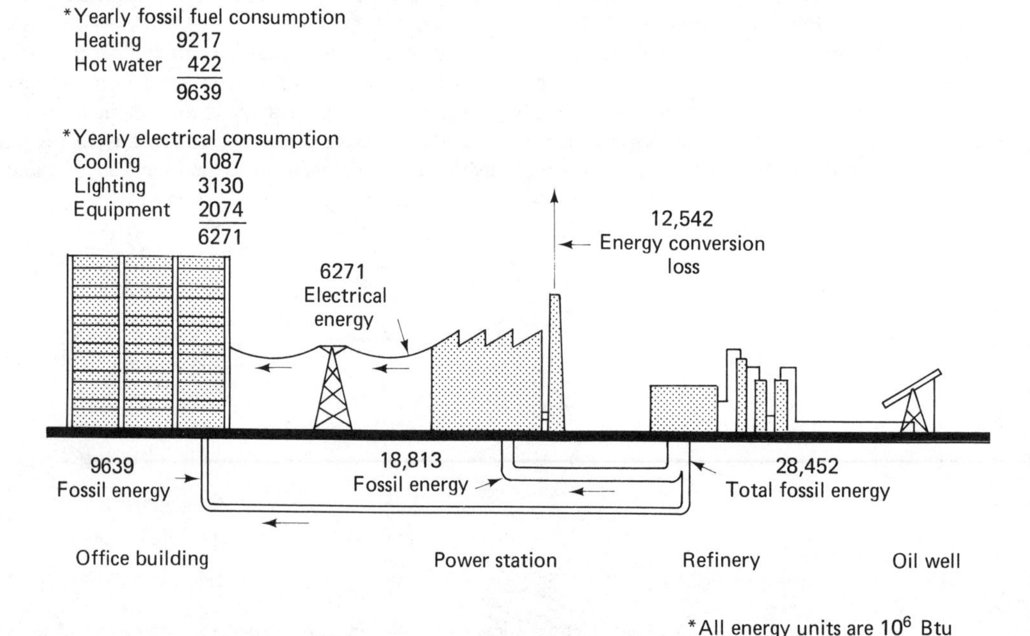

Figure 2.2. Yearly energy requirements for office building in New Hampshire (conventional modern design) in terms of primary, raw source, fuel requirements. [GSA (2)]

Energy Concepts

Energy in all forms, except for heat, is the capacity to do work.

All matter and all things have energy.

Energy is never created or destroyed, but is only converted from one form to another.

Work is the product of force and the distance moved along the direction of force, or power multiplied by time.

Work is energy in ordered form; it is the highest quality form of energy, consequently it is the most valuable.

Power is the rate of doing work or the rate of converting from one form of energy to another.

Power Rating is the normal maximum rate at which energy can be converted by a device. For an electric device such as a motor or heater, it is the power input. For fuel consumers it is usually the nominal maximum output at, for example, the shaft of a diesel motor, or output from a boiler.

Primary energy is a form of one of the earth's natural energy resources.

Crude oil, natural gas, solar radiation, uranium, coal, etc., are primary forms of energy.

Secondary energy has been converted from primary energy to one of the forms of energy more suitable for intermediate or end use.

Electricity is a secondary form of energy, as are gasoline, steam, hot water, hydrogen, coke, etc.

A distinction is made between secondary and primary energy because of the (sometimes large) loss in converting to secondary forms such as electricity or synthetic gas.

Raw source energy is the amount of primary energy used to produce a unit of secondary energy.

Including thermal and conversion losses, and energy used internal to the power plant, the raw source energy of 1 kWh of electricity is approximately 3 kWh of primary energy input at the power plant.

Thermodynamics

Heat is the energy that flows between two bodies which are at different temperatures. Heat, therefore, may be defined as the flow of energy between two systems which causes the temperature to rise, or causes the phase state (solid, liquid, or gas) of the colder one to change. The British Thermal Unit (Btu) is the work or heat which must be transferred to one pound of water to raise its temperature by one degree Fahrenheit. The kilocalorie is the heat transferred to raise the temperature of one kilogram of water one degree Celsius.

Quality of energy is a measure of its versatility and ability to do work. Energy in mechanical or electrical form has the highest quality. The quantity of thermal energy decreases as the difference in temperature to its surroundings decreases and it becomes zero at the ambient temperature.

Carnot efficiency is the maximum theoretical efficiency that a real heat engine could achieve. It is based on a simple reversible engine that results in the same heat transfers and work when operated as a heat pump as it does when operated as an engine, except that the signs of the quantities are reversed. As a heat engine it takes in energy at a higher temperature T_h, converts some of this to work, and rejects the remainder of the energy at a lower temperature T_1. The Carnot efficiency depends only on these two temperatures expressed on the absolute (Kelvin) temperature scale, $e_C = 1 - T_1/T_h$. Since heat is usually rejected at ambient temperatures, the higher the source temperature, the higher the Carnot efficiency. Efficiencies for actual heat engines are typically a factor of 2 smaller.

Coefficient of performance, COP, is the usual figure of merit for a Carnot engine operated as a heat pump. It is the ratio of the heat, which has been moved from a lower temperature source at T_1 to a higher temperature output at T_h, to the work used to operate the Carnot heat pump. The maximum theoretical COP $= T_1/(T_h - T_1)$. Practical heat pumps can move 4 to 8 times the amount of energy required to operate them.

Absolute temperature: The Kelvin scale for temperature is one in which the freezing point of water is 273 K. This absolute-temperature scale is used for calculating COP of efficiencies of Carnot engines, and is one in which zero is $-273°$ on the Centigrade scale, or $-460°$ Fahrenheit. One unit of the Kelvin scale is equal to 1°C or 1.8°F.

Available work, also called exergy, or simply availability, has the dimensions of energy.

It is the upper limit of the work that could be obtained from energy at a temperature T_h with respect to a final temperature T_1 under ambient conditions.

According to the first law of thermodynamics, all energy is conserved; it is never lost but merely converted from one form to another. However according to the second law of thermodynamics, it can be dispersed, or its temperature brought closer to ambient temperature so that it cannot be used to perform further work.

Available work is a quantity that is consumed, or used up, as the quality of energy decreases from flame to ambient temperature.

Loss of available work is irreversible.

It is the natural irreversibility of all real processes which proceed spontaneously that causes a loss of available work.

Fuel has a high availability because of the high flame temperatures it is capable of producing, or the high level of available work it can yield in a fuel cell.

Enthalpy has the dimensions of energy per unit mass and is a term used in combustion and in air conditioning.

For an ideal gas, enthalpy is proportional to temperature.

In addition, for air which contains some percentage of moisture, the energy per unit mass will depend on the latent heat contained in the mass of water vapor.

A change in enthalpy will depend on *(a)* the change in temperature, and *(b)* a change in the moisture content due to the heat of vaporization of water, 970 Btu/lb. (0.627 kWh/kg).

Efficiency

Definitions of efficiency take two forms: the conventional measure and the second law efficiency. The concepts of the minimum energy required for a given purpose and second law efficiency have not received much attention previously. Consequently, present second law efficiencies of most end uses are low (cf. Table 2.1). Because energy is never created nor destroyed (the first law), the conventional measure of efficiency is useful in keeping track of flows into and out of a system. The usual measure of efficiency (first law efficiency) is the ratio of output to input:

$$\frac{\text{Energy provided or converted into desired form}}{\text{Energy input to the system}}$$

It is this definition of efficiency (sometimes called the first law efficiency) that is used when describing a boiler as 65 percent efficient, meaning that 65 percent of the energy content of the fuel was converted to heated water. This is the measure being used when the efficiency of a power plant is given as 33 percent, or when the coefficient of performance (COP) of a heat pump is given as 3.2. However, this definition is inadequate as an indicator for possible improvement.

Table 2.1. The Second Law Efficiencies of Some Important Uses for Energy in the United States. (3)

Application	*Second-law efficiency (%)*
Residential and commercial space heating	6
Residential and commercial water heating	3
Air conditioning and refrigeration	5
Automobile propulsion	10
Steel production	21
Petroleum refining	9
Cement manufacturing	10
Paper production	Less than 1

Note: The weighted average second-law efficiency for the applications shown here is about 8%, yet these uses account for about 60% of the U.S. fuel use. This suggests that there is considerably greater potential for stretching fuel supplies by improvements at the point of end use than there is at the point of converting fuels to secondary energy.

SOURCE: Thomas F. Widmer, "Accelerated Conservation: An Alternative National Energy Plan" (Waltham, Mass: Thermo-Electron Corporation, Feb. 15, 1977). (3)

A much more meaningful definition compares the energy actually used with the minimum possible energy use for a specific purpose.

$$\frac{\text{Minimum possible energy use for a given purpose}}{\text{The energy actually used}}$$

This is called the second law efficiency because the minimum quantity of energy required for a given task is determined by the second law of thermodynamics. It provides a yardstick for all fuels and all energy-using tasks. The second law efficiency refers to the task to be performed rather than to the particular piece of equipment used. The results for some important examples using collectively about 60 percent of energy in the United States are shown in Table 2.1. (3) The weighted efficiency for the uses shown is about 8 percent, a figure thought to be representative of the entire United States economy. Although this suggests potentially enormous opportunities for savings, creativity in redefining tasks or combining them in cascade offers the potential for even greater improvements.

Energy Quality and Available Work

The second law tells us that the quality of energy can only change in one direction and that energy loses its capacity to do useful work—eventually reaching the point of zero usefulness. Actually, when energy is used, we do not "consume" energy, but rather *available work.* As available work is consumed, the quality of energy is degraded; the *quantity* of energy remains the same. The second law allows us to calculate the available work required to convert metals from ores, to cool a building, to propel a car, or to convert the energy in natural gas to a high temperature flame, for example. (See References 3, 4, 5, 6 and 7.)

Good energy-saving practice strives to harness energy at the highest quality or temperature possible; that is, to avoid unnecessary degradation from friction, or from large temperature or pressure drops, or through mixing of different temperature energy flows.

The following example shows that it is wasteful to burn fuel solely to obtain low-quality energy required at low space of process heating temperatures. Consider, for example, two ways of producing electricity and steam or hot water to use for district heating, process heat, etc. The more efficient of the two methods considered in Figure 2.3, is to first produce high pressure steam at a temperature of 930°F (500°C).* The available work in the steam is used to drive a back pressure turbine, where it is converted to mechanical energy that drives an electric generator. The steam is cooled to a temperature which may be 300 to 400°F (150 to 175°C) at the output of the turbine, as required for its intended purpose. Once some high quality work has been extracted, the inexorable increase in entropy guarantees that lower-quality heat will be available as energy flows along its paths to disutility. With this scheme, it is possible to convert approximately 30

*The difference between the flame temperature of the burning fuel (2,500°F or 1,400°C) and the high-pressure steam also represents a considerable potential loss in available work. The use of a gas turbine-waste heat boiler (or so-called "topping cycle") helps to recover some of this available work that is otherwise lost due to the large temperature difference between the flame and steam temperatures.

Case A: Separate production of electricity and heat

(Combined first law efficiency η = 52%)

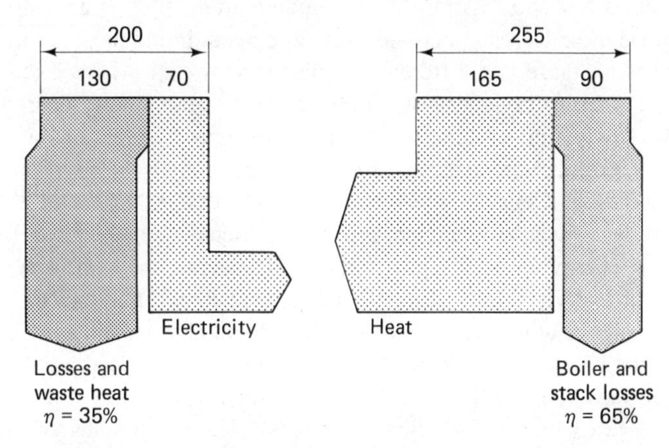

Total fuel requirement for separate production is 455 units/h

Case B: Cogeneration of heat and electricity

(Combined first law efficiency η = 85%)

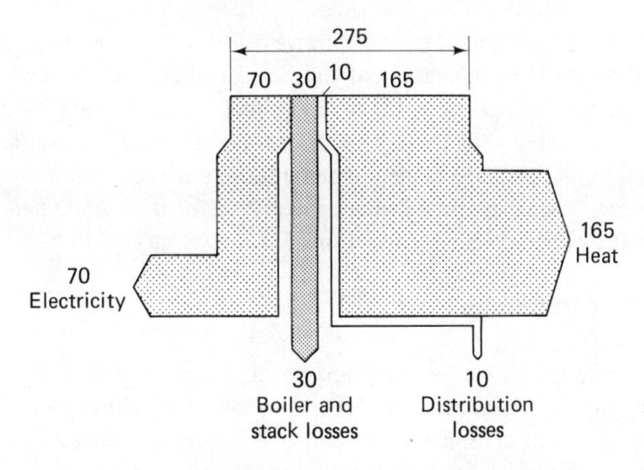

Total fuel requirement for cogeneration is 275 units/h

Rate of fuel saving is 180 units per hour, or 40% less than separate production with same useful output.

Figure 2.3. Typical power balance for separate heat and electricity production compared with cogeneration.

percent of the high-quality energy in the fuel to electricity and 50 percent to useful low-temperature heat. The conventional method of producing the electricity and heat separately loses almost 50 percent of the energy content of the fuel. With cogeneration, it is possible to reduce these losses to perhaps only 20 percent in the most favorable cases. The power balances for these two different schemes is shown in Figure 2.3.

Not surprisingly, because of the former cheapness of energy, our supply system emphasizes quantity. Consequently, too many opportunities are missed to reduce operating costs by better matching the quality to the need, and to take advantage of other possibilities for cascading. Consider the mismatch, for example, in using costly and high-quality electricity to heat low-temperature water. Electrical energy can be converted into mechanical energy using an electric motor. However, the same amount of energy in 85°F (30°C) is not very useful for producing mechanical motion. Nor can the energy in the warm water be upgraded in quality so that it would be capable of melting iron.* However, all of the energy supplied to the electric motor or from the casting of iron eventually becomes available in the form of lower temperature heat for purposes such as warming water (see Chapter 6 for actual examples). These examples are not limited to thermodynamic utopias, but they do require that users understand the principles of matching quality of supply with quality of use, and that traditional resistance to adopting new ideas is eliminated. Now that energy is more expensive, better organization and management of energy use will be worthwhile.

An example of the practical matching of the right quality of energy to the task is the curing of resin coatings with ultraviolet irradiation. Process heat is typically used for drying and curing. The curing of resin, however, is a chemical process which can proceed more rapidly with the application of high-quality heat. A conventional source of heat applied at the quality optimum for the chemical process would burn the surface before the interior was cured. Short-wavelength ultraviolet light corresponds to a temperature source of high quality more closely matched to the energy requirements of the chemical curing process. In one industrial plant, the replacement of direct heating with ultraviolet light reduced the rate of energy use from 3.5 megawatts to 45 kilowatts. (4)

Principles for Buildings

Many people pay twice for using too much energy, and then end by throwing it away before receiving full value. Figure 2.4 shows the energy flow proportioned for an office building in the north central United States.† A similar Sankey diagram could be constructed for any building, such as a retail store or a manufacturing facility. This diagram shows two points which must be understood in order to make full use of the energy brought into a building. First, and contrary to popular opinion, it shows that the energy

*In theory, 3 percent of the energy in the warm water could be raised in quality for the iron-melting task, given a perfect heat pump and an environment capable of receiving 97 percent of the remainder at a temperature of 68°F (20°C).

†The flows have been proportioned for the annual energy use given for the ASHRAE 90-75 modified design. Although the proportions will be different for each building, the principles remain the same. (See Ref. 30, Chap. 1.)

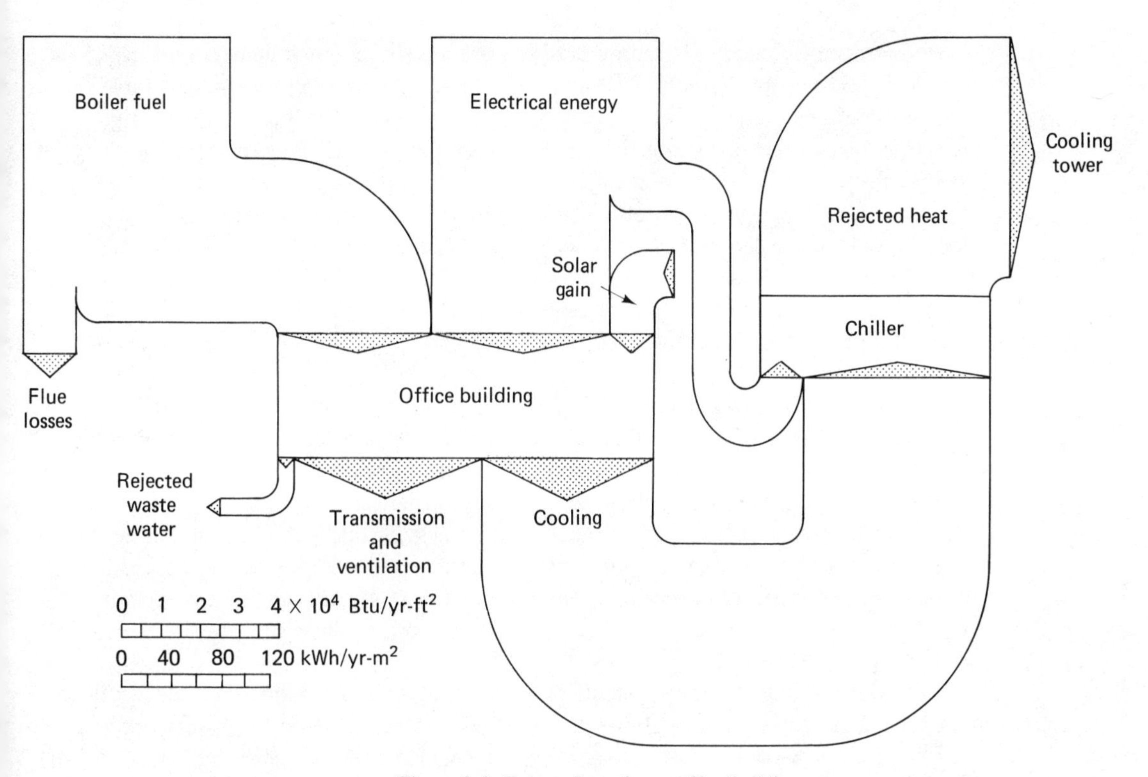

Figure 2.4. Energy flows for an office building.

used for air conditioning is primarily determined by the need to remove energy brought into the building for lighting, fans, and other devices, and only secondarily for removing heat from the sun! This explains why air conditioning is used for modern buildings even in cold climates. Cutting down on lighting and unnecessary energy in manufacturing will also cut air conditioning costs.

The second point is that even in this relatively cold climate, three times as much heat is rejected as is required from the heating system to maintain comfort. This means that heat should be transferred from areas of surplus to areas of need. If the heat can be captured at a high enough temperature, it may be recoverable for comfort heating (or preheating) directly. To improve the usability of waste heat, an attempt should be made to increase its temperature at the source of the energy flow. Perhaps the rate of flow of ventilation air, cooling water, etc., could be reduced, and mixing with other, colder flows could be avoided so that the rejected heat is more useful. The principle is to use the energy that would otherwise be rejected from the cooling tower and through exhaust ventilation to replace fuel. Furthermore, thermal storage can be a powerful energy-saving strategy. Energy can be stored when it is in abundance (or when it is cheaper) for use later, as for example during the subsequent unoccupied period. Figure 2.4 shows that one of the problems is the overabundance of energy, and that by recovering it for reuse, operating costs may be reduced.

One principle, then, is based on cutting unnecessary or superfluous energy use. Let us use lighting as an example. The annual cost for lighting is given by the simple but fundamental relationship: cost per year $= P \times t \times c$. Thus the annual cost for lighting is determined by the cost charged for each unit of energy c, the power or rate of energy use P (also called demand), and the annual hours of operation, t. Annual costs can be cut by reducing P with more efficient lighting or with lower illumination levels. By lighting only while the area is occupied, t can be shortened. For example, the cost of \$150 to install one switch per 1,000 ft^2 could be recovered in less than a year in many cases, if it were used to switch off the lights except when the area was occupied or being cleaned. Annual costs could also be reduced by choosing a cheaper energy service—in this case available daylight.

A similar relationship involving the product of demand, operating time and cost per unit of energy gives the essence for the cost of any energy use. The essential relationships for heating (and for some aspects of cooling) are given below.

Costs can be reduced by cutting the intensity of demand and the time of operation. Reducing temperature differences to the outside, improving insulation, eliminating cooling of stairwells and storage areas, and cutting superfluous ventilation would be examples of reducing the intensity of demand. Perhaps the heating system can be shut down in the summer to reduce t. The cost of heat could be lowered by increasing boiler efficiency and by other means.

Be aware of hidden waste systems that are out of sight and running needlessly or at low efficiency. Unnecessary lighting receives considerable attention from amateur energy savers because of its visibility. Other equipment which does not call visual attention to its unnecessary or inefficient operation may be easily overlooked.

Basic Functional Relationships for Heat and Building Heating, and Their Annual Costs

1. Annual cost for heat, C_{Anh}

$$C_{Anh} = c_h \times Q_h \times t + C_h \times (1/A) \qquad (1)$$

c_h = cost of heat (see below)

Q_h = average heat load in Btuh

t = hours of operation per year

C_h = capital investment in heat producing facility

$1/A$ = capital recovery factor[a]

$$c_h = \frac{P_f}{e \times h_f} \times I_f \times (1 + C_{om}) \qquad (2)$$

P_f = annual average price of fuel, electricity, etc., in $ per unit of purchase

h_f = energy value of fuel as obtained from supplier in Btu (or kWh) per unit of fuel

e = efficiency of heat-producing facility in percent

I_f = fuel inflation factor (for computing future costs)[a]

C_{om} = cost of operation and maintenance (from accounting department or estimate 10 percent)

2. Building heating (cooling) load

$$Q_h = (AU + h_a n)T_{avg} - Q_{ext} + nq_{hum} \qquad (3)$$

AU = sum of conductive heat loss rates through the walls, windows, floor, etc., of the building. This is the sum for each area, A, times its respective heat transmission coefficient, U (typical average values for existing construction are 0.25 Btuh/ft^2°F or 1.5 W/m^2°C)

$h_{air}n$ = heat rate to heat n air changes per hour, h_{air} is the heat capacity that is typically 1.08 Btuh/°F ft^3/min, or 0.34 W/°Cm^3/h (neglecting changes in humidity)

T_{avg} = average temperature difference to the outside

Q_{ext} = average rate at which heat is supplied from external sources such as the sun, and for the purposes of some analyses, heat from lights and other equipment in the heated space

nq_{hum} = heat load for humidifying n changes per hour of dry, outside air, in Btuh (or kW)

[a]Refer to "Expressions used for financial analysis of energy savings," p. 28.

The Order of Approach

Another principle concerns the order of an efficiency improvement program: first, reduction of demand; second, reduction of distribution losses; and third, improvement in production efficiency. Reducing the demand for heating or lighting, for example, required at the point of use, may make it possible to apply measures to the distribution system which would not be possible if the order were reversed. As an example, after reducing the power needed for lighting, air flow rates can be lowered to the extent that there is less internal heat which must be removed. This order of approach to energy saving and some of the general methods to apply are shown in relationship in Figure 2.5.

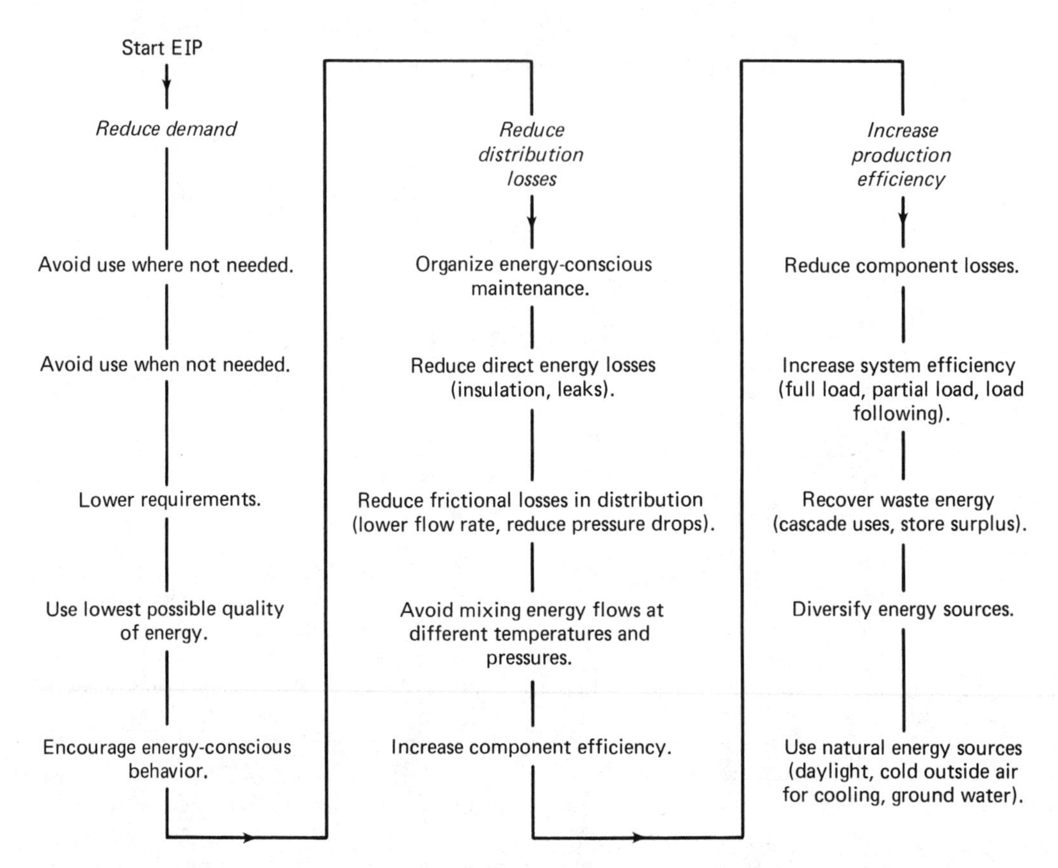

Figure 2.5. Order of approach and general principles for an energy efficiency program (EIP).

2.2 FINANCIAL CONSIDERATIONS

The quantity of energy to be saved by operational measures, and those that can be included under maintenance is impressive. However, capital expense will eventually be required to achieve additional savings. This section explains some financial considerations for use in evaluating proposed energy-saving projects. Government and nonprofit organizations will also find these methods useful to determine which efficiency improvement projects can save enough money to justify borrowing money to do them.

Life-Cycle Accounting

The method of accounting is crucial to formulating an efficiency improvement program, since design options are usually compared on the basis of costs to the owner. Before the energy price shock, most design and investment decisions were dominated by the object of minimizing first costs, virtually ignoring operating costs. First costs are not, however, an accurate indication of the cost of a project nor its after-tax profitability, since operating and maintenance costs are not included. Life-cycle costs reflect the cost of a project over its lifetime and allow trade-offs to be evaluated between a system that is inexpensive to build versus one that is inexpensive to operate. Unfortunately, the large increase in building and equipment costs which has occurred over the last few years, has made it difficult for the owner to shed her or his traditional preoccupation with the size of the initial investment. Those aware of the economic and environmental consequences of unbridled energy use should strive to convince owners to extend their time horizon at least to the economic lifetime of the project or, preferably, to consider life-cycle costs.

Life-cycle costs are the sum of the yearly owning and operating costs for the period of life expectancy of the project. The owning and operating costs consist of amortization and interest, based on the initial costs; taxes; insurance; labor and materials for operation and maintenance; and energy costs. The comparison of the life-cycle costs of two projects examines the difference between the initial costs and operating and maintenance costs. The methods for including the escalation of energy prices over the life of the project will be discussed later. The reduced operating, maintenance, and other long-term costs of a more efficient design will be important to the owner who places significance and value in the future. The opportunities from life-cycle costing may be missed by developers who build for speculation or owners who believe that they will always be able to pass increased operating costs along to their tenants. Government agencies, non-profit organizations, and developers who will also be the long-term owners are examples of those in a position to take advantage of the savings achieved by a more efficient design throughout the lifetime of the project.

The Analysis of Financial Investments in Energy Saving

The business economist knows that the dollar savings from energy efficiency improvements are equivalent to the income earned from other investments. They will be taxed accordingly and they will be evaluated for their investment performance in comparison with other opportunities. The purpose of a financial analysis of a project for the capital budget is to provide a quantitative measure of its value under present and reasonably foreseeable future circumstances. The following methods for analyzing an investment proposal are based on the time value of money, called its *present value*. According to this concept, an amount of money that is available for investment today is more valuable than an equal amount received at some point in the future, because of its capability for earning interest during this period.

The value of money used to save energy is compared to the cost savings that are expected to accrue over the economic lifetime of the improvement, discounted to reflect the fact that these savings will be received in the future. The economic lifetime of an investment (as distinguished from the physical or useful lifetime of a capital improvement) is the period during which the improvement will be amortized. The merit of an investment, then, is based on the present value of the savings received during this period.

Listed below are three expressions useful for the financial analysis of energy savings.

Equation (1) is the expression used to discount, or to determine the present value of a dollar to be received in the future.* Equation (2) is the present value of one dollar per year received from energy savings for a period of n years (an annuity).†

Expressions Used for Financial Analysis of Energy Savings[a]

1. Present Value (discount rate) $$PV = \frac{1}{(1 + i)^n}$$

   i = interest rate
   n = number of years

2. Present value of an annuity $$A = \frac{[(1 + i)^n - 1]}{i(1 + i)^n}$$

2.1. Capital recovery factor = $1/A$

3. Energy inflation factor $$I = \frac{[(1 + r)^n - 1]}{nr}$$

   r = rate of annual energy price increase
   n = number of years over which the average energy inflation is calculated

[a]Helpful tables will be found in Appendix 3: Present value (Equation 1) in Table 1; present value of an annuity (Equation 2) in Table 2; present value of energy savings with fuel inflation in Tables 3a through 3k.

However, because the costs of energy will increase, the size of future savings will also increase. Since the rate of inflation for energy is expected to be considerably higher than the rate of inflation for other commodities, it will be worthwhile to take this into consideration. Equation 3 is an inflation factor used to evaluate the *average* saving during a period of n years when the cost of energy is escalating at an average annual rate r. For example, if energy prices increase by 12 percent per year, during the twenty-year life of a project, the *average cost* of energy during this period would be a factor $I = 3.60$ greater than present energy costs. Tables for discounting (Equation 1), for the present value of an annuity (Equation 2), and for the present value of energy savings with energy price inflation will be found in the Appendix.

*If money can be invested to earn a return of 10 percent per year, then the present value of one dollar to be received 5 years in the future is 62.1¢ (see Table 1 in Appendix 3). If the one dollar were to be received 10 years in the future, then its present value today would be only 38.6¢. If the discount rate were 15 percent rather than 10 percent, the present value of one dollar received 5 years from today would be 49.7¢.

†If $1,000, invested at 10 percent interest, is to be repaid in five equal installments, the annual payment would be $1,000 ÷ 3.791 or $263.8. (see Table 2 in Appendix 3). The present value of these five payments received at one, two, three, four, and five years into the future would be $239.8, $217.9, $198.3, $180.2 and $163.8. Hence, although a total of $1,318.9 would have been received from payments on the investment at various times in the future, their present value sums to $1,000.

Analysis of an Investment

To calculate the present value of a capital investment that will generate a stream of future savings, the following data is required:

1. The cost of the investment, C
2. The saving during the first year, S
3. The economic lifetime of the investment, n
4. The interest, i, at which the money can be obtained
5. The estimated, average rate of fuel inflation, r

The cost of the project is the total outlay (including any non-capital expenses) that will be made if the course under evaluation is taken. Therefore, it should include shipping, installation, costs for training operators, and other factors, even though these may not normally be considered capital expense items. The annual savings are the differences between the gross energy savings and the annual operation and maintenance expenses that would be required because of the new investment. The longest possible lifetime which can be justified should be chosen. As explained earlier, a policy which requires that a project pay for itself in a period much shorter than its service life tends to focus design criteria on minimizing first costs, with the consequences of higher life-cycle costs and greater energy use.

According to the present value method, the project should be financially acceptable if the present value of the savings exceed the net cost. Their present value is the initial value of the savings, times the factor from Tables 2 or 3 for the corresponding discount rate and lifetime years. The following examples will serve to explain the method. Consider an investment of $1,000 to save fuel worth $260 today, but expected to increase in value at 12 percent per year. Furthermore, assume that this is a non-taxpaying organization which can borrow money at 8 percent over a period of 6 years. The factor 6.828 (obtained from Table 3e in the column under 8 percent and in the line for 6 years), times $260 gives $1,775. Since the present value of the savings ($1,775) over the 6-year lifetime exceeds the project's cost, it would be acceptable.

To compare the merit of one project with another, it may be useful to express the results of the analysis in one of the following terms—the benefit/cost ratio, the internal rate of return, or the discounted pay-back period. The benefit/cost ratio is a comparison of the present value of the benefits (savings) generated by a project during its economic lifetime, with its costs. The benefit–cost ratio of 1.78 is greater than 1, indicating that the project would be acceptable. The financial priority among projects with equal risk and other factors would be for the one with the highest benefit/cost ratio.

The internal rate of return on investment (ROI) is that discount rate which results in a present value for the savings just equal to the expenditure. Table 3e can also be used to obtain an approximate value for the ROI. The ratio of the cost, ($1,000), to initial savings ($260) is 3.85, which lies between 25 percent and 30 percent in the line for 6 years. (The exact ROI must be solved by an iterative process and is 28 percent.) Since this rate of return on the investment is much higher than the interest rate of 8 percent, it would be a very satisfactory investment.

The discounted pay-back period may also be estimated with the aid of Table 3e in the Appendix. The ratio of the cost to initial savings, 3.85, can be found under the 8 percent interest column between the third and fourth years. (The actual payback period is slightly over three-and-a-half years.) Since the investment can be recouped with interest in less than its economic lifetime, the project is justified. These present value methods put projects which may have different lifetimes, costs, and savings all on the same footing for comparison with each other, as well as with income-producing investments.

Had fuel inflation been neglected, the benefit–cost ratio, rate of return, and the discounted pay-back period would have been 1.2, 14.4 percent and 4.8 years, respectively. The analysis for a tax-paying firm is somewhat more involved because depreciation and income tax credits enter into the calculation of the after-tax, net present value of savings.

Present value analysis for a profit-making firm: Consider an investment of $1,000 to save fuel worth $260 with 12 percent fuel inflation. Assume that the project has an economic lifetime of 6 years, and that the interest rate is 15 percent. In computing the after-tax present value with a 50 percent tax on profits and a 20 percent investment tax credit, sum-of-digits depreciation will be used to obtain the net profit. The analysis is displayed in Table 2.2, where one should note that due to round-off, sums may not agree to within $1 or $2.

The net, after-tax present value is $1,078, and with the 20 percent tax credit, the net present value is $1,278.

The benefit–cost ratio is 1.28.

Table 2.2

	Year							
	0	*1*	*2*	*3*	*4*	*5*	*6*	*Total*
Saving	$260	$291.2	$326.1	$365.3	$409.1	$458.2	$513.2	$2363
Depreciation factor (Note: 1 + 2 + 3 + 4 + 5 = 15)		5/15	4/15	3/15	2/15	1/15	0	
Depreciation		$333.3	$266.7	$200	$133.3	$67.67	0	
Profit		42.1	59.5	165.3	275.8	390.5	513.2	1362
Tax @ 50%		21.1	29.7	82.6	137.8	195.3	256.6	
After-tax cash flow[a]		$312.3	296.4	282.6	271.2	262.9	256.6	1882
Present value factor (at i = 15%)		.8700	.7561	.6580	.5720	.4972	.4323	
Present value		$271.5	$224.1	$185.8	$155.1	$130.7	$110.9	$1078

[a]Depreciation expenses are added back to obtain the total cash flow, e.g., $645.6 for year 1, $563.1 for year 2, etc.

The discounted payback period is somewhat less than 3 years.

The internal rate of return on investment (ROI), as determined by iteration, is 27.8 percent.

Had the fuel inflation been taken as zero, the above factors would be $1,059 for the after-tax, present value. The benefit–cost ratio would be 1.06. The discounted payback period would be in somewhat over 5 years. The ROI would be 18 percent.

The Bottom Line

As noted at the beginning of this section, the method of accounting is important to the success of an efficiency improvement program, since different options are usually compared on the basis of cost to the owner. When a more efficient alternative also requires a higher initial investment, the comparison of alternatives should be referred to differences in the total annual costs of ownership, including amortization of investment and energy costs. The bottom line for the decision-making process is: will the additional expense for the more efficient option be returned with interest from the savings in the annual costs of ownership?

2.3. SAVING, NEGATIVE SAVING, DO-NOTHING, AND SAVING

One hears and reads of many suggestions to save energy and cut operating expenses. Some can be helpful, some appear to offer saving but achieve nothing in reality, and some are actually erroneous—even harmful. To ensure that the measures applied have a beneficial and long-lasting effect, it is essential that do-nothing and harmful measures be separated from constructive efficiency improvements.

The following categories can be identified:

1. Real saving
2. Negative saving
3. Do-nothing
4. Multiple savings which we call SAVING

1. Real savings are founded upon understanding how energy is used and how energy-using equipment functions to maintain comfort conditions and to supply other needed services and processes. An example would be the saving achieved by reducing winter heating temperatures from 78°F to 70°F (26°C to 21°C).
2. Negative savings are the results of the misapplication of energy-saving methods (ESM) which causes energy use to increase, or which is made at the expense of a pleasant and productive environment and occupant morale. Examples would be the lowering of thermostats from 78° to 70° in interior rooms, consequently requiring additional refrigeration to maintain this lower temperature in these zones;* or the removal of lamps, resulting in complaints from employees who require high levels of illumination for tasks with low contrast.

*Lowering interior zone temperatures could very well lead to a double loss if air is cooled for interior rooms, then heated up again for perimeter warm air supply.

3. Do-nothing is the description of some actions that achieve negligible saving, if any, despite the expenditure of considerable time and money in some cases. Examples would be removing lamps to decrease energy used for lighting, only to have the equivalent energy supplied in the form of reheat elsewhere in the system; or expenditures for such devices as vapor injection equipment for burners, fuel oil additives, transient suppressors, or unnecessarily complicated control systems. These actually save no energy in and of themselves, but merely function as expensive placebos to explain any savings observed which result from closer attention to energy use or some unidentified improvement in operation.
4. SAVING is the multiple saving which results from decreasing energy use in one place, resulting in decreased energy use elsewhere as well. For example, if excessive lighting is decreased, the cooling load on the chiller is also reduced in many systems. As another example, several benefits were achieved by a manufacturer of high voltage equipment who cut back on water use. Water meters were installed and valves, which had been set full open since they were installed, were adjusted for the actual flow required. Three benefits resulted from the 80 percent reduction in annual water use:

1. In some cases, the output temperature was sufficiently high that the hot water could be reused for space heating.
2. The water system now had sufficient capacity to serve a new addition to the plant, and plans to expand the water system were no longer necessary.
3. Yearly savings of $12,500 were achieved which reflected reduced impact on the water supply and the sewage systems.

Only a fraction of the methods given in the checklists will be applicable to any one situation. Methods from the checklists can be misused if they are applied under the wrong conditions, with consequent do-nothing or even negative savings.

An example from the checklist which is not universally applicable would be, in some cases, the addition of insulation, an apparently innocuous energy-saving activity that could backfire. If the area already has a surplus of heat, this could lead to uncomfortable conditions and to the expenditure of extra energy for ventilation fans. Or, incorrect application of insulation could lead to moisture problems or fire hazards. Another example would be raising the room temperature during the air-conditioning season; if the building uses reheat for humidity control, this could result in additional energy requirements for reheating.

We will try to point out some of the pitfalls as we go along. However, the only way effectively to avoid these pitfalls, is to understand local conditions and to have full knowledge of how your energy-using equipment functions. Consult the manufacturer's literature with each piece of equipment being considered and use it along with this book. There may be special characteristics that have to be taken into account.

Difficulty can also occur by trying to optimize too small a microsystem at the expense of something else in the overall system. A general principle is that one should strive to make the system to be optimized as large as possible and yet still have it manageable (see Figure 2.6).

Some examples will serve to illustrate the interaction between different systems and show that optimization of the larger system may require a compromise, or less than minimum energy adjustment for some of the sub-systems. With careful planning and

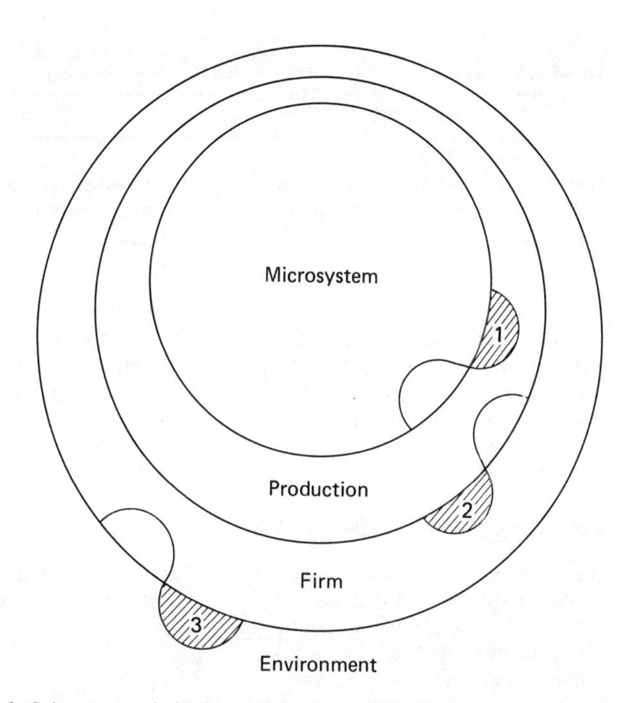

Figure 2.6. Subsystem optimization at the expense of the larger system. Examples would be: (1) minimizing energy use at the expense of production output; (2) minimizing energy use in production at the expense of quality, thereby shifting the burden onto the firm and its reputation; (3) minimizing first costs through installation of a resistance electric heating system, thereby shifting the burden onto future tenants and onto society at large because of its impact on fuel and capital supplies.

operation, a larger window area may result in sufficiently less electricity use for lighting to offset some heat transmission problems. On the other hand, reflective window coatings, which help reduce the summer air-conditioning load, may preclude even larger savings that might be achieved by using daylight for lighting and the winter sun for warmth.

In a purely technical approach to energy efficiency we may forget to take advantage of some of the environmental resources at hand. Consider that well water may be a very satisfactory heat sink in summer and a good heat source in winter. In a broader perspective, optimization means reducing deleterious effluents, with local recycling if possible. Heat can be recovered from air and waste water discharges. The heat content of garbage and other solid wastes can be captured with a waste-burning boiler, thus retrieving another energy source from a landfill. As the system concept is enlarged, more opportunities for multiple savings become apparent.

In summary, negative consequences can be minimized with sufficient knowledge of existing conditions and the effect of each proposed action. The opportunity for multiple

Table 2.3

Some resistance to EIP	*Response*
Saving face	
For 20 years we have already been doing everything possible to save energy.	Increased energy prices may make new approaches necessary.
Pride	
Our system is essentially brand new, and it was designed by one of the largest engineering firms in the country; besides, we have no money for any additional investment.	There is a lot of energy-saving potention in the practical small details of operation.—By the way, not all engineers have been energy conscious in their planning.
Can old dogs learn new tricks?	
We can order an EIP study to prove to upper management that we appreciate the problems, but there will be little practical chance to introduce any of the proposals.	Let us try a realistic, not overly enthusiastic approach together.
Not me	
We would be interested in ESA but there is nobody in charge.	Management should designate a coordinator.
Too much risk	
Our system is finally running without any complaints from the tenants. It is not worthwhile to risk any disruptions.	Do the owner and tenants actually understand the magnitude of the opportunity being missed?
Progress (??)	
This is a return to the dark ages.	We have to learn again what reasonable energy use is without decreasing necessary comfort or production.
Not here	
The advantages to us are too insignificant compared to our other problems and opportunities. That is an interesting idea but it won't work here.	A lot of small savings add up to a big sum. Let's try it.

savings can be enhanced by making the boundary of the system as large as possible. Many of our present difficulties stem from not fully understanding how things work and/ or from trying to optimize too small a system.

Resistance to Energy Saving

From our experience, the most successful efficiency improvement programs are based on a positive attitude of those involved. When management provides encouragement, and operating personnel are willing and informed, energy use can be improved. Some of the different forms of resistance to starting an EIP are summarized, with responses, in Table 2.3.

REFERENCES

(1) DOE, "Electric Utilities: Percentage Produced from Each Source, data for the month of August, 1977," *Monthly Energy Review* (October, 1977), p. 33.
(2) GSA, "Energy Conservation Guidelines for New Office Buildings" (Washington, D.C.: Public Buildings Service, General Service Administration, July 1975).
(3) Thomas F. Widmer, "Accelerated Conservation: An Alternative National Energy Plan," (Waltham: Mass.: Thermo-Electron Corp., February 15, 1977).
(4) C. A. Berg, "A Technical Basis for Energy Conservation," *Technology Review* (February 1974), pp. 14–23.
(5) W. Carnahan, K. Ford, et al., "Efficient Use of Energy: A Physics Perspective," in *Efficient Use of Energy*, (New York: American Physical Society, 1975).
(6) M. H. Ross and R. H. Williams, "Energy and Economic Growth," A report prepared for the Joint Economic Committee of the U.S. Congress, in *Achieving the Goals of the Employment Act of 1946—Thirtieth Anniversary Review*, vol. 2, paper 2 (Washington, D.C.: U.S. Government Printing Office, 1977).
(7) E. Gyftopoulos, L. Lazaridis, and T. Widmer, *Potential Fuel Effectiveness in Industry*, (Cambridge, Mass.: Ballinger, 1974), p. 89.

3

Energy Planning

In view of the improvements in the regional economy and employment to be obtained from greater energy efficiency, and the growing concern for the environmental impacts from energy use, state and local governments are taking increased responsibility for energy management. It is the purpose of this chapter to encourage community and regional planners to assume more responsibility for improving energy efficiency. District heating should be considered a basic utility, along with power and gas, which might be efficiently integrated in some existing areas and in the planning of new communities. City codes can also be effective aids in improving building efficiency and in restricting the use of resistance heating where this would help to stretch existing electrical capacity and to lessen the increase of electrical rates. Land-use planning should strive not only for improved living and commercial amenities, but should reflect the need for less energy-intensive living patterns and the need to integrate commercial and industrial facilities for greater efficiency.

In the future, to assure an adequate long-range energy supply, larger regions should develop energy plans similar to the planning that has been traditional for water supply and sewage disposal. Some of the supply systems which offer greater efficiency would integrate district heating, integrated waste disposal, and the generation of electricity. In outlying areas, as population density decreases, the distribution of district heating eventually becomes uneconomical and the distribution of fuels (gas and oil) would then be favored. In the cities, utilization of a central energy plant to provide heating as well as cooling for both commercial and residential use can achieve significant savings, especially when heat production is combined with an electricity-generating plant.

Plants burning solid waste to produce heat* and electric energy have been constructed in many European countries (and to a limited extent in the United States and Canada), usually where the costs of both solid waste disposal and fuel are high. For example, typical wet municipal refuse from which glass and metals have been separated has a heating value (even though wet) of about one-third that of high-value bituminous coal. In the past, incineration or burial in landfills have been the most economical methods for disposal of solid wastes. Although for various reasons it has not been generally worthwhile to attempt to capture the heating value of the disposed wastes, it has been estimated that the heat potential in industrial, commercial, and residential wastes from which solids have been removed is equivalent to 160 million tons of coal. As the population increases and our economy grows, significant amounts of solid wastes will be generated; compounded with higher fuel prices and increasing waste disposal costs, there is now an increasing interest in the United States (as there has been historically in European countries) to use these wastes for generating electricity and hot water for process use (Figure 3.1).

Among other objectives, residential areas should be planned to minimize the need to use the automobile for local shopping and commuting purposes. Access to recreational

*Steam, or preferably super-heated water.

Figure 3.1. Photograph of the waste incineration plant with heat recovery for the district heating system of Basel, Switzerland.

open spaces for walking, exercise, sports, and other activities should also be planned so that the quality of life is less dependent on the use of the car. Such organization represented by these unconventional planning criteria may be found in some of today's most livable communities.

Commercial and industrial complexes can be aggregated to share total energy facilities, and to implement the use of heat rejected from high-temperature requirements. Such industrial complexes, which require a collection of users with compatible operating schedules and flow requirement, will place unusual demands on planning and funding.

By assuming responsibility for energy planning at the regional or local level, local economic conditions can be improved and action can reflect local conditions. In California, to cite an example, new government buildings are designed to innovative energy-saving standards, and active assistance is provided citizens in reducing their energy needs. As a consequence, unnecessary construction of power plants has been avoided and the use of expensive liquified natural gas has been reduced. Since local labor and local construction firms are involved in the energy-saving business, they have both helped to create and benefited from the economic prosperity in California, while other parts of the country were still seeking more conventional solutions. (See Reference (35) in Chapter 1.)

3.1 PROJECT FOR THE ACCELERATED DEVELOPMENT OF DISTRICT HEATING IN WEST GERMANY (1)

This example discusses a plan for the accelerated development of large-scale district heating as a measure to help reduce this country's heavy dependence on oil imports. Data for this project are summarized below. Essentially, the study shows that it would be possible to provide economically about 60 percent of the space heat demand for the entire Federated Republic of Germany (in those areas densely enough populated to be economically supplied with district heating) from large thermal power plants. These plants, which are located relatively close to densely settled areas, could simultaneously produce useable heat and electricity (cogeneration). It is interesting to note that the capital cost for each unit of primary energy saved is relatively low as compared with medium-return-rate energy-saving actions. The specific capital productivity for energy saved compares favorably with other, medium-return energy-saving measures.

The study gives the following conclusions:

1. The electric generating district heat plant must be located close to consumers.
2. Plants supplying heat only (thermal plants) are needed during the first phase of development.
3. The upgrading of thermal plants to the simultaneous production of heat and electricity should become more economical as the system is developed section by section.
4. Enforcement of air pollution laws would help promote the development of district heating by encouraging users to hook up to the system.
5. Space for the district heating lines will require acquisition of rights-of-way in the streets, etc., which can be implemented in cooperation with municipal authorities and the utilities.

6. Subsidies should be offered to encourage existing buildings to connect to the district heating network.
7. Subsidies will be required during the start-up phase of the accelerated development of district heating.
8. Long-range planning for the district heating system and for other metropolitan energy supply needs should be carried out to assure an adequate, future energy supply.
9. A public information campaign would be required to acquaint the potential users with the features of the district heating system and the possible advantages of hook-up.
10. Additional research and development of some engineering problems may be able to lower future costs.

Data from the West German Study

Listed are some typical data from the economic and technical study of the expansion of district heating in West Germany, as presented at the Fernwärme International. (1) There are:

1. 104 co-generating power plants
2. 366 district heating plants
3. 465 heating systems, total 5,300 km (33,000 mi)
4. a total connected thermal power of 23,000 MW (78,000 MBtuh)

Proposed accelerated development:

A connected thermal power of 40,000 to 70,000 MW (136,500 to 239,000 MBtuh) by the year 1990. (This would correspond to one-fourth of the total low-temperature heat demand below 200°C (395°F).

Total investment up to 1990 would amount to between $7.5 billion and $14 billion (about $200/kW or $15,000/BPDE).

This would result in a saving of primary energy resources of 80×10^6 to 135×10^6 MWh per year (50 to 80 million bbl of oil).

This district heat system saves fuel at the rate of 1 MWh per $100 of investment, which represents a specific capital productivity of $875/kW or $62,000/BPDE. This compares favorably with some of the other energy-saving improvements discussed in Chapter 5.

Hook-up subsidies of $40,000/MW ($11,700/MBtuh), or $400 per apartment, would be required in the initial phase.

Nuclear or fossil fuel power plants near cities can supply heat to a surrounding region within a radius of 30 km (20 mi).

Smaller fossil-fueled co-generation plants with capacities of 100 MW (340 MBtuh), and thermal power plants with capacities of 30 MW (100 MBtuh) or less, can be used with a district heating network to serve smaller and medium-size communities.

With simultaneous production of heat and electricity, there is a small reduction (about 15 percent) in electrical energy produced in existing thermal nuclear power plants.

3.2
PLAN FOR THE METROPOLITAN AREA OF LUCERNE (2)

Next we describe an energy-planning study for an urban area with a population of 120,000 and the development up to the year 2000 of a publicly owned distribution network for district heating and natural gas. (2) The planning objectives were to improve safety of supply, reduce the influence on environment, and achieve good economy.

At present, the heat supply in the region consists almost entirely of oil, which is burned in individual furnaces; the reliability of supply depends on the availability of oil. District heating is not in use and the development of a distribution system for natural gas is in its infancy.

In terms of the objective—to improve long-range safety of supply, especially during possible periods of international crisis—the following must be considered:

The growth of primary energy demand should be minimized, even if there is a population or industrialization increase. This is to be achieved by such energy-saving measures (ESM) as improvements in ventilation and heating systems, heat recovery, improved insulation of buildings, etc.

More rational use of energy should be achieved. This means a reduction in production, distribution, and end use losses with, for example, the simultaneous production of heat and electricity or the use of heat pumps.

The heat supply should be diversified to include several additional primary energy forms, with about equal contribution to the total supply. This means reduction in the present use of oil and gas. Refuse and industrial-waste heat should be used wherever possible.

In Lucerne, it is finally being recognized that with continued increase in energy demand, a deterioration of the environment must be expected if no corrective measures are taken. A major goal of the energy plan is to reduce this impact on the environment, particularly in the area of air pollution. The emission of soot, particulates, and sulphur dioxide, as well as peak values of emission at critical locations, and auto emissions, must be reduced.

Any new heat supply system based on distribution networks for district heat and natural gas should be capable of operating without deficit at each step throughout its development.

Deficit during the start-up phase should be held to a minimum and publicly subsidized.

Heat prices paid by the users should be competitive with equivalent costs for operating individual home oil-heating systems.

Description of the Region

The area considered in this study can be divided according to the following distribution characteristics:

Zones which are suitable for district heating, and where good economy can be expected because distribution costs will be low. In such zones there is a high population density, the topography is advantageous (no large differences in geographical height), and the distance between the heating plant and the main users is small.

Zones which are suitable for the further development of the natural gas supply network.

Zones with low heat density or with large distances between users and the heating plant. Land use planning should emphasize the construction of small housing communities. Other heating systems should be promoted such as electric, heat pumps, and solar energy.*

For the whole region, vigorous energy-saving methods must be implemented to assure that today's heat demand is stabilized despite further increase in population.

The cheapest part of the energy supply system is for that which does not have to be supplied because of the avoidance of waste.

The Heat Supply in the Year 2000

The mature system could serve the city and several neighboring communities with one or two main heating plants. Assuming the price of heat from the district heating system is competitive, a high proportion of the users would be connected to the system without requiring connection by ordinance or building code, as users who can be supplied economically with district heating will voluntarily want to connect to the system. The price for district heating would be pegged to the equivalent price for residential oil heating, and is expected to be about 5–10 percent higher than with individual oil furnaces (including capital costs of both systems). The energy sources to be used include natural gas, oil, waste, and if available in the distant future, heat from nuclear power stations. A nuclear power plant situated about 20 kilometers (12½ miles) from the city had been discussed for construction during the 1990s. However, the concept of district heating is such that it would be economic in the year 2000 even if no nuclear power station were built.

Typical data for the district heating system are given below. The investment required is $80 million to supply a connected thermal power of 300 MW (1000 MBtuh). This corresponds to $4,000 per apartment. About half of this cost must be provided with some type of investment subsidy to cover the deficit during the starting phase.

Today the supply of natural gas is operated at a deficit. The cost of the existing network is too high as compared with the amount of gas sold. The use of natural gas should be increased to the point that the system can be operated without a deficit. The use of natural gas should be extended to:

large users who could switch to oil during peak periods (when outside temperatures are low), such as industrial plants and large apartment complexes.

new areas which could not be supplied with district heating (i.e., the old, medieval section of town), through expansion of the existing natural gas network.

*The thickly settled regions at lower elevations receive little sunshine during the heating season because of the climatic tendency for temperature inversions. The cold, thick fog which frequently covers Central Europe has a high concentration of industrial and urban pollutants.

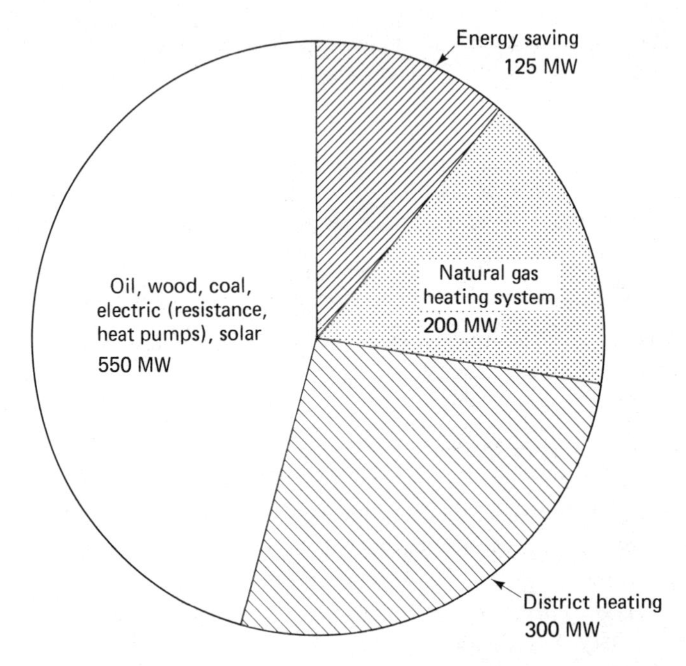

Figure 3.2. The expected distribution of energy for the city of Lucerne and its outlying suburban regions in the year 2000. Individual heating systems fired with wood, coal, and (mostly) oil, which account for over 90 percent today, should be reduced to about 50 percent. The increased use of solar heating is expected; however, because of uncertainties in predicting its technical development, solar has been included with oil, etc.

Electric heating systems would be able to cover only a small portion of the total energy demands, depending on the available capacity of the distribution network. Heat pumps could be used in heavily insulated buildings. Heat pumps should be used to upgrade waste heat (i.e., from ventilation exhaust) and to take advantage of the solar heat in well water. Sources of warmer subterranean water should be identified. With heat pumps using subterranean water, the efficiency of delivering comfort heat can be very high compared with electric resistance heating.

Individual oil heating systems would still have to cover about half the total heat demand, predominantly in areas with a lower population density. Renewable heat sources such as solar heating systems could achieve considerable importance, especially in these regions (see Figure 3.2).

Typical Data of the District Heating System

Technical Data. Power station with the co-generation of heat and electricity 100–200 MW thermal power. Heating plants of 30–100 MW designed to cover peaks in heat demand. Distribution network with 130°C supply/70°C return (270°F/160°F) water piped through 100 to 400 mm-diameter lines (4″ to 16″).

Energy Saving. The production of electricity in a thermal power plant (efficiency 30 percent) and heat production in individual heating systems (efficiency 65 percent) will be replaced by co-generation (efficiency 85 percent). The saving during the first phase when only heating plants (without co-generation) would be used would be 5 percent per MW and 15 percent for the combined system with heat and electricity produced by a nuclear power plant. The specific capital productivity would initially be $3,700/kW of saved thermal demand, decreasing to $1,300/kW for the saving achieved with co-generation. This corresponds to an investment of $425 ($150 with co-generation) to save 1,000 kWh per year.

Costs. Heating plant $40 to $60/kW heating capacity. Distribution network $80 to $120/kW ($400–$800 per meter, $650,000 to $1.3 million per mile). Total system $120/kW to $180/kW (in special cases up to $280/kW) of heating capacity.

Price of Heat. Equivalent to 10 percent more than the cost of private oil heating (heat cost + amortization + 10 percent), $8 per 10^6 Btu ($28/1000kWh).

Operating Assumptions. No mandatory connection to district heating should be required. In the district heating zones, 30–60 percent of the consumers would connect to the system.

Economically Viable Areas for District Heating. Heat demand greater than 250 MBtuh/mi^2 (30 MW/km^2) or greater than 13,000 persons/mile^2 (5000/km^2), which corresponds to a town of 10,000 to 15,000 inhabitants with 30 percent connected.

Development Phases for District Heating

Phase 1: Construction of a heating plant in the city and smaller district heating systems in the outlying communities.

Phase 2: Connection of district heating systems of some of the communities and the city to a new, larger heating plant (possibly with the simultaneous production of heat and electricity). A plant to be constructed at the location of the refuse incinerator to also allow use of the heat from waste.

Phase 3: Enlarge distribution network increasing the economics of the operation.

After the start of Phase 1, the final stage would be achieved in 20 years. The size and schedule for construction of the district heating network would depend on energy cost escalation and availability of capital, as well as the urgency for substituting oil and reducing sulphur dioxide emission. The maximum possible share of the heat load by the

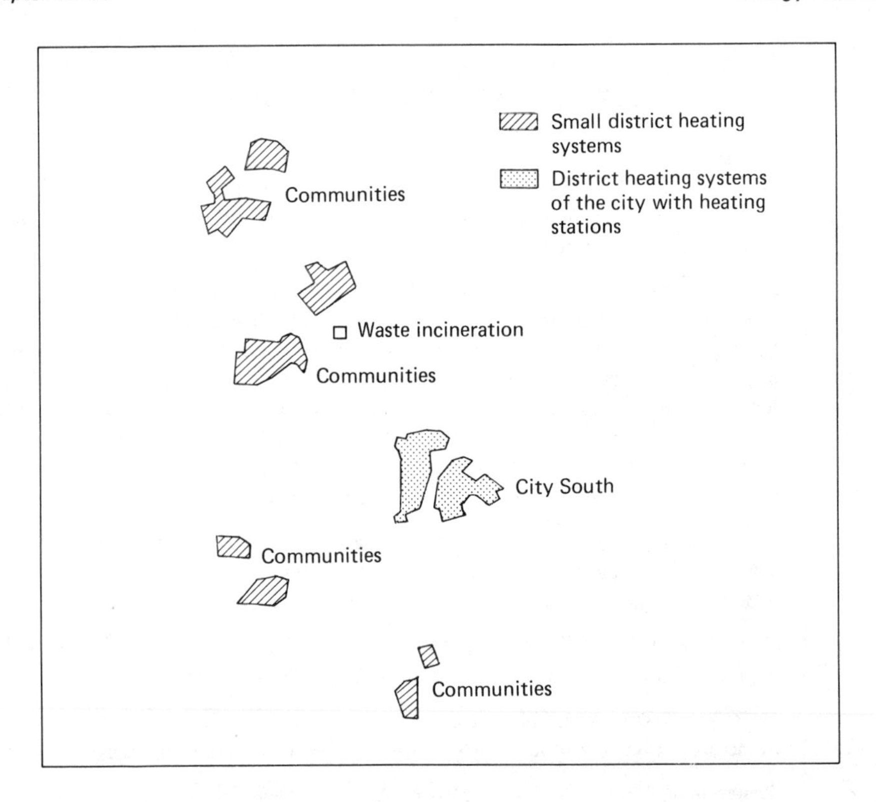

Figure 3.3. First phase of a proposed district heating development for the city of Lucerne, Switzerland.

year 2000 would probably be somewhat less than 50 percent. The first and third development phases are shown in Figures 3.3 and 3.4.

3.2 CONVERSION OF AN EXISTING POWER PLANT FOR DISTRICT OR PROCESS HEAT

The next example shows that it is also possible to convert an existing power plant to simultaneous production of electricity and heat. While such a conversion will result in lower production of electricity, higher overall efficiency will be achieved. Figure 3.5 shows the relationship between turbine efficiency and the temperature of the heat output. In the figure, the cycle efficiency corresponds roughly to the ratio of the electricity output to the total fuel input.

Although the efficiency of the plant is lower, the higher temperature of its rejected

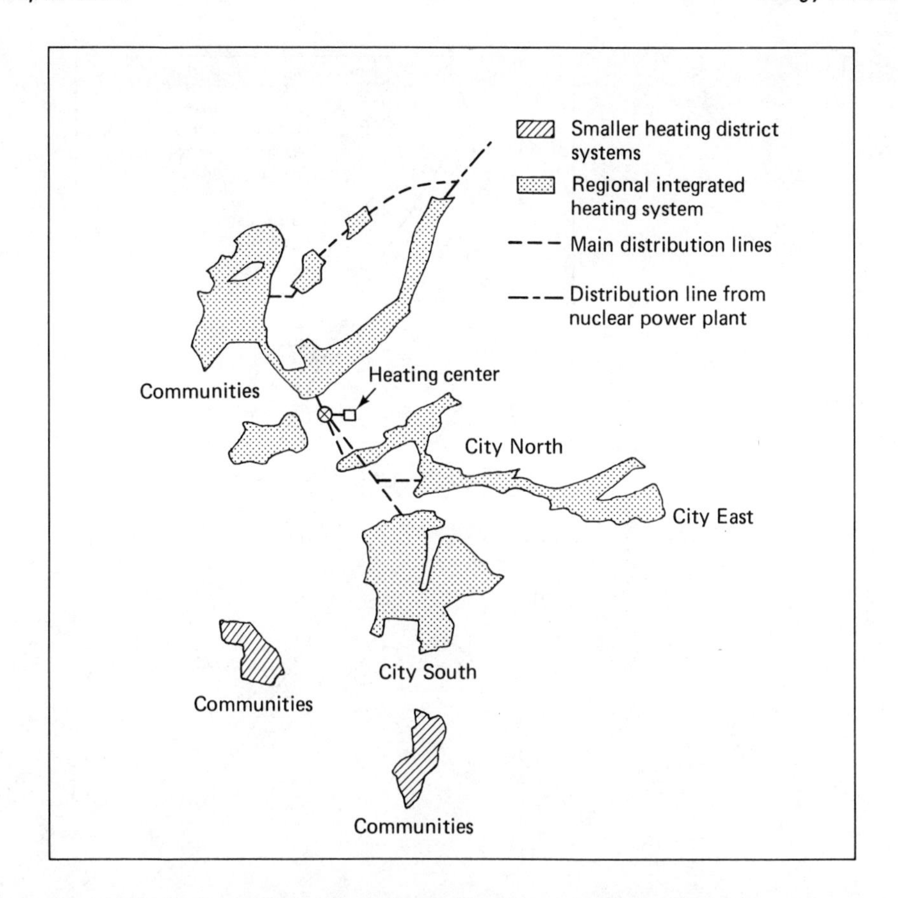

Figure 3.4. District heating Phase 3, about the year 2000.

heat means that it is worthwhile to transport it to where it can be put to use. In the following example, the conversion of an existing nuclear power plant is illustrated. (3)

Present Situation

The city of Berne has a population of 160,000. The central part of the city is already served by a district heating system with a capacity of 150 MW (maximum 250 MW). For heat production, a waste burning thermal plant is used which supplies the base load. A fossil-fired (oil and natural gas) district heating plant with simultaneous production of heat and electricity covers the peak demand in winter. In addition, on the west side of the city there are several communities with large apartment developments having their own smaller district heating systems with a total capacity of almost 100 MW. West of

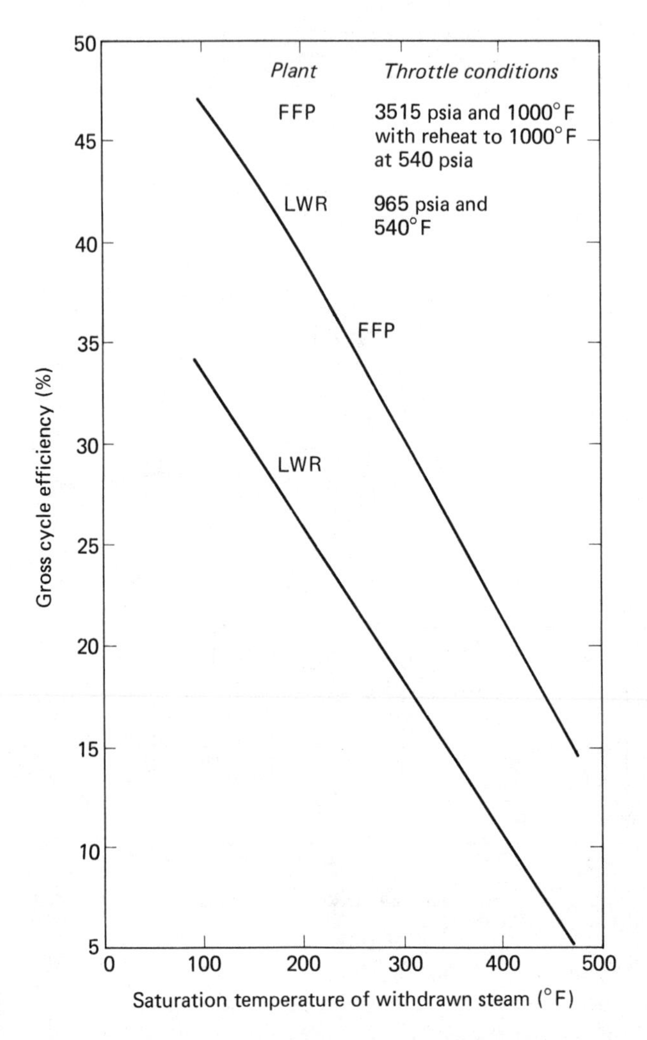

Figure 3.5. Effects of exhaust conditions on efficiencies of large fossil-fueled plant (FFP), and of light water reactors (LWR). [Oak Ridge National Laboratories (4)]

the city, about 13 kilometers (8 miles) away from the district heating station, there is an existing nuclear power station that has been in operation since 1972.

This boiling water reactor has a thermal power of 950 MW. It supplies saturated steam at a pressure of 70 atmospheres (1,000 psi) and 515°F (268°C) to two groups of steam turbines, only one of which is shown in Figure 3.6. Each turbine group consists of a high-pressure extraction turbine series connected with a low-pressure turbine to the

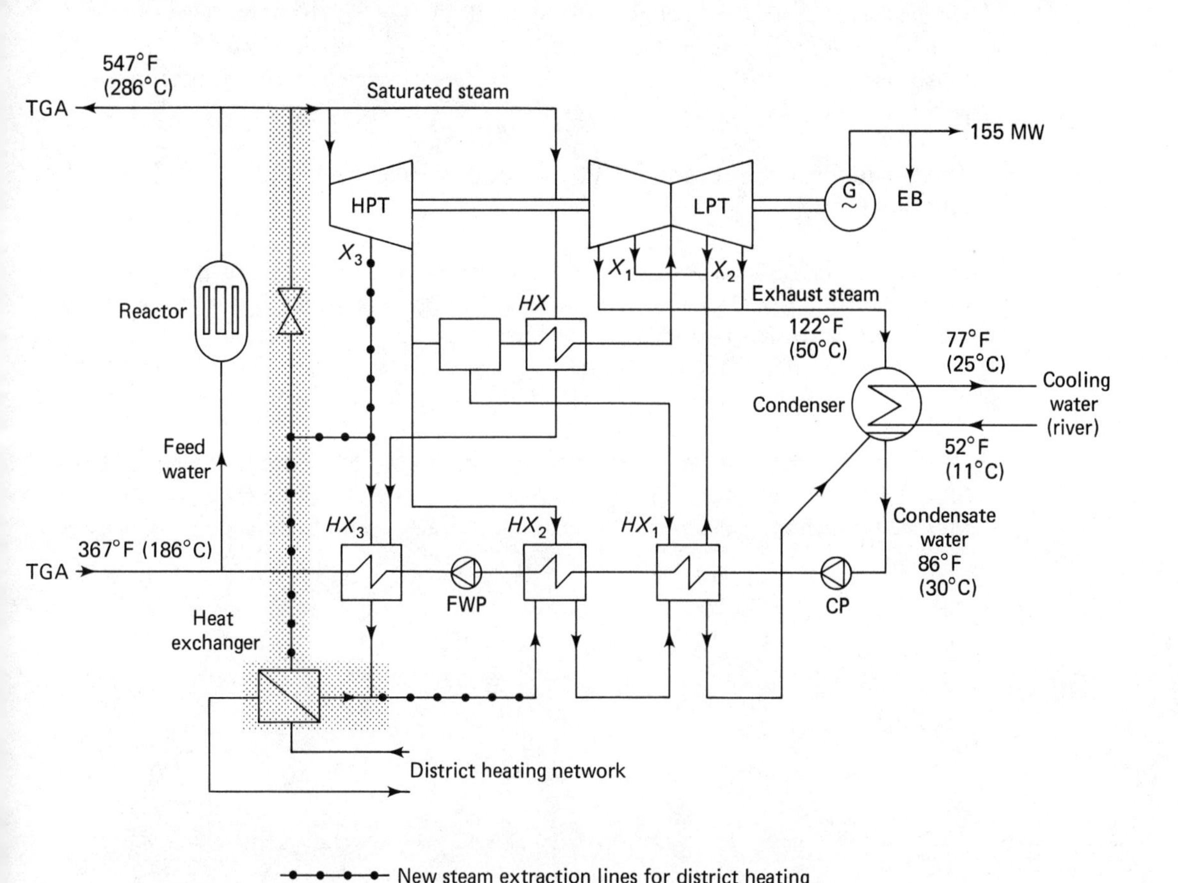

Figure 3.6. Nuclear power plant showing turbine group B with simultaneous production of heat and electricity. The nuclear reactor has a thermal capacity of 947 MW (3232 MBtuh) and the net electric output of each generator is 155 MW. In order to avoid erosion while supplying the required temperatures from extraction point 3 (x_3), the preheater (heat exchanger Hx_3) was taken out of service, causing a slight decrease in electric generating efficiency. For situations requiring greater power, or for operation of the turbine group B at part load, there is also a provision for supplying the district heat requirements from the reactor-saturated steam line.

X:	Extraction point	FWP:	Feedwater pump
HPT:	High pressure turbine	CP:	Condensate pump
LPT:	Low pressure turbine	G:	Generator
TGA:	Turbine Group A	EB:	Electricity for use at the station
HX:	Heat exchanger		

electric generator. The full electric power output of both turbogenerator groups, when operated with condenser cooling water of 52°F (11°C), is 310 MW. The difference between the thermal input and the electrical power (640 MW) is rejected as waste heat to the river. With this temperature for the condenser water, the cycle efficiency is 33 percent.

The System for Simultaneous Production of Heat and Electricity

Nuclear heat is used to supply steam to the existing district heating system at a temperature of 355°F (180°C) during the entire year to meet the needs of some industrial process heat users. Steam from the nuclear reactor goes to the high-pressure turbine, from which it is extracted at the tap x_3 shown in Figure 3.6. If necessary, steam can also be drawn directly, though with some loss of efficiency, from the saturated steam line coming from the nuclear steam supply system at a temperature of 535°F (280°C).

For the conversion, it was necessary to construct a new addition to the turbine building. Since the steam from the boiling water reactor is radioactive, the radioactive steam is separated by a heat exchanger to prevent contaminated steam from entering the district

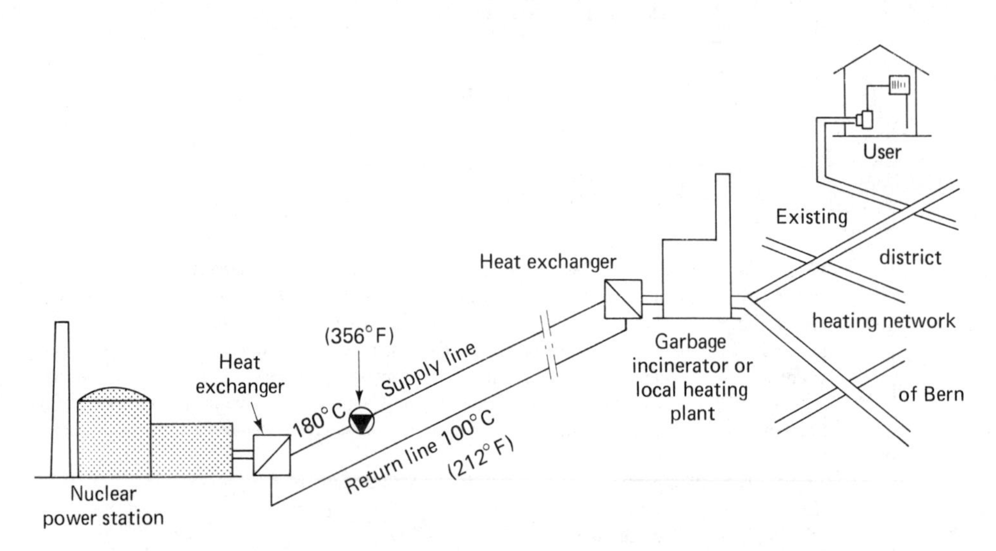

Figure 3.7. District heating from a nuclear power plant located 13 km (8 mi) from the local heating plant (garbage incinerator with heat recovery). The local heating plants continue to be a very necessary part of the system for handling peak loads and for use when the electric power plant is down for repairs or refueling. (3)

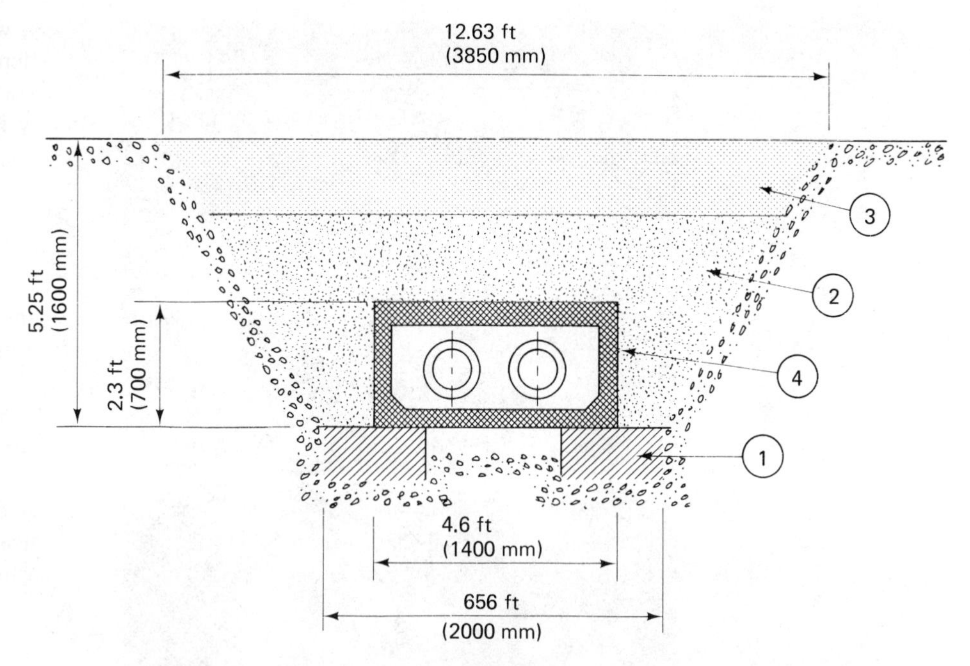

Figure 3.8. Cross section through a district heating line trench. The pipes are about 220 mm (8½ in.) in diameter and are covered with a 50 mm (2 in.) thickness of insulation.

1 Foundation
2 Filling material of gravel
3 Covering material
4 Concrete channel for district heating

heating system. The heating system is maintained at a higher pressure than the steam from the nuclear reactor. This means that in case there is a leak, no radioactive steam should be able to enter the system. In addition to this heat exchanger, there are two additional barriers. The city distribution network is separated by a second heat exchanger from the heat transport pipes to the nuclear station. Furthermore, an exchanger at each house separates the individual user from the city distribution system (see Figure 3.7). The insulated transport lines are installed underground, parallel to the freeway in a protective concrete channel (see Figure 3.8). An example of the installation of district heating pipes in an existing neighborhood is shown in Figure 3.9.

In operation, the heat from the garbage incinerator is used, as previously, for base loads. The existing, fossil-fueled neighborhood boilers are used as a reserve (during repair or refueling of the nuclear power station) and for covering the peak demand above the capacity of the transport line and the garbage burning incinerator (see Figures 3.10 and 3.11).

Figure 3.9. The installation of the pipes for neighborhood distribution of district heat in the city of Basel, Switzerland.

Economics

The cost includes the additional installations and changes in the nuclear power plant and the existing district heating station, the new heat exchanger stations, installation of the transport lines, heat exchangers, etc. Not included are any further possible developments to extend the district heating to other parts of the city. The investment costs:

Modifications to the nuclear power plant	\$ 5.6 million
Transport lines and heat exchanger with 200 MBtuh capacity (60 MW)	\$10.8 million
Total costs	\$16.4 million

Additional data are given in Table 3.1.

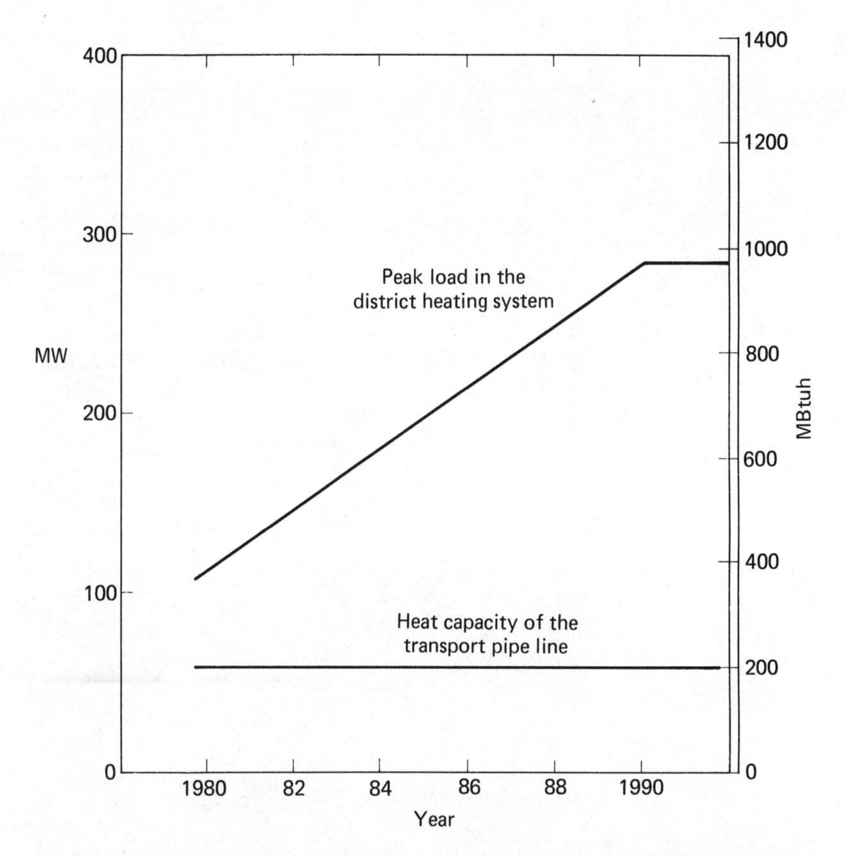

Figure 3.10. Future development of the heat load in the district heating system. The garbage incinerator supplies the base load. Then the heat is used from the nuclear power station, and finally oil is used to cover peak loads and periods when the nuclear reactor is down for refueling.

Table 3.1. Data for Calculating the Cost of Heat from the Nuclear Power Plant

System data		
Optimum pipeline capacity	60 MW	(200 MBtuh)
Annual nuclear heat energy supplied	356,000 MWh	(1,215,000 MBtu)
Annual costs		
Yearly fixed cost	$1,654,000	
Cost of energy for pumping the super-heated water	$93,200	
Forfeited nuclear electricity production[a]	$2,739,000	
Total yearly costs	$4,486,000	

[a]The cost of the nuclear electricity production forfeited for the opportunity to supply energy to the district heating system is calculated as follows:

Reduced power capacity of turbine group B	19.24	MW
Reduced electricity production	124.5 million	kWh
Cost per kWh	$0.022/	kWh

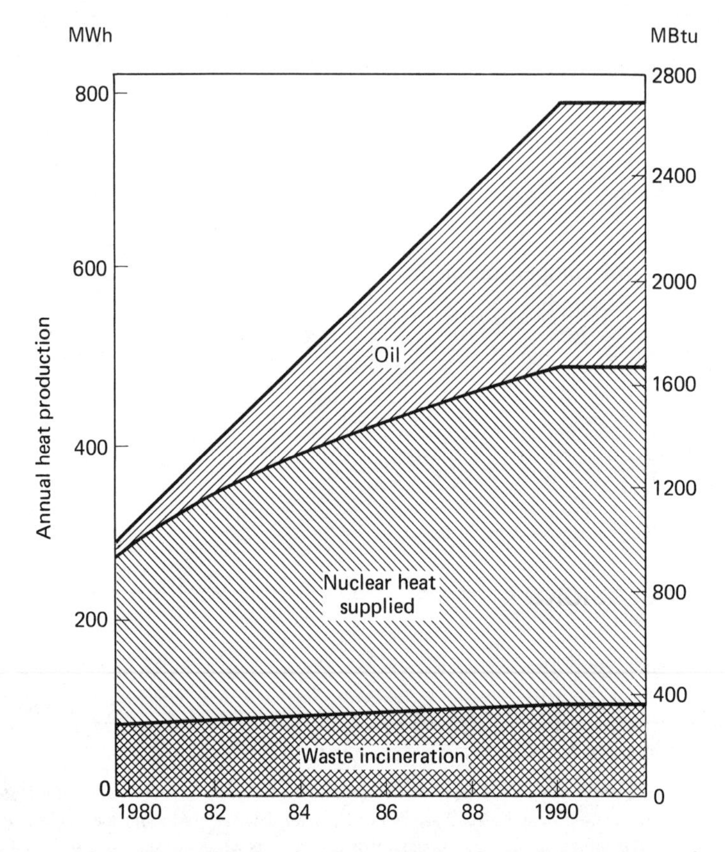

Figure 3.11. Expected development of the heat supplied to the district heating system from oil, uranium, and from burning garbage.

Heating Costs

The yearly heating costs include:

Capital costs for changes in the nuclear power plant, transport line, and necessary alterations to the existing district heating systems

Costs of operation and maintenance

Costs for energy to pump the super-heated water

Compensation for the decrease in electric production of the power plant (about 20 MW)

The value of the corresponding fraction of the nuclear fuel for the 425,000 MBtu of heat energy supplied (about 125,000 MWh) from the nuclear steam supply system

The costs of heat including the deficit during the start-up phase are $3.70/MBtu ($12.50/1,000 kWh). See Table 3.2.

Economic Comparison

A cost comparison of the new composite system must assume continued operation of the existing heating plants and the garbage incinerator, since even if all the heat were to be supplied from the nuclear plant, the existing plants must be ready with a reserve of fuel to take over the load from the nuclear power plant if necessary.

Production costs for district heating are $3.05/MBtu ($10.40/1,000 kWh) using 1976 data—with residual oil at $13.80/bbl ($0.33/gal) burned with an efficiency of 80 percent. Compared to oil, the cost of nuclear heat at $3.70/MBtu ($12.65/1,000 kWh) is somewhat higher.

CONCLUSIONS

Simultaneous production of heat and electricity will increase the overall efficiency of the nuclear power plant by 33 percent to 37 percent, but at the expense of a 6 percent reduction in the yearly production of electricity. Because of the higher efficiency, there is a total saving of 222,000 MBtu (65,000 MWh) in primary energy. In addition, 260,000 bbl of oil can be replaced with nuclear fuel. The specific investment to save 1,000 kWh of energy saving is $255. This corresponds to a capital productivity of $2,200 per annual average kW of energy saved. See Table 3.2 for a summary of the district heating examples discussed in this chapter and a comparison of their specific capital productivities.

If it is desired to use nuclear energy from an existing nuclear power station for a district heating system, the first step should be the construction of the distribution network. The minimum size to obtain economical operation of the combined system, assuming a distance of 10 miles (15 kilometers) for the transport of heat, is 135,000 MBtu (400 MW); this corresponds to 40,000 apartments, each with a demand of 34,000 Btuh (.01 MW).

3.3 THE ACTIVITY OF AN ENERGY COMMISSION IN A SMALL TOWN

Possibilities to realize the goals of energy plans exist not only in large cities but also in small communities. As an example, we choose a community with a population of about 10,000. (4) An energy commission was formed which consisted of seven members (the

Table 3.2. Summary of District Heating Examples

System	Cost ($)	Savings ($/year)	Energy savings MBtu (kWh)	SCP[a] $/BPD ($/kW)	Comments
Accelerated development of district heating with co-generation in West Germany	7.5×10^9	1×10^9	2.73×10^8 (80×10^9)	62,000 (875)	Development of district heating network in West Germany would supply a connected heat load of 1.4×10^5 MBtuh (40,000 MW) by 1990 at a cost of $14,200/BPD ($200/kW), that would serve 25% of the total annual heat load under 390°F (200°C). Cost of heat is $8.20/10^6 Btu ($28 MWh). Example is from Ref. (1).
District heating for Lucerne and its outlying communities	51×10^6	8.7×10^6	1.2×10^6 (35×10^7)	92,000 (1300)	A district heating system for Lucerne, Switzerland, started from scratch, could supply 30% of the heating needs at a cost of $8,500/BPDE ($120/kW) to $12,700/BPDE ($180/kW) of heat capacity by the year 2000. This would save 15% per connected MW of heat supply over the use by individual residential boilers. Cost of the heat is $3.70/10^6 Btu ($12.50/MWh). Example is from Ref. (2).
Nuclear power plant adapted for district heating of Bern	16.4×10^6	See comments	2.2×10^5 (65×10^6)	156,000 (2200)	A nuclear power plant with 310 MW electric capacity can be adapted to feed 355°F (180°C) super-heated water into a district heating system 8 mi (13 km) away. This reduces electric capacity by 6.2%, but raises overall efficiency from 33% to 37%, resulting in a saving of primary energy shown. Replacement of oil with uranium is 26×10^4 bbl (445×10^6 kWh) per year. The cost of heat is $3.00/10^6 Btu ($10.40/MWh). Example is from Ref. (3).

[a]The specific capital productivity (SCP) is the cost of reducing the annual rate of energy use by the energy equivalent of one barrel of oil per day (BPDE) or by one kW.

head of the utility system, the manager in charge of operation of the town electricity and water works, two specialists for heating systems, etc.).

The following areas were treated:

Energy-saving methods for existing buildings

The production of heat and distribution to new developments

Evaluation of previously untapped energy sources such as methane from sewage treatment, recovery of heat from exhaust ventilation

Use of waste energy

Use of solar energy

Increased insulation of buildings

For the existing buildings of the community, a small district heating plan was evaluated. Because the community consists mainly of one-family houses, it was definitely concluded that district heating would not be practical with acceptable economics. The possibility of district heating for new apartment developments was examined.

Other projects considered were the use of gas from the sewage treatment plant, in combination with a heat pump, to supply heat to the surrounding area. Solar collectors were examined for heating the open-air swimming pool.

For new public buildings in the community, a building code was recommended spelling out minimum values of insulation. Recommendations were made for private residences. The personnel responsible for the operation of the buildings in the community (schools, church, administration buildings, and residential complexes) were instructed about energy-saving methods and improved boiler operation. The owners of single-family houses were informed about the possibilities of heat pumps, solar energy, and district heating.

This example shows that there are practical steps that can be taken by small communities to achieve energy savings at different levels.

3.4 LAND-USE PLANNING

Sooner or later, state and local governments must address the potential efficiency improvements possible with land-use planning. Of the many issues which come to mind, alternative uses for the waste heat from existing and future power plants deserve emphasis.

Heated discharge water has many potential uses in addition to heating and cooling residences and commercial and industrial establishments. This rejected heat can be used for the warming and cooling of greenhouses to improve crop growth and yield and to reduce fuel costs. It may be used in open-field agriculture (where the farm acts as the direct-contact heat exchanger for the power plant), allowing a rapid plant growth, improving crop quality, and permitting control over temperature extremes and growing seasons. Another potential use is in aquaculture. Extensive thermal aquaculture facilities could revolutionize fish and seafood production, similarly to what has been done in the poultry industry. Data indicate that fish grow three times faster at 83°F than at 76°F, and that shrimp growth is increased by 80 percent when water is maintained at 80°F instead of 70°F. (6,7)

3.5 MODULAR INTEGRATED UTILITY SYSTEMS (MIUS)

The amount of heat normally wasted during conventional electricity generation, and the conservation of energy which would result from the recovery and use of that heat, are widely recognized. Unfortunately, at present and into the near future, the cost of transmitting and distributing steam or hot water to residential developments limits the economic feasibility of utilizing waste heat from central station plants. As with total energy systems, the MIUS concept places electrical generation near consumers in order to economically utilize waste heat for domestic heating and cooling. (8)

A basic problem of the residential environment is waste disposal. Tons of solid waste are produced daily by every community. Disposal of that waste is a continuous burden. Undisposed of, that waste suffocates decent housing and the living environment with junk and sewage.

The MIUS concept is an expansion of the total energy system. One example of an integrated facility concept is shown in Figures 3.12 and 3.13. A total energy facility provides two of the basic utility services: electricity and space heating and air conditioning. Three additional services are included in a MIUS facility: solid waste processing, liquid waste processing, and water purification. Up to this point, the process is relatively conventional; it is at this point that MIUS takes a most unconventional turn.

One part of the system concerns reuse of waste heat from the heating system, air conditioning unit, incinerator, water heater, and clothes dryer for water reclamation in a low-temperature evaporation unit. Such units have the proven capability to produce water of higher quality than municipal supplies. This water could be reused for bathing and washing clothes. Effluent from these operations poses only minor treatment problems and, with filtering and disinfecting, can be used again to flush toilets.

Drainwater from the kitchen sink, lavatory, and dishwasher combines with that of the toilet for treatment. Part of the treated water is recycled to the reclamation unit for processing, with the remainder reused for watering the lawn. In winter, only this small amount would go into the sewer.

Sludge from the waste treatment process and refuse are fed into the incinerator. Scrubbers remove pollutants from the incinerator gases, and the waste heat is again used to reclaim water for another trip through the system.

As discussed above, there is energy in solid waste. Much of it can be burned and the resulting heat can be combined with heat reclaimed from the engine generators. If solid waste combustion is added to a total energy system, the thermal efficiency of the plant can be improved and the quantity of solid waste to be disposed of can be reduced to incinerator residue.

Liquid waste contains reclaimable water which can be used as the process fluid of

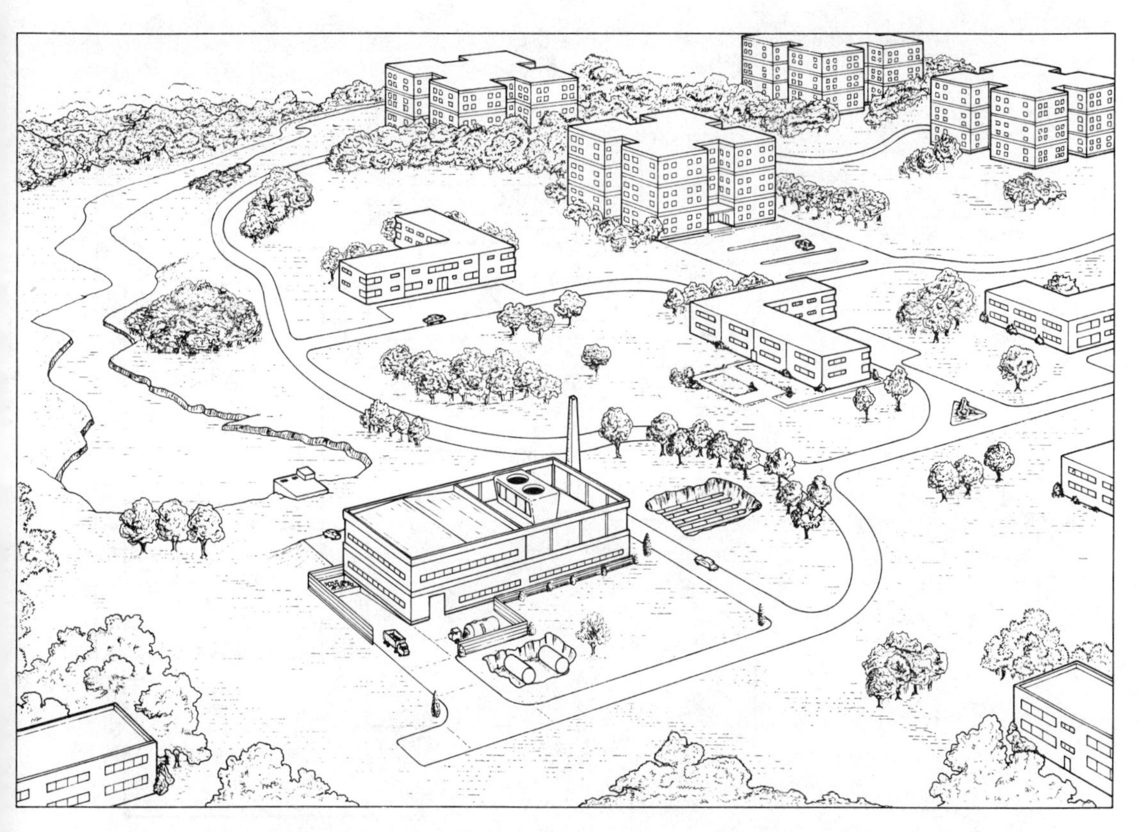

Figure 3.12. Artist's concept of a MIUS and building complex. [Oak Ridge National Laboratory (8)]

the total energy system. At present, losses of water from the system—by evaporation, drainage of lines, or leakage—must be replaced by purchased city water. Processed liquid waste can provide this water resupply from within the system. Excess water can be discharged into the natural environment without ecological damage. Sludge can be treated and sold as raw material for fertilizer or soil treatment, used as land fill, or burned to produce additional heat. By providing liquid waste treatment in the MIUS, technical problems of central treatment plant capacity limitations are avoided.

The final element of MIUS as an integrated utility system is the addition of a water purification unit if potable water is not already available. Raw water can be purified on-site and used as potable water within the complex served.

Not one of the services included in the MIUS concept is unusual; all of them are being provided throughout the country today by equipment and material long in use. What is unusual is combining them in such a way as to produce a community that is almost self-sufficient in utilities services except for the basic fuel supply. The community's

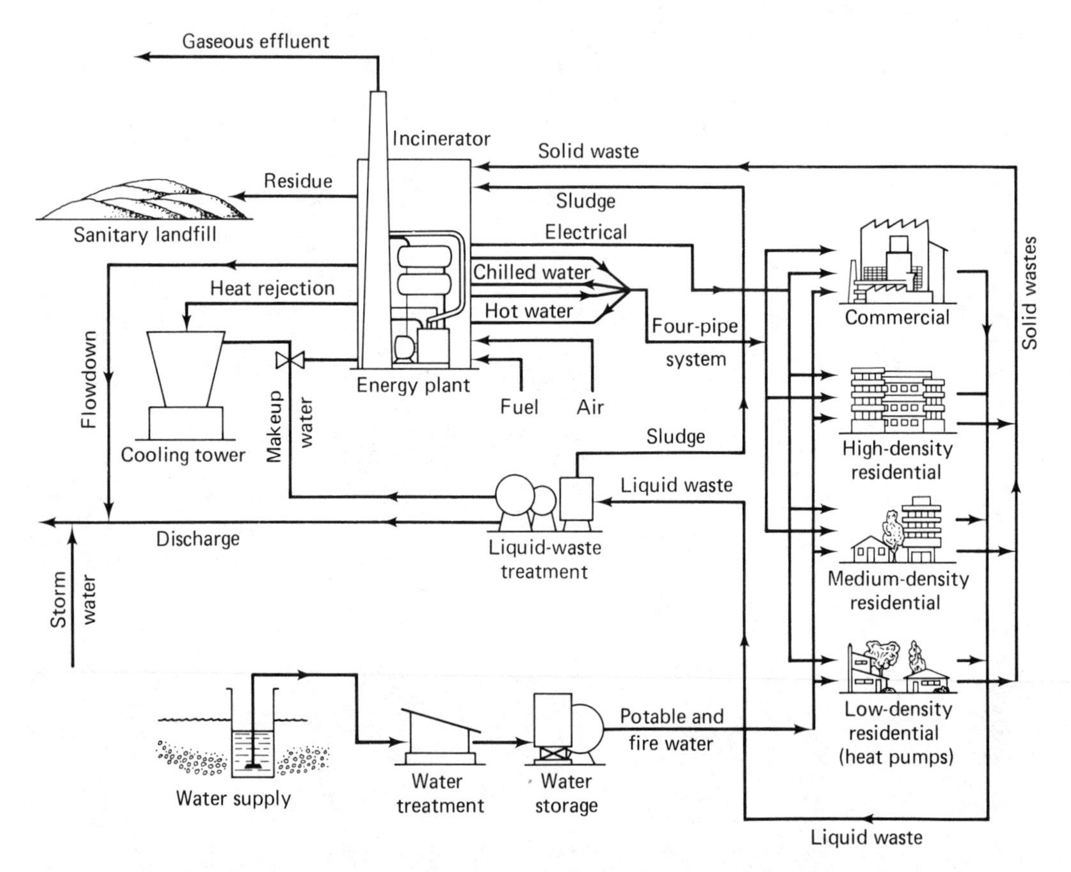

Figure 3.13. A typical MIUS concept. High- and medium-density population areas are supplied with district heating and chilled water. Residential units in the low-population area are heated and cooled by air-to-air heat pumps and auxiliary electric heaters. Heat recovered from the total energy plant as a by-product of electricity generation for these heat pumps and other uses of electricity is available for the district heating network serving the high-density commercial, industrial, and residential areas. [Oak Ridge National Laboratory (8)]

energy requirements, waste disposal, and water purification are integrated within a single functional core.

Energy conservation is a high-priority national goal; but developing communities face additional and increasingly serious problems related to the adequate treatment and disposal of liquid and solid wastes. Many communities, faced with the task of upgrading existing waste disposal facilities and with the growing need for new, environmentally acceptable waste treatment capacity, have resorted to building moratoriums to limit growth. The MIUS concept addresses these problems by integrating the solid and liquid waste treatment systems with a total energy system.

The MIUS can employ gas- or oil-powered internal combustion engines to generate electricity for a building complex. The engine exhaust and coolant heat can supply the buildings with some of their air conditioning from heat-absorption-type water chillers, some of their hot water, and some of their space heating. Liquid waste-water can be purified to a degree acceptable for recycling as cooling water, irrigation, or—perhaps some day—drinking water. Inexpensive waste heat from the prime movers can be utilized in the purification process. Additional gas or oil and solid waste can be burned whenever the "waste heat" from generating electricity is inadequate to meet the demands for building services. A pleasing arrangement of the utility service building within the consumer complex could be achieved as shown by an artist's concept in Figure 3.12. (8) A schematic diagram of one of the many possible MIUS concepts based on present-day technology is shown in Figure 3.13.

One example of the application of these concepts is in Jersey City, New Jersey. With HUD assistance, a model integrated residential and commercial complex is being evaluated; it includes almost 500 residential apartments, a school, shops, offices, and places for recreation. Electricity, space heating and air conditioning, and the energy for hot water are provided by a total energy plant operating on No. 2 diesel oil.

REFERENCES

(1) Gesamtstudie: "Die wirtschaftlichen und technischen Ausbaumöglichkeiten der Fernwärmeversorgung in der Bundesrepublik Deutschland," Fernwärme International (February, 1977).

(2) M. Kiss, H. Mahon, and H. Leimer, *Energiesparen Jetzt*. (Berlin: Bauverlag, 1978).

(3) Arbeitsgruppe bestehend aus Vertretern des Elektrizitätswerkes und des Kernkraftwerkes Mühleberg unter Mitwirkung der Firma Gebrüder Sulzer Aktiengesellschaft, "Nukleare Fernwärme für die Stadt Bern" (Winterthur, Switzerland, 1976).

(4) A. J. Miller, et al, "Use of Steam-electric power plants to provide thermal energy to urban areas," ORNL-HUD-14 report to the U.S. Department of Housing and Urban Development by the Oak Ridge National Laboratory (January 1971), p. 9.

(5) F. M. Bachmann, "Bericht über bisherige Erfahrungen in der Gemeinde Küsnacht/Zurich," Seminar Bauliche Massnahmen zum Energiesparen in der Gemeinde (Gottlieb Duttweiler-Institut, Ruschlikon, 1976).

(6) M. M. Yarosh, B. L. Nichols, E. A. Hirst, J. W. Michel, W. C. Yee, "Agricultural and Aquacultural Uses of Waste Heat," ORNL Report No. 4797 (Oak Ridge, Tenn.: Oak Ridge National Laboratory, July 1972).

(7) Eric Leber, "Waste Heat Utilization: Promises and Problems," *Public Power* (March/April, 1979), pp. 28–30.

(8) W. R. Mixon, et al., "Technology Assessment of Modular Integrated Utility Systems," (Oak Ridge, Tenn.: Oak Ridge National Laboratory, December 1976).

4

The Ten Steps of an Efficiency Improvement Program

This chapter explains the four phases for an energy efficiency improvement program (EIP). The first three phases involve the application of energy-saving methods (ESM), which are differentiated by the financial and time requirements needed to carry them out. Each of these phases can be divided into ten steps. The fourth phase is one of consolidation, in which the results obtained in the first three phases should be maintained and expanded.

Implementation of an Efficiency Improvement Program

A successful EIP results from the application of many saving methods and is an interactive process. This means that the experience from previous actions will be used to determine new energy-saving actions. As understanding of the energy use is improved, more and more possibilities for energy saving will be discovered. Normally, the simplest and cheapest ESM are carried out first (see Figure 4.1).

Phase I: Operational Changes. In the first phase of an energy-saving program, the operation of the plant is checked. This initial examination is sometimes referred to as a **Walk-Through Energy Audit**.Through contact with operating personnel, energy-saving actions can be determined which will require no investment and can be carried out immediately. These changes will depend on the improved control of energy using equipment and informing users of improved operating methods such as turning off idle equipment and unneeded lighting. They will relate also to maintenance personnel. With operation and organizational changes, and maintenance improvements, energy use can be reduced.

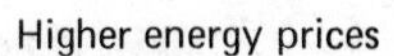

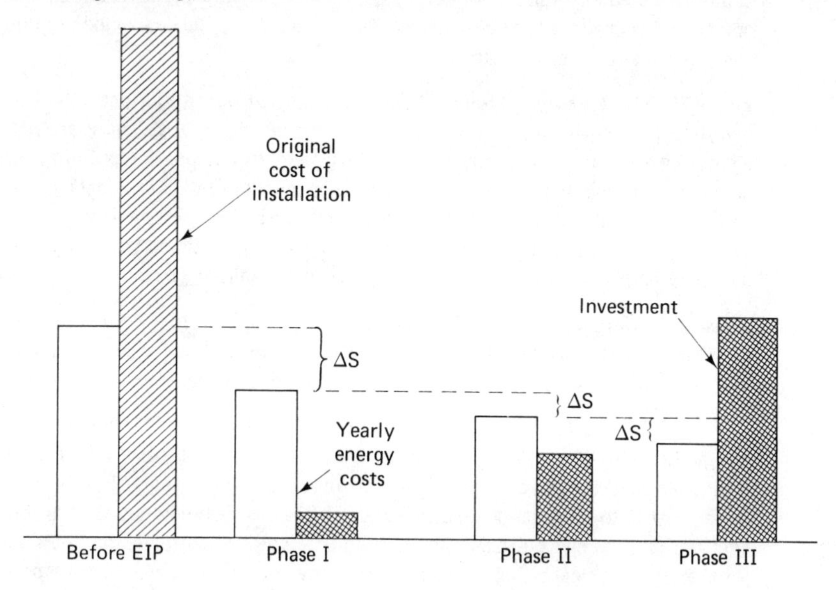

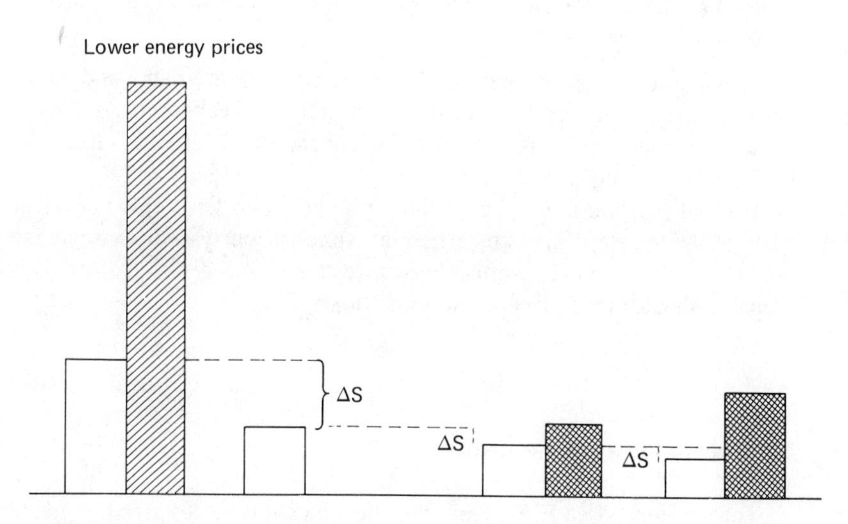

Figure 4.1. Example of typical energy cost savings (ΔS) and investment needed during Phases I, II, and III of an Efficiency Improvement Program (EIP). With higher energy prices, measures requiring a capital investment may already be considered in Phase I to achieve energy cost reductions ΔS.

A more detailed examination of energy use (detailed energy audit) will often uncover additional operational changes. A conprehensive energy audit should be conducted preliminary to making extensive changes and investments.

Phase II: Short Return Changes. The ESM carried out in this phase will require a certain investment—a small one, however, compared with the results. These are defined by having a short rate of return, normally less than one year. In this phase, special attention is given to automatic actions, that is, actions that are independent of the user or operating personnel, such as improved boiler controls, tightening up outside air dampers, and improving their control. The basic data indicating the choice of energy-saving methods is usually determined during the analysis of energy use (energy audit) conducted during Phase I.

Phase III: Medium Return Changes. ESM in this phase require investment. The return of investment will be normally less than 10 years. The preparation for these methods—for example, hearings, check list, cost estimates, proposals, and decisions—can be carried out in Phase II.

Phase IV: Consolidation. This phase should follow each EIP. After the actual decision to carry out the EIP, this phase is the most important. The savings obtained with the initial application of a given measure have to be maintained in the following years. In other words the follow-up action has to be organized such that the measures instituted and all participants of an EIP continue to save energy. Campaigns to enlist employee participation could use such means, for example, as stickers, or a barometer showing savings achieved, or perhaps an internal news letter on energy saving that would report on results and new proposals, and have a suggestion column for reader ideas.

The possibilities for personnel involvement in an EIP are many and there is no limit to the imagination. At the Massachusetts Institute of Technology, for example, energy-saving proposals can be made over the telephone by dialing the 5 letters in the acronym ENCON (energy conservation).(1)

Each of the first three phases can be divided into ten steps. Experience shows that some of the ten steps may be carried out simultaneously—for example, steps 6, 7, and 8—estimation of saving, estimation of investment, and decision of priorities—may (and probably should) take place at the same time.

Summary for Implementation

The ten steps of an EIP are outlined next and will be described in greater detail later.

1. **Start EIP:**
   Management decision to start.
   Select suitable boundaries to avoid negative or ineffective savings.
   Define program objective, capital recovery factor.

2. **Select Team:**
   Project leader and engineers.
   Operating management and users.
   Outside help from utilities, manufacturers, and expert consultants.

3. **Analyze Conditions—The Energy Audit:**
   Get data on energy use. Get equipment manuals.
   Understand how people and equipment work.
   Define comparative energy statistics—energy per unit area of floor space, per unit product, per unit weight or time, etc.—or define efficiency of energy use.

4. **Estimate Goals:**
   Set short- and long-term goals.
   Re-define amenity levels—temperatures, lighting, etc.—and exceptions.

5. **Check ESM:**
   Use checklist to survey possibilities.
   Place all in questions.
   Focus on largest users.
   Take every small saving for big results.

6. **Estimate Savings:**
   Make realistic estimates of saving. *Remember:* energy savings add non-linearly.

7. **Estimate Investments:**
   Follow up all consequences.
   Evaluate effects on production, occupants, and on the environment.
   What investment is required?
   Be conservative, include costs in time and space.

8. **Decide Priorities:**
   Management decides priorities.
   Coordinate renovations and equipment modifications with production schedules and maintenance.
   Define the short- and long-term programs.

9. **Implement ESM:**
   Implement the saving program.
   Be empathetic with users' adjustment to changes. However, do not overreact to complaints.
   Be flexible, if necessary, in adjusting ESM.

10. **Evaluate Savings:**
    Monitor actual savings and costs.
    Compare with goals.
    Go to the next phase of the program.
    Perform maintenance for continued savings.

4.1 MANAGEMENT DECISIONS

The decision of management to carry out an EIP will include the definition of goals. To achieve these goals, it is necessary that management fully support the program and show its conviction. A responsible member of management should be closely informed with the planning and with the problems during the EIP.

It is essential to define the boundary conditions, for example, the limits of the system to be analyzed, the time available for analysis, the allowable effect on production, and the costs for analyses. When deciding priorities it is important to start with actions that have a substantial effect. This way, those involved in EIP will be motivated by early success for further work. The leadership qualities of the project manager will be important in determining the success of the EIP. He or she must be able to learn from each phase and be able to motivate her or his collaborators and keep management and personnel informed on the effects of the EIP. It would be an advantage if a full-time project leader could be delegated for the EIP; otherwise a task force can be formed.

4.2 SELECTION OF THE TEAM

The first task of the project leader is the selection of the team for the energy-saving program. In a large company this team should be an interdisciplinary team with experience in all aspects of the operations and systems involved (see Figure 4.2). For smaller companies, a simpler organization is needed. The members will not be occupied full-time on the energy-saving project. If you want to reduce the financial risk and work involved with an EIP, it is also possible to hire outside experts to carry out the program for an honorarium based on the program's success. For example, the fee might be based on a percentage of the energy cost savings achieved in the first year after the implementation of the EIP.

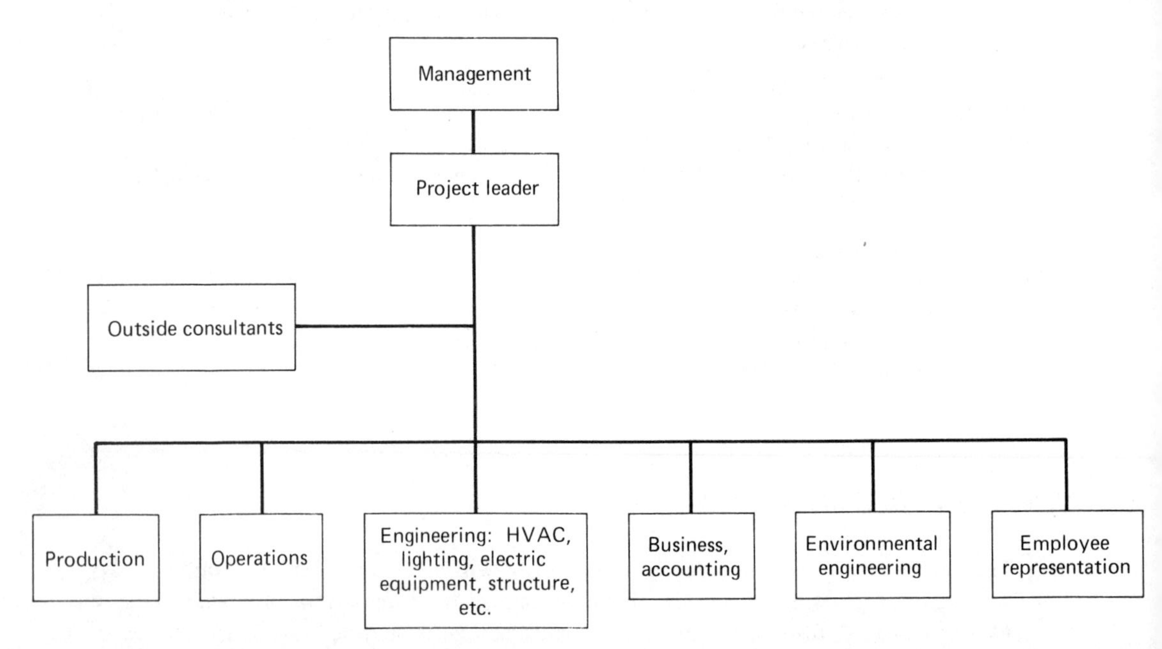

Figure 4.2. Organizational structure for the task force for an energy efficiency improvement program (EIP).

The selection of experts should be made very carefully. Experts should be competent and should know the technical risks involved in the EIP. In some cases bids can be solicited from equipment suppliers for *performance* based contracts to improve the efficiency of equipment.

4.3 ANALYSIS OF PRESENT CONDITIONS: THE ENERGY AUDIT

The basis of a successful energy-saving program is a thorough understanding of the present situation. Therefore a detailed analysis of energy use is necessary. This will require familiarity with applicable blueprints and with the operating and maintenance documents for the systems and equipment. (In many instances, it will be necessary to contact the designer and equipment manufacturers for replacement of missing copies as without this information it would be like working in the dark.) However, of primary importance is the goal of reducing energy waste. It is not enough merely to work on the analysis of existing conditions for several years, without actually implementing energy-saving programs. The analysis of present conditions can be carried out in four steps:

1. Determine energy use in the forms of oil, gas, coal, electricity, etc.
2. Determine where and when this energy is used.
3. Determine why energy is used this way.
4. Compare energy use with good design practice or with the energy use in similar companies.

Normally at the beginning the data available are insufficient for a detailed analysis. Energy use will be available only in units of dollars. There will be no differentiation between primary and secondary energy forms, and data will be available only for the entire company. It will be useful at this point in the analysis to organize an energy bookkeeping system which shows energy use by fuel type, electricity (incl. demand), steam, etc., and such uses by departments as a function of time and production. The energy use should be shown in Btu (or kWh) and not in dollars. This will allow comparison of energy use over time despite changing energy prices.

At the time an energy bookkeeping system is being organized, a commensurate method of determining energy use should be installed. This will ensure billing according to actual energy use for each department.

The energy bookkeeping should make it possible to summarize where energy is wasted. This can be achieved by comparing energy use with good design practice and by calculating efficiencies. This will permit the potential for energy saving to be defined.

If you want to understand changes in energy use and compare your energy use with other companies, the energy use should be normalized per unit area of office space, per unit weight product, or per person served (e.g., per student, etc.). The energy use per $ft^2(m^2)$ in several office buildings can be compared if certain characteristics are considered: situation of the building, type of air conditioning system, special requirements, equipment used for production, etc. In this way the energy use of similar users can be compared. Standard norms can be defined and used to determine deviations of energy use from those values, and can be the basis for monthly reporting. A necessary condition for this is to use the same form of energy bookkeeping for all departments.

4.4
GOALS

When the present situation is known in detail the amount of energy wasted should be clear to you. Before energy savings are discussed in detail, it is necessary to estimate realistic goals.

A good early goal is 10 percent saving in the first phase. Long-range savings on the order of 30 percent or greater may be possible. Comparison with other companies should help you determine the goals of your EIP.

Goals should include the amount of energy to be saved, the maximum capital investment allowable, and effects on users and on production. It is also important to keep in mind, that despite efficiency improvements already achieved, each time that there is an increase in energy prices, new and additional measures become financially feasible. In 1973 IBM set out to reduce its energy use in 34 major facilities in the United States by 10 percent. Pressured by increasing energy prices and using hardheaded analysis, by 1980 IBM had reduced energy use by 50 percent. (2) Savings well in excess of $100 million were achieved with surprisingly little investment.

4.5
CHECK ENERGY-SAVING ACTIONS

In Chapter 5 energy-saving actions are summarized in checklists. They have been compiled from our own experience and the experience of many different companies and municipalities. Use these checklists to determine the ESM for your establishment. There are three major aspects:

Look for energy savings with large savings potential. This will help to convince people that energy saving is worthwhile.

Don't forget that large energy savings represent the sum of many small actions.

Work out all the ESM together with the departments to be affected. This will help them to understand and support the energy-saving actions.

Add to the checklists from your own experiences. As you work through the lists of possible energy-saving measures take time to ask, "Under what conditions could this apply?" or "How could I improve on each measure listed?" Everything should be placed in question.

4.6
ESTIMATION OF SAVINGS

In the early stages it is sufficient to have a rough estimate of savings in energy and costs. At this point classification according to small, medium, and large is sufficient.

In addition to energy savings, any effects on production quality, production quantity, the environment, and on working conditions should be estimated. In calculating the dollar savings, it is necessary to use the incremental costs for the last unit of heat and the last units of electrical energy and demand, rather than the average costs, which are usually higher.

It is important to understand that the savings from different measures are not simply additive, i.e., 3 + 3 is not 6, but 5 or less. For example, if weatherstripping of doors and windows would result in an energy saving of 10 percent and a reduction of room temperature would also result in an energy saving of 10 percent, the sum of the two saving actions if carried out simultaneously would be $1 - .9 \times .9 = 19$ percent. Moreover, if heat recovery from an abundant source of waste heat were also to be added, the value of the 19 percent savings might be lower, because of the possibly lower value for the recovered heat. Perhaps it is better to err on the low side when estimating the expected savings.

4.7 ESTIMATE INVESTMENTS

To complete the classification of ESM the cost for implementing each measure must be estimated as well. The costs for all direct and indirect consequences of ESM should be included. Furthermore, the investments for several different measures may be larger than their individual sums (i.e., 3 + 3 will be 7 or more). One should also consider that the combination of several energy-saving actions may result in unexpected indirect costs.

If the energy savings and the costs are known, the rate of return can be estimated according to Chapter 2. It is essential to include the effect of energy price inflation. Once the rate of return for each ESM is determined, then the priority of the energy-saving measures can be determined.

4.8 DECIDE PRIORITIES

It will be helpful to organize energy-saving actions according to Table 4.1, and assign a priority to each ESM.

Energy-saving measures can be prioritized according to the following criteria:

Rate of return

Automatic or non-automatic measures

The measure's effect on the occupants or on production

Normally, during the first phase operational changes achieve 10 percent savings, usually without investment. These actions are psychologically essential because the quick success after ESM may help stimulate people to follow up on the EIP. Another 10 percent savings can be achieved with small investments. It would be helpful to apply the savings achieved in these first two phases toward the larger investments required in the third phase.

Table 4.1. Form for Checklist

No.	Energy-saving measure	Who is Responsible	Approximate energy saving[a]	Approximate investment cost[a]	Other consequences	Priority

Note: Large energy savings are greater than 1 percent of yearly energy consumption; large investment costs would be greater than 1 percent of yearly energy costs.

[a]First estimate, without doing a calculation, denotes if large, medium, or small.

Measures which do not rely on user or operator intervention for their success deserve special attention. Without such measures, there is a danger that the program will lose its effectiveness after a short introduction period. Such (automatic) energy-saving measures can continue to be effective long after their introduction. An example of an automatic measure would be heat recovery, which may provide a more predictable and lasting saving than such non-automatic measures as turning off unneeded lighting.

Measures having some risk of affecting productivity should be assigned a correspondingly lower priority. It would be imperative to select the participation of the occupants in any decision on measures which could adversely affect them.

4.9 IMPLEMENT ENERGY-SAVING METHODS

In implementing the energy-saving program there are two main points to be considered:

Be flexible, but consistent.

Go nearly to the limit of complaints.

This means you should not be afraid to propose something new. If complaints occur, you should handle them seriously and adjust the EIP if necessary.

4.10 EVALUATE SAVINGS

If an EIP, or a phase of it is completed, the effective energy saving should be compared with calculated values and goals. In addition to dollar savings other effects of the energy-saving program should be evaluated as well. It would be advantageous to publish the results.

At the end of an energy-saving program it is most essential to ensure that old habits and waste of energy cannot regain their previous importance. You should check possibilities for the follow-up phase of the energy-saving program and see that energy use is continuously monitored, and that operation and maintenance remain energy conscious. Ensure that the momentum gained during the energy-saving program continues.

The following suggestions can help to maintain the momentum gained during the energy-saving program.

1. Continue periodic meetings of the energy-saving team.
2. Review energy use. Encourage new suggestions. As energy prices rise, re-evaluate measures listed in Chapter 5.
3. Emphasize energy saving in employee training.
4. Improve energy-use and energy-cost accounting.
5. Keep up the campaign of making people aware of the need to conserve.
6. Publicize results with the names of the people working to save. Reward employees for efficiency improvements achieved.

7. Hold technical courses on economics of insulation, steam traps, lighting, air conditioning, etc., energy saving at home and while driving.
8. Take management through the facility and demonstrate how energy is being saved. Introduce management to personnel who are helping to make improvements.

The organization of EIP is summarized in the following checklist.

CHECKLIST	ORGANIZATION OF AN EFFICIENCY IMPROVEMENT PROGRAM
1. Management decisions.	Outline the boundary conditions and set goals for the EIP. Define the budget. Set the dates for the completion of the analysis of energy use and the implementation of the energy-saving program. Delegate a member of top management and the project leader who will be responsible for the EIP.
2. Select team.	Production Operation Engineering Accounting Environment Consultants Suppliers Employees
3. Analyze present conditions.	Get data, blueprints and maintenance instructions. How much primary energy is used for each department (Energy audit). Use Btu (or kWh) not dollars. Include also water and rejected wastes. Where and when is energy used? (Daily and monthly recording of data.) Coordinate use with weather, and with deviations from design conditions. Why is energy used this way? How is primary energy transformed to secondary energy? Calculate efficiency. Identify areas with large energy saving potential and areas of large energy use.

	Compare original design conditions for the system with the actual existing conditions. Compare energy used with other systems of known good design. (Define energy budget. Relate data to unit area or unit of production.) Is a change to new requirements possible?
	Coordinate energy use in different areas and with respect to time. Is it possible to transport energy from where it is in surplus to where it is needed? Can storage or multiple use avoid rejecting high quality energy?
4. Goals.	Set short- and long-term goals.
	Primary energy saving.
	Saving in electrical energy and demand costs.
	Reduction in the use of oil.
	Saving other resources (water, etc.).
	Set capital recovery criteria.
	Define allowable influences on the users.
5. Check ESM.	Use the checklists in Chapter 5.
	Work out a checklist of ESM for your company or organization.
6. Estimate savings.	Estimate savings for each ESM: (a) energy saving; (b) energy cost-saving; (c) effect on production quality, output and scheduling, etc.
	Estimate total energy saving.
	Be conservative with estimates.
7. Estimate investments.	Estimate investments for each energy-saving measure in dollars and in percent of the annual energy saving.
	Check all of the consequences of ESM.
	Eliminate unrealistic ESM.
	Estimate the total investment needed.
	Calculate the rate of return for each and for the package of ESM.
8. Decide priorities.	Operational changes.
	Large savings with small influence on users and on production.
	Other savings.
	Long-range program.

9. Implement ESM.	
10. Evaluate savings.	Compare with goals. Decide further ESM. Re-examine performance periodically and after equipment is serviced.

REFERENCES

(1) Department of Physical Plant, Massachusetts Institute of Technology, "Energy Conservation at MIT" (Cambridge, Mass.: April 1975).
(2) Daniel Yergin, "Conservation energy key '80s source," The Portland Oregonian, January 27, 1980, p. 8, Sect. 6.

5

Methods for Improving Energy Efficiency

5.1 AN OVERVIEW OF ENERGY SAVING

In this chapter, methods for saving energy are described for public and commercial buildings, businesses, manufacturing, and commercial transportation. Although this chapter is divided into subsections dealing with specific energy systems, it would be a mistake to approach energy saving from a compartmentalized approach. Successful energy saving must be done with full knowledge of the many factors which affect energy use. Unless these factors are known and used to reduce the energy use of the whole system, it is possible to reduce energy use in one place only to find that it has increased somewhere else. Without full knowledge and a systems approach, changes made may result in achievements far short of the potential, or they may actually do more harm than good.

At the end of each section in this chapter is a checklist with places to look for energy-saving opportunities and methods to be applied. The reader should consult the equipment manual for each piece of equipment examined for efficiency improvement and use it along with this book. The checklists are essentially a menu of practical methods from which options may be selected according to one's own particular appetite. Not all of these methods will be cost-effective in every situation. It will be necessary for you to choose those suited for your application. We also provide examples of how others with similar problems have used some of these methods.

There are many similarities between a poorly run building, a poorly run business, and a poorly run manufacturing operation. In poorly managed buildings we observe excessive illumination, poor control over temperatures, and unoccupied space which is nevertheless ventilated, and heated, and cooled. In industry we see poor scheduling or product flows and idle equipment left running at full power. In poorly run businesses we see loss due to poor maintenance, spoilage of merchandise, and overhead out of control.

In energy-wasting buildings huge exhaust fans draw out vast amounts of heat right next to huge intake fans that replace heat with outside air—which then must be heated or cooled, filtered, and dehumidified or humidified. In inefficiently run industry there are plumes of steam and rows of empty ovens, awaiting an occasional load, maintained at operating temperature throughout the shift. In business we see poor routing of deliveries and lack of employee orientation. Vigorous, informed energy management can improve efficiency and reduce waste.

The main idea is to keep an overall perspective and not to focus only on one piece of equipment, or one aspect of the building, or one particular product, or one part of the business. To best apply these methods one must first understand the function of each piece of equipment, how energy is used, and the effect of climate and other factors. One must know how the user's needs are met, and how a particular change would affect the users and relate to subsequent, possible improvements. With the systems approach, improvements are integrated with the whole system. Then, when the system is used, it requires the least amount of energy to get the job done.

Energy Use in Buildings

Since most of our activities take place in buildings, we consequently devote considerable attention to obtaining more efficient energy use in them. The energy used in buildings is also an important factor for industry, since Department of Energy data indicate that between 20 percent and 50 percent of the fuel used by industry is used for plant heating. (1)

In the United States, a little over one-third of the total energy is used in commercial and residential buildings (see Figure 5.1). (2) Because approximately 70 percent of that energy is used for space heating and cooling, it deserves a great deal of attention.

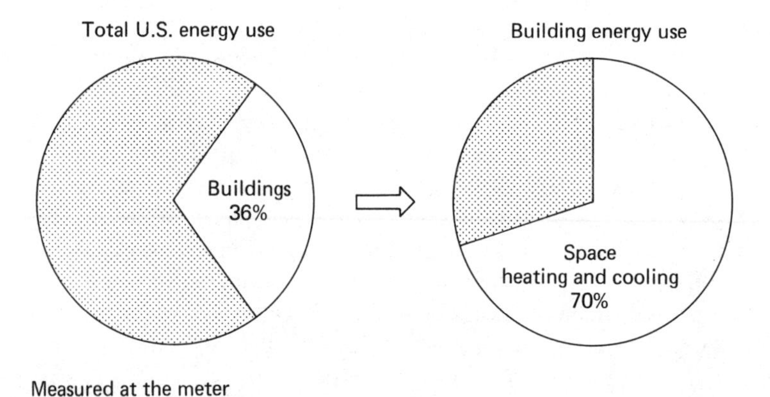

Figure 5.1. The importance of space heating and cooling in residential and commercial buildings.

The total floor area of non-residential buildings in the United States has been estimated by the Rand Corporation to be 29 × 10^9 ft^2, (2.7 × 10^9 m^2), of which it is estimated that 5 to 5.8 billion $feet^2$ (460 million to 540 million $meters^2$) of this floor area is devoted to office space. (2,3)

For new buildings built in conformance with the very modest requirements of the ASHRAE 90-75 guidelines, there would be significant reductions in energy use as compared with the results of traditional construction practice.* (4) For example, with windows and insulation in accordance with ASHRAE 90-75, heating energy use for the average single-family residence could be reduced from the present 142 million Btu (41,600 kWh) per year to below 50 million Btu (15,000 kWh). (3) According to the "minimum performance standards" of ASHRAE 90-75, the potential saving for commercial buildings is approximately 40 percent relative to 1973 practice. Although more would be spent for insulation, double glass, and improved controls, there are cost reductions for HVAC systems. For each 30 percent reduction in system capacity there is a 10 percent to 15 percent cost reduction. New buildings built to the ASHRAE Standard 90-75 may cost less to build, as well as less to operate.

*ASHRAE. The American Society of Heating, Refrigeration, and Air Conditioning Engineers is a source of data, standards, and handbooks for this industry which are available in many libraries.

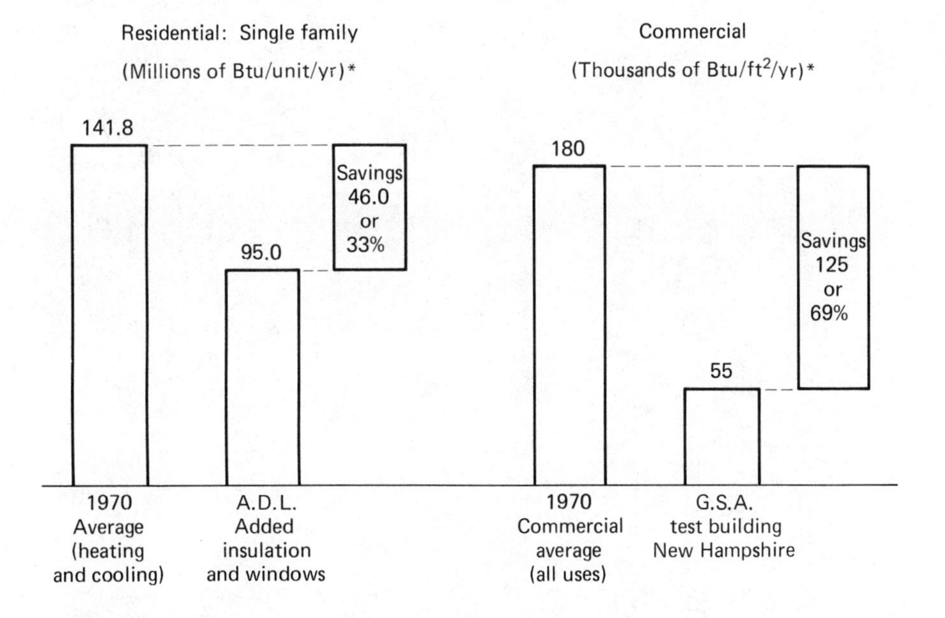

Figure 5.2. The long-term potential for increased energy efficiency using improved standards for building construction. The average energy use shown is that measured at the envelope of the building and does not include the raw source equivalent of electricity. [U.S. Department of Energy (6)]

Building Energy Performance Standards (BEPS) proposed by the U.S. Department of Energy represent more stringent measures to reduce building energy requirements. (5) The standards require that the design energy requirements of the building, including all installed components and systems, not add up to more than the designated budget. The use of a building energy budget, rather than perscriptive guidelines such as those found in ASHRAE 90-75, is an attempt to give the designer greater latitude in selecting design methods from improved energy utilization. The BEPS budgets for different buildings do not require the use of advanced technologies. Many buildings built prior to BEPS have exceeded these standards simply with the use of careful design methods.

For example, with the methods used in the design of the General Services Administration Building in Manchester, New Hampshire, and with today's technology the design energy was reduced from 180,000 Btu/ft^2 (570 kWh/m^2) to 55,000 Btu/ft^2 (175 kWh/m^2) referred to gross area at the building boundary, as shown in Figure 5-2.* (6)

*Referred to *primary* energy and *net* area, the design energy use of the new Manchester building would be 200,000 $\frac{\text{Pri Btu}}{\text{net ft}^2}$ $\left(580 \frac{\text{Pri kWh}}{\text{net m}^2}\right)$ per year.

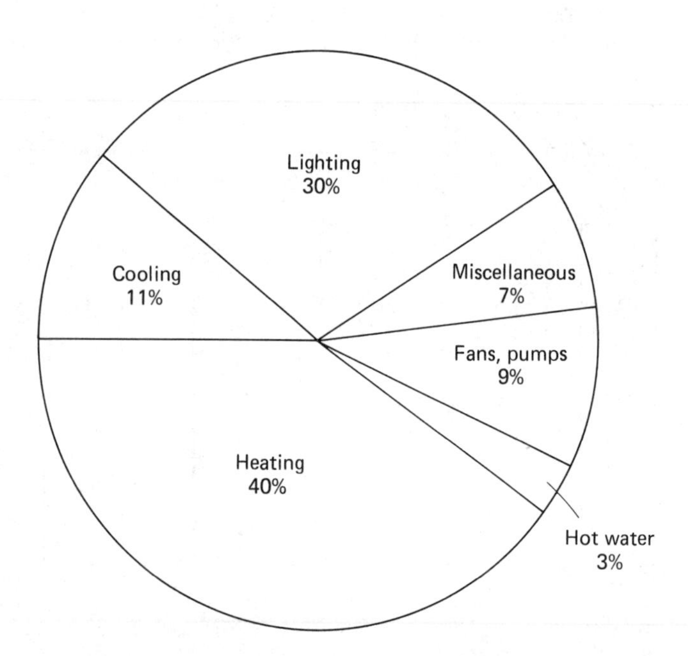

Figure 5.3. Energy use in a typical office building. As much as 30 percent of this energy can be saved, in some cases more. A survey of the energy use of 180 buildings in Manhattan by the New York City Interdepartment Committee on Public Buildings found the median energy use for this group of buildings was 161,000 Btu/ft^2 (510 kWh/m^2). The annual use varied between a low of 62,000 Btu/ft^2 (200 kWh/m^2) to a high of 545,000 Btu/ft^2 (1700 kWh/m^2) before accounting for energy use in the generation and transmission of electricity. (7)

If we look at an example of energy use in an existing air conditioned office building operated according to traditional methods (see Figure 5.3), we see that about 30 percent is used for lighting, the rest for heating, air conditioning, fans and pumps, and miscellaneous uses including office equipment and elevators. Typically, this use can be trimmed in existing buildings by about 30 percent, depending on specific equipment and local circumstances. The factors influencing the efficiency of energy use in buildings are shown in Figure 5.4. The interrelationships in this user-structure-equipment system must be understood so that changes can be integrated into the system to achieve the most efficient operation. Examples of annual energy per unit area and per occupant are shown in Table 5.1. This compares typical and average use with energy efficient and maximum energy

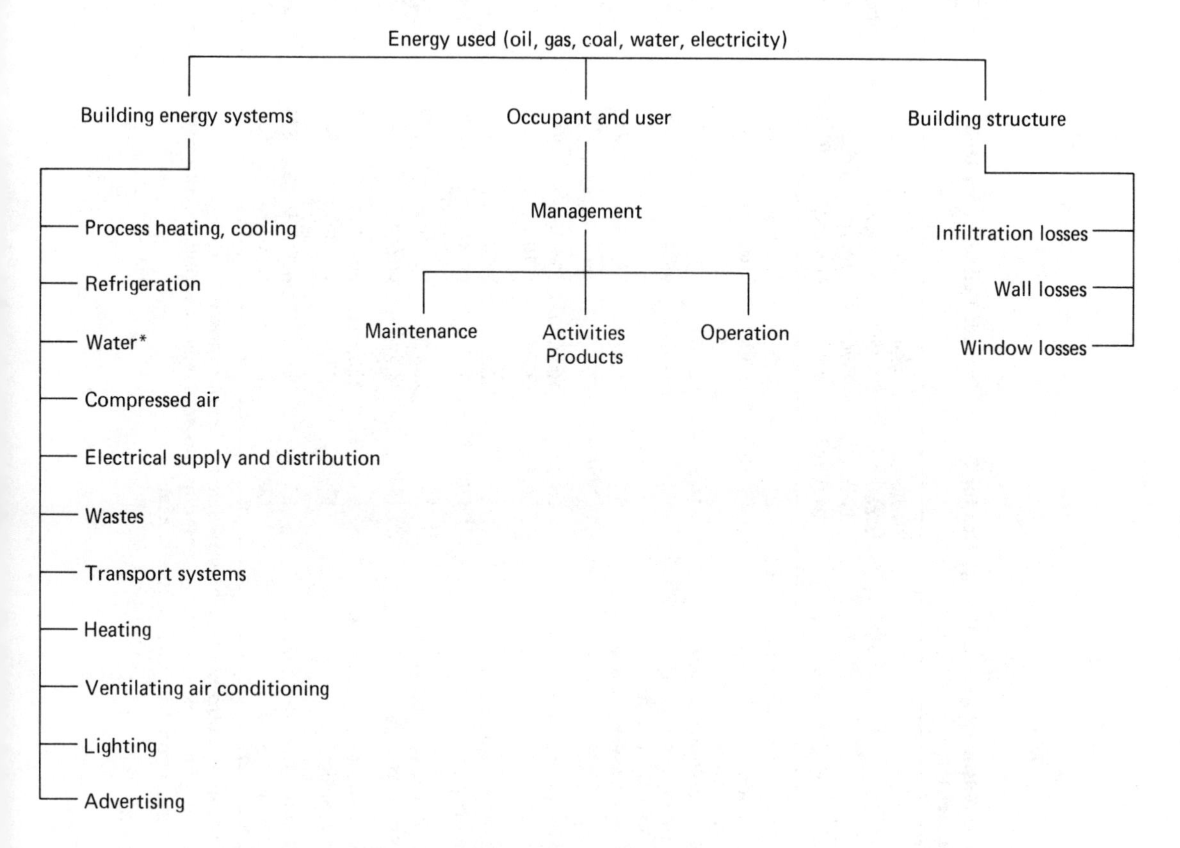

Figure 5.4. Building energy use is determined by energized systems, nonenergized systems, and human systems. An effective efficiency improvement program requires that the entire pattern of energy use be analyzed so that changes made will be integrated into the system in full light of the interrelationships which will occur.

Table 5.1. Comparison of Annual Energy Use for Some Different Buildings per Unit Area and Occupant

Reference	*Example*	*Nonheat electricity kWh/ft² (kWh/m²)*	*Heat Btu/ft² (kWh/m²)*	*Total energy Btu/ft² (kWh/m²)*	*Area per occupant ft² (m²)*	*Energy per occupant MBtu (MWh)*	*Primary[a] Btu/ft² (kWh/m²)*	*Oil per occupant,[b] gallons (Kg)*
R.G. Stein (8)	Typical design, NY office building (all electric, 1 million ft²)	21 (226)	53,925 (170)	125,600 (396)	225 (21)	28.3 (8.28)	268,000 (850)	436 (1,528)
NBS (9)	Base building design Manchester, N.H. (11,000 m²) 126,000 ft²	14.3 (154)	55,880 (176)	104,700 (330)	210 (19.5)	22 (6.44)	202,300 (638)	306 (1,073)
NBS (9)	Energy saving design	9.7 (104)	20,870 (66)	54,000 (170)	210 (19.5)	11.3 (3.32)	120,300 (379)	182 (637)
DOE (10)	Average of 307 building survey	26.4 (284)	104,000 (327)	194,000 (610)	242 (22.5)	47 (13.76)	374,300 (1,181)	653 (2,287)
DOE (10)	Building with minimum energy	6.5 (70)	13,876 (44)	36,000 (114)	184 (17)	6.6 (1.94)	80,430	107 (374)
DOE (10)	Building with maximum heat energy	31 (335)	759,395 (2,895)	865,000 (2,725)	624 (58)	540 (160)	1,077,000 (3,400)	4,845 (16,970)
DOE (10)	Building with maximum electricity	69 (740)	133,461 (420)	368,000 (1,160)	186 (17.3)	68.5 (20.0)	840,000 (2,649)	1,126 (3,945)
DOE (10)	Building with maximum total energy	37 (395)	755,401 (2,380)	880,000 (2,775)	314 (29)	276 (81.0)	1,134,000 (3,577)	2,568 (78,932)

[a]Electricity converted to primary energy using a factor of three.

[b]Primary energy per occupant represented as the equivalent gallons of No. 2 fuel oil. (Use has been converted to kilograms of oil using the convention 1 kg oil per 10,000 kcal.)

using buildings. The energy use per occupant is related to a measure of its cost during a period of rapidly changing prices by representing primary energy as the equivalent gallons of oil per person per year.* These data are illustrative only, because of differences in climate and use. Nevertheless, the difference of a factor of 45 between the building for which annual primary energy use is 4,845 gallons of oil-equivalent per occupant and the one for which it is 107 gallons must also represent important differences in operation.

Improved Management and Process Control in Industry

In manufacturing, significant energy savings have been achieved by tuning up equipment, improving management practices in plant operation, careful use of lighting and air conditioning, recycling scrap in metal industries, and improved process control. These improvements have neither interfered with production nor compromised worker safety or productivity, and yet some measures have achieved significant savings at little expense. In the production and processing areas, energy waste is often the result of long-unquestioned practices that, upon analysis, can be shown to have no effect on employee productivity or product dependability. Several examples will illustrate the type of savings possible from reexamination of existing practices.

A small electronics plant operated a line with four operators, for cleaning complex assemblies. It was the practice to run this line, like the rest of the plant, for two shifts per day. Although it had the capacity of 2 to 2½ times the capacity of the rest of the plant, it had been operated at below capacity so that it was always available, so that units needed for a rush order would not be held up. Rescheduling permitted the operation of this cleaning line for only one shift, thus avoiding the need to operate equipment drawing 28.5 kW during the second shift. Careful scheduling insured that no unit needed in a hurry was held up, and resulted in saving 57,000 kWh per year.

Re-examination of process specifications can also achieve savings. For example, the published specifications for the manufacture of an electronic unit called for heat treatment lasting one hour at 125°C. As a result, the manufacturing plant purchased two new ovens. Subsequent review of the specifications discovered that this was an over-specification (to allow for the contingency that the oven was started cold), rather than an actual requirement, which was only for a twelve-minute treatment. Since no one had questioned the specifications, both capital funds and energy were wasted.

New plant practices to save energy can have a ripple-through effect on other scheduling. An example of careful equipment turn-on as the result of rescheduling is described in the reprinted article from the Raytheon Corporation's "Energy Bulletin #10." (11)

Better planning to maintain a full, continuous manufacturing schedule will avoid the start-and-stop losses of energy and the resulting product defects. This can also avoid the inefficiency of operating only small segments of the processing facilities during weekends.

*In some cases, energy costs of 40 percent of annual rentals have been reported following a 300 percent increase in prices.

ENERGY BULLETIN #10[a]

Don't Be an Eager Beaver!
Save by Turning on Equipment as Late as Possible.

In Raytheon plants energy consuming devices are shut off at nighttime. Temperatures in process equipment are lowered. Obviously, it takes time to bring this equipment up to operating temperature again. Therefore, steam or electricity must be turned on ahead of the start of the morning shifts allowing for "heat up." To be on the safe side, often equipment is turned on much too early and unnecessary energy losses occur between the time that the equipment is ready for use and production begins.

At Caloric Corporation, "Andy" Andersen noted that the steam charts for a dryer indicated completion of the heat up period more than an hour before start of the shift. He calculated the steam use during this hour and noted that $2,000 could be saved annually by delaying steam supply to the equipment. He immediately arranged for a change in routine, with the result that his plant can increase its savings. It may be worth your while to check whether some of your equipment is not turned on too early!

Electrical equipment should not all be brought to "heat up" at the same time. By staggering starts, peak demand can often be lowered. Remember that heat up loads frequently are much larger than running loads.

For example, improved scheduling can avoid idle equipment being maintained at operating temperature and reduce cool down between processes. According to the Department of Energy, in one steel reheating plant the application of on-line computer controls on stock scheduling yielded a 25% reduction in fuel consumption. An additional benefit gained in this instance was an increase in production of somewhat more than 12%. The control system enabled the plant manager to eliminate several bottlenecks in the process, which had not yielded to manual control procedures. (11)

[a]Example of bulletin distributed to energy managers in the manufacturing plants and subsidiary firms of the Raytheon Corporation.

The Checklists

Potentially applicable energy-saving methods (ESM) are summarized in checklists at the end of the sections in this chapter. The ESM are organized by cost with operational and maintenance measures first, followed by short and medium pay-back measures (see Table 5.2.). The checklists are also organized according to a sequence: end use, distribution, and production. It makes sense to follow this sequence, as opportunities to reduce distribution and production losses do appear in a completely different light, once end use losses have been cut.

Practical energy-saving does need to ask: "How can I cut energy loss in this building? How can the efficiency of this piece of equipment be improved?" The topically organized checklists will provide help. Organization of ESM by specialized topic, however, has one drawback: measures cannot be considered isolated from their impact on other parts

Table 5.2 Organization of the Checklists with a Typical Example for the Different Systems of an Office Building. The Relative Cost Savings for Measures Given in the Checklists Would Depend on the Particular Details for Each Building.

	1. Operating changes	*2. Short-return measures*	*3. Medium-return measures*
Building envelope	Close window blinds at end of day and during abnormally hot and cold periods.	Weatherstrip, install low-leakage, outside air damper.	Install double-glass windows.
Electrical systems	Switch off unused equipment.	Add time switches to cut off some electrical units at time of peak loads.	Replace oversize motors.
Heating	Control heating to rational levels, set temperature back after hours.	Install automatic combustion controls.	Convert to heat pump.
Ventilation Air conditioning Refrigeration	Turn off AC one hour before quitting time.	Install automatic start-up controls.	Recover heat from exhaust air.
Lighting	Remove unnecessary lighting. Switch off in unoccupied rooms.	Install time switches for irregularly used rooms.	Replace incandescent lighting systems with fluorescent or other more efficient lighting.

of the larger energy system and the environment, as for example the effect of reduced lighting on the HVAC system, or on occupant comfort. Each measure may cause problems or opportunities to appear somewhere else. A successful program combines an aggressive approach to cutting losses with a broader viewpoint that assures that the productivity of the system and the quality of the occupants' working environment is not sacrificed. A comprehensive efficiency improvement program (EIP) for your particular situation will consist of a collection of specific measures from different sections in this book, plus your own measures together with those collected from elsewhere. At the risk of some repetition, we have included as examples some integrated checklists put together by others as examples of their saving-energy programs.

Our checklists include measures for the utility systems of large office and manufacturing buildings, which have more elaborate equipment than smaller commercial buildings and single-family residences. However, many of the ESM can be applied to smaller residential and business buildings as well.

The physical and financial conditions vary greatly between one building and the next. Only a few of the measures listed will be practiced in any one building. Some of the measures listed make sense only under very special circumstances. The checklists also contain methods for reducing energy *costs*, even though the action may not result in actually reducing energy use. Such methods generally result in improved load management to better utilize the energy supply system, as for example in the reduction of peak electrical demand or a better use of the tariff structure.

The lists were compiled from our experiences with successful efficiency improvement programs, some of which are described in Reference 12, and from the general References 13 through 35, and from references specifically cited elsewhere.

Our experience has shown that considerable energy is saved from simple operating changes. It is a common belief that energy-saving actions consist of adding insulation to buildings, of returning to the dark ages by reducing room temperatures below comfort levels, of adding expensive computer control, or of making costly changes to existing air conditioning systems. As the organization of the checklists already shows, there are many energy-saving actions that exist without the need of large investments. Adequate comfort can be maintained even after executing an energy-saving program.

Some energy-saving actions can even result in an improvement in comfort conditions. For an example, reorganizing the lighting system can improve the quality and usefulness of lighting, while decreasing glare and energy cost. High lighting levels are the causes of eye fatigue for many people working eight to ten hours in over-illuminated offices. Task-oriented lighting will result in significant improvement in lighting comfort.

Examples of changes with large savings and small effects on the occupants would be upgrading weather stripping or sealing of buildings, reducing air flow rates, using heat recovery, improving partial load operation, and improving adaptation to load changes. High energy efficiency improvements can be achieved especially when building extensions are planned, or when components are replaced or systems overhauled.

Care Is Required in Using the Checklists!

The checklists contain a number of measures. In your case maybe

5 out of 100 are both possible and economical now,

10 out of 100 should be re-examined later,

and some of the remainder may only be applicable in the more distant future.

It is definitely your job to separate those actions which are applicable from those which are not, because you are the one who knows the specific problems in your situation. Of course, any changes should be made in conformance with local codes and safety requirements.

No checklist can be complete. It is intended that the list be extended by each user according to his own experience. No checklist can be used as a cookbook. Read our comments, and also ask the advice of specialists, if necessary, before applying the recommended actions. In this way you can avoid negative savings (see examples given in Table 5.3). In one office building, as the result of an energy-saving program, the outside air flow rate was reduced. At the same time, we noticed that the cooling units were running in the winter. It was obvious (to us) that the reduction of outside air flow rate had resulted in increasing the cooling load. Some other examples of possible consequences of the misapplication of measures from the checklists are given in Table 5.3.

Table 5.3. Examples of Possible Negative Savings Due to Incorrect Use of the Checklists

Action	*Positive aspects*	*Negative aspects*
Close off unused windows.	Heat losses in winter and heat load in summer are reduced.	Need for artificial lighting is increased.
Reflecting windows.	Summer heat load is reduced.	Solar contribution to winter heat load is reduced. Lighting need is increased.
Install double-glass windows.	Summer heat load/winter heat loss is reduced.	In peripheral areas with a surplus of internally generated heat, even in winter, additional cooling may be needed.
Install external sun blinds, or internal draperies.	Summer heat load reduced.	If sun blinds are closed all of the time, the need for artificial lighting is increased.
Apply insulation.	Heat losses reduced.	Cooling load increased.
Turn off heat in unused areas.	Reduced heating requirements.	Possible freezing of drains and water pipes; humidity problems (condensation, rusting).
Use electric resistance heaters for spring and fall local heating.	Avoids use of central heating system during the temperate seasons.	The overall efficiency of an electrical heater is less than half that of a central heating system with an oil- or gas-fired furnace.
Open windows to eliminate artificial ventilation in winter and temperate seasons.	Energy for air transportation saved.	Heat loss through uncontrolled natural ventilation may be much higher than what it would be with controlled artificial ventilation. Possible problems with dirt in areas with heavy air pollution.
Discontinue humidity control.	Energy saving both summer and winter.	Possible increased illness; dry conditions may cause annoying shocks; electrostatic problems with computer equipment.
Reduce outside air flow rate to minimum requirements.	Energy demand reduced.	Possible complaints from occupants if there is heavy smoking.
Replace filters with ones having a lower pressure drop.	Pressure drop reduced.	In certain cases, if speed of fan motors is not adjusted, a higher energy demand can result.
Reduce excessive lighting.	Cooling load, energy use decreased.	Effort is wasted if dual duct or air conditioning systems with reheat simply add equivalent energy elsewhere.

(continues)

Table 5.3. Examples of Possible Negative Savings Due to Incorrect Use of the Checklists *(cont.)*

Action	*Positive aspects*	*Negative aspects*
Lower level of lighting.	Less electrical energy use.	A double or triple saving could be achieved if at the same time air flow rates were reduced as well. The demand for heating in winter may be somewhat increased. Poor morale and productivity due to inadequate illumination.
Insulate hung ceilings.	Reduced heating and cooling.	Possibility of shortened equipment lifetime, perhaps even fire hazard caused by covering electrical fixtures, ballasts.

Note: The positive and negative consequences must be weighed against each other before it will become clear for your particular circumstances if a measure would result in a real saving, a negative saving or do-nothing, or a SAVING as defined in Chapter 2.

Energy-Saving Experts

You can call upon outside consultants for advice in helping to avoid negative savings. We would like to caution you, however, about the danger of speaking only with a lighting expert, or only with an air conditioning expert, or only with a specialist for an isolated manufacturing process. Modification of one part of the system generally affects other parts of the system. Unfortunately, the complexity of much of the equipment encourages specialization. A solution focusing on a single part of the system may produce less than optimum results, or even negative savings. Such is typically the case when a salesperson attempts to convince you of the solution to your energy problem with his particular piece of equipment or product. An example is the overpromotion or misapplication of costly computer-controlled energy management systems (EMS). Be wary of testimonials. All too frequently these are expensive examples of savings which could have been achieved by simpler and less costly means. Testimonials referring to a well-known name do not imply that the expenditure was made by a well-informed person. Get advice from several independent sources. However, in the end, there is no substitute for your becoming informed about how your equipment functions, and where, when, and how energy is used.

5.2 ENERGY-SAVING MEASURES FOR THE BUILDING ENVELOPE

The structure of a building has a direct relationship to the energy requirements for heating, cooling, lighting, and other needs. The envelope of the building is the boundary between the inside, where some attempt is made to control the environment, and the outside which is in contact with the elements. The envelope, therefore, includes the walls with its windows and doors, the bottom floor, and the upper roof. These elements of the structure influence the amount of uncontrolled air infiltration into and out of the building, as well as the heat transfer through the envelope as the result of the sun and wind, and the temperature differences between the inside and the outside.

A checklist for energy-saving measures should begin with measures to reduce unwanted heat transfer and air flow through the envelope.* However, many people, even including some "experts," seem to believe that energy-saving consists mainly of adding additional insulation to existing buildings. This may be the case for some buildings. For smaller buildings without insulation, adding insulation can lead to cost-effective savings. Usually, however, controlling air flow through and around windows and doors, cracks, vents, and other openings should have priority, because this leads to larger savings with less cost. Some improvements may be justified by improved tenant appeal—better sound isolation, reduced drafts, and improved appearance.

The HVAC systems (e.g., burner, ventilation) frequently offer opportunities with smaller cost and greater savings. For this reason, energy-saving measures for the building envelope should not be considered alone, but rather in concert with improvements to be made in other areas.

An example illustrating the relative merit of different combinations of structural improvements to an apartment house is shown in Tables 5.4 and 5.5. (36) Considering the low energy costs at the time of this study (1975), such improvements often indicate unattractive rates of return based on energy savings alone. Shorter rates of return may be expected if the energy improvement is combined with other renovations to the buildings. These renovations were justified for their improvement to comfort, property value, and as a long term investment against much higher energy prices.

The cost of upgrading the insulation and glazing for a 50-story office building, as modeled for the climate conditions of New York City, is shown in Table 5.6. (37,38) The analysis of the value of added insulation to a large office building is more complex than for a residential building. The heat produced by lighting, mechanical systems, and other equipment becomes a factor, and the energy required to condition outside air for ventilation is an important quantity for large buildings. Better insulated walls and windows do not necessarily lead to reduced energy consumption in larger office and manufacturing buildings with large internal heat generation. (Note this as a caution!) (39) Some buildings must be cooled during the entire year. To trap more of that heat inside with increased insulation would add to the air conditioning load. Therefore, the analysis shown in Table 5.6 is one from which caution must be used in drawing any generalization. The effect of increased insulation, according to this analysis, showed a small net increase in annual costs using 1973 prices. This analysis was done before the increase in oil prices and in electricity rates. Two years later, this investment would have yielded an annual return of $70,000.

In smaller buildings, heat loss through the surface represents a greater proportion of the energy use. For single-story buildings, heat loss through the roof is relatively more important, and restaurants, shopping centers, and large manufacturing buildings, for example, may offer good opportunities for upgrading the roof insulation, and covering flat roofs with white marble chips.

*In addition to those references specifically cited, we have found References 31 to 35 helpful.

Table 5.4. Comparison of Energy Saving Actions Related to the Building Described in Ref. 36.

Example: Three residential buildings with 48 apartments, 32,500 ft^2 (3000 m^2) of heated space built in 1962.

Construction: Pitched roof; stucco finished light concrete block walls with plaster interior finish; single glazed, wood framed windows. U factor of the walls is 0.25 Btuh ft^2°F (1.4 W/°C m^2).[a]

Specific heat requirement per unit of living space 40 Btuh-ft^2 (126 W/m^2) at 10°F (−12°C).

ESM	*Outer wall*	*Exposed ceiling*	*Basement ceiling*	*Windows*	*Reduction of specific heat demand, Btuh-ft^2(W/m^2)*[a]	*Cost,*[b] *\$/ft^2 (\$/m^2)*
1	Insulation of the exterior surface (with ventilation space between wall and insulation boards)	Insulation with finished plaster	Insulation	Old windows	14 (40)	1.20 (13)
2	—	—	—	New wood-aluminum single glazed windows with the old sash	11 (35)	2.75 (30)
3	Same as ESM 1	—	—	New wood frame with insulating glass	12 (37)	4.00 (45)
4	—	Insulation finished with plywood	Insulation	Triple glazing with new all-wood frames and sash	16 (50)	4.00 (45)

Notes: 1. Two design approaches were used to achieve a reduction of the specific heat demand by 30 to 40%:(a) improvement of the insulation of the outer walls together with improvement of the windows, or (b) no improvement of the outer walls, but improvement to critical areas having large heat losses, as for example the exposed ceilings of the basement and upper apartments, and the use of high performance windows.

2. The renovation costs for an apartment with an area of 1100 ft^2 (100 m^2) amounted to about \$4,500. Even if one takes a heating cost of \$400 per year, the saving in heating costs would amount to only about \$120 to \$150 per year.

3. For each building, the financial analysis must be worked out individually, as special circumstances do not permit generalization. General experience does lead us to expect:

Reduction of infiltration using better weather stripping yields a very good pay-back.

The use of storm windows results in attractive savings.

The addition of more insulation to basement and attic surfaces, and improvement of, e.g., doors, may result in relatively good economy.

Addition of insulation to the outer walls, if done at the time of overall renovation of the roof and facade, may yield acceptable returns.

[a]For an explanation of U-values, see page 96. [b]Net area.

Table 5.5 Economic Analysis of Structure Improvements to a 50-Story Manhattan Office Building

Initial incremental cost of adding insulation to oil fired, boiler heated, electrically cooled office building	
Glazing: single to double: \$2.50/sq ft × 240,000 sq ft	\$600,000
Wall: from R3 to R19: \$1.00/sq ft × 240,000 sq ft	240,000
Roof: from R5 to R37.5: \$2.00/sq ft × 37,500 sq ft	75,000
	\$915,000
Initial savings in	
Heating Plant	\$ 50,000
Refrigeration Plant[a]	500,000
	\$550,000
Net cost for additional insulation	\$365,000
Annualized cost for additional insulation, at 8 percent mortgage and 6.5 percent property taxes	55,881

Annual operating savings	*Pre-embargo energy prices*[b]	*Post-embargo energy prices*[c]
Reduction in oil consumption from 394,000 to 170,000 gal	\$27,000	\$ 85,500
Reduction in electricity for air conditioning	16,500	26,400
Reduction in accessory equipment operating in refrigeration plant	9,000	14,400
	\$52,500	\$126,300
Net change in annual costs	\$ 3,381 loss	\$ 70,419 profit

Note: The 50-story office building was modeled as a "typical" structure of steel and concrete construction. The glass was assumed to have a U-factor of 1.13; walls 0.2; and roof 0.15. The improved insulation would result in U-factors of 0.58 for the glass; 0.048 for the walls; and 0.026 for the roof. (Data in Btu/hr°F ft^2, multiply by 5.7 convert to W/m^2°C. See p. 96.)

[a]Solar heat gain was reduced 36%, justifying a 10% reduction in the size of the refrigeration plant.

[b]Oil at \$0.12/gal; electricity at \$0.025/kWh.

[c]Oil at \$0.38/gal; electricity at \$0.04/kWh.

SOURCE: *Energy Conservation in Buildings: The New York Metropolitan Region* (37,38).

Table 5.6. Typical Costs for Improvements to the Building Exterior

Doors, Windows (Cost per Door or Window)[a]	*Costs ($) for 10*	*Costs ($) for 200*
Reflective foil applied to windows	24	20
Sun protection with venetian blinds, installed	10	
Add single glazing with aluminum frame to existing window[b]	130	95
As above but with openable sash	315	230
Replace windows with double glass in new aluminum frames	275	200
Adding opaque insulation to unneeded windows	65	
Weatherstrip double hung window	33	26
Weatherstrip door, metal (wood) frame	95	50
Install door closer (min. 10 doors)	90	
Add vestibule to building entry	3100	
Insulation	*Cost per unit area, $/ft²*	*($/m²)*
2 in. (5 cm) insulation applied to exterior of building	3.25	35
1.5 in. (3.8 cm) insulation applied to interior of building	2	22
5 cm insulation sprayed onto interior of roof (with reflective aluminum overcoat)	26	23

Notes: These cost data are given to provide perspective as to the relative costs of different measures. Costs have changed since these studies were done for the U.S. Government in January, 1975 and in any case they would vary from one region to another.

[a]Per 10 windows (200 windows) of 3 ft × 4 ft (1.1 m²) each. Doors are 21 ft² (2 m²).

[b]Without removing existing windows.

SOURCE: Department of Energy, 1975. (17)

When Should Improvements to the Building Envelope Be Considered, and Where?

In contrast to the operational measures considered with air conditioning systems or furnaces, etc., there are fewer significant savings achievable without some investment. Many of the actions may be deferred to that time when repairs or new additions are planned. With additions to existing buildings and, of course, with new buildings, many of the recommendations can be achieved, at no significant additional cost. Nevertheless you will still find in the checklist a number of advantageous methods that can be used now, for example, caulking and the weatherstripping of windows.

Typical costs for energy-saving improvements to existing buildings are given in Table 5.6. In Table 5.7, cost savings are compared with investments for nine different examples of improvements to municipal buildings in Massachusetts. The specific capital productivity (SCP) of the improvement has been calculated for each of these examples in order to develop a figure of merit for policymakers. The SCP is the investment required to reduce the rate of energy use, averaged over the year, by one 42-gallon barrel of oil per day (BPD), or by one kW (i.e., saving 8,760 kWh per year). For example, an SCP of $46,000/ BPD corresponds to investing $3 in a measure that would save one gallon of oil per year. Many of these improvements show that it is cheaper to invest in energy saving than to produce new energy. Some industrial examples of investments to block the flow of heat, and the savings that resulted, are shown in Table 5.8.

The Importance of Infiltration Is Not Generally Appreciated

Air flows between the outside and the interior around windows, doors, and through cracks in the facade. The resulting heat exchange can be remarkable, and represents an important waste not only for residences (see Table 5.4.) but also for commercial buildings.

This need not apply only to older buildings, as we illustrate with the following example. In an air conditioned office building, experiments were conducted during the spring and fall in which an attempt was made to shut down the air conditioning system. This did not prove to be a satisfactory measure. Although this building was only a few years old, the infiltration during these tests was so high that paperweights were necessary to keep letters from blowing off the desks. As long as the air conditioning system had been overpressurizing the building, the outward flow of air had masked this problem. It appears to be a common defect of many buildings constructed recently, both in the United States and Europe, that they were not constructed tightly enough for good energy economy.

It would not be reasonable to start with other measures, e.g., insulation of the outer walls, without first reducing the infiltration to acceptable levels. In many buildings (without air conditioning or without artificial ventilation) air flows of 2 to 4 air changes per hour have been measured. An air change of 1 is more than sufficient to supply adequate fresh air, and in many cases lower amounts are specified. An air change of 12 cubic feet per minute (cfm) or 20 cubic meters per hour per each person corresponds to an air change every 4 hours for typical residential space, and to an air change every 75 minutes

Table 5.7. Comparison of Investment and Energy Saving for Some Building Improvements (40)

ESM	*Example (yr. of constr.) [gross floor area]*	*Cost #(1); [SCP] (2)*	*Savings*	*Assumptions (3)*	*Comments*
Roof insulation	School, Concord (1951) [39,400 ft^2 (3700 m^2)]	$22,000 [$41,400/BPD ($585/kW)]	$2,900	Roof area 39,400 ft^2 (3,700 m^2). Cost for oil before renovation $14,000[b]	U-value improved from 0.22 to 0.08 Btuh°F ft^2 (1.25 to 0.454 W/°C m^2).
Roof insulation	Fire Station, Attleboro (1969) [2,100 ft^2 (195 m^2)]	$1,900# [$17,000/BPD ($240/kW)]	$715	Roof area 2,100 ft^2, (196 m^2). Cost of oil before renovation $913.[a] Building is 30% air conditioned in the summer.[c]	U-value improved from 0.67 to 0.08 Btuh°F ft^2 (3.8 to 0.454 W/°C m^2) by spraying the roof with foam insulation.
Wall insulation	Davol School (1892) [15,964 ft^2 (1500 m^2)]	$9,400 [$34,300/BPD ($485/kW)]	$1,750	Wall area 8,392 ft^2, (780 m^2). Cost for oil before renovation $6,185.[a]	Uninsulated brick wall improved from a U-value of 0.48 to 0.08 Btuh°F ft^2 (2.72 to 0.45 W/°C m^2) with fiberglass and covered with a gypsum board interior wall.
Bubble window insulation	Highland School, Fall River (1901) [16,968 ft^2 (1580 m^2)]	$600 [$20,600/BPD ($290/kW)]	$170	Window area 1,400 ft^2, (130 m^2). Cost for oil before renovation $6,500.[b]	Plastic, bubble type window insulation on upper half of all windows improves U-value from 1.13 to 0.65 Btuh°F ft^2 (6.4 to 3.7 W/°C m^2).
Storm windows	Davol School (1892) [15,964 ft^2 (1500 m^2)]	$2,900 [$25,800/BPD ($365/kW)]	$425	Window area 1,700 ft^2 (160 m^2). Heating degree days (3,000°C-days) 5,408°F-days.[a]	U-value of single windows improved from 1.13 to 0.54 Btuh°F ft^2 (6.4 to 3.1 W/°C m^2).
Automatic door closer	Firehouse, Acton (1950) [5400 ft^2 (502 m^2)]	$800 [$8145/BPD ($115/kW)]	$585#	Indoor temperature 60°F (15°C). Door had been left open 2 hrs/day. Infiltration rates: 4 air changes/hr with door left open and 2 air changes/hr with door closed. Cost for oil before installation $2,430.[a]	After a fire alarm the doors are opened and then automatically closed after the trucks have left.

ESM	Example (yr. of constr.) [gross floor area]	Cost #(1); [SCP] (2)	Savings	Assumptions (3)	Comments
Wall insulation	School, Fall River (1901) [16,968 ft^2 (1577 m^2)]	$13,350 [$31,900/BPD ($450/kW)]	$2,400	Wall area 11,916 ft^2 (1,107 m^2). Cost for oil before renovation $6,500.[b]	U-value of uninsulated brick wall improved from 0.62 to 0.08 Btuh°F ft^2 (3.5 to 0.45 W/°C m^2).
Existing windows blocked up	Firehouse, Attleboro (1959) [8608 ft^2 (800 m^2)]	$1,200 [$16,300/BPD ($230/kW)]	$460#	Window area 430 ft^2 (40 m^2). Cost for oil before renovation $3,862.[b]	The existing windows were replaced by cement blocks, 3 inches of insulation and plaster. The U-value was improved from 1.0 to 0.06 Btuh°F ft^2 (5.7 to 0.34 W/°C m^2)
Door control of heating	Dept. of Public Utilities, Acton (1965) [1735 m^2 18,644 ft^2]	$750 [$9900/BPD ($140/kW)]	$465#	Door was open 2 hrs/day for 2 200 days of heating season. Interior temperature 60°F (15°C). Exchange reduced from 4 to 2 air changes/h.[b]	Switches on the door now shut off the heating system when the door is open.

Notes:
(1) The # examples have payback periods of less than three years, and hence represent very attractive examples.
(2) The specific capital productivity (SCP) of the conservation investment is the cost divided by the annual energy saving.
(3) [a]gas—$3.00/ccf ($0.016/m^3)
[b]oil—$0.36/gal ($0.11/kg)
[c]electricity—$0.055/kWh

Source: Massachusetts Dept. of Community Affairs, *Energy Management in Municipal Buildings,* 1977.

Table 5.8. Examples of Costs and Savings from Industrial Improvements to Block the Flow of Heat

Energy saving measure	*Cost ($)*	*Savings ($/year)*	*Energy savings, MBtu (kWh)*	*SCP,[a] $/BPD ($/kW)*	*Comments*	*Reference*
Improved furnace insulation	1,750	924	1950 (570,000)	1560 (22)	By increasing insulation on a Behrenburg Glass Co. bending furnace gas consumption was reduced 19% and warm up time was reduced from $2\frac{1}{2}$ hours to $\frac{1}{2}$ hour. Production increased by 10% due to more uniform heat distribution and the working environment was more comfortable. Ceramic fiber replaced refactory brick.	(41)
Loading air lock for railroad cars	20,000	3,600	1840 (540,000)	23,000 (325)	An air lock was installed to restrict air flow when railroad box cars were admitted to a Westinghouse Pittsburgh factory. Rate of fuel saving is computed on an annual basis.	(42)
Truck loading dock shelters	650	580	290 (85,000)	4750 (67)	Devices to restrict the loss of heated air around warehouse doors when trucks are loading also make working conditions more comfortable. Cost is per shelter. Annual saving is computed for New York climate.	(43)
Heavy fuel oil tanks insulated	20,000	9,000	4100 (1.2×10^6)	10,000 (145)	At the Micafil plant in Zurich, the absence of any snow buildup on 40,000 gal. (150,000ℓ.) tanks was a clue to their high surface temperature. After insulation, significantly less steam was required to heat heavy, residual oil to pumping temperature.	(44)
Insulate oil storage tank	15,000	20,000	10,000 (3×10^6)	3200 (45)	Monsanto Fulton plant.	(45)

[a]The specific capital productivity (SCP) is the cost of reducing the annual rate of energy use by one barrel of oil per day (BPD) or by one kilowatt.

in an office building. Excessive infiltration is a problem for many buildings both old and new, and may be one of the most common factors causing wasteful fuel consumption.

The influence of energy saving measures on the reduction of building heat loss will be important nationally, especially in the residential area for which owners can cut costs by undertaking much of this work themselves. The effect of selected improvements to an older, 3-family wood-frame dwelling in New England is shown in Figure 5.5. (Additional examples can be found in Reference 46.) There are many older buildings without storm windows and without ceiling insulation. Not infrequently, rising energy costs have resulted in owners deferring maintenance to pay for fuel. This has further exacerbated heat loss and has led to the abandonment of much needed housing. The energy efficient rehabilitation of existing housing is essential to help control rising energy costs. Invest-

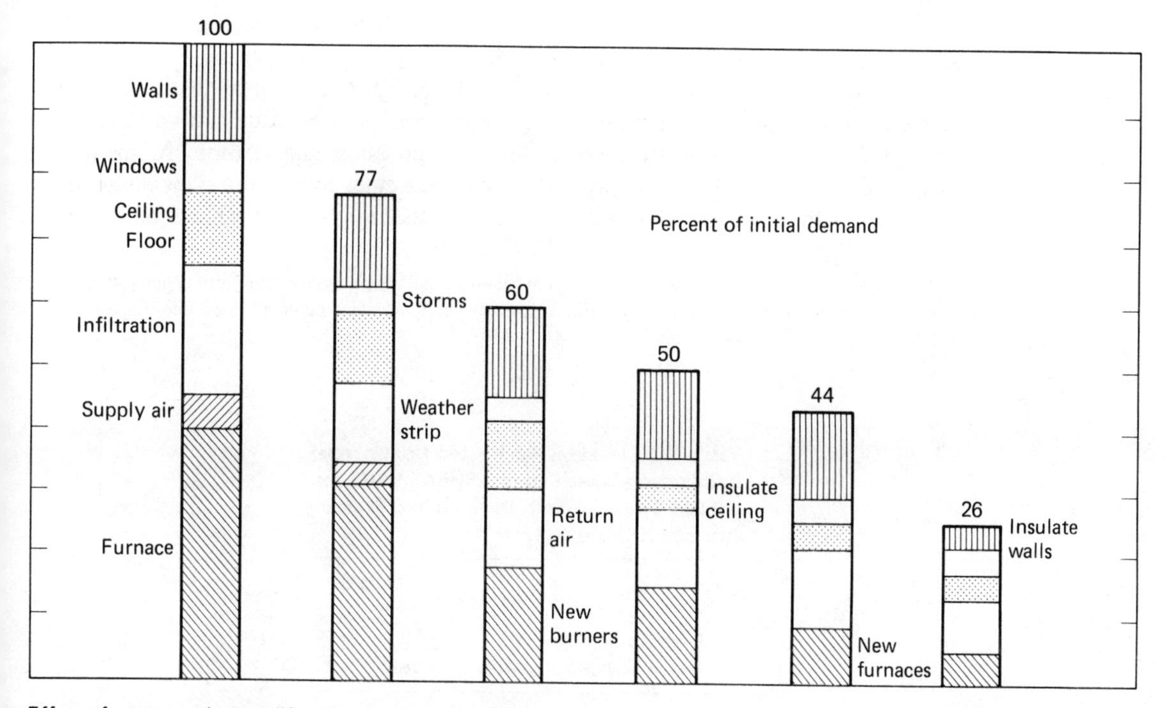

Figure 5.5. The effect of selected modifications on space heating requirements for a low-rise, multiple-family house (triple-decker). Adding storm windows and weatherstripping reduces demand to 77 percent of its initial level. With new burners and adding a return air duct for the forced, warm air heating system, the level is reduced to 60 percent. Adding ceiling insulation to the upper apartment saves another 10 percent. With wall insulation and by moving the heating equipment (new, instead of adding new burners) inside the spaces to be heated, demand is reduced to 26 percent of its initial level. Energy efficient rehabilitation of older, multifamily dwellings can lead to lower monthly payments by the occupants (including both fuel and loan amortization).

ments in energy efficient rehabilitation, using measures such as those shown in Figure 5.5, can result in lower monthly costs (both fuel and amortization) for the occupants.

Insulation of the Walls and Roof

Standards have been established in many countries for the improved insulation and energy efficiency of new buildings. In the United States the ASHRAE standard 90-75 was used widely following its introduction in 1975. (4) Different states have their own versions of these building energy performance standards. In Switzerland the SIA 180/1 provides a new norm for building construction; and in Germany the DIN standards have also been revised. (47) The national average heating season in the United States (30-year mean) is 212 days with 5,156°F-days (2,864°C-days).* The ASHRAE 90-75 standard for thermal transmission coefficients (U-values) applicable to commercial (type B) buildings for this heating climate are given in Table 5.9.

The energy-saving potential of additional insulation must be weighed against the cost of installation in a particular building. The optimum amount of insulation will depend on energy prices, on the cost of the insulation, interest rates, and so forth. As the cost of fuel is expected to rise more sharply in the future than other commodities, the minimum in the cost-benefit analysis is going to occur for increasingly better insulated buildings.

*Degree-day: According to long-established convention in the United States, there are as many degree-days in a 24-hour period as there are degrees difference between the average outdoor temperature for the day and 65°F (18°C) (2500°C-days = 4500°F-days).

Table 5.9. U-Values for Heated Commercial Buildings as Specified in ASHRAE 90-75 for a 2,864°C-Day (5,156°F-Day), United States Average Heating Season (4)

	U-value	
Building element (in contact with heated space)	$\frac{Btu}{°F\ h\ ft^2}$	$\frac{W}{m^2\ °C}$
Walls (incl. windows & doors)		
Under 3 stories	0.29[a]	1.7[a]
Over 3 stories	0.35[a]	2.0[a]
Floors (over unheated space)	0.08	0.45
Roof	0.082	0.47

[a]The U-value of the walls is an average that includes the effect of wall, windows, and doors.

For those who do not take advantage of the minimum lifetime cost obtained with more efficient construction, but who instead emphasize minimum first costs, this may require more stringent codes, federal government intervention, and demands by a more aware clientele for lower operating costs. Examples of guidelines which set higher standards are the General Service Administration's "Energy Conservation Guidelines for New Office Buildings," and the U.S. Department of Energy Building Performance Standards (BEPS). (48,5)

The energy use of a 34-story apartment house in the New York metropolitan region has been examined as a function of different amounts of insulation, method of operation, and other factors. (49) The study indicated that if oil- and gas-heated apartment houses were insulated to the same degree as electrically heated apartment houses, the fossil-fuel-heated buildings could reduce heating energy consumption by 15 to 20 percent. Table 5.10 compares the construction characteristics assumed for the fossil-fuel-heated and electrically heated apartment houses in the study. In fossil-fuel-heated apartment houses with sealed windows, the study suggested that increased thermal insulation may increase the total kilowatt hours needed for cooling. However, if kilowatt hours for air conditioning are converted into fuel consumed at the power plant, the reduction in fuel consumption for heating far exceeds the increase in fuel consumption for cooling. Moreover, with added insulation, peak air conditioning load (maximum kilowatt demand) would be reduced.

Table 5.11 shows the economic effect of adding substantial amounts of insulation to the apartment house discussed above. (38) The results show that while the net annual cost of much greater insulation was $5,300 more than savings, using 1975 energy prices, the subsequent increase in energy price would have caused this investment to be fully offset by annual savings.

Infrared cameras have been used to spot areas with insufficient insulation. (34,50) With this equipment, parts of the building with large heat losses (warmer areas of an otherwise cold exterior surface) become visible to the eye. This shows up leaks around windows and "heat bridges" in masonry or concrete buildings. Heat bridges are contacts between the inside and the outside of the envelope (e.g., concrete floors, structural members that extend through the walls) which lead to cold floors and heat losses.

Table 5.10. Comparison of Previous Apartment House Insulation Levels (49)

Building element	*Fossil fuel*	*Electric heat*
Walls	4" brick, 8" block, 1" insulation	4" brick, 8" block, 2" insulation
Roof	10" concrete, 2" glass fiber insulation	10" concrete, 2" glass fiber insulation
Glazing	1/4" clear plate	double glazing (1/2" air space)

Table 5.11. Economic Analysis of Adding Substantial Amounts of Insulation to a 34-Story Apartment House in the New York Metropolitan Region

Initial *incremental* cost of additional insulation to apartment house with oil-fired boiler, electric air conditioners.	
ESM:	
Glass: single to double glazing @$2.50/sqft	$120,900
Walls: from R4 to R24 @$1/sqft	112,700
Roof: from R6 to R37.5 @$2/sqft	18,000
	$251,600
Initial savings in:	
Reduced heating plant	15,000
Reduced cooling plant (100 ton reduction in capacity)	60,000
	$ 75,000
Net additional investment	176,000
Annualized net additional investment[a]	26,946
Annual operating savings:	
Fuel oil consumption reduced from 70,560 gal to 30,480 gal	
40,080 × $.38/gal for No. 4 oil[b]	15,230
Air conditioning electricity consumption reduction @$.04/kWh[c]	2,400
Reduction in electric accessory operation	4,000
	$ 21,630
Net annual cost	5,316

[a]Mortgage term of 30 yrs.; 8% interest rate; 6.5% property tax rate; excludes tax credit considerations.

[b]No. 4 oil cost based on mid-July, 1975, quotation from Public Fuel Service Company, New York, N.Y.

[c]This analysis fails to account for initial savings that would be attributable to reduced heating plant size.

SOURCE: Environmental Law Institute (38).

U-Value

To reduce heat loss through the building envelope, or through insulated equipment, the U-value must be reduced by upgrading insulation. The U-value for a given building section or thickness of insulation is the amount of heat transferred per unit time and area from the warm side to the cold side per degree temperature difference. Low U-values indicate good insulating properties. Thus, U-values are expressed in the following units:

Table 5.12. Examples of the Evaluation of U-Values

Example 1. The U-value for single and insulating glass

	Single glass British (SI)[a] units	*Insulating glass British (SI)[a] units*
Outer air film	0.17 (0.03)	0.17 (0.03)
Glass, $^1/_4$" thick (6.35 mm)	0.04 (0.01)	2 × 0.04 (0.01)
Air space $^1/_2$" thick (13 mm)	None	0.88 (0.15)
Interior air film	0.68 (0.12)	0.68 (0.12)
Total resistance ($R_{tot.}$)	0.89 (0.16)	1.81 (0.32)
U-value (= $1/R_{tot.}$)	1.12 (6.4)	.55 (3.13)

Note: The air films contribute the major portion of the insulation value and that the insulation value of the "glass alone" is 0.05 percent of the single window.

Example 2. The U-Values for a low heat transfer wall section and an uninsulated wall with a 4" (100 mm) air space

	Uninsulated Wall		*Low heat transfer*	
	British units	*SI units[a]*	*British units*	*SI units[a]*
Outer air film	0.17	0.03	0.17	0.03
Brick veneer (4", 100 mm)	0.44	0.08	0.44	0.08
Air space	0.91	0.16	0.88	0.15
Insulation	None	—	11.0	1.94
Concrete block (12", 300 mm)	2.27	0.4	2.27	0.4
Insulating board (½", 13 mm)	None	—	1.52	0.27
Interior decoration	0.31	0.05	0.31	0.05
Interior air film	0.68	0.12	0.68	0.12
Total resistance	4.78	0.84	17.27	3.04
U-value (= $1/R_{tot.}$)	0.21	1.19	0.058	0.33

Note: This illustrates the significant impact of adding insulation during construction. (The importance of tight joints during construction should also be stressed.)

[a]Data may not add due to rounding off.

Btu per hr-ft^2-°F (British) or watts per m^2-°C (SI Units). The U-value is dependent upon the resistance (R-value) of several components, as explained in the examples shown in Table 5.12.

R-Value

The resistance to heat flow through any part of the envelope is dependent on the several elements in its path between the warm side and the cold side:

1. The outside surface traps a film of air which resists heat flow. The resistance of this thin film depends upon the wind turbulence.
2. Each separate layer of the building section resists the flow of heat. The resistance (R-value) of any material is simply a measure of how good an insulator it is. The R-values for some insulating materials are given in Table 5.13. The higher the R-value, the better the insulator.
3. Each measurable air space adds to the overall resistance.
4. The inside surface also traps a film of air. (Smooth, reflective surfaces also retard heat transmission by radiation.)

The sum of these resistances gives the overall resistance, whose inverse is the U-value. The lower the U-value of a given building section, the more it resists heat transfer. The comparative value of different materials can be judged with the help of Table 5.13. Certain foam insulations can be applied through small holes to fill wall cavities.

New materials should be used with caution, as their properties have only been evaluated under limited conditions and their suitability and characteristics after aging have not been completely determined. The characteristics with respect to settling, shrinkage, corrosion, toxicity, and general suitability of application should be checked before installation. Following manufacturer's instructions with regard to the proper method of application (as, for example, the suitable temperature range for the application of foams and appropriate fireproofing), can help to avoid problems.

Where To Insulate

The greatest heat transfer (up to 50–65 percent) occurs through the roof of low buildings with no insulation. In this case, roof insulation should be high on the list of priorities. Rigid board insulation can be used above the roof deck, batting insulation can be placed underneath the roof level, but this can cause difficult condensation problems. When re-roofing a flat roof, the rigid board insulation can be laid down over the old, built-up roof. The replacement roof can be laid on top of this. Another alternative is to use perlite expanded aggregate with an asphalt binder. This can be applied in thicknesses up to 5 in. (13 cm) and sloped to avoid water ponding.

Underinsulated, one-story commercial buildings with dropped ceilings are easy to insulate. The ceiling panels can be lifted and the batting placed atop the adjoining panel. This type of insulation can only be effective where air exchange between the heated space and the uninsulated space above the dropped ceiling has been minimized. It would be

Table 5.13. R-Values of Some Insulating Materials

Material	Thickness in. (mm)		Resistance[a] $\left(\frac{ft^2\ h°F}{Btu}\right)$	Resistance $\left(\frac{m^2\ °C}{W}\right)$
Calcium silicate[b]	1	(25)	1.7[a]	0.3
Cellular glass	1	(25)	2.5[a]	0.44
Cellulose fibers	1	(25)	3.7[a]	0.65
Ceramic fiber[b]	1	(25)	2.2[a]	0.39
Corkboard, 8 lb/cf (130 kg/m^3)	1	(25)	3.7[a]	0.65
Diatomaceous silica[b]	1	(25)	1.4[a]	0.25
Expanded polystyrene, 3.5 lb/cf (55 kg/m^3)	1	(25)	5.26[a]	0.93
Expanded polyurethane, 2.5 lb/cf (40 kg/m^3)	1	(25)	6.25[a]	1.1
Glass fiber	1	(25)	4.0[a]	0.7
Insulating boards (wood fiber)	0.5	(13)	1.52	0.18
Mineral fiber bat	2.75	(70)	7.0	12.35
	3.5	(90)	11.0	1.95
	6.5	(165)	19.0	3.35
	10.5	(265)	33.0	5.8
	12.0	(300)	38.0	6.7
Mineral fiber board, 18 lb/cf (300 kg/m^3)	1	(25)	2.86[a]	0.5
Mineral fiber	1	(25)	1.9[a]	0.33
Mineral (or glass-fiber) wool	1	(25)	3.7[a]	0.65
Organic urea foam (Tripolymer)	1	(25)	4.4[a]	0.78
Polyurethane foam	1	(25)	7.1[a]	1.25
Shredded wood fiberboard	2	(50)	3.5	0.62
	3	(75)	5.25	0.93
Styrofoam	.75	(19)	4.17	0.73
Urea-formaldehyde foam (Borden)	1	(25)	4.2[a]	0.74
Vermiculite, 4 to 6 lb/cf (65 to 100 kg/m^3)	1	(25)	2.27[a]	0.4
Air spaces				
Air space (vertical)	.75	(19)	.92	0.16
Aluminum foil on both sides (vertical)	.75	(19)	2.64	0.46
Horizontal, with foil, winter	.75	(19)	1.84	0.32
Horizontal, with foil, summer	.75	(19)	3.6	0.63
	4	(100)	8.9	1.57
Three horizontal foils, summer	8	(200)	22.3	3.94
Air film, interior (vertical)			.61	0.11
Air film, exterior (vertical, turbulent)			.17	0.03

Note: Resistances given decrease with age and moisture content.

[a]Resistivity for 1 inch thickness (25 mm).
[b]High temperature insulation, value given for 500°F (260°C).

Source: Egan (51), Kinsey & Sharp (52), and manufacturer's literature

particularly effective over an air-conditioned area in the summer. However, use a vapor barrier next to the warm space and be certain that the area above the ceiling does not fall below freezing as a consequence of the insulation, if water pipes are run above the ceiling. Insulation should not be placed on top of electric junction boxes, lighting fixtures, or fluorescent ballasts. Insulation can cause overheating and will shorten the life of ballasts, negating any saving from reduced heat flow.

A similar treatment can be extended to the floors. Consider placing 2-in. (5 cm) rigid board insulation around the edges of the building. Conductive heat loss through 2 ft. (60 cm) of exposed concrete block in the foundation wall equals that of 8 ft. (2.5 m) of insulated wall. The insulation should be placed under the slab and extended 40 in. (1 m) down the side of the foundation; this can result in as much as a 90 percent reduction in energy loss.

The method of insulating walls will vary greatly. For example, some improvement may be obtained by filling the cavities of concrete block walls with vermiculite. However, this may not always be practical. Since the exterior walls of buildings are either completely solid or have difficult access to cavities, the insulation must often be applied to the exposed exterior surfaces so as not to disrupt the productive, inner building space. A styrene foam laminate with fiberglass mesh is available in 1-in. and 2-in. (2.5 cm and 5 cm) thicknesses for retrofitting insulation on existing masonry walls. Applied with an adhesive and covered with stucco, this system can improve the U-value of a lightweight concrete (0.33 Btu/°F hr ft^2 or 1.9 W/°C m^2) to .08 or 0.45 W/°C m^2 respectively. Other solid board insulations can be similarly applied; however, attention must be directed to ventilating the back surface if local conditions could lead to moisture buildup. Surfaces of stucco, aluminum, vinyl, fiberglass panels, or wood can provide an attractive exterior finish over the insulation at increased costs.

Usually it is easier to apply insulation on the interior surface of the walls, although where long thermal time constants are desired, exterior insulation may be preferable. This may require putting up new studs to provide a support for the insulation. The surface may be paneled or covered with drywall and painted. Often attention to small problem areas needing insulation or weather-stripping will make life much more comfortable for the occupants. There are many wasteful but otherwise valuable buildings where wall insulation, weather-stripping, and/or new windows are a sound investment that add years of useful and more comfortable life.

There are other important parameters besides the U-value. The heat storage capacity of the building is also an important factor in climates with warm summers and extreme day–night temperature changes. For new buildings with low heat capacity, insulation added to the outer side of the facade will not only reduce heating of the exterior skin in the summer, but it will also cause a longer thermal time constant. However, this will reduce the savings possible from nighttime temperature setback. The decision to insulate the outer surface or the inner surface of the envelope should be made with the help of a specialist. Although inside insulation may be cheaper, there can be important moisture problems. These can be minimized with a vapor barrier on the surface closest to the interior occupied space (using a paint formulated for this purpose or plastic film).

Insulation on the exterior of the building is frequently better suited to blocking heat bridges, and also results in better control of dynamic heat flow. However, in most cases, this will involve a total facade renovation, with the consequently higher costs. In any case, while the insulation of a building is also an opportunity to improve its acoustic characteristics, to reduce air infiltration, and to obtain better humidity control by eliminating moisture diffusion through the envelope, due regard should be given to the fire rating of the materials to be used.

In concluding, we note that according to insulation manufacturers, 1 unit of energy is required to produce sufficient material to save 14 units of energy per year—an excellent trade-off!

Windows

The windows in use today have both assets and deficiencies that we have summarized below. Basically their advantages outweigh their disadvantages. However, it is the way in which the window is used that determines whether or not it is an asset.

In exceptional cases, larger windows can lead to lower energy costs. For example, in an air-conditioned building with induction registers, larger windows allow the reduction of artificial lighting. The increased heat losses result in additional energy costs for the HVAC system, but these cost increases are usually lower than the decreased lighting costs. The same reasoning is not generally valid for other air-conditioning systems. This is a good example of why windows and air-conditioning systems, or any other part of the energy system for that matter, should not be considered alone.

Advantages and Disadvantages of Windows

Advantages

Permit the use of natural light, and the reduction of artificial lighting to a minimum.

Reduce winter heating requirements as the result of solar energy admitted on the sunny side of the building.

Provide visual contact with the outside world.

Allow the use of natural ventilation (with openable windows).

Provide attractive element of architectural design.

Disadvantages

Cause higher heat losses in winter.

Cause additional cooling load in summer through heat conduction and through solar radiation.

Peak solar radiation through unshaded windows can greatly increase the size and capital expense of an air conditioning system required to remove it.

The Correct and Incorrect Use of Windows

More energy efficient	*More energy intensive*
Windows are sized to permit optimum penetration of daylight. (This assumes that lighting is automatically shut off when daylight is sufficient.)	All glass curtain walls are used.
Moveable solar blinds are installed on the exterior of the building where wind conditions permit.	Office buildings are too deep to permit use of daylight, or some offices in building have no windows.
Occupants use window blinds or other moveable solar protection to minimize the solar cooling load.	Window blinds are used on the interior rather than the exterior of low-rise buildings.
Window blinds or drapes are closed at night to reduce winter heat loss.	Window blinds are incorrectly adjusted so that artificial light is needed when they are used for solar screening during the cooling season.
Insulating glass and low heat loss frames and sash are used.	Single glazed windows are installed.
Effective gasketing and weather-stripping on windows are used.	Windows are opened in winter to reduce the room temperature or raise the humidity.
Window sash is effectively caulked against infiltration.	Reflecting windows are used in place of exterior moveable blinds on the sunny sides and where the value in the sun's warmth rejected during the heating season is greater than the saving in cooling during the air conditioning season.

Much can be done with windows in existing buildings to improve their energy characteristics while retaining their functional and esthetic qualities. Possibilities range from attractive methods of uniformly blocking up a fraction of the window area in glass-curtain-walled, modern buildings to replacing the broken glass in warehouse windows. Considering the extremely widespread use of single glass in the United States and elsewhere, the great potential for the reduction of heat losses becomes apparent; double glazing has become the new standard of expectation.

Sometimes, simply closing the blinds at night is a good energy-saving measure. At the Whitmore Administration Building of the University of Massachusetts at Amherst, a projected annual saving of $10,000 a year in energy costs could be achieved closing the

Solar Radiation

A basic principle is to keep the sun out in the summer, but to let it in in the winter. To achieve this effectively, unprotected glass on the east and west walls should be avoided where possible. Unprotected glass facing in these directions receives little solar heat in the winter and causes heavy heat loading in the summer. The glass should be shaded with external shading devices such as awnings or overhanging canopies. The most effective shading, possible on buildings up to 30 stories high (wind conditions permitting), can be obtained with moveable blinds, mounted on the outside of the windows. Some examples of window shielding are compared in Table 5.15. Exterior moveable blinds dissipate heat more effectively than any other shielding system. Horizontal louvers are needed on the southern windows, while blinds with vertical louvers are best for east or west exposures.

Basic to the assumption that moveable blinds are worthwhile, is that they will be

Table 5.15. Comparison of Different Window Systems

Window type	Form of sun shielding	Heat transmission coefficient		Total radiation transmission (%)	Artificial lighting required[a] (%)	Light trans-mission (%)
		Btu / h °F ft²	W / m² °C			
Double glass	Without blinds	.57	3.2	80–90	60[30]	80
Double glass	Louvered sun shades (mounted outside)	.57	3.2	15–20[b]	60[30]	20[b]
Double glass	Venetian blinds (inside)	.57	3.2	50–60[b]	60[30]	20[b]
Reflecting glass	Venetian blinds (inside)	.30	1.7	15–30[b]	100[90]	30[b]
Absorbing glass	Venetian blinds (inside)	.46	2.6	30–50[b]	100[70]	40[b]

Notes:

Double glass: U-value is higher than reflecting double glass.

Reflecting glass: Offers poorer use of daylight, external reflections can be annoying.

The "ideal window system": Optically transparent, double or triple glass with low emissivity, infra red coating (U-value of 1.5) combined with exterior blinds automatically controlled so that they could be raised and lowered, and the angle of the louvers adjusted remotely according to the weather conditions.

[a]Percent of the yearly operating hours in which artificial lighting is required in the exterior zone extending 10 ft (3 m) from windows 5 to 6½ ft high (1½ to 2m) in order to achieve an illumination of 100 ft candles (1000 lux candles) in an open plan office or 60 ft (600 lux) in a private office respectively.

[b]Blinds drawn and adjusted for the optimum angle.

existing vertical venetian blinds at night. (34) This apparently
closure was made by an infrared-sensitive TV [sensitive to tem
small as 0.35°F (0.2°C)], used to disclose heat leaks at the U
the saving is that the combined radiation barrier and insulation
raised the interior temperature at their surface by an average
unit area through single glass, regardless of its thickness, is 1
wall. If the windows in an office building with 60 percent gla
with double glass, the heat losses through the walls would be
high cost, however, this is not generally done. If the windows
any other reason, then double glazing may become cost effecti

Good glazing should never be used with poor frames. Con
7 times greater for solid aluminum—as compared to woode
example, given a 10-ft.^{2} window, edge loss will be about 25
total if the frame is aluminum, as compared to a loss of 12 perc
If aluminum is required, then window frame assemblies with th
of rubber, plastic, or a similar material having a substantially low
be used. A thermal break can cut in half the loss through the fra
systems, the U-value of the frames is at least as good as the U-va
U-values for the frame and glass in turbulent air (windy) condit
5.14.

As the technology of selective window coatings improves, spu
of improved solar collectors, higher performance windows with
effectiveness will appear. As can be seen from Table 5.14, the U
with a high visual transparency, low heat emissivity surface (
current technology already exceeds that of standard triple glazin
esthetic value of windows, we look forward to improved, relativ
windows in the future.

Table 5.14. Typical U-Values of Frame and Glass under Air Conditions.

		$\frac{Btu}{h\ °F\ ft^2}$
1.	Single glass	1.1
2.	Double glass (or storm window) with (12 mm) 1/2-in. spacing or greater	.55
3.	Triple glazing, each spaced 1/2-in. (12 mm)	.36
4.	Double glazing, low emissivity surface	.32
5.	Double glazing, low emissivity surface and filled with heavy gas	.24
6.	Quadruple glazing with special frames	.18

used by the occupants to minimize the solar loading of the air conditioning system. However, experience indicates that manually operated blinds are infrequently operated effectively. Photographs of buildings show that an average of only 30 percent of the venetian blinds are used on a typical sunny day. The indication is that the users close the blinds only when their offices become uncomfortably warm. The result is a large amount of thermal energy stored in the building, which must later be removed by the cooling system. If proper use of blinds or drapes and lighting cannot be assured, then automatic controls should be considered.

Moveable solar shading systems allow the possibility of using available solar energy in the winter for heating by reducing the heat demand on the southern and western parts of the building. For this purpose the installation of thermostats (or thermostatic radiator valves) may be necessary. Calculations show that for walls with a sunny exposure and good windows (double glazing), the heat gain from the sun is greater than the heat loss during the night and cloudy winter days. For optimum energy balance, including lighting, the windows should not be too small. The reflecting windows that have been heavily promoted by the glass manufacturers do not take advantage of these free energy sources. Whether the value of the solar heat in the winter exceeds the cost of removing it in the summer depends on the specific local conditions and the efficiency of the air conditioning system. However, with moveable, exterior solar blinds and non-reflective windows, you can essentially have your cake and eat it too!

For buildings less than 30 stories high, insulating glass with exterior venetian blinds is preferable to the reflecting glass. According to Table 5.15, both systems have the same total radiation transmission; however, the availability of natural lighting is much poorer with reflecting glass. In the exterior zone of an office building, adequate daylighting is available up to 70 percent of the time with non-reflecting, insulating glass. In comparison, the theoretical availability with reflecting glass is only 10 percent which, translated into practice, means that artificial lighting is used 100 percent of the time.

CONCLUSIONS

The positive characteristics of windows can be exploited, while their negative effects can be minimized with good design and conscientious use. With appropriate form, size, and location, windows can have a positive effect on the building's yearly energy balance for heating and lighting. Because of this, we consider windows to be a desirable element for the purpose of energy saving. Also for this reason, one should avoid extreme solutions in new construction, such as all glass or windowless office buildings, or cubic buildings.

CHECKLIST	MEASURES WITH SHORT RETURNS
	Building
Caulk cracks, apply weather-stripping.	Seal around window sash, exterior venetian blind openings, building foundation, and along roof line, skylights.
Plug unused vents and openings.	Ventilation shafts, unused furnaces and fireplaces, chimneys, ventilation grills. Repair leaky exhaust dampers.
Insulate.	Underneath the roofs of low, poorly insulated buildings.
Permanently seal unused windows.	Especially in cold weather climates, where the effect of infiltration is large. Install plastic sheet over poorly fitted windows and skylights.
Apply weather stripping to outside doors.	Apply to windows also.
Seal elevator shafts.	Check that doors close under all weather and pressure conditions. Automatic vestibule doors should be adjusted to minimize the times when they are simultaneously open.
Repair weak door closers.	
Reduce infiltration to refrigerated spaces.	Use door drapes, narrower doors; install buzzers with delays on doors to signal a door which has been left open.
Reduce losses through walls Storage on cold walls.	Organize storage cabinets, filing cases along north-facing or other cold walls. Choose these walls to hang wall displays, bulletin boards, etc.
Insulate heat leaks or cold bridges through walls.	On external surfaces, insulate unheated basement and garage ceiling under occupied space. Insulate walls toward unheated areas (e.g., storage or warehouse).
Create openings to avoid the need for artificial ventilation.	For example, in garages or areas with internal heat loads like photocopying machines. Do not insulate buildings, manufacturing areas, laundries, where there is a problem getting rid of the heat.
Examine insulation.	Repair damaged insulation and provide mechanical protection if needed (especially in high-traffic areas or with low garage roofs); replace wet insulation; eliminate source of water; provide vapor barriers.
Choose white paint.	When repainting, use light colors to reduce lighting requirements.
Use reflective roof finishes.	When reroofing, choose white or light-colored roofing surface.
Reduce winter heat losses and summer heat gains through windows.	Close the blinds or drapes on unused north-facing windows in the winter. Urge personnel to adjust blinds to reduce summer heat gain, adjust angle of louvers for comfortable natural lighting levels.
Reduce window heat loss.	Lower window blinds at night. Insect screens reduce air turbulence and hence reduce heat loss.

CHECKLIST	MEASURES WITH MEDIUM RETURNS
	Building
Install vestibules	or rotating doors. Install windbreakers to shelter doors without vestibules. Modern, plastic vestibules look attractive, are easily retrofitted, and are economical.
Lower ceiling height.	To reduce the heated volume and the volume of air that must be handled by the air conditioning system, where this would not interfere with fire sprinklers.
Reduce losses through walls	
Upgrade insulation.	Walls U = 0.08 to 0.05 Btuh°F ft^2 or less (k = 0.5 to 0.3 W/°C m^2 or less). Insulate under roofs (ventilate space between roof and insulated ceiling), use multiple radiation barriers for more effective ceiling insulation. Also protect walls against moisture penetration (diffusion) to maintain good R-values, avoid rot, and maintain humidity control. Apply radiation barriers (double or triple aluminum foils) as insulation on the ceiling of refrigerated room ceilings to reduce summer AC load.
Install windows with low heat loss.	Install storm windows. If replacing windows, consider double glass, triple glass (also in glass doors), add storm doors. Consider the suitability of interior storm windows.
Install window shades.	Consider insulating window quilts to block heat loss at night. Use insulating shades with multiple layers of reflective foil. Consider motor-driven, remote-control units.
Install window frames with low heat loss.	Replace metal window frames with heat-barrier type.
Install moveable, external sun blinds.	To reduce summer heat gain, replace internal sun blinds. Consider installing more effective sunshading. Install remote-controlled, external venetian blinds.
Reduce solar radiation in summer.	If no shading is available, at least install simple roll blinds. Venetian blinds allow much better utilization of natural light. Use vertical louvers for east- and west-facing windows; horizontal blinds are best for south-facing windows.
Reduce wall losses.	Plant trees and shrubs to reduce winter winds. Use deciduous trees for reduced solar gain. Consider berming windowless walls.

Reduce losses through walls (cont.)	*Building (cont.)*
Analyze energy saving by reflecting windows.	Use reflecting windows only if external sun blinds are not applicable (e.g., for high-rise buildings). Consider reduced winter heat gain to decide on net effect. Also consider effect on lighting of reflecting solar film on windows (check overall efficiency first).
Decrease room size.	In warehouses, large machine halls, shipping and receiving areas, build smaller enclosures that are practical to heat and cool to comfortable levels.
Isolate spaces or group rooms.	According to temperature and humidity requirements, and to reduce the need of artificial ventilation.
Size chimney to heating unit.	To improve draft, raise chimney height to clear surrounding roofs, tanks, and other obstacles. Use low thermal mass construction.

5.3 PROCESS HEAT AND PROCESS COOLING

One major area common to most industries is the use of process heat, the production of which requires almost one-half of the fuel consumed by industry. This section treats methods for more efficient *use* of process heating and cooling.* The *production* of industrial heat in boilers and furnaces is covered, together with production of comfort heat, in the second part of Section 5.4. The production of process cooling is likewise treated with the production of comfort cooling in Section 5.5. Cogeneration and industrial heat recovery will be discussed in Chapter 6.

The value of process heat as a function of different fuels is shown in Figure 5.6. For example, process heat at $10/MBtu ($34/MWh) corresponds to residual fuel oil at $1/gal, natural gas at $6.65 per 1,000 cubic feet, and coal at $145 per short ton, assuming an annual average production and distribution efficiency of 65 percent. At $10/MBtu for heat, one short ton of steam would be worth about $20.80 ($22.95 per metric ton) assuming the condensate is returned at about 180°F (80°C).†

Improvements to the process heat system can help to hold down fuel costs. A complete

*References 53, 54, 55, and 56 are general references for this section.

†At 65 percent efficiency, 10 pounds of steam = 1 gallon of oil, and one kilogram of steam = 1 kWh of raw fuel energy.

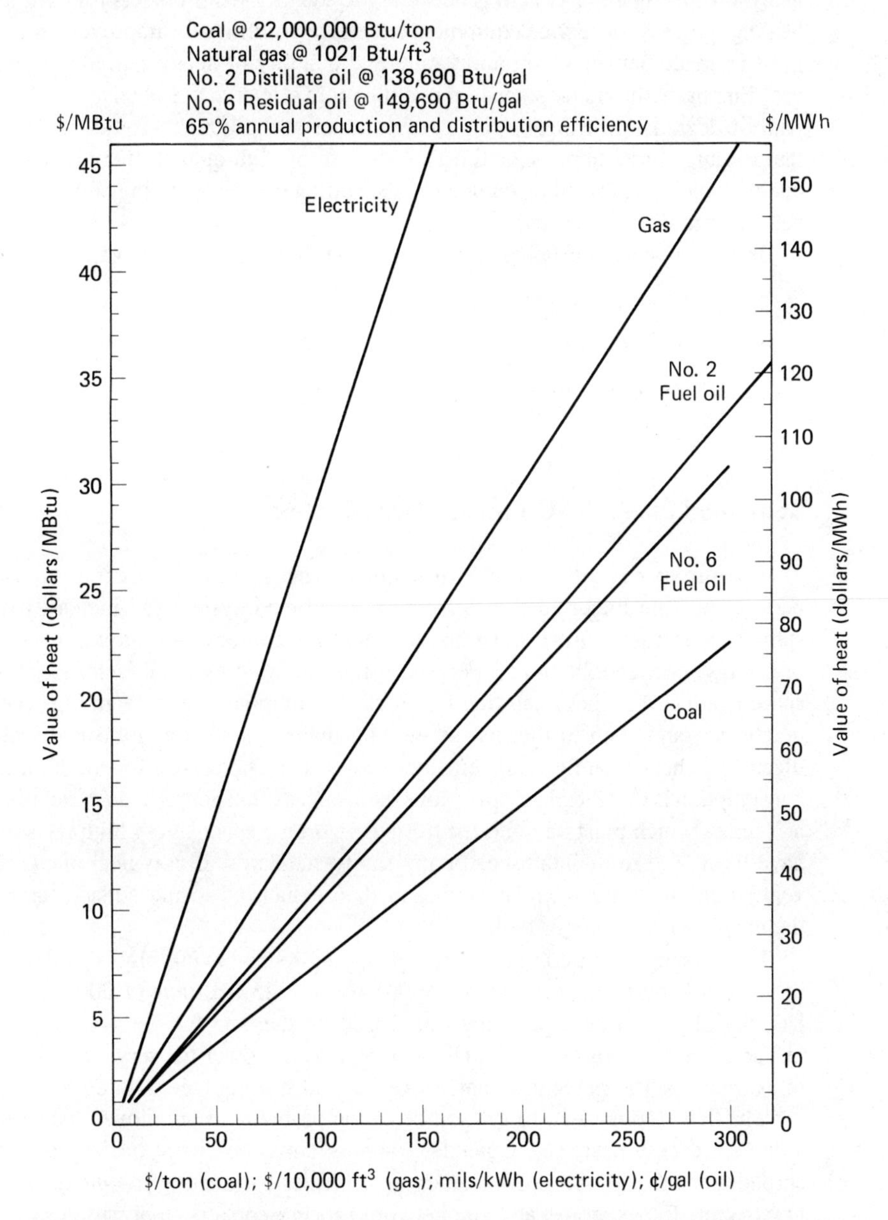

Figure 5.6. Value of heat as a function of electricity cost, and of fuel cost with 65% production and distribution efficiency.

analysis of the heating system is necessary to identify requirements for each item of space heating, process, or service equipment and the possibilities for improvement. A distinction must be made between the temperature and quantity of heat required at the point of end use. Emphasis should be placed on making more effective use of energy in its unavoidable path of degradation from the high temperature at the boiler through process to ambient temperature. In addition to optimizing the use of high-quality energy, fuel costs can be cut by reducing leaks, improving controls, setting proper operating conditions, recovering heat, and reducing line losses.

Improved system efficiency involves the following sequence of action:

Reduce demand.

Improve maintenance.

Reduce distribution losses.

Improve production efficiency.

Combined Plant Heat: Process Heat Systems

Because space heating requires low-quality energy, and because it is often an important part—if not the major load—of the total combined system (as indicated by a winter-summer load variation), it must be taken into the strategy. According to the Department of Energy, between 20 and 50 percent of the fuel used by manufacturing plants goes for space heating. This has been true for electrical equipment manufacturing, chemical manufacturing, and even in thermal processing industries such as rubber and plastics manufacturing, heat treating, and semiconductors. (1) The reason for the high level of fuel consumption is that the plant operations themselves often generate dust, volatile chemicals, and fumes which must be removed from the working space. Very high rates of ventilation have been used to maintain health and safety standards. In a typical plant, cold outdoor ventilating air is drawn in, heated, and then exhausted to the outside, carrying with it the energy of the heating fuel.

For perspective in comparing space heating loads having high ventilation rates with those in non-manufacturing areas, approximately 25 MBtu/yr (7,000–8,000 kWh/year) has typically been used per occupant, based on the data for the average building given in Table 5.1. This varies from a factor of 50 percent lower for very large office buildings, to as much as 150 percent higher for residential heating (see Section 5.4).

The first step is to go to the source of the problem and minimize the release of dust, volatiles, excess heat, etc. Consider trapping emissions with local suction, hoods with shrouds, or other devices. Also study possibilities for ducting in untempered outside air to carry off fumes from vats, smoke from process cookers, lubrication mists, and other undesirable effluvia. It may be possible to use precipitators or charcoal filters to remove the remainder and perhaps recover a useful byproduct. One basic technique which has been successful, is to seal the building and then make use of a mechanical ventilation system, so that all the air entering and leaving the plant space can be controlled. Further savings can be made by using a heat exchanger to allow thermal energy from the warm, outgoing air to heat the incoming cooler air. Plant space heating savings of 30 percent or more, with some savings as high as 80 percent, have been attained by these methods.

The question is whether space heating should be combined with process heat or separated from it. We list some of the pros and cons for this question below.

Pro:

It may be convenient (already in existence).

Process heat may be always available anyway when space heat is required.

No separate lines are required if space heat is needed where process heat is being delivered. A separate space heat production facility is avoided.

Con:

It is a mismatch of energy quality, and a loss of the opportunity to extract valuable mechanical work before the energy is degraded to ambient temperature.

Separate lines may be required anyway if office space is separate from the production area.

Space heating may be required only when outside temperatures are lower than 55°F (13°C) in manufacturing areas, 60 to 63°F (15 to 17°C) in office areas. Consequently, a process heat system supplying a large space heat load may be oversized, and have higher losses than necessary during the spring/summer/fall.

In the converse system, when the manufacturing lines may be shut down in winter, an oversized process system with oversized losses must still be operated for space heating.

The higher temperatures and pressures required for process heat mean that even if the same energy is ultimately supplied, more expensive equipment is involved than if a separate system were designed to supply the space heat load alone.

Using process steam to heat the office buildings of the plant would be an example of an inappropriate use for the process heat system.

Perhaps an opportunity is being missed to use rejected heat from the processes that is now going to waste.

The answer to the question depends upon the timing and size of demands for space and process heat, and alternatives which govern the economics of the individual situation. The answer must also look ahead to the possibilities of changes in demand; demand might increase because of future plant expansion, or it might decrease because of improved efficiencies. If the option to change exists, keep an open mind and calculate the annual fuel costs for the alternatives. It may be that the most efficient system will be separate space and process heating boilers.

An example of the inefficiency of combined space and process heat is illustrated by the system of a large, modern chemical firm. In June, their two 350 MBtuh* (100 MW) natural-gas-fired boilers were operating at less than 20 percent of capacity; in addition, their control personnel had adjusted the excess air so the CO_2 concentration was down to 5 percent! This senseless waste could have been avoided with proper instruction of the operators, or with the installation of an automatic flue gas-analyzing controller. This process heating system was entirely too large to meet the rarely occurring space heating conditions in the firm's sparsely occupied manufacturing areas and in their numerous laboratory and office buildings. Furthermore, fewer than 10 percent of their process heat users actually required the 340°F (170°C) temperature supplied. Much of the load was due to distribution losses to office buildings where no heat was needed and to lower temperature heat users. Rather than 350°F (180°C) hot water, space heating requires temperatures lower than 195°F (90°C), and 140°F (60°C) temperatures are quite satisfactory if air heat exchanger surfaces are large enough. Lower temperatures mean lower heat losses from the distribution lines. This has to be balanced against additional energy for pumping a greater volume of low-temperature water.

Reduce Demand

Can the process be changed to one with less energy need or lower energy quality (lower temperature)? Can users be separated or better grouped according to their temperature and pressure needs, their mode of operation and operating time?

A certain strategy is required to minimize fuel costs:

1. Know the efficiency and warm-up time of each piece of equipment. Know which ones require highest temperature—there may be only a few which lack adequate heat exchange surface or for which there might be another solution, such as locating them closer to the source of production or improving the heat exchanger.
2. Lower the supply temperature to determine which users complain first. Improvement of these few users may enable the supply temperature to be lowered permanently; this will cut distribution losses and unnecessary losses at each one of the other process heat users.
3. Coordinate product flow and warmup and operating times to optimize through-put at minimum operating costs.

Process design changes also offer opportunities for energy savings. Distillation requires significant amounts of energy, generally in the form of steam to the boilers. By vapor recompression distillation, overall energy requirements for a given separation can be reduced by as much as 25 to 75 percent. (53) The following passage shows an example of a simple change in the steam flow in presses for molding phonograph records which cut steam demand by a factor of more than 2.

*Million Btu per hour (MBtuh)

Steam Traps for Steam-Heated Cyclic Processes

EXAMPLE

The installation of steam traps on a group of presses for molding phonograph records was the principal factor in an energy saving of 87,800 MBtu per year.

Phonograph records are compression molded from thermoplastic resins using a steam heating and water cooling cycle. Formerly, the molds were heated by blowing 150 psig steam through the mold channels directly into the drain line. It was felt that traps were not practical, since they offered too much restriction to the flow of cooling water, and lengthened the molding cycle to a degree which was not acceptable.

It was found possible, however, to use traps during the heating part of the cycle and to have no trap during the cooling portion by the installation of both a trap and a solenoid controlled by-pass. The by-pass valve is open during the water cooling cycle, and is closed by a timer about four seconds after the water is shut off and the steam turned on. These traps and by-passes reduced the steam consumption from 6.1 pounds of steam per record to 3.0 pounds, and actually shortened the molding cycle instead of lengthening it.

Other refinements, such as a newly designed mold with more efficient heat transfer, ultimately reduced the steam consumption to 2.11 pounds per record. At the same time the production rate (theoretical) was increased from 40 records per hour per press to 90 records per hour.

If 20,000,000 records are produced per year and the steam used in the presses provide 1100 Btu per pound,

$$\begin{aligned}\text{Annual energy saving} &= 20{,}000{,}000 \text{ records/yr} \\ &\quad \times (6.10 - 2.11) \text{ lb steam saved/record} \times 1100 \text{ Btu/lb steam} \\ &= \underline{87{,}800} \text{ MBtu per year}\end{aligned}$$

If energy cost is $1.20 per MBtu,

$$\begin{aligned}\text{Annual cost saving} &= 87{,}800 \text{ MBtu/yr} \times 1.20 \text{ \$/MBtu} \\ &= \underline{\$105{,}000} \text{ per year}\end{aligned}$$

SUGGESTED ACTION

Check all steam heated equipment for the existence of efficient traps. They can be used even when the steam is turned on intermittently for a time as short as 10 or 20 seconds. Also, consider returning steam condensate to the boiler feedwater tank.

SOURCE: National Bureau of Standards. (57)

Reduce Distribution Losses, Improve Maintenance

The energy losses through leaks, inadequate insulation, and poorly maintained traps can be considerable. Is the maintenance of your plant adequate? To cut distribution losses, consider better organization of the process heat flow by situating high temperature users closer to the source. Coordinate users with progressively lower temperature requirements (and in simultaneous operation) to use rejected heat from the previous process. Product output from one process may be more efficient to transport to the next production step than is low-temperature heat.

The use of special heat transfer fluids to replace water in process steam installations may help to avoid flash and condensation losses. Unlike steam, the same fluid can both heat and cool. Closer temperature control can be obtained and unneeded heat can be returned to the heater. (58) In some process steam plants, certain of these fluids have been used in conjunction with some other straightforward measures, such as plugging steam leaks and exact metering, to reduce heating fuel consumption by as much as 50 percent. (1)

Insulation applied during the former cheap oil era is now insufficient, and may be wasting 30 to 40 percent or more of the energy that could be saved with cost-effectively increased thicknesses. (56) The cost-benefit analysis to determine the proper thickness on process heat and process cooling lines is described with extensive graphs and charts in the DOE publication *Economic Thickness of Industrial Insulation*.* (56) Listed below are some representative heat losses for process heat lines due to insufficient insulation.

The pipe in the following example is 330 ft. (100 m) long, 4 in. (100 mm) in diameter, and carries 10,000 MBtu (2.9 million kWh) of hot water per year at a temperature of 340°F (170°C):†

1. For a bare, uninsulated line, the annual heat loss would be 1,100 MBtu (3.2×10^5 kWh) per year. This is 11 percent of the heat transported and corresponds to a loss equivalent to 9,100 gallons of oil.
2. For a poorly insulated line with 0.8 in. (20 mm) of insulation, the annual heat loss is 330 MBtu (9.6×10^4 kWh) per year. This is 3.3 percent of the heat transported and corresponds to a loss of 2,750 gallons of oil.
3. For a better insulated line with 2.4 in. (60 mm) of insulation, the annual heat loss is 100 MBtu (3×10^4 kWh) per year. This is 1 percent of the heat transported and corresponds to a loss of 800 gallons of oil.

Improve the Maintenance of Steam Lines and Traps

Because of the conditions of temperature and thermal expansion, steam distribution systems require a continuous maintenance program to replace gaskets, packing and insulating materials, and to repair traps and other leaks. Defective fittings, rupture of corroded parts, and worn gaskets that release steam are usually no problem to locate.

*Available from the U.S. Government Printing Office (Washington, D.C. 20402), stock number 041-018-00115-8.

†The line is assumed to carry process hot water with a 55°F (30°C) temperature drop for 2,100 hours per year and to remain at its temperature of 340°F (170°C) for 5,500 hours during the year.

This is not necessarily the case for components of the system that are out of sight in darkened tunnels or in distant trenches, or in defective steam traps.

Steam leaks are tolerated because personnel have become accustomed to the clouds of steam and the magnitude of the loss is not appreciated. It is easier to rationalize that the trouble and cost of repair would not save much steam. As additional leaks develop, they are accepted with the same rationalization. That is why the plumes of steam seen rising from small laundries and large manufacturing plants alike have become accepted as a normal symbol of industrial activity.

Although steam leaks are expensive, the cost of a particular leak, and therefore the justification for its repair, may be a problem to quantify. The loss for some typical pressures and hole sizes is given in Table 5.16. However, the diameter of a leak is usually not known, and mainline meters seldom have adequate resolution to determine the leak rate. Estimating loss thus becomes a matter of judgment. One oil company facilitated estimation by photographing the clouds issuing from different hole sizes under typical weather conditions. (59)

Table 5.16. Costs of Steam Leaks Represented in Gallons of Oil per Month and Kilograms of Oil per Month

Size of leak		*Temperature of saturated steam*			
in.	*mm*	*240°F (115°C)*	*310°F (155°C)*	*380°F (195°C)*	*465°F (240°C)*
1/16	1.6	34 gal/mo (128 kg/mo)	100 gal/mo (374 kg/mo)	259 gal/mo (961 kg/mo)	627 gal/mo (2,357 kg/mo)
1/8	3.2	137 gal/mo (515 kg/mo)	399 gal/mo (1,500 kg/mo)	1,031 gal/mo (3,872 kg/mo)	2,507 gal/mo (9,418 kg/mo)
5/32	4.0	215 gal/mo (807 kg/mo)	623 gal/mo (2,340 kg/mo)	1,631 gal/mo (6,612 kg/mo)	3,921 gal/mo (14,735 kg/mo)
7/32	5.6	421 gal/mo (1,582 kg/mo)	1,223 gal/mo (4,594 kg/mo)	3,162 gal/mo (11,880 kg/mo)	7,686 gal/mo (28,879 kg/mo)
1/4	6.35	550 gal/mo (2,067 kg/mo)	1,598 gal/mo (6,005 kg/mo)	4,131 gal/mo (15,522 kg/mo)	10,044 gal/mo (37,739 kg/mo)
5/16	8.0	856 gal/mo (3,216 kg/mo)	7,188 gal/mo (27,008 kg/mo)	6,454 gal/mo (24,251 kg/mo)	15,686 gal/mo (58,938 kg/mo)

Assumptions:

1. Steam produced and distributed at 65 percent production and distribution efficiency from No. 6 residual oil.
2. 720 h/mo.
3. Energy loss converted to kilograms oil using the convention of 10,000 kcal per kg oil.

Example of Savings Resulting from Steam Leak Survey

At the Sun Oil Company's Toledo refinery, two engineering summer students used photographs of calibrated steam leak plumes to make a steam leak survey. As a team, they determined which leaks would be cost effective to repair. They found 247 leaks totaling 14,040 lb/h, and ranging in size from 3 lb/h to 800 lb/h. The average leak was 57 lb/h. Thirty-six leaks with losses larger than 95 lb/h were scheduled for immediate repair. The remainder were incorporated into the regular maintenance schedule. Each leak was tagged and a maintenance work order was initiated which included the size of the leak, the cost of the leak, and the cost of repair. (59)

At the time of this study, the cost of these leaks amounted to $300,000 per year. Some firms contract for outside specialists to repair steam leaks and then continue periodic inspections to insure a tight, leak-free system.

Condensate Should Be Recovered

Fresh feedwater required to make up for lost condensate must be heated to the process temperature. This requires 2 percent additional fuel for each 20°F (10°C) additional feedwater heating. Fresh feedwater, despite demineralization and anticorrosion processing, still causes additional boiler scale and corrosion, all of which are minimized by returning condensate to the boiler.

Part of the condensate will flash to steam when the pressure on it is reduced. Steam venting from any source is a waste of one kWh of fuel value for every kilogram of steam vented. Steam escaping from the condensate tank flash vent may be evidence of one or more malfunctioning traps blowing through. The following example illustrates the value in recovering condensate.

One plant in California had not been able to justify condensate recovery lines at the time the facility was constructed because of cheap energy and water prices. Although freshwater treatment had always been relatively expensive, the fear that the condensate might become contaminated if the process went out of control helped rationalize the omission of a condensate recovery system.

This design was eventually modified because of the high cost of fuel ($4.85/mBtu or $16.70 mWh) and the rising cost of water and water treatment (35 percent of the cost of condensate replacement). By returning process condensate, make up water costs were reduced 53 percent, at a saving of $200,000 per year. The cost of additional collecting facilities and a water quality control system with a process failure alarm were repaid in less than a year. (25)

Steam Traps

In a properly operating distribution system, steam is trapped on one side of a valve, while condensate is allowed to flow through and return to the boiler. The failure of a trap to operate properly can have a significant effect on the overall efficiency and fuel consumption of the system. Below is a summary on traps for steam conservation similar to the one at the Dow Chemical Co. (53)

Steam Conservation

Check all steam traps

What? Steam traps are used to separate air and non-condensable gases that, by their insulating effect, reduce the efficiency of process heating equipment.

Why? To save steam and to prepare your tracing and heating systems for winter operation.

By whom? Plant operators, mechanics, and fitters from the field service units.

Frequency of testing: Once per shift by plant operators is not unreasonable, or daily, weekly, etc.

Safety: Check all steam and condensate lines for insulation. When testing steam traps protect yourself from possible thermal burns.

Testing: Bucket traps—Strong No. 141 or Armstrong 811.

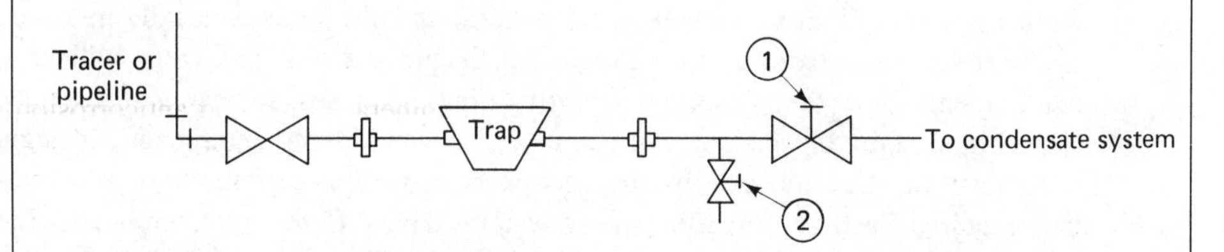

Testing: Thermodynamic or disc trap. "Sarco" 52, Yarway 29 Series, or strong DD-70.

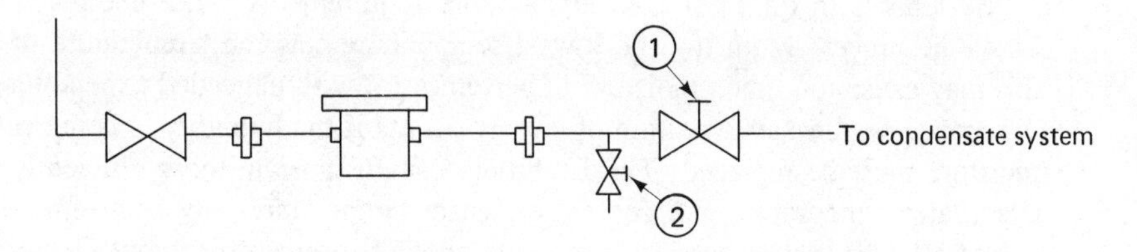

Close valve no. 1; open valve no 2; trap should cycle and close. Should trap fail to operate, close valve no. 2 and open valve no. 1. Red Tag for further check. Should trap blow continuously, it could be worn out, undersized, on a cold system, or process start-up.

A properly operating trap should be closed to hold back live steam, and open occasionally to pass condensate. Thus there should be a difference in temperature between the steam and condensate sides of the trap. If other noise in the piping is low enough, the opening and snap closing may be heard from disc-type valves. The periodic opening and closing may also be heard from mechanical (bucket) valves, although this may not be audible from certain thermostatic valves. When this sound can be heard, it is evidence that the valve is not stuck in the open or closed position. An industrial stethoscope is helpful in picking up the sounds of valve cycling, while a metal bar or screwdriver may be improvised to transfer sound from the pipe to the ear.

Heat Recovery and Cogeneration of Process Heat and Electricity

Heat recovery from industrial processes and cogeneration of electricity and process heat offer significant opportunities to reduce overall energy costs. A variety of methods and examples are discussed in Chapter 6. Additional information is given in References 60 through 77.

Process Cooling

Many of the methods for improving the efficiency of the process heating system would serve equally well for process cooling, allowing for the usually lower temperature differences. The equipment, however, for generating cold fluids is usually more expensive to buy and more costly to operate. Frequently the process cooling system, with its different requirements for temperatures and operating times, costs less to operate if it is not coupled to the air conditioning system. For example, air conditioning equipment is designed both to remove moisture released by the occupants as well as to remove or add moisture to their required fresh air. Process cooling systems have different moisture loads. Electronic equipment, instruments, and computers, which require heat removal, generate little or no moisture and require little fresh air. Plating rooms or refrigeration systems handling food may have high moisture loads.

Systems with different cold temperature requirements waste energy in three ways. First, the process requiring the lowest temperature sets the temperature of the system; this may cause too much moisture to be removed, with unneeded expenditure of energy. A second, needless expenditure of energy occurs if the humidity is being controlled and moisture must be replaced. Third, chillers usually operate more efficiently with higher evaporator temperature and cooler condenser temperature; any heat removed at a temperature lower than necessary represents another loss of efficiency. Combined systems may be less expensive to build, but they certainly do not operate less expensively.

Could there be unnecessary run-around cycles in your process cooling systems? Check the flows. Is water being removed and added later? Raise the heat exchanger (cold deck) temperature by cutting the chillant flow rate if its temperature cannot be raised above the desired dew point. Humidity sensors can accomplish this control automatically. For additional methods see Section 5.5.

Sometimes a process must be cooled after heating. Evaluation of the thermal balance of a given process should also pinpoint places where excessive sub-cooling of process streams can occur because of low cooling water temperatures. If the process stream must subsequently be heated, steam consumption will be increased. Modified operating conditions are normally able to correct this problem.

Do not overlook the possibility of using rejected heat from a process to generate cold fluids for cooling. Heat rejected at temperatures between 250°F (120°C) and 160°F (70°C) can be used to produce chilled water with an absorption chiller. Production efficiency of absorption chillers increases with increased input temperatures. At 220°F (105°C), the coefficient of performance* would be approximately 0.5. Although this is low compared with large centrifugal chillers, it may represent a use for condensate at this temperature that is not otherwise being recovered. The hot water rejected from absorption chillers is still hot enough for drying, heating domestic hot water, or space heating. A cascade arrangement using a steam-turbine-driven centrifugal chiller and an absorption unit as a steam condenser can pump $2\frac{1}{2}$ kWh of heat per kWh of steam input at typical process heat temperatures.

The overall efficiency of some process cooling systems producing chilled air with a central refrigeration machine, which by itself has a fine coefficient of performance (COP), is often quite low when the energy for the fans and other auxiliaries is included. In Chapter 8 we discuss one system, for example, with a COP less than 1.5 even with free cooling (i.e., using cold, outside air). One secret to saving energy is to keep the volume of air and the distance it must travel as low as possible.

*The coefficient of performance (COP) is the ratio of the heat energy removed to the input energy required to operate the equipment.

CHECKLIST	OPERATING CHANGES: PROCESS HEAT/COOLING
Reduce Demand	
Is unused heat turned off?	Shut off steam to unused equipment, also during lunch or after working hours. Improve product flow to reduce idle time during which equipment is up to temperature.
Improve product flow.	Reduce storage time for hot or cold products awaiting next process. Avoid letting products cool down between stations. Use existing computer to program flow and minimize both energy and unproductive waiting by operators.
Reduce operating time of equipment.	Can start up times be delayed? Run only the equipment needed. Avoid holding standby boilers and equipment in hot or running condition except when their immediate use may be needed.
Review process.	Can steps or processes be eliminated? Can heating or cooling temperatures be altered to save time and energy? (E.g., can a drying process be accomplished without any heat?) Evaluate temperature reductions in process equipment, i.e., washers, dryers, ovens, dip tanks, etc. A few degrees drop may not affect the product, but will definitely produce fuel or electricity savings.

Eliminate Unnecessary Demand

Reduce temperatures in process equipment.	Steam injection heating of domestic hot water is a misuse of higher quality energy. Heat domestic hot water with heat recovered from boiler exhaust, compressor heat, etc. Or use flash steam from high-pressure condensate as low-pressure steam for heating hot water, boiler feed water preheating.
Conserve high pressure steam.	Obtain use steam for low-pressure uses, cleaning, etc., that is exhausted from high-pressure equipment.
Use stored heat.	Shut down boilers and other heating equipment sufficiently early so that the energy stored in the system can be utilized before the end of the working day.
Record steam demand.	Meter and record steam use per unit product output. Compare metered consumption at regular intervals to identify unusual changes. Check accuracy of recording meters.
Eliminate unnecessary demand.	Shut off or disconnect unnecessary steam heating and trace lines, for example during the summer months. Shut off lines to building space heat during summer months.
Reduce preheat time to minimum.	Also reduce temperatures when production stops for a longer period.

Reduce Distribution Losses

Reduce losses in condensate return system.	Check condensate return lines for leaks and damaged insulation.
Eliminate steam venting.	Condensate receivers should be inspected for overflow and steam exhausting to waste. Inspect condensate tank vents. Plumes of steam are an indication of one or more defective traps. Determine which traps are defective and adjust, repair, or replace as necessary.
Improved heat flow may permit reduction of supply temperature.	Correct sluggish or uneven circulation of steam. This is usually caused by inadequate drainage, improper venting, inadequate piping, or faulty traps and other accessory equipment. Check vacuum return system for leaks. Air drawn into the system causes unnecessary pump operation, induces corrosion, and causes the entire system to be less efficient.

Reduce Distribution Losses (cont.)

Service chilled water system.	Proper maintenance of chilled water piping will improve the efficiency of the piping system.
Improved distribution efficiency may allow flow rates to be reduced, or supply temperature to be increased.	Inspect all controls. Test them for proper operation. Adjust, repair or replace as necessary. Also check for leakage at joints. Check flow measurement instrumentation for accuracy. Adjust, repair, or replace as necessary. Inspect insulation of chilled water pipes. Repair or replace as necessary. Be certain to replace any insulation damaged by water. Determine source of water leakage and correct. Inspect strainers. Clean regularly.
Keep heat exchanges clean, free of scale, fouling.	Inspect heat exchangers in cooling system. Large temperature differences may be an indication of air binding, clogged strainers, or excessive amounts of scale. Determine cause of condition and correct.
Improve maintenance.	With future costs of new plant and operating expenses and equipment increasing faster than labor costs, it pays to upgrade maintenance schedules. Proper maintenance of steam piping will, among other things, prevent unnecessary wastage of steam.
Repair leaks.	Repair steam leaks—repack steam valves, stuffing boxes, add insulation if necessary to steam or cooling water lines.
Schedule routine inspection.	A thorough inspection of trenches and remote areas should be made for leaks and open discharge to waste from traps and steam lines. Special attention should be given to blow-down and blow-off valves. Any losses in the distribution system require greater energy input for the same end use. Inspect insulation of all mains, risers and branches, economizers and condensate receiver tanks. Repair or replace as necessary. Inspect vents and remove all obstructions. Clogged vents retard efficient air elimination and reduce efficiency of the system. Correct any excessive noise which may occur in the system to provide more efficient heating and to prevent fittings from being ruptured by water hammer.
Provide for periodic maintenance checks on controls, valves, and accessories.	Inspect all pressure-reducing and regulating valves and related equipment. Adjust, repair, or replace as necessary.

Reduce Distribution Losses (cont.)

Service boiler feed water heater.

Repair any steam and water leaks at packing—piping joint cracks, flanges, gauge glasses, etc., in boiler-feed water heaters. Internal inspections should be scheduled to clean or repair spray heads and trays in the heater to assure proper deaeration and efficiency. Float valves should be free acting; sticking or binding may cause warm water overflow to waste. This type of waste usually goes unnoticed. The steam pressure control valve to the heater and the atmospheric relief valves should be tested periodically to reduce waste and improve safety.

Inspect and repair steam traps.

Worn out and leaking steam traps and traps with open bypasses anywhere in the plant can cause an excessive return of steam to the feed water heaters or receivers, which would escape to waste through the atmospheric relief valve. A trap testing and repair program will help to avoid one of the most costly and common problems of wasted energy encountered in many plants.

Use IR heat detectors.

Use infrared heat detector to check temperature drop across properly functioning trap and to check for location of hidden steam leaks, faulty insulation.

Listen to trap to determine if it is opening and closing.

Operating Sounds of Various Type Traps

Type	*Proper operation*	*Failure*
Thermodynamic (disc-type)	Opening and snap-closing of disc several times per minute	Rapid chattering of disc as steam blows through
Mechanical (bucket)	Cycling of sound of the bucket as it opens and closes	Whistling sound of steam blowing through at high velocity
Thermostatic	Sound of periodic discharge if on medium to high load; possibly no sound if light load, throttled discharge	Whistling sound of steam blowing through at high velocity

Improve production efficiency.

Do not allow condensate to be wasted.

Reduce Distribution Losses (cont.)	
Increase condensate recovered.	Reduce energy costs for feed water heating, and costs for corrosion and scale control by getting more users to return condensate.
Monitor water quality.	If your raw water quality is good, you may not need to blowdown according to schedules established for worse conditions.

CHECKLIST	SHORT-RETURN IMPROVEMENTS: PROCESS HEAT/COOLING
Reduce Demand	
Insulate.	All dryers, tanks, and process equipment should be well insulated. All surfaces too hot to touch should be insulated for savings, safety, and comfort. Cover open tanks to reduce heat losses. Cover the surface of hot liquids in open tanks with plastic foam particles to reduce evaporative losses.
Improve seals on loading doors.	Replace seals and gaskets of steam equipment, pressurized cookers, and dryers. Is steam escaping and increasing demand on the air conditioning system?
Install shut-off valves.	Install valves on steam and condensate lines to allow unused, individual pieces of equipment, or sections of the plant that are not in use during certain periods, to be shut off.

CHECKLIST	MEDIUM-RETURN CHANGES: PROCESS HEAT/COOLING
Reduce Distribution Losses	
Insulate steam condensate and hot water and chilled water lines; upgrade existing insulation.	Properly install and maintain insulation on steam lines. One foot of noninsulated 8″ steam line could result in an annual loss equivalent to one ton of coal, 179 gallons of fuel oil, or 25,000 cubic feet of natural gas. All high-temperature piping, ovens, dryers, tanks, and processing equipment should be covered with suitable insulating material.
Shorten waterlines.	Shorten hot and chilled water lines, regroup users of high temperature closer to heat source. Repair leaks in any lines. Shut off or remove unused lines.

Reduce Distribution Losses (cont.)	
Use non-freezing heat transfer medium.	This can eliminate the need to heat trace water pipes in winter.
Reduce Demand	
Upgrade steam controls, valves.	Replace valves which allow live steam to be passed directly through to exhaust. Use proportioning steam supply valves and controls to reduce or eliminate abnormal sudden steam demands.
Improve equipment cooling.	Do not allow equipment heat to increase room heat load; use separate cooling systems for components (preferably water cooled). Use spot or local cooling for personnel rather than cooling entire area.
Separate contaminated exhaust air	and recycle the clean air.
Use low pressure steam turbines.	To drop pressure for low-presssure users, extract mechanical work from steam (or air lines) using steam-driven motors to operate pumps, air compressor fans, shaker beds, cool transport systems, etc.
Check installation of simultaneous heat and electricity production.	Recover more of the high-quality energy available in fuel.
Cascade users for multiple use of energy.	Since very little high-quality energy goes into the chemical bonds of a product, with better insulation, heat supplied at high temperature to first process would be available with only a slight temperature drop for a second and third process. Although drying operations do require heat to change phase from the liquid state to the gaseous state, this can be done with heat at temperatures between 40°C and 120°C; hence drying should be able to use heat rejected from other processes.
Segregate users of low quality heat.	Separate energy users which can be satisfied with a low temperature level. Also segregate users according to operating mode and time.
Replace inefficient equipment.	Install forced circulation fans in convection ovens.
Upgrade restaurant and kitchen equipment.	Replace unpressurized food steamer (compartment, deck, or convection steamers) with pressurized steamers or steam jacketed kettles when modernizing kitchens. Replace older and surprisingly wasteful steam tables and bain maries with newer, better insulated units.

Reduce Demand (cont.)

Apply process heat only where required.	Induction heating, because it is able to apply the heat more nearly to the actual volume required, can be more energy-efficient than, for example either convection or radiant furnaces. For the same reason, microwave ovens can overcome the inefficiency in microwave energy production by the efficient, localized application of heat to the food to be cooked.
Use automated processes for better energy and quality control.	For example, automated welding can produce more uniform quality welds with significantly less energy.

Increase Production Efficiency

Meter condensate	If all condensate is returned, a meter on the condensate lines can serve as a measure of steam use.
Install a meter on the make-up water to boilers.	Strive to improve the percentage of condensate returned by recording daily make-up water or condensate flow. Excessive make-up indicates a loss of boiler water, increases treatment, and is wasteful.
Improve feedwater chemical treatment.	Deaerate boiler feed water at high temperature 205°F (95°C) for more efficient oxygen removal to reduce corrosion.
Generate process heat from wastes.	Burn office paper, shipping crates, industrial wastes to generate steam and reduce air and water pollution.
Recover process heat from existing incinerators.	Fumes, vapor, and industrial wastes which require incineration, especially at high temperatures, may provide opportunities for significant heat recovery.
Use high pressure condensate.	Utilize flash steam from high pressure condensate as low pressure steam for heating hot water or boiler make-up water.
Use storage to operate process heat boilers at higher loading for improved efficiency.	Store off-peak energy by the use of steam or hot water storage. Use insulated storage tank to receive heated water produced at optimum boiler efficiency for use during low-use periods at night, weekends.
Install small steam boilers for more versatile and efficient production.	Installing a steam boiler to serve peak demands or to use for light loads may eliminate the necessity for running a larger boiler to handle a relatively small load in the summer. Boilers shold be operated at near full load capacity for most efficient operation. Multiple units of the proper size may

Increase Production Efficiency (cont.)	prove extremely efficient for peak-shaving and flexible efficient operation. A small boiler at a distant site may eliminate large distribution losses. Would this be a suitable service for a waste burning boiler?
Improve recovery of waste heat at low temperatures.	From water/air over 85°F (30°C) with heat exchangers, from water/air under 85°F (30°C) with heat pumps. Use condensation heat from process refrigeration units.
Use seasonal storage to improve process refrigeration economics.	Consider seasonal storage of cold (ice) collected during the winter. Use for summer cooling, and for improving the COP of food freezers, etc.
High temperature heat transfer fluids.	Consider thermal heating fluid to avoid condensate and flash losses, and make-up.

5.4 HEATING SYSTEMS

This section covers improvements first in the use and distribution of heat in buildings, and second in the production of heat from boilers and furnaces. While it also refers to saving measures that apply to the heating systems of buildings with climate control, buildings without air conditioning or forced ventilation are given more emphasis in this section. Buildings with HVAC systems receive more attention in Section 5.5. While process heat for industrial use has been treated in Section 5.3, measures for the improvement of boilers and furnaces for both comfort heat and process heat are included in this section.

Winter space heating systems are a necessity for human health and comfort. It is possible, however, to greatly reduce wasted heating energy while still maintaining adequate working and living conditions. The task of people implementing an Efficiency Improvement Program (EIP) will be to obtain a consensus of what adequate comfort should mean, and perhaps to convince people that comfort requirements, as they have increased in recent years, have become too high.

Heating Use and Costs

Based on the data given in Table 5.1, the energy for average heating use per unit area from the 307-building survey was 104,000 Btu/ft^2 (327 kWh/m^2). (10) This is a factor of five higher than the energy-saving design for the United States Government Office Building in New Hampshire, despite the relatively harsh New Hampshire winters with 7,380°F-days (4,100°C-days). (78) For large buildings, in which heating fresh air represents a larger heat load than conduction through the envelope, the annual heat use may be 38,000 Btu/ft^2 (120 kWh/m^2) or less. Particular heat use, of course, will vary

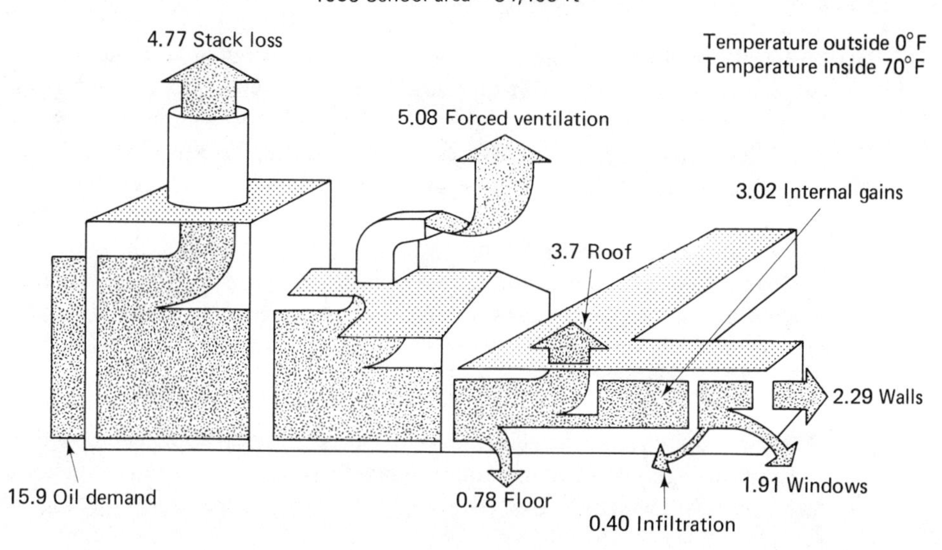

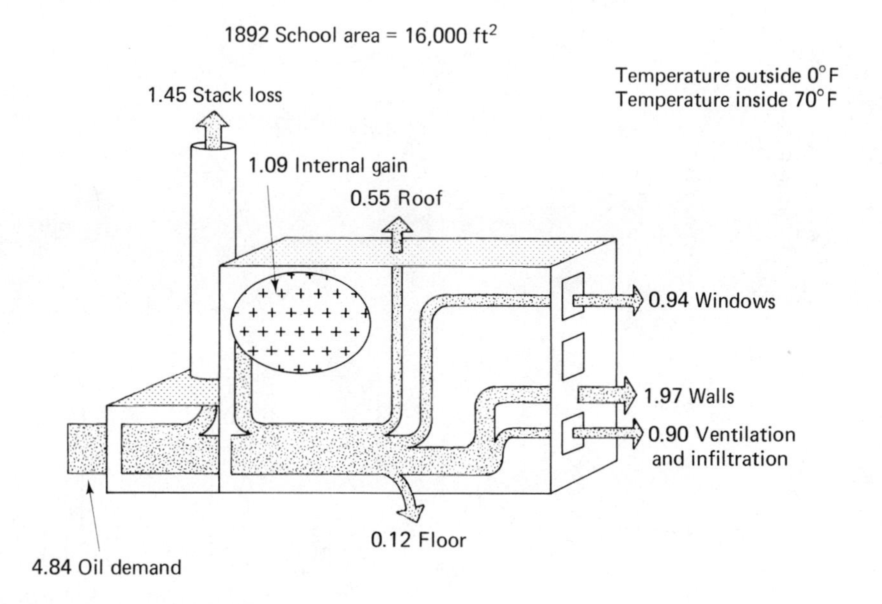

Numbers refer to gallons of oil per hour

Figure 5.7. Heat flows at 0°F for schools of two different eras. [Massachusetts Department of Community Affairs (40)]

greatly with climate, details of the building and equipment, and the care used in its operation.

For the purposes of comparison, the per person energy used for heat as derived from the average heat use in the 307 commercial building survey listed in Table 5.1 is 25,000,000 Btu per year (7,360 kWh/yr). This can also be compared with the per capita energy for residential heating in the United States of approximately 37,000,000 Btu/yr (10,700 kWh/yr). The annual cost for heat per person can be obtained from Figure 5.6 in the previous section. This figure gives the cost of heat produced and distributed with an annual efficiency of 65 percent, as a function of the cost of different fuels.

Examples of heat flow in buildings are illustrated in Figure 5.7, showing representative school buildings of two different eras (1956 and 1892). The energy flows are expressed in gallons of oil per hour. The flows have been evaluated for an outside temperature of 0°F (−18°C). (40)

The 1956 school loses energy at the rate of 18.9 gallons of oil per hour at 0°F (−18°C). Offsetting this loss are 3 gallons per hour in heat released internally by the occupants and from the lighting. In comparison, the pre-1900 school loses a total of almost 6 gallons of oil per hour, which is offset by 1.1 gallons from occupants and lighting. If the older school were scaled up to the same floor area of the more modern building, a comparison would indicate that the newer building uses 50 percent more fuel. The reason for this is the use of forced ventilation in the newer construction. These two examples represent

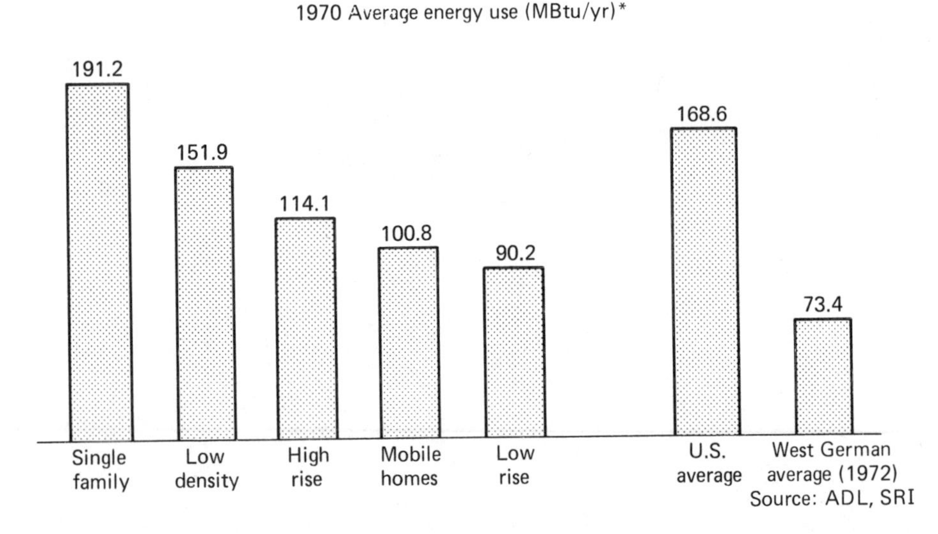

Figure 5.8. Energy use per unit area in different types of buildings. These examples show that there are more efficient types of dwelling units than those represented by the U.S. average. [U.S. Department of Energy (6)]

previous design and construction practice. Once we have identified where the losses occur, it will be evident that there is an immense opportunity for saving.

Figure 5.8 shows energy use per unit area for different types of buildings. Also shown is a comparison between the average energy use in buildings in the United States and in Germany. The lower average use in the German buildings is the result of different energy management in these buildings.

Losses and Potential Savings

Even if adequate comfort is still maintained, the energy used for heating is in many cases 20 percent higher—and in some cases as much as 80 percent higher—than it should be. The opportunity for reducing energy wastes in existing heating systems can be summarized as follows:

While still maintaining adequate living and working conditions, the *demand* for heat can be reduced by 10 percent to 20 percent, and even more in some cases.

Reduction of *distribution losses* will achieve an additional saving.

Improvement in the efficiency of *production* (or conversion of fuel to heat) can lead to substantial savings. In addition, additional heat sources, such as the heat from solar radiation, lighting, or other internal sources, can be used to better advantage, perhaps with the help of some form of heat storage.

A comparison of costs and savings from a study of improvements to heating systems in municipal buildings in Massachusetts is shown in Table 5.17. (40) These examples illustrate some of the practical savings from investments in heating system improvements. The SCP is an indication of the productivity of the conservation investment, i.e., the specific cost to save energy of one 42 gallon barrel of oil per day (BPD). An equivalent measure is the cost to reduce the rate of energy use by one kW (i.e., saving 8,760 kWh of fuel per year). For example an SCP of $46,000/BPD corresponds to investing $3 in a measure that would save one gallon of oil per year. As can be seen, some of the most effective improvements required only a small investment.

Energy-Saving Measures: Reduce Heating Demand

Promote policies which provide an incentive for saving. Meter, and if possible bill users for the heating energy used. In one department store located in a large shopping center, a proposed heat recovery system with short payback was not installed because the heating energy was billed according to the floor area of the store, rather than to the actual amount used.

Consider heating to lower room temperatures. (79, 80) Monitor the temperatures in a representative sample of rooms to gain a better idea of their levels and fluctuations. The vertical temperature variation and the variation in temperature between where people are working (sitting) and the location of the thermostat should be considered before making any temperature changes.

It is surprising how frequently one finds temperatures that are actually too warm, even uncomfortably so. Lower temperatures are healthy and comfortable once people have become used to them. Complaints concerning low humidity will also be fewer at lower temperatures.* The important element is perceived comfort. This has partly to do with minimizing the effects of exposure to surfaces that are cold (such as exposed windows or cold walls), and the avoidance of drafts (from leaky windows or mechanical ventilation). People with different personal requirements can more easily adjust to cooler rooms with a light sweater, than can people find relief in rooms that are too warm. However, since lowering temperatures from their customary setting without discussion with the occupants and their concurrence may result in a negative impact on morale and productivity—far outweighing any energy saving—a generous measure of diplomacy amd sensitivity is required.

Due to heat storage in the building and its contents, some heating systems can be shut off before quitting time. Consider setting back thermostats after working hours and on weekends.†

Many buildings are allowed to fall back to 55°F–60°F (12°C–15°C). Attention should be given to condensation problems and the possibility of unprotected water pipes freezing, especially if temperatures are allowed to fall back to lower temperatures. The cost of supplying portable heaters for those who must work after hours may be more economical than maintaining the whole building at normal temperature. Or, only a portion of the building could be heated, with better organization of after-hours activities. Automatic controllers are available that will start up the heating system before the working day starts, and shut it down early in the evening, depending on the outside temperature. These units will control the time of start-up according to how far back the temperature has fallen, and to the capacity of the heating system to supply heat. (Note that boilers usually operate more efficiently during periods of continuous operation under full load, as during the early morning warm-up period, than during intermittent operation.)

Take advantage of intermittent sources of heat, such as the sun, a kitchen, or business machines or manufacturing processes. Areas such as lecture halls and stores, which at times may be heavily occupied, should have automatic controls to shut down heating when internal heat sources are present. Install local thermostats, thermostatic valves on radiators, which will take advantage of these other sources by reducing unneeded heat flow. This will also save on cooling energy.

Look for areas where the heating can be shut off completely (with no danger to water pipes, etc.). Stairwells, entrance halls, storage rooms, utility areas with already sufficient internal heat production, out-buildings, covered walkways, unused portions of buildings

*Indoor temperatures in the mid-60s are healthier than those in the mid-70s, according to the American Medical Association. Heating homes and offices in the winter removes moisture from the air, which could aggravate respiratory problems and contribute to dry throat and nose, dry and itchy skin, and coughs. Respiratory problems can also result from moving from overly heated interiors to the cold outdoors, because the body adjusts only gradually to sudden temperature changes. (81)

†Because night temperatures are lower, the fuel consumption can be as much as 30 percent greater during the period after normal working hours than during occupancy, despite setback of temperature.

and manufacturing areas, and interior halls may be areas where heating goes to waste. The temperature, along with the lights and air conditioning, should be turned down in other intermittently occupied areas such as cafeterias, lecture halls, hearing rooms, churches, and shipping and process areas.

Unnecessary heat loss may occur in the domestic hot water system supply for washrooms, toilets, and shower rooms. Energy, water, and money can be saved by reducing water flow to the required levels. Alternately, a pressure regulator for each floor may be used to achieve reduced flow. Faucet nozzles are available which distribute a reduced flow of the water more uniformly. For motels, hotels, and dormitories, effective, reduced-flow shower heads are available (see Section 5.8).

It may be possible to heat or preheat domestic hot water with cooling water from condensers, compressors, or water cooled process equipment. A heat exchanger in the line to the cooling tower may be able to supply wash water for equipment, rinse water for electroplating, etc. Perhaps the flow may be safely reduced to bring the cooling water up to acceptable temperatures and to reduce water costs as well.

Reduce Distribution Losses in the Heating System

Long heat lines and ducts, and higher-than-necessary temperatures, may lead to avoidable distribution losses. Use of local heaters will allow shutting off of lines to distant locations. According to a Department of Energy study on the economical thickness of insulation, which treats examples in the distribution portion of heating and cooling systems, insulation specified for 1975 economic conditions saves an average of 30 percent to 40 percent more than the insulation specified before the oil price shock. (56) Insulation specified for current economic conditions would save correspondingly more. In past years, when first-cost analysis was the criterion for capital expenditures, using more than minimal amounts of insulation, if any at all, in the heating distribution system was not justified. Today, the reasoning that said that the heat lost in distribution lines ends up in the heated space anyway, is no longer acceptable to those looking for cost effective improvements. The heat loss from ducts and heat piping can be 30 percent of heat supplied, in some cases even more. This heat does not end up in the areas where it is most needed, but rather in basements, high ceilings, hallways, and service chases. The economical thickness of insulation to apply will have to be computed for each case. The DOE reference above, and References 82 and 83 may also be helpful. Insulation should be protected from moisture and possible mechanical damage.

The stand-by losses of the distribution system can be determined during a period when no heat is being supplied to end use. Consider for example a system with a fixed-firing-rate burner and all of the valves to space heaters and radiators closed. By dividing the number of minutes of burner operation required to maintain the water temperature by the total time during the test interval, the percentage of the burner's input used to supply stand-by and distribution losses can be determined.

Check radiators, heating vents, and unit ventilators for cleanliness, especially dust buildup (clean filters), and obstructions from office equipment, drapes, etc. Are these units in the right position and correctly sized and controlled?

Table 5.17. Comparison of Cost and Savings for Heating System Improvements

ESM	Example (yr. of constr.) [gross floor area]	Cost # (1); [SCP] 2)	Savings ($)	Assumptions (3) [yearly cost of fuel before improvement]	Comments
Demand reduction					
Room thermostats	Fowler School, Fall River (1897) [16,343 ft^2 (1,520 m^2)]	$400# [$7080/BPD ($100/kW)]	$270	Second floor 5°F (3°C) warmer due to the heat rise from the first floor [$6,000].[b]	Second floor was too warm, and had no thermostats. 4 thermostat controlled valves were installed in steam pipes.
Thermostat controlled ventilators	Fitzpatrick School, Pepperell [32,000 ft^2 (2,974 m^2)]	$1,000 [$11,300/BPD ($160/kW)]	$460	Temperature was reduced 5°F (3°C) [$12,000].[b]	Ten room thermostats and valves control hot water heat to local ventilators.
Air flow reduced	Junior High School, Acton (1960) [66,600 ft^2 (6,130 m^2)]	$800# [$5000/BPD ($70/kW)]	$780	40°F (4°C) average outside temperature. Flow of outside air from 15 cfm to 10 cfm (25 m^3/h to 17 m^3/h) per occupant with 25 occupants per ventilator. [$25,900].[b]	40 unit ventilators were readjusted. The new adjustment conforms with Massachusetts code.
Air flow reduced	Willard School, Concord (1958) [39,500 ft^2 (3,670 m^2)]	$400# [$4250/BPD ($60/kW)]	$480	Outside air reduced from 15 cfm to 10 cfm (25 m^3/h to 17 m^3/h) per occupant, with 25 occupants per ventilator. 40°F (4°C) average outside temperature. [$14,500].[b]	25 room ventilators were readjusted. Readjustment used one hour of labor per ventilator.
Production improvements					
Heat recovery	Administration building, Concord (1960) [32,193 ft^2 (3,000 m^2)]	$3,500 [$15,900/BPD ($225/kW)]	$1,000	Flue gas temperature lowered from 600°F to 300°F (300°C to 150°C). [$7,200].[b]	Heat exchange "economizer" installed at exit of flue from boiler. CO_2 is 8%.

Hot water heater	Fire Station, Fall River (1951) [9,576 ft^2 —(890 m^2)]	$240# [$12,000/BPD ($170/kW)]	$96	1,900 gallons of fuel @ $632 used only for domestic hot water May through September.[b] Efficiency raised from 0.5 to 0.75 with separate water heater.	Heating system boiler could be shut down with a separate, gas fired water heater.[a]
Heat recovery	Fowler School, Fall River (1897) [16,364 ft^2 (1,520 m^2)]	$5,000 [$31,900/BPD ($450/kW)]	$735	Flue gas was reduced from 550°F to 300°F (290°C to 150°C) [$6,000].	Economizer installed in exhaust gas.
Heat wheel	High School, Attleboro (1962) [430,000 ft^2 (40,000, m^2)]	$20,000 [$36,000/BPD ($510/kW)]	$2,600	10,000 cfm (17,000 m^3/h) of air at 80°F (27°C) is exhausted from the swimming pool when the average outside temperature is 34°F (12.4°C). [$87,600].[b]	Heat Wheel air to air heat exchanger transfers 60% of heat in exhaust air to incoming air.
Boiler controls	Town House, Concord (1860) [32,193 ft^2 (1,820 m^2)]	$150 [$6400/BPD ($90/kW)]	$115	The second boiler need only operate under extremely cold weather conditions. Heavier loading of the operating boiler is improved 5%. [$2,300].[b]	Two boilers, each sized for 66% of the maximum load had been operating simultaneously.

Notes:
(1) The # examples have payback periods of less than three years, and hence represent attractive examples.
(2) The specific capital productivity [SCP] of the conservation investment is the cost divided by the annual energy saving.

(3) [a] gas—$3.00/ccf ($0.016/m^3)
[b] oil—$0.36/gal ($0.11/kg)

SOURCE: Massachusetts Department of Community Affairs (40).

The adjustments of the heating system should be checked periodically. Bleed the air from lines, radiators, and fan coil units. Measure and correctly set the air flow rates, as these are often excessive. (See examples shown in Table 5.17.)

1. The initial adjustment after the commisioning of a building is often not ideal. Readjustment after EIP is also needed.
2. Yearly examination by qualified personnel should be done.

By reducing the temperature of domestic hot water to 100°F (38°C) for lavatories, distribution losses are reduced. Distribution losses are also reduced by shutting off the circulation pump for domestic hot water after working and cleaning hours. It may prove more efficient to provide local boosters for kitchens, etc., than to tolerate the distribution waste when the other requirements cannot use it at a temperature much higher than 115°F (45°C).

Example of the Selection and Evaluation of ESM from the Checklist

In the following example we illustrate the use of the checklist for a modern office building with heating only and openable windows in all of the offices (no forced ventilation or air conditioning). This example will demonstrate that of the measures that look possible, fewer turn out to be actually practical, and yet they save approximately 25 percent. This office building was built in 1965. It has 5 floors and is occupied by about 100 persons (engineers, designers, and staff). There are no rooms other than offices (except conference rooms which are treated in the example as office space). The window area is 30 percent of wall area, the conductivity of the walls is 0.2 Btu/hr°F/ft^2 (k = 1W/m^2°C). The windows are double glass. The building is used 1,500 hours per year, or about 50 percent of the working and cleaning hours (7 a.m.–7 p.m., five days a week, or 3,000 hours per year).

The energy for heating during working hours and during cleaning is the same as the energy for heating after working hours. This is due to the lower nighttime outside temperatures.*

Energy-saving actions considered:

1. Lower room temperatures from 72°F to 68°F (22° to 20°C).
2. Lower room temperatures after working hours from 64°F to 59°F (18° to 15°C).
3. Start after-hour set-back at 5 P.M. instead of 7 P.M.
4. Reduce domestic hot water temperature from 140°F to 105°F (60° to 40°C).
5. Reduce domestic hot water consumption by 20 percent.
6. Use portable electric heaters in the five coldest corner rooms. Reduce duration of central heating operation by 30 days a year during the temperate seasons.
7. Reduce air/fuel mixing ratio, improve burner efficiency by 3 percent (average over a year).

*There are about 6,500°F-days (3,600°C-days) during the heating season.

Table 5.18. Summary of EIP, Phase I, for a 100-Employee Office Building

	Measure	*Calculated savings (%)*[a]	*Practical savings (%)*[b]	*Comments*
1.	Low room temperature	12	8	Reduction in average room temperatures may be in practice only 2°F (1°C).
2.	Night temperature	19	10	Due to heat capacity of building after hours temperatures may be greater than 59°F (15°C).
3.	Night set back	2	—	Not used for first phase.
4.	Hot water temperature	0.3	—	Domestic hot water use is low in this office building.
5.	Hot water conservation	1	—	
6.	Local service at low load	6	3	A reduction in the use of central heating system of only 15 days may be more realistic. Local electric heaters are already used when very cold.
7.	Boiler efficiency	3	—	We assume that burner service is already good.
8.	Electric boiler	1	—	First phase is only with no investment.
9.	Insulation blankets	2	—	Disturbing people in first phase is to be avoided.
10.	Furnace insulation	3.6	2	Conservative value used for first phase.
11.	Automatic shut down	5	—	Investment too large (50% of yearly heating costs) for such a small building, 5% saving would be in addition to 10% of No. 2.
12.	Thermostatic valves	6	—	Investment for replacing existing valves would be too high for 1st phase (additional costs for a new building would be much lower).
13.	Circulation off after hours	4	4	Circulation losses are equal to the energy actually used for domestic hot water consumption!

[a]Energy savings (in percent) estimated from analysis.

[b]Energy savings thought to be achievable in actual practice (a lower value is more realistic because of human factors, etc., not included in analysis).

8. Install separate electric boiler for domestic hot water preparation during summer and part of the temperate season.
9. Apply super insulation blankets behind radiators.
10. Add insulation to furnace, reduce surface temperature from 160°F to 105°F (70° to 40°C).
11. Install automatic control units for optimum shut-down.
12. Install thermostatic valves on 50 percent of the radiators (the other 50 percent are used for base load) and reduce temperature variations. Make use of solar energy through windows, keep sun-shades open during winter, use heat of lighting and internal energy sources. The average temperature is assumed to be reduced by 2°F (1°C).
13. Shut down domestic hot water circulation after hours.

The calculated energy savings are given in Table 5.18.

CONCLUSIONS

Preliminary estimates add up to more than a 50 percent saving in energy (!). In the first phase, 25 percent was achieved using a very conservative evaluation. The first phase was accomplished without significant investment nor with any disturbing consequences for the occupants.

For this small building there were no measures requiring an investment which could yield an adequate return. This could be different for a larger building because some of the costs would be approximately the same for a much larger saving (e.g., automatic start-up equipment).

This is an example of how today, in order to save energy, some motivation is needed other than simply the desire to save money. The actual yearly energy cost had been so low, even with present energy prices, that for years nobody had looked at this building's energy system. On the other hand, it did not cost any money to save 25 percent of the yearly energy use, and therefore these measures were among the best ventures the owner made.

An example of EIP by the Home Office of the Connecticut General Insurance Company is described below. By 1974 this program, which combined improvements to the heating system with a number of other improvements, had already shown a 29 percent reduction in energy use.

Connecticut General Home Office Energy Conservation Measures Which Have Been Implemented

Boiler de-rated to 22,000 lbs. of steam • draft fans reduced in speed and fitted with inlet vortex damper controlled by flue gas oxygen analyzer • burners fitted with automatic turn-down ratio of 4 to 1 replace inefficient combustion control system • all boiler output control at lower capacities is manual at present • (the original control system provided for automatic changeover from heating to cooling at an outdoor temperature of 40°F for the perimeter zones, and energy was wasted because the system operated either with reheat or with recooling:) the induction changeover controls were disconnected and hot water or chilled water to the coils programmed manually. (Experience now shows that secondary cooling is not required now until outdoor temperature rises to 60°F) • space temperatures during occupied periods have been reduced from 74°F in the winter to 68°F when heating is required and raised from 72°F to 78°F in the summer when cooling is required • no reheat is used • ventilation requirements reduced from 15 to 6 cfm per person • all down escalators turned off during the day except for heavy traffic periods • lighting which had been uniform at 60 ft candles was reduced in corridors, cafeterias, lobbies, equipment rooms and in other areas with less demanding tasks by removing lighting bulbs and disconnecting ballasts • when relamping occurs, lamps with greater efficiency (more lumens per watt) are installed • one additional deep well installed • spray coils in 9 HVAC units modified to provide adiabatic cooling and spray coil humidification • one of four additional units has been similarly equipped and is under test before completing modifications of all units • modifications made to permit cooling with well water for return air and outdoor air in place of cooling minimum outdoor air only, and cooling load reduced by about 700 tons enabling centrifugal units to be removed from service for a large percentage of the time thus reducing both electrical consumption and electrical demand • the additional well provides sufficient capacity for condenser cooling to replace the cooling tower for the 900 ton chiller when in operation • enthalpy controller has been installed on one air handling unit for testing and evaluation • domestic hot water temperature reduced from 140°F to 100°F, but raised to 120°F when complaints arose.

The total consumption of energy for the main building in 1972 prior to conservation was $201{,}224 \times 10^6$ Btu's, the equivalent of 223,580 Btu's per sq. ft. per year for the main building.

After the first year of conservation, the energy consumption for the main building in 1973 was reduced to $188{,}323 \times 10^6$ Btu's per year, but an extra load of 544×10^6 Btu's was added for processing. The gross consumption including the added process load was the equivalent of 209,850 Btu's/sq ft/yr.

In 1974, with the continuing conservation program, the yearly consumption for the main building dropped to $164{,}138 \times 10^6$ Btu's, but extra process load and the print shop loads raised the total to $165{,}131 \times 10^6$ Btu's/yr. or the equivalent of 165,130 Btu's/sq ft/yr. The actual savings in energy conservation amounted to 63,580 Btu's/sq ft/yr. or a reduction of 29%. (84)

Improve Production Efficiency

The remainder of this section is concerned with the efficient production of heat for comfort, domestic hot water, and for manufacturing processes. A summary of methods for improving production and heat recovery will be followed with a discussion of combustion and a more extensive explanation of how to improve the efficiency of furnaces and boilers (see References 85 and 86).

Although this discussion is intended for small- and medium-size combustion equipment, the basic ideas would apply to units of any size. The efficiency of utilization ranges from below 25 percent for lightly loaded equipment in intermittent operation to 85 percent and higher for fully loaded burners in good adjustment. The inefficiency is due to overfired and oversized installations operating at less than optimum adjustment. The proper adjustment and operation of combustion equipment as described in this section can result in significant reduction of unnecessary fuel costs. For example, in 1975, the 78 boilers of the Raytheon Company consumed 30 percent less fuel than in the preceding year. As much as 10 percent was saved after boiler operators adjusted oil temperature for correct viscosity, carefully tuned the burners, and reduced firing rates to a minimum. (86)

Approximately 40 percent of the fuel consumed in the United States is burned in boilers and furnaces for space heating, process heat, and heat treating and drying. A 5 percent to 10 percent improvement in combustion efficiency on a national scale could reduce consumption by the equivalent of a million barrels of oil per day, and result in significant improvement in air quality. However, realistic and effective standards will be required for new and existing boilers* if the potential for efficient operation is to be achieved.

For example, the ASHRAE standard 90-75 set a goal for minimum combustion efficiency of only 75 percent. This was already the minimum efficiency standard for new equipment 25 years ago. (4) (Such a boiler would consume 12 percent more fuel than a slightly more expensive boiler that operates at 85 percent efficiency.) A more efficient boiler has, among other improvements, a greater or more effective heating surface so that more heat can be extracted from the products of combustion.

Heating equipment operates below potential efficiency, primarily because of overfiring and poor regulation of combustion air. The following measures may improve the operating efficiency of a boiler.

1. Reduce excess air (compatible with acceptable smoke), to obtain high CO_2 content in the exhaust gas (low O_2 content).
2. Reduce firing rate compatible with heat requirements† and lower exhaust gas temperature.‡
3. Optimize the combustion in the burner, heat transfer in the boiler, and operation of the chimney as a system.
4. Keep fireside and waterside heat transfer surfaces clean to reduce energy lost up the flue.
5. Improve heat transfer with spinners (turbulators) in the fire tubes or by adding heat recovery to reduce stack gas temperature.
6. Upgrade boiler insulation and decrease other losses.
7. Keep records of load (degree-days, process demand, etc.), boiler operating conditions (e.g., stack temperature), and fuel consumption. This can help in identifying abnormal operation and in monitoring the effectiveness of your actions.

Heat recovery to lower excessive exhaust gas temperature may be attractive if the boiler has too little heat exchanger surface. Heat recovery can be used to heat combustion air, to provide space heat, to preheat return water to the boiler, or to heat domestic hot water. Conditions for favorable heat recovery economics are: high flue gas temperature, high annual boiler load factor, and use for the lower temperature heat obtainable, for example, to heat large quantities of feedwater. A summary of heat recovery examples for space heat and domestic hot water is given in Tables 5.19 and 5.20. Heat recovered

*Remarks made for boilers apply similarly for furnaces and radiant heaters. The term "boiler" is used here to represent all such fuel-burning equipment as well as the heat exchanges in such equipment.

†See reference 85, pp. 22 and 23

‡The exhaust gas should not fall below a minimum temperature of 350°F (175°C) without using special precautions to avoid corrosion.

Table 5.19. Comparison of Costs and Savings for Improvements to Boilers and Furnaces

Energy-saving measure	*Cost ($)*	*Savings ($/year)*	*Energy savings MBtu (kWh)*	*SCP[a] $/BPD ($/kW)*	*Comments*	*Reference*
Heat wheel	5,600	4,100	6500 (1.9×10^6)	1840 (26)	Heat from drying oven exhaust can be used for space heating with 78% recovery when fresh air is at 37°F (4°C). Exhaust temperature is 300°F (150°C). Energy recovered replaces 6500 MBtu (1,900,000 kWh) of natural gas fuel.	(87)
Flue gas to water heat exchanger	17,340	5,360	5500 (1.6×10^6)	6700 (95)	Space heat for a dairy and boiler feedwater preheating was obtained by placing a finned coil in the exhaust of a gas fired boiler. Cost includes total cost of heat distribution system. It is estimated that 75% of the heat formerly wasted up the stack is recovered.	(88)
Paint curing oven exhaust	2,000	933	410 (1.2×10^5)	1700 (24)	Paint oven exhaust is 1000 cfm (1,700m^3/hr) at 480°F (250°C). Heat is recovered with a glycol loop connected pair of heat exchangers. Exhaust gas exits at 300°F (150°C) to avoid corrosion and condensation problems. Recovered energy is used by Raytheon for space heating.	(89)
Air to air heat exchanger	5,000	2,500	560 (1.65×10^5)	3100 (44)	Heat exchanger installed in 1200 cfm (2,000m^3/hr) oven exhaust at 350°F (175°C) saves 6,100 gallons of propane/year at Raytheon plant in California.	(89)
Combustion controls	10,000	20,000	10,000 (3×10^6)	2100 (30)	Monsanto Fulton plant, excess air reduced from 30% to 20%.	(45)
Combustion controls	10,000	20,000	10,000 (3×10^6)	2100 (30)	Monsanto Fulton plant, savings expected as the result of installing oxygen analyzer on a second utility boiler.	(45)

[a]The specific capital productivity (SCP) is the cost of reducing the annual rate of energy use by one barrel of oil per day (BPD) energy equivalent, or one kilowatt.

Table 5.20. Comparison of Costs and Savings for Heating System Changes and for Heat Recovery

Energy-saving measure	*Cost ($)*	*Savings ($/year)*	*Energy savings MBtu (kWh)*	*SCP[a] $/BPD ($/kw)*	*Comments*	*Reference*
District heating from a casting foundry	160,000	84,000	12,000 (3.5×10^6)	28,000 (400)	The Sulzer iron foundry sends water heated in the cooling system for their casting furnaces to the district heating system in Winterthur.	(90)
Space and process preheat from an iron foundry	220,000	240,000	31,750 (9.3×10^6)	15,000 (210)	Warm (117°F or 47°C) cooling water from a casting furnace is used for preheating boiler feed water for the space and process heating systems. Energy from the district heating network supplies this heat when the cooling furnace is not in operation.	(90)
Factory entrance control for space heat	4,000	5,600	785 (2.3×10^5)	10,000 (152)	Heating near the entrance is actuated by door closure switch. Space heaters operate if temperature is below set point and if doors are closed. Cost is for ten doors. Example from a casting foundry in Winterthur, Switzerland.	(90)
Air compressor heat used for domestic hot water	20,000	140,000	34,000 (10^7)	1,275 (18)	The Sulzer factory in Winterthur, Switzerland, uses the cooling water from their air compressors for domestic hot water at a temperature of 122 to 131°F (50 to 55°C). The compressors are oil free so that no heat exchanger is needed. Costs were for a pump and a storage tank.	(90)

[a]The specific capital productivity (SCP) is the cost of reducing the annual rate of energy use by one barrel of oil per day (BPD) or by one kilowatt.

from the exhaust air of buildings and other sources may provide a way to extend the capability of an existing boiler to provide heating for an extension to the building. In this category would also be the additional unused capacity of the boiler remaining after completion of an EIP throughout the building and after improvement of the operating efficiency of the boiler itself.

Combustion

Combustion is the chemical oxidation of fuel components with the 21 percent oxygen in the combustion air; this results in the release of heat and in the formation of CO_2, H_2O, and other unwanted components (pollutants). Solid and liquid fuels must be converted to gas before they can burn. The role of the burner is to vaporize the fuel and mix it with the combustion air in the proper proportions for efficient combustion, as given in Table 5.21.

Liquid fuel burners must vaporize or atomize the fuel into droplets sufficiently small to be completely vaporized by the time they have traveled through the combustion zone. Atomization may be achieved by spraying the oil at high pressure through a nozzle, or by a spinning cup, or (in larger units) by air or steam injection. Air atomizing burners are more efficient and avoid the introduction of undesirable, additional H_2O.* (Steam atomization causes a loss of boiler capacity and lowers the dew point of corrosives in the exhaust gas.)

Bulk coal and solid waste burners must heat the fuel so that enough volatiles are vaporized to sustain combustion, and sufficient heat must be retained in the combustion zone above the fuel bed of coal or waste fuel burners to completely oxidize the vaporized fuel. The heat required to convert liquid and solid fuels to the gaseous phase comes from absorption of radiation in the reaction zone of the flame and from recirculating burning gases. For complete oxidation, each atom of the fuel must come in contact with oxygen before it is cooled below the combustion temperature by contact with the surface of the combustion chamber or with the colder excess combustion air.

The color of the combustion flame is blue for the burning of lighter hydrocarbons, carbon monoxide, and hydrogen; a luminous yellow flame is produced from glowing carbon particles from solid fuels or from the end result of the cracking of liquid fuels.

For complete stoichiometric combustion† the quantity of air given in Table 5.21 must be intimately mixed with the vaporized fuel while it is above the combustion temperature. Even with already gaseous methane (natural gas), excess air above the theoretical minimum is required because of the difficulty of achieving uniform mixing and complete chemical reaction. Figure 5.9 shows the calculated equilibrium flame temperature for No. 2 fuel oil and the composition of the combustion gas as a function of the stoichiometric

*Do not be misled by persistent advertising for vapor injection equipment which falsely claims that the addition of water (in any form) or the addition of other fuel oil additives can improve efficiency or save money. Such claims of savings, supported by clever but erroneous testimonials, can be traced simply to the use of better boiler management practice as described in this section.

†Stoichiometric is complete burning with no excess oxygen or unburned fuel.

Table 5.21. Approximate Combustion Characteristics of Various Fuels

Fuel characteristics				Combustion characteristics							
	Higher heating value[a]			CO_2 content by volume in dry exhaust gas						Excellent combustion	
	Btu	kWh	Air fuel[b]	100%[c]	120%*	140%*	160%*	180%*	200%*	Excess air (%)	Smoke no.
No. 2 Fuel Oil (low sulfur)	140,000/gal	12.9/kg	15.0:1	15.4%	12.7%	10.8%	9.4%	8.3%	7.5%	12–13%	1
No. 6 Fuel Oil[d] (low sulfur)	144,000/gal	12.6/kg	15.4:1	16.5%	13.6%	11.6%	10.1%	8.9%	8.0%	15–16%	2–3
Bituminous Coal	14,000/lb	9.0/kg	10.3:1	18.4%	15.3%	13.0%	11.4%	10.1%	9.1%	20%	3
Anthracite Coal	13,300/lb	8.6/kg	9.6:1	20.2%	16.8%	14.4%	12.6%	11.2%	10.1%	20%	3
Methane (natural gas)	1030/ft^3	10.7/kg	10.0:1	11.7%	9.6%	8.1%	7.0%	6.2%	5.5%	14–18%	0
Approximate oxygen content by volume in dry exhaust gas					*3.6%*	*6.1%*	*8.1%*	*9.5%*	*10.6%*	—	—

[a]Fuel oils and coal have a wide variation in their density, heating value, and chemical composition, depending upon their origin and on their treatment at the refinery.

[b]Air to fuel mass ratio for stoichiometric combustion.

[c]100% corresponds to stoichiometric combustion.

[d]Some types of No. 6 oil require preheating for pumping and a higher temperature for atomization.

*Excess air.

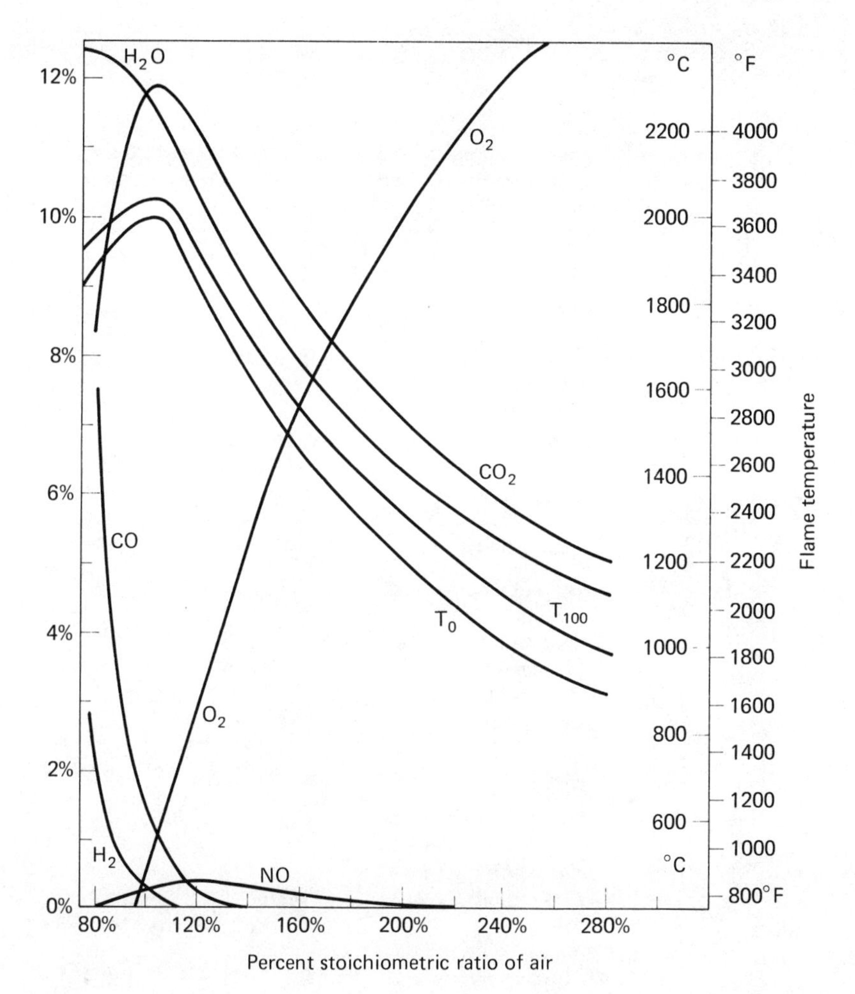

Figure 5.9. Calculated equilibrium flue gas composition and flame temperature for No. 2 distillate fuel oil as a function of the stoichiometric ratio of air. The flame temperature is shown for oil and air inlet temperatures of 32°F (0°C) and 212°F (100°C). The gas composition shown includes the water from combustion. Uniform mixing has been assumed.

ratio of combustion air. For this example ideal, uniform mixing of the combustion components has been assumed. In practical burners such uniform mixing is not achieved, and a good balance between high flame temperature and acceptable unburned fuel occurs at about 15 percent excess air. Older burners may require as much as 33 percent more air than the stoichiometric ratio to obtain acceptable smoke emission. The correspondence between excess air and carbon monoxide, carbon dioxide, and oxygen is shown also in Figure 5.9. Soot or carbon monoxide, and carbon dioxide or oxygen, are easily measured parameters that permit optimization of combustion (as discussed with Tables 5.22 and 5.23) and minimization of air pollutants from incomplete combustion.

Table 5.22. Examples of Savings from Improved Boiler Adjustment and from Reduced Firing to Satisfy a Lower Demand for Heat after an EIP. (89)

	Example A	*Example B*
1. *Initial conditions:*		
Carbon dioxide in flue gas (excess air)	10% (50%)	5½% (160%)
Smoke number	0	½
Net stack temperature	528°F (276°C)	650°F (345°C)
Output	2.6 MBtuh (760 kW)	1 MBtuh (295 kW)
Input (No. 2 fuel oil)	23.6 gal/h (75.6 kg/h)	12 gal/h (38.4 kg/h)
Efficiency (initial)	78%	65%
2. *After reducing excess air:*		
Carbon dioxide in flue gas (excess air)	13.8% (12%)	9.5% (55%)
Smoke number	1 to 2	1
Net stack temperature	514°F (268°C)	550°F (280°C)
Efficiency	83%	76.5%
3. *After reducing firing to meet actual demand for heat after EIP:*		
Output required	2 MBtuh (585 kW)	0.8 MBtuh (235 kW)
Carbon dioxide in flue gas (excess air)	13.8% (12%)	10% (50%)
Smoke number	1 to 2	1 to 2
Net stack temperature	437°F (225°C)	500°F (260°C)
Efficiency	84%	79%
Fuel input	17.2 gal/h (55 kg/h)	7.2 gal/h (23 kg/h)
4. *Savings:*	6.4 gal/h	4.8 gal/h
Operating hours and cost of fuel	8250, $0.48/gal	6300, $0.45/gal
Annual dollar savings	$25,000.	$13,600.
Percent improvement	27%	40%

Table 5.23. Comparison of Costs and Savings for Improvements to Boilers and Furnaces.

Energy-saving measure	*Cost ($)*	*Savings ($/year)*	*Energy savings MBtu (kWh)*	*SCP[a] $/BPD ($/kW)*	*Comments*	*Reference*
Automatic flue gas analysis	30,000	23,170	33,000 (9.7×10^6)	1900 (27)	Continuous flue gas analysis at paper mill enables oxygen content to be reduced from 5.3% to less than 2%. This reduces excess air from 30% to 10% or less.	(91)
Automatic boiler tube cleaner	9,500	8,350	4100 (1.2×10^6)	4800 (68)	American Can Co. in Baltimore installed an automatic boiler tube cleaning system in two boilers and reduced fuel use by 7.46% based on use over two seasons.	(92)
Combustion air preheater	90,000	67,000	56,000 (1.64×10^7)	2340 (33)	Kaolin powder (a raw material for manufacturing paint) is produced in a gas fired furnace at a temperature of 1140°F (615°C). The cool-down heat of this product is used to preheat furnace combustion air.	(93)
Combination controls on dual fuel boiler	29,000	40,090	44,400 (1.3×10^7)	640 (9)	A South Carolina textile mill replaced a two year old manual control system on a dual fuel, oil or gas fired boiler. This reduced oil use by 9% and gas use by 1%, because of closer control of air for efficient combustion.	(94)
Replace boiler baffle	1,500	10,130	11,300 (3.3×10^6)	215 (3)	Check of boiler efficiency showed it was down from specifications which was traced to a deteriorated flame baffle. Repair saved fuel with an energy value of 45 MBtu (13,200 kWh) per day.	(95)

[a]The specific capital productivity (SCP) is the cost of reducing the annual rate of energy use by one barrel of oil per day (BPD) (energy equivalent) or by one kilowatt.

Recovery of Useful Heat

The initial effect of introducing air in excess of the stoichiometric ratio is to improve combustion efficiency with the reduction of carbon monoxide and soot. Overall efficiency is, however, the product of combustion efficiency and the efficiency of extracting heat from the combustion (gas) in useful form. Since excess air must be heated in the combustion chamber, and since it decreases the temperature of the hot gases and their residence time in contact with the heat exchanger surfaces, too much excess air decreases boiler efficiency.

The larger volume of gas flowing up the flue at an elevated temperature represents an unwanted energy loss. Too little air can result in higher concentrations of dangerous carbon monoxide (especially with natural gas) and increased soot formation that must be cleaned from the boiler in order to maintain heat transfer efficiency.

Boiler efficiency can be improved by cleaning heat transfer surfaces; removing soot and boiler scale will increase capacity and lower stack temperature. Oil- or gas-fired boilers converted from coal service may be improved by the insertion of baffles, or spinners or turbulators in oversized fire tubes if sufficient draft is available. More draft may be obtained by adding additional height to the stack, or an induced draft unit may be added. Forced draft and induced draft operation are less sensitive to wind and temperature changes. Recovery of heat from hot flue gases will improve overall efficiency when a use for the heat recovered at this temperature can be found. Water-cooled heat exchanges are frequently the more economical because of the smaller surface areas and easier installation in the exhaust flue, and are suitable for boiler feed water, preheating domestic hot water, and other purposes.

Preheating combustion air with the low temperature heat recovered from the flue can improve combustion efficiency (fuel vaporization) and lead to fuel savings of 5 percent to 8 percent for combustion air preheated to 300°F to 440°F (150°C to 225°C) respectively. (96) Hence, the possible advantages of drawing higher temperature combustion air from near the flue or from the ceiling of the boiler room should at least be considered. Since combustion air can amount to as much as 5 percent of the building heat load if drawn from the heated comfort space, using building exhaust, exhaust air from toilets, chemical processes, etc., could reduce this loss.

The losses up the flue increase as the temperature of the exhaust gas increases and are inversely related to the CO_2 content. The most efficient operation of burner and boiler corresponds to the maximum CO_2 (minimum O_2) and the minimum stack temperature that can be obtained with acceptable burner stability and smoke number. The lower the stack temperature, the less loss of energy up the flue. Frequent flue gas analysis and control of combustion with periodically checked instruments helps maintain high efficiency.

Due to sulfur-oxides formed from the combustion of sulfur compounds contained in the fuel, the temperature of the flue gas should be maintained above the dewpoint for condensation of sulfurous and sulfuric acids.* Usually temperatures are held above 350°F

*The water from combustion combines with sulfur compounds, vanadium, and other trace constituents in the fuel to form acids. Their corrosive action depends upon the type of steel in the boiler, the material in the stack, and the origin of the fuel (its chemical constituency), as well as the temperature of the exhaust gas.

(175°C) to avoid corrosion. The lower the sulfur content of the fuel, the lower can be the exhaust temperature. With the lower sulfur fuels now prescribed, the temperature can be lower than conventional practice. Expert advice should be obtained to determine the minimum temperature for your particular fuel and combustion conditions. However, unless it is possible to cool the exhaust gas below the condensation temperature of the water vapor produced from combustion, only the lower heat value of the fuel can be recovered as a theoretical maximum. Hence the maximum possible efficiency available without recovering the heat of condensation would be 90 percent for gas, 93 percent for the lighter, No. 2 oil, and somewhat higher percentages for fuels with higher ratios of carbon to hydrogen.

There are some new units which recover the heat of condensation by extracting heat from the flue gases until they are below the condensation temperature of the water products of combustion. One method subjects the exhaust gas to direct contact with a water spray. A plastic lined flue ducts away the exhaust. Overall efficiencies of 95 percent are reported. (97)

Boiler Losses

As we mentioned above, combustion equipment is frequently oversized as a result of over-design. If it is also overfired as the result of the same over-design, then the efficiency will be lower than at optimum loading. For equipment fired for peak conditions but operating under a lower load, the energy not withdrawn by the load simply goes up the stack in the form of higher temperature exhaust gases. The boiler efficiency under average conditions will be lower than for continuous operation at optimum loading and combustion.

Radiation and conduction losses occur continuously as long as the boiler is at an elevated temperature, even when the boiler is not being fired. Hence, during the off-cycle intermittently fired boilers continue to lose energy proportionally to the boiler temperature. (84) These losses, which may be as much as 4 percent for some older, non-insulated equipment, are usually less than 1 percent for units with good insulation. Loss can be reduced by reducing temperature to the minimum required by demand and with insulation to reduce conduction and radiation losses. Where codes permit, the use of an automatic stack damper that closes during the off-cycle can reduce flue loss.*

Oversized equipment has a larger heat transfer surface that generally results in more efficient heat transfer to the load. However, since the larger equipment also loses a greater amount of energy through radiation and convection, a point of diminishing benefit obtains as a result of an oversized boiler. The optimum firing and loading occurs for older, less well insulated equipment at between 60 percent and 80 percent of rated capacity.

Adjustments for Better Efficiency

For a given boiler and fuel, there are three basic variables which determine the overall efficiency: the load, the firing rate or gallons per hour of fuel consumption (including

*Reliable off/on fuel control must be interlocked with the damper so that poisonous and explosive hazard is not caused by gases from leaking fuel trapped by a closed damper.

the frequency of operation), and the draft. After the load has been reduced as the result of reduced demand and distribution losses, it may be possible to reduce the firing rate. As the result of improved efficiency by means of the adjustments described below, it may be possible to reduce the firing rate still further.

For intermittently fired equipment, and boilers with two or more firing rates, the annual efficiency of operation depends not only on the efficiency during combustion, but also on the number of starts and the idle time while at an elevated temperature. For example, boilers operated during the summer for the production of domestic hot water only, might have reasonable efficiency during firing, but have a seasonal term efficiency of only 20 percent. The best strategy may be to shut down the large unit and produce summer hot water and steam needs from a much smaller unit sized to this load. Given the choice of intermittently firing a partially loaded boiler at high rate, or firing for a longer period at a lower fuel rate, you may be able to obtain somewhat higher annual efficiency by adjusting the firing rate for the longest operating time of the burner.

Operating records should be carefully examined to determine that changes made do improve average fuel economy and that these improvements are maintained.

Measurements

The basic instruments needed to optimize combustion and to measure efficiency are simple, easy to use, and not very expensive. A carbon dioxide sampler, a smoke tester, and a stack thermometer will allow one to determine if the equipment needs readjustment and to adjust for the correct air input. With larger equipment, instruments which allow one to make these adjustments, or preferably, which automatically control the draft and firing rate, should be permanently mounted. A kit, which also includes a draft gauge and very helpful instructions, is available from one manufacturer.* A similar outfit including an oxygen concentration tester and a carbon monoxide tester is available for optimizing gas firing; any instrument may also be ordered separately. It is worthwhile to use these instruments even if the adjustments are left to a boiler or burner mechanic. In many cases, except for some conscientious ones, burner mechanics may simply adjust for flame color and no visible smoke, and claim that according to their experience the burner is now well tuned. Frequently verifying proper boiler efficiency may consume some time, but considerably less fuel and the resultant improvement in operation should return the cost of the kit in much less than a year for all but the smallest of installations. According to combustion experts, because of former low energy prices, few people were available who were concerned enough to take the time to become trained and fully equipped to properly adjust combustion equipment.

The measurement of carbon dioxide and oxygen is made with an Orsat-type sampler in which both the carbon dioxide and sulfur oxides are dissolved out in the solution used. Hence the carbon dioxide indicated will be higher according to the percentage of sulfur in the fuel. Since the water of combustion is also condensed by this method, the concentrations of carbon dioxide and oxygen measured are with respect to the *dry* exhaust

*One manufacturer is the Bacharach Instrument Co., 625 Alpha Drive, Pittsburgh, Pa. 15238.

gas. In Table 5.21, therefore, we have given the concentration of these gases as the percentage of dry exhaust gas. Since the excess oxygen does not depend very strongly on the type of fuel combusted, this is a more reliable indication of excess air for waste and combination fuel burners. A fuel cell detector for excess oxygen, although more expensive than the Orsat devices, gives a convenient electric signal that can be used for effectively controlling the combustion air.

The instruments to measure firing conditions are simple to use. The Orsat-type CO_2 sampler draws a volume of exhaust gas across a solution of potassium hydroxide which changes volume according to the percent of carbon dioxide, and this percentage is read off a scale according to the height of the liquid column. The smoke tester pumps a sample of the stack gas through a small area (spot) of filter paper. The spot darkens according to the presence of both visible and invisible concentrations of smoke and is compared with a standard on a scale of 10. These gas samples as well as the flue temperature are taken near the outlet in an area where the gases are well mixed. With the results of these measurements, the efficiency of equipment can be quickly determined by use of a handy calculator provided with the kit. For larger installations, these portable instruments should be used to confirm the proper operations of permanently mounted instrumentation. Orsat samplers require maintenance and periodic calibration, both of which are frequently omitted. The dollar savings possible by operating at maximum combustion efficiency may justify the installation of measurement and automatic combustion control equipment that is already available on relatively small units. (Examples of the potential savings will be illustrated later.) Installations of a few million Btu/hr or less may be able to repay the cost of heat load, draft, and oxygen sensors with automatic firing rate and draft controls. Temperature, pressure, and flow should be metered both for fuel as well as for the boiler or furnace output.

The following are procedures for adjusting oil-fired, commercial boilers. Similar remarks would apply to other combustion equipment (see also References 86, 98, 99, and 100).

1. Clean boiler heat-transfer surfaces, flue passages, and chimney. Clean the fuel line filter, and check the line and valve for leaks. Clean the burner, especially the nozzle of the atomizer and the air handling parts. Seal any air leaks into the combustion chamber or heat exchangers.
2. Confirm (a) that the fuel is suitable for the burner, and (b) the fuel's firing properties—preheat temperature, CO_2 for stoichiometric combustion, sulfur content—with each delivery. (Current supply sources of oil may change rapidly, which means that viscosity, heat value, corrosive sulfur, vanadium content, etc., may vary from one delivery to the next.) New, low-sulfur fuels have unusual viscosity–temperature relationships and different corrosion characteristics. Obtain from your supplier the value of the temperature for the oil viscosity you require.
3. Use the proper pumping and atomization temperature. Smoky combustion can result from either too high or too low preheat temperatures. In general the temperature required for proper atomization is different than for pumping. The correct preheat temperature (firing viscosity) depends on whether your burner uses pressure, steam or air, or rotary cup as the method of atomization. Confirm the correct viscosity for each burner. If the viscosity of a particular fuel is unknown, start at 250°F (120°C) and reduce the preheat temperature until the best combustion is obtained.

4. Adjust the atomizing pressure to the burner manufacturer's recommendations. If the burner has been modified, make sure that the instructions for the latest modification are confirmed. For smaller burners with fixed firing, seasonal changing of the nozzle size to better correspond to the seasonal load will save fuel.
5. The object of this procedure is to obtain the highest possible CO_2 concentration (lowest O_2) without exceeding the smoke limits given in Table 5.21. The ideal sampling point would be in a straight section of the duct midway between the furnace exit and the stack draft; however, the sampling hole should be at least 8 diameters from the exit and 2 flue diameters from the stack draft opening. The sampling tube should draw a representative volume of the exhaust gas from the center of the duct.

   Using the air setting for best visual adjustment for clean combustion, operate the burner at full firing rate until equilibrium flue temperature is reached. This should be under load. A sufficient number of smoke and CO_2 readings should be made to determine the characteristics for your equipment at different settings of excess air, similar to the example shown in Figure 5.10. This can be done by adjusting the control linkage and/or the damper setting to obtain different amounts of excess air, and then taking smoke and CO_2 readings at each setting. The

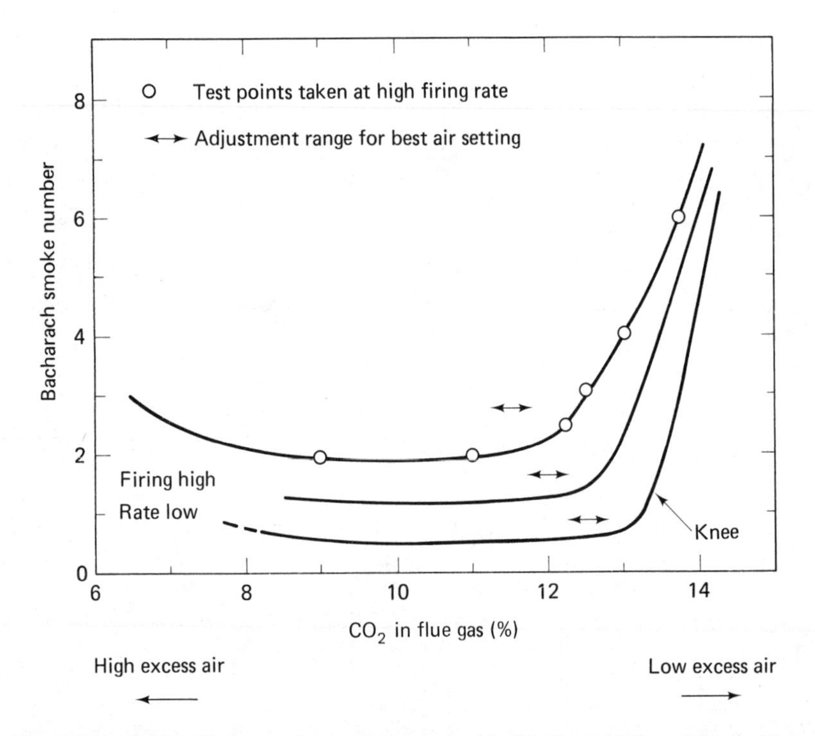

Figure 5.10. Smoke-CO_2 curves for a typical commercial oil boiler at high, medium, and low firing rates. A sufficient number of tests should be made to define the knee of the smoke-CO_2 curve. At lower firing rates the knee of the curve is sharper, and reliable operation can be obtained at higher CO_2 settings (lower excess air). With variable-firing-rate burners the opposite condition may be found at low rates. [After EPA (93)]

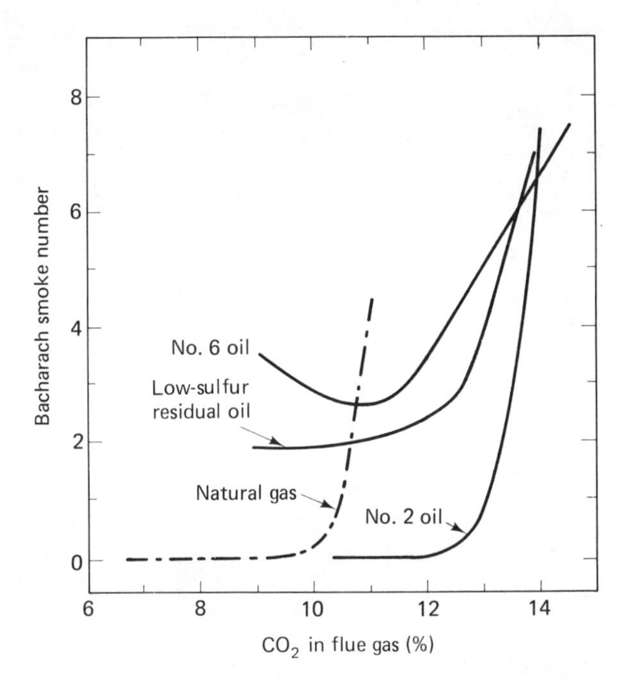

Figure 5.11. Smoke-CO_2 characteristics of a typical commercial boiler firing three different grades of fuel oil and natural gas. [EPA (93)]

object is to accurately determine the knee, or point at which the smoke starts to rise rapidly. The exact shape of this curve will depend upon your particular equipment, and upon the type of fuel, as shown in Figure 5.11.

6. Adjust the excess air setting for about 1/2 percent CO_2 below the knee. At this point it should be possible to obtain the smoke level shown in Table 5.21. The object is to adjust for highest possible CO_2 without excessive smoke, as this will be the setting for highest thermal efficiency. Check that atmospheric flue dampers can move freely and have sufficient range to open or close to obtain proper combustion draft in extremely cold or hot weather. Modern, forced draft or induced draft boilers are less dependent on weather conditions, but they must be kept free of soot. It should always be possible to obtain smoke numbers less than 4 with No. 6 oil. A smoke number of 3 will lengthen the time between cleanings of soot fouling from boiler surfaces and should meet the requirements of most air quality districts, while 2 will cause only slight sooting and will not increase stack temperature appreciably. The stack temperature on a modern, properly sized and fired boiler operating at full load will be about 100°F to 150°F (55°C to 85°C) above the steam or water temperature. Older boilers may have higher stack temperatures of up to 575°F to 670°F (300°C to 350°C). The effect of stack temperature and CO_2 on boiler efficiency is shown by the curves in Figures 5.12, 5.13, and 5.14. The greatest savings from reducing excess air occur with older boilers with higher stack temperatures. See example in Table 5.22.
7. As a result of reducing excess air and cleaning and sealing the equipment, efficiency will be higher and it may be possible to meet the design load with a reduced firing rate. As the interval

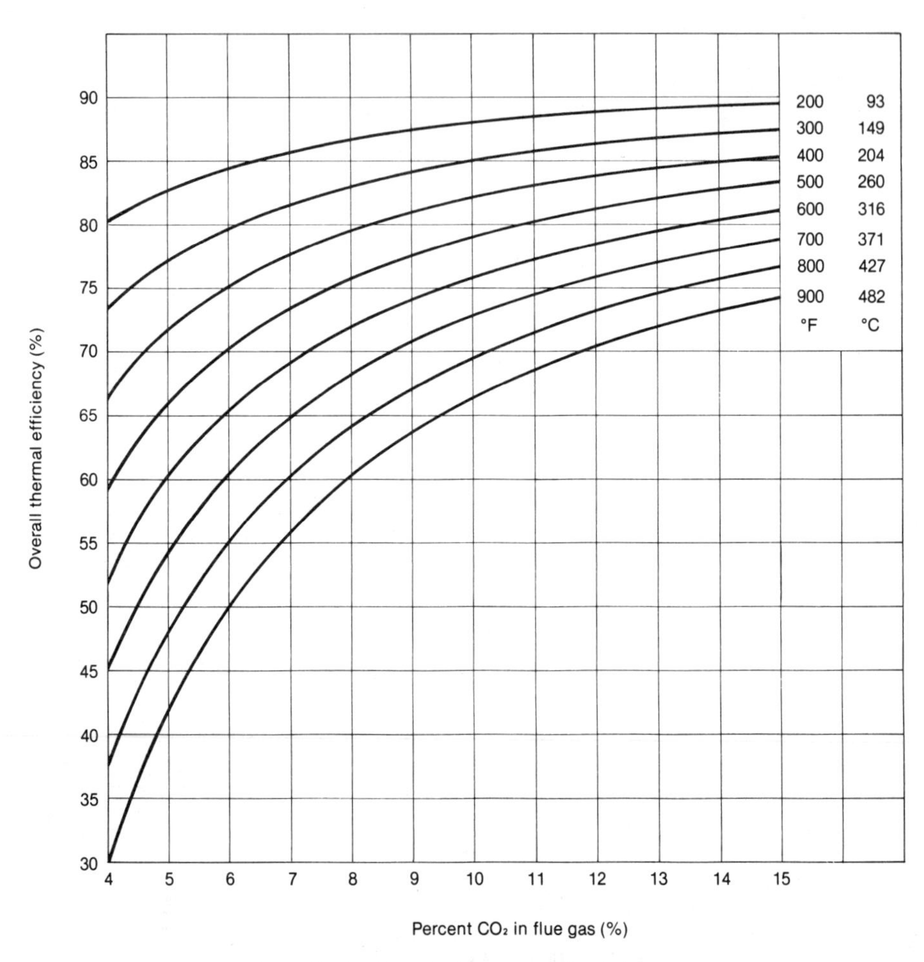

Figure 5.12. Thermal efficiency of No. 2 fuel oil.

between firings increases, the cycle-efficiency decreases, and falls rapidly for units firing less than 10 minutes per hour. The percentage firing should correspond to the percentage of the design load supplied. For space heat, for example, the percentage firing time should correspond roughly to the ratio of the difference in temperature between room and outside, to the maximum difference for the coldest weather expected. Domestic hot water and process heat must be added to the space heat loads.* Another method, for constant loads, is simply to reduce the supply temperature and firing rate by steps until there are complaints. Locally applied ESM may satisfy the complaints and permit additional reduction.

For some burners, due to less turbulent mixing at lower firing rates, it may not be possible to decrease the combustion air in the same proportion and maintain acceptable smoke. Thus it may be necessary to accept slightly decreased CO_2 (decreased combustion efficiency) to obtain improved cycle efficiency and lower fuel consumption.†

*Without adequate storage, large peak loads (e.g. "tankless heating coils") restrict the amount that the fuel nozzle can be reduced.

†For example, Brookhaven tests show a $5^1/_2$ percent reduced fuel consumption as the result of reducing the firing rate from 1.28 gal./hr to 0.678 gal./hr, despite a decrease in the CO_2 in the stack gas. (Reference 85, p. 22)

8. Under maximum loading, it should be possible to achieve the manufacturer's specifications for CO_2 levels and acceptable smoke numbers. If not, possibly fuel atomization or air mixing need improvement. Flame impingement on cold surfaces will give poor combustion; check for proper or dirty nozzle, or fouled air-handling parts. Check for clean insulators and properly positioned electrodes, and for prompt ignition to avoid oil deposits, smoke, and boiler fouling. Leaking oil after shut-off will cause furnace sooting. For installations relying on natural draft by a stack, the atmospheric damper may not be following changes in the weather. If the burner was tuned on a cold day, and if it smokes when the weather is hot, there may be insufficient draft to provide enough combustion air. Check the atmospheric damper if there is poor efficiency during cold weather. If the temperature of the stack rises more than 50°F (30°C) above the clean value, the flue passages should be cleaned and the water-side boiler surfaces should be checked for calcium scaling. If the temperature is still too high, consider reducing the firing rate, or possibly using exhaust heat recovery.
9. For burners with variable firing rates, set the high firing rate as low as possible. The measurements listed under steps 5 and 6 above should be repeated at low and intermediate firing rates. Typically, the CO_2 is lower than for maximum firing.

   The control system should be checked at several intermediate firing rates for proper proportioning of the combustion air by testing smoke numbers and CO_2. If the system uses a mechanical control linkage, this should be checked for backlash on both increasing and decreasing firing rates. Use a felt-tip pen to mark linkage settings so as to retrace the direction

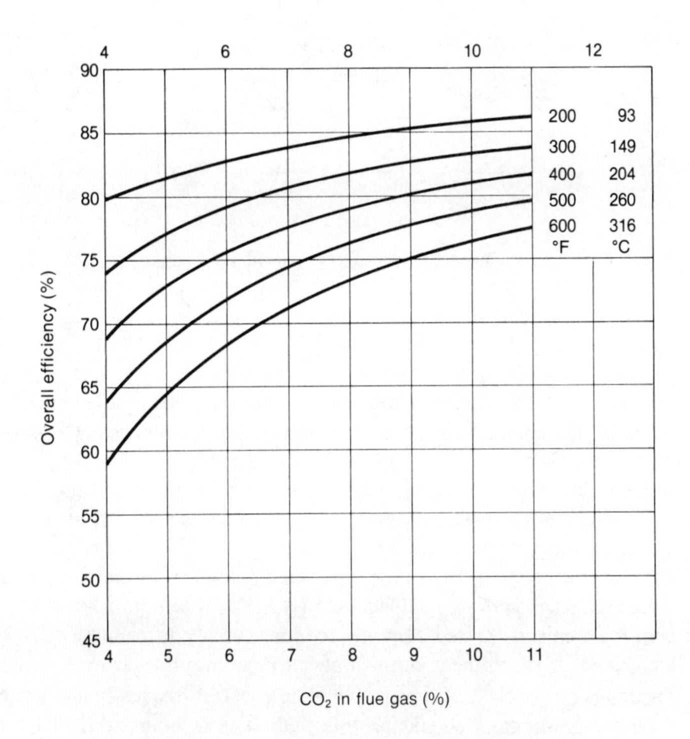

Figure 5.13. Thermal efficiency of natural gas.

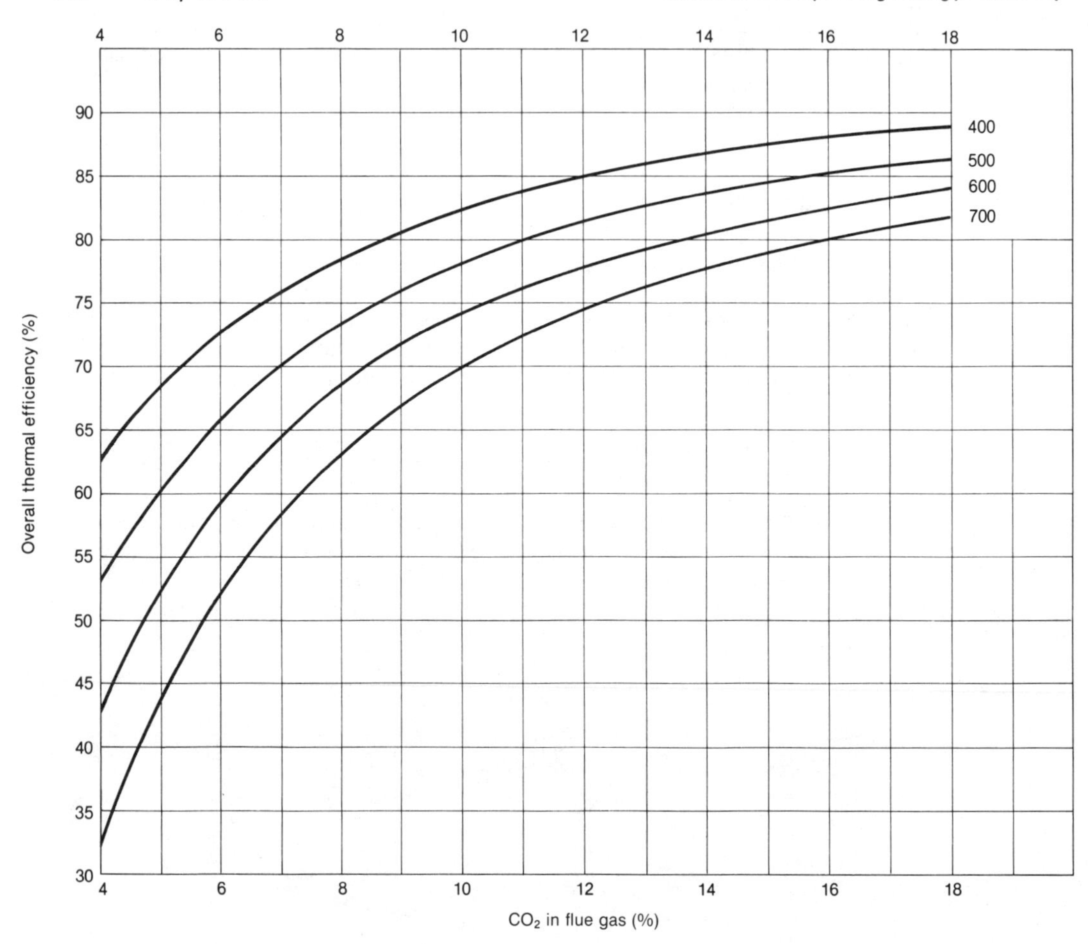

Figure 5.14. Thermal efficiency of anthracite coal.

and position of the adjustments. If the combustion control cannot maintain efficient conditions, it may pay to replace it with a modern system with continuous flue gas analysis. Some examples of investments for new controls are shown in Table 5.23.

On dual fuel equipment, settings should be repositioned after fuel switching. If this is not possible, a compromise is usually required in the air settings to obtain the maximum average efficiency for both fuels, and over the average firing rates. On gas/oil equipment, preference must be given to adjustments for gas at the highest and lowest firing rates to reduce the carbon monoxide hazard to acceptable limits (typically 400 ppm); however, check EPA and local code requirements. Since air settings are considerably different for oil, the supply of an adequate amount of air should be confirmed after the settings for gas have been made.

10. The efficiency of the equipment should be checked frequently and a periodic (at least annual) maintenance program should be followed. It is recommended that a stack thermometer be permanently mounted and the temperature recorded in a log along with the ambient temperature

and other indications of load. A temperature higher than normal is a quick indication that the equipment is not at its top efficiency and may require some remedial treatment. CO_2 and smoke should also be regularly recorded. It is important that operating data be recorded when the equipment is new to compare both with (a) the manufacturer's specifications, and (b) later operating data, for an indication of specific maintenance required.

Gas-Fired Burners

The operation of gas-fired equipment is usually considered to require less maintenance, and for this reason gas-fired equipment may be needing attention to bring it up to present day standards. If your boiler is operating outside of the carbon dioxide range specified

Table 5.24. Comparison of the Specific Capital Productivity (SCP) of Improvements to Small Boilers

No.	*Improvement*	*Estimated cost ($)*	*Savings ($/year)*	*Yearly energy savings, MBtu (kWH)*	*SCP ($/kW)*	*SCP ($/BPD)*
1	Reduced burner nozzle size	20	54	18.7 (5,480)	32	3,270
2	Retrofit controls					
	(a) Variable aquastat for boiler water temperature	150	54	18.7 (5,480)	240	17,000
	(b) Timers for thermostat control	100	48	16.6 (4,865)	180	12,750
	(c) Zone controls	150	30	10.4 (3,050)	430	30,450
	(d) Continuous circulator operation	30	30	10.4 (3,050)	86	6,100
3	Burner steady-state efficiency adjustment	20	15	5.2 (1,525)	115	8,140
4	Retention-head burner with reduced firing rate	150	72	25 (7,325)	180	12,750
5	Variable firing-rate burner	300	90	31.2 (9,140)	290	20,500
6	Convective stack damper to reduce downtime losses	150	60	20.8 (6,100)	216	15,300
7	Purgeable, low mass boiler	400	120	41.6 (12,200)	290	20,500

Notes:
Savings from retrofit actions are not additive.
Savings have been based on oil at 40¢/gallon.

SOURCE: Brookhaven National Laboratory. (85)

for your equipment, it should be checked by an expert. Atmospheric gas burners do need to be cleaned, and this and any adjustment should be done by an experienced technician. Sufficient oxygen is much more critical in gas firing than for oil, where insufficient oxygen is indicated by smoking. A substantial loss of unburnt combustibles from gas may occur before smoking occurs. In addition to the possible buildup of dangerous levels of carbon monoxide, oxygen-poor firing incurs a danger that the flame will extinguish and then reignite explosively. Furthermore, the fuel loss from only $^1/_2$ percent of carbon monoxide in the flue gas could amount to as much as 5 percent. Atmospheric burners may be replaced by forced draft units for better efficiency over wider firing rates. Pilot lights can be replaced with spark ignitors to eliminate this stand-by loss.

Examples of some investment to improve the efficiency of boilers and furnaces are given in Table 5.23, together with the resultant savings and the specific capital productivity. The costs for some improvements to smaller oil-fired units are shown in Table 5.24.

Conversion to Coal or Wood

In the past, oil and gas have been more convenient and, in many cases, cheaper alternatives to coal and wood. Unfortunately, the fuel situation is changing. The cost of oil is rising sharply, and the government may ultimately decide to restrict the use of fuel oil while more actively promoting the use of more abundant coal and renewable wood supplies. It is suggested that new facilities which lend themselves to future conversion to coal or wood operation be considered.

Such boilers are available and are known as dry base, fire tube boilers. They must be elevated sufficiently above the ground to permit eventual addition of a stoker. They must also be located in sufficiently large boiler rooms to provide space for ash collection, and in some cases a scrubber and induced draft fan. Such a boiler could also operate on oil or gas. When firing oil, turbulators in the smoke tubes will provide for combustion efficiency equal to those of other boilers designed for oil-firing only.

If such convertible boilers are to be used, fuel storage space must be provided. It is customary to carry a one-month supply of fuel in reserve. For instance, a 200 HP* boiler fires 747 pounds (340 kg) of coal per hour maximum. If the average consumption is 270,000 pounds (120,000 kg) of coal per month, the volume occupied will be 5,400 ft^3 (150 m^3) and the floor space 500 ft^2 (50 m^2) when the bunker is about 11 feet (3.5 m) high. Larger boilers will require more coal storage. Coal bunkers should be situated close to railroad sidings, if additional transport costs are to be avoided.

The conversion back to coal may be possible if the original coal and ash handling equipment is still available and if local air quality requirements can be satisfied. As recently as 1978, some operators who obtain their coal by truck directly from the mine have been able to produce heat for between $1 and $2 per million Btu net, including costs for ash disposal.

Conversion of existing oil- and gas-fired boilers to wood, wastes, and/or coal may

*6.7 MBtuh or 2000 kW.

be possible where not precluded by lack of space. More combustion volume is needed than for oil or gas, and if greater excess air is required to fire coal, the output will be lower because of lower heat transfer. One method for conversion, if space in front of or beside the existing equipment permits, would be to fire the coal externally by adding a water-jacketed coal fire box. With this scheme, the hot combustion gas is led through the opening previously used for the oil burner so that the existing boiler is used only for heat transfer. Consider adding height to the chimney if additional draft would be required.

Technology developed in the wood-rich Northeast has overcome traditional problems of low efficiency and pollution. Using low cost wood chips, stick wood, compatible solid wastes, and forest and mill wastes, this new technology can provide low-cost heat without the former problems of pollution.

CHECKLIST	OPERATIONAL CHANGES
Reduce Heating in Selected Areas	*Heating System: Eliminate Unnecessary Demand*
Turn off heat in irregularly occupied or non-occupied areas.	Turn off or reduce heat to minimum in stairwells, entrances, utility rooms, outbuildings, basements, garages, storage rooms. Check for possible danger of freezing; drain water pipes.
Isolate heated areas within larger areas.	Isolate or close off particular sections within large areas that require higher temperatures than the rest of the area. For example, reduce the heating of a large manufacturing, shipping, or work area with the use of a small room or heated booth.
Replace general area heating.	Replace general area heating in workshops, storage areas, warehouses, and other high-ceiling areas, or where traffic-induced infiltration occurs; use local heating where and when it is needed. Use task-oriented spot heating. Use IR or radiant heat rather than air heaters.
Do not heat ramps.	Turn off heat used for melting snow and ice on steps and driveway ramps and remove snow and ice manually. Roof over these areas, or use waste heat not usable otherwise.
Shut off heating units in overheated areas.	Either move heating units from overheated areas to cold spots or shut them off.
Eliminate unnecessary domestic hot water outlets.	Avoid unnecessary use of hot water, e.g., for washing, where cooler water would suffice.
Turn off circulation during non-working hours.	For example, domestic hot water. Also turn off distribution to temporarily unoccupied areas and save the heat loss from the lines to these areas, as well as reduce their needless heating.

Reduce Heating in Selected Areas (cont.)	*Heating System: Eliminate Unnecessary Demand (cont.)*
Minimize low load operation.	During the temperate seasons, allow temperatures to vary over a wider range in order to minimize use of heating system. For exceptionally cold areas, use local electric heaters if overall primary fuel consumption would be less than if operating complete heating system.
Shut down earlier.	Shut down system earlier and take advantage of the heat stored in the system and the building; start up as late as possible.
Allow Adequate Temperatures	
Heat to lower level during occupancy.	68°F (20°C) in winter for office space, and 60–65°F (15–18°C) in stores, service areas, and production plants (where customers and employees are wearing heavier, outside clothing, or where more active physical work is being performed). Each Fahrenheit degree reduction will save approximately 3% on heating fuel (5% per °C reduction).
Check critical areas.	Monitor temperature and occupants' perception of comfort in selected rooms in order to assure that required working conditions are actually being maintained. Areas with chronic problems of being too cold, too hot, or too drafty are frequently areas with excessive energy consumption.
Heat to lower level during non-working hours.	Reduce room temperatures during nighttime, weekends, vacations, as low as possible without damaging material, or endangering the freezing of water pipes. The maximum temperature could be allowed to fall as low as 50–55°F (10–13°C).
Lower water temperatures during non-working hours.	Lower heating system and domestic water temperatures during the night, on weekends.
Reduce domestic hot water temperature.	Reduce temperature to handwarm level (reduce storage, mixing, and insulation losses). Use water- (and energy-) saving spray heads on faucets.
Improve Energy-Conscious Occupant Behavior	
Create incentive for energy saving.	Bill tenants or departments for energy used. Inform management of heat used in different departments. Set goals, post results.

Improve Energy-Conscious Occupant Behavior (cont.)	*Heating System: Eliminate Unnecessary Demand (cont.)*
Meter energy consumption.	Record fuel consumption and heating system production and boiler efficiency vs. influences of weather, occupancy, and product output. Compare with data for previous year and with norm for standard production conditions. Use record to correct faults, or to identify excessive demand.
Inform occupants on energy saving.	Keep windows or outside doors closed as much as possible. Lower temperature by reducing heat input instead of opening windows.
	Leave thermostats alone and set at the predetermined level, to avoid forcing system to readjust to new setting. For long periods of absence, notify operating personnel to turn heat off.
	Use local service only during temperate season and only if heavier clothing does not provide adequate comfort.
	Pull office drapes or blinds at night to reduce heat loss. Lower external window blinds or close shutters to reduce heat losses due to air turbulence and radiation. Let the sunshine in during winter.
	Do not leave doors to the outdoors open longer than needed. Repair door closers.
Achieve Energy-Conscious Maintenance	*Heating System: Reduce Distribution Losses*
Repair insulation.	Repair insulation, especially on circulation lines and distribution ducts in unheated or utility areas and basements where lost heat is not useful or only seldom desired. Replace damaged or wet insulation, cure source of moisture. Protect insulation from further damage.
Repair leaks.	Repair leaks in valves and distribution system.
Adjust thermostats.	Periodically reexamine the setting of thermostats. Adjust thermostats on radiators and heaters to take advantage of temporal availability of heat from other sources such as process equipment and business machines, lighting, people, solar radiaton through the windows.
Check and readjust radiators.	As a result of EIP, radiators may be oversized. To avoid overheating, water temperature and distribution may need readjustment. Avoid blocking of radiators and wall heaters by objects which might disturb flow or radiation patterns.
Remove air.	Bleed air trapped in lines and from radiators.

Achieve Energy-Conscious Maintenance (cont.)	*Heating System: Reduce Distribution Losses (cont.)*
Repair leaking domestic hot water taps.	In many cases summer fuel consumption is surprisingly high, and it can be greatly reduced by repairing dripping domestic hot water taps.
Organize Frequent Furnace Maintenance	***Heating System: Increase Production Efficiency***
Check and adjust burner frequently.	Check CO_2 content of furnace exhaust regularly. If CO_2 content is too low, reduce stack losses by reducing air input. Make sure that burner operation is stable also at start-up (but avoid too-frequent start-ups). Adjust burner to lower heat demand after EIP.
Clean burner.	Clean air filters, fuel line filters, check for nozzle clogging. Check oil tank and fuel lines for condensation water, and fuel shut-off valve for leaks. Check electrodes and insulators, also flame distribution—flame should not impinge on any cold surface.
Check chimney and atmospheric damper.	Check for soot buildup, water damage, or corrosion. Atmospheric damper should move freely with sufficient range to open or close for extreme weather conditions.
Clean combustion surfaces.	Periodical cleaning will greatly increase efficiency. Are soot blowers maintained and used regularly?
Remove boiler scale.	Follow a regular water treatment and maintenance (blow-down) schedule for water-side heat exchanger surfaces. Check feedwater treatment for minimizing boiler scale and corrosion. Also check anti-corrosion electrodes (and power supply if one is used). Fuel consumption can be reduced in some cases if a continuous blow-down cycle can be used with heat recovery.
Seal furnace doors, insulate.	Close draft doors and dampers with automatic controls to non-operating furnaces, reduce boiler cool-down, and avoid loss of heated air up chimney. Repair furnace and oven doors so that they seal effectively.
Seal leaks.	Seal leaks into combustion, heat exchanger which waste hot air up stack, spoil flue gas analysis, and quench combustion temperatures. Replace leaking coverplate and casing head gaskets. Seal around doors and burners. Use self-sticking aluminum tape up to 250°F (120°C), sheets of incoloy foil for higher temperatures, or fibrous cements. Check furnace arch for leaks.

Organize Frequent Furnace Maintenance (cont.)

Heating System: Increase Production Efficiency (cont.)

Calibrate CO_2 and O_2 meters.	Repair CO_2 and O_2 analyzers, calibrate. Use portable analyzers to check stack analyzers and to help balance burners.
Repair draft gauges.	Confirm that sufficient draft exists to obtain proper fuel–air ratio. A good target is 0.1 in (2–3 mm) water column at the entrance to the convection section.
Adjust firing viscosity.	Check that oil temperature is giving the viscosity required. (Some low sulfur oils may unusual viscosity-temperature characteristics.) Confirm oil viscosity characteristics with each delivery.
Use dry steam.	Check that atomization steam is dry, and that it is supplied at minimum rate for acceptable combustion. Any extra water carries lost heat up the flue and adds to corrosion problems.
Schedule maintenance.	Schedule regular maintenance and efficiency checking, rather than waiting for a fault condition.
Balance burners.	Balance air flow on multiple burner boilers to obtain proper fuel/air mixture at each burner. Control air to all burners simultaneously by enclosing them in a plenum.

Improve Part-Load Efficiency

Make optimum use of storage.	Where storage of steam or hot water exists, larger variations in storage temperature can be tolerated and still maintain the required temperatures at the point of use. This would enable the storage tank to be charged less frequently, but with the boiler operating for a longer interval and at higher loading and hence at higher efficiencies.
Make optimum use of furnace units.	Where several furnaces exist, shut down one or more to enable remaining furnace(s) being fired to operate at higher loading, and for longer periods for higher efficiency.
Shut down large units in summer.	If it is possible to separate loads which are continuous throughout the year from loads that vary due to weather or production, then it would be more efficient to supply the former from a smaller furnace operating more closely within the range of its maximum efficiency. Or, for example, seasonal switching of domestic hot water to a boiler sized for this load would allow shut-down of the heating system. The larger system may otherwise operate with less than 30% efficiency.

Improve Part-Load Efficiency (cont.)	*Heating System: Increase Production Efficiency (cont.)*
Adjust firing to load.	Change nozzles to correspond to seasonal load.
Use exhaust air.	Use building exhaust air, or air from exhaust hoods, or take air with combustibles (e.g., from paint spray booths) through flame arrestor. Obtain double saving by ducting in exhaust air from roof ventilator around boiler flue to preheat combustion air.
Lower heat to oil storage.	Some new, low-sulfur residual fuels have lower viscosity and therefore storage tank and fuel line heating may be reduced accordingly.
Reduce External Loads	*Heating System: Eliminate Unnecessary Demand*
Make use of external heat sources.	Use sun energy transmitted through large south or west windows by installing thermostatic valves on this side.
Install automatic door closers.	Also carefully seal off rooms with different temperature levels.
Reduce Internal Loads	
Reduce cooling loads.	Cover refrigerators, ice machines, and display cabinets, at least during non-working hours. Increase their insulation.
Limit hot water consumption.	Install outlet device to limit flow of hot water to 1/2 gal/min (1.9 ℓ/min) for lavatories.
	Install device to limit outlet temperatures to 105° F (40° C)
	Install self-closing valve [1 qt. (1 ℓ) in lavatories per use]. Use spray or aerator heads for more efficient distribution with less water.
Make use of internal heat sources.	Relocate photocopier and other heat-producing process equipment to areas where heat is welcome. Box up or wall up heat-producing equipment. Duct heat where it is wanted in winter, outside in summer.
Reduce heat tracing.	Heat tracing of cooling water lines to prevent freezing in cold weather can be reduced by substituting liquid heat transfer media (e.g., ethylene glycol-water). Sprinkler system lines can be pressurized with air to hold back the water until a spray head is opened by fire.
	Insulate fuel lines to residual air storage tanks. Heat load to smaller tanks can be reduced by insulating them. (Check rate of winter snow melting to estimate significance of loss.)

CHECKLIST	SHORT-RETURN SYSTEM CHANGES
Reduce Internal Loads	*Heating System: Reduce Distribution Losses*
Redistribute heat.	Install ceiling fans to circulate hot air near ceiling to lower, occupied levels.
Use Optimal Insulation	
Insulate distribution lines in unheated areas.	Increase insulation. Insulate uninsulated valves. Increase insulation on distribution lines in unheated or utility areas.
Apply insulation behind radiators.	Apply additional insulation behind radiators ("space blankets" or aluminum foil) where temperature difference to outside is greatest.
Improve Control and Regulation	
Relocate thermostats.	Look for optimum thermostat location for better regulation and comfort. Add thermostats, reorganize zones according to load.
Use external sensors for water temperature regulation.	Use external sensors (sun and wind in addition to temperature) to control start-up, shut-down, and to determine optimum water temperatures.
Reduce Furnace Losses	*Heating System: Increase Production Efficiency*
Insulate furnace.	Insulate exposed surfaces, doors. Replace insulation on a more frequent schedule with thicker layers that reflect economies to be obtained with higher fuel prices. Add insulation to surface over 125°F (50°C). Use infrared photography to locate hot spots.

CHECKLIST	MEDIUM-RETURN SYSTEM CHANGES
	Heating System: Reduce Distribution Losses
Install time switches.	Install time of day or time interval switches on local heaters.
Install shut-down/start-up unit.	Install units for optimum shut down/start-up using as parameters: outside conditions, building heat storage and insulation, occupancy, and internal heat loads, night or weekend, room temperature required.
Install a clock thermostat	to achieve consistent night set-back to the desired level.

Adjust to Effective Load	*Heating System: Reduce Distribution Losses (cont.)*
Install thermostatic valves.	Use solenoid valves to shut off heat when window switch is open. Install thermostatic valves (to reduce the number of valves required, use some of radiators in a room as base load units).
Use zoning.	Use suitable zoning according to load on radiators. For example south-north side, average load areas, or high load areas (entrances, corners).
Install heat storage.	If heat demand and production is not simultaneous, use heat storage, increase size of existing storage tanks.
Increase System Efficiency	
Replace obsolete radiators.	Also check for optimum location. Do losses to outside occur because of low insulation at location of radiators? Is the air flow path obstructed due to furnishings? Readjust distribution system after EIP.
Distribute recovered low-quality heat.	Use lower water temperatures 125°F (50°C) if available from heat recovery or heat pumps. Reduce temperature level in room (which is possible with a more uniform envelope temperature distribution). Install piping to reuse heat content in waste or cooling water. For example, can water from a washing process be reused in another wash cycle, or for floor or equipment washing, or for flushing toilets?
Replace resistant electric heaters.	Replace electric resistance heaters with more efficient systems: hot water system with radiators or heat pumps.
Use catalytic burners.	Use catalytic heaters to produce clean heat where it is needed without production and distribution losses.
Increase Furnace Efficiency	*Heating System: Increase Production Efficiency*
Lower exhaust gas temperatures.	Install heat recovery units (heat pipes, extended surface coils). Lower exhaust gas temperature to 340°F–400°F (170°–200°C). With oil heating: to avoid condensation and formation of H_2SO_4, use appropriate stack construction to allow for lowest possible exhaust gas temperature. Measure stack wall temperature.
Use automatic combustion control.	Use linearized O_2 and load sensors to continually optimize firing rate and excess air.

Increase Furnace Efficiency (cont.)	*Heating System: Increase Production Efficiency (cont.)*
	Add extra heat transfer surface to boiler to improve efficiency and increase output, with finned or studded (easier to clean) tubes.
	Preheat combustion air with waste heat or flue gas—air preheater to improve fuel vaporization and combustion, as well as reduce stack temperature, or take air from top of boiler room.
	Check furnace exhaust for use with gas turbine, or use exhaust from gas fired turbine for process heat production.
	Install efficient forced draft burners for easier excess air control, better efficiency. Use new burners which can efficiently burn the most difficult and aggressive fuels (high-metals, catalyst, high-viscosity pitch, etc.)—previously considered unsuitable as boiler fuel.
	If draft is insufficient for all-weather conditions, add additional stack height or use induced draft fan to avoid positive draft conditions.
	Use automatic draft doors and draft damper controls to eliminate heat loss up flue and reduce boiler cool-down during non-operating period of boiler.
	Add heat recovery to waste incinerators.
	Recover energy from exothermic reactions. Examples would be the oxidation of sulfur, ethylene, ammonia, or methane to produce respectively sulfuric acid, ethylene oxide, nitric acid, or carbon black.
Monitor furnace behavior.	Meter CO_2, O_2 and temperature of chimney exhaust, energy production.
Avoid low-load operation.	Use special units for domestic hot water preparation during summer and temperate season. Use: gas-fired boilers, waste heat (hot water drains, warmed cooling water as input to boilers, etc.), solar energy.
Replace pilot lights.	Replace continuously burning gas pilot lights with electric ignition.
Replace obsolete burner.	When furnaces need to be replaced, change number and size of furnaces to avoid operation at particularly low load.
Improve heat transfer.	Install metal strips in oversized fire tubes (turbulators) to increase gas turbulence, heat transfer to yield lower stack temps.

Reduce Use of Non-Renewable Energy	*Heating System: Increase Production Efficiency (cont.)*
Use internal heat sources.	Use internal heat sources (cooling water; exhaust air from transformer rooms, utility rooms, or air conditioning; condenser heat from refrigeration units). Recover condensate from process steam and use it to augment office heating, or return condensate to boiler feed water.
Use waste heat.	Use waste incineration. Use waste gas streams containing combustibles for fuel rather than flaring.
Use heat pumps.	Use electrical or diesel-driven heat pumps preferably with waste heat recovery or, for example, use groundwater, lake water, or river water as a heat source. Replace small inefficient heat pumps with larger, more efficient units.
Optimize the Use of Production Units	
Produce heat and electricity simultaneously.	Connect to district heating (preferably using simultaneous production of heat and electricity).
Produce heat, electricity, and cooling simultaneously.	Consider total energy units if life cycle costs are favorable.
Connect to larger units.	Connect to larger units in neighborhood (buy or sell energy).
Change, Diversify Energy Source	
Use oil and gas.	Diversify (for example use oil and gas) to improve energy supply under all circumstances.
Keep coal.	Keep coal-burning facilities (if environmental controls permit) as a reserve.
Use wood.	Burn low-cost wood chips.
Use wastes.	Use paper, packing crates, and other waste rather than paying to have these energy resources hauled away to a land fill.

5.5 VENTILATION AND AIR CONDITIONING

Methods for improving energy use in buildings and for processes which require forced ventilation and artificial cooling equipment are discussed in this section. An energy-saving program for this area should be coordinated with other ESM throughout all related areas to achieve maximum impact. Saving programs should also integrate measures for the building structure (Section 5.2) with attention to reducing building infiltration and moisture penetration, improving windows, etc., and heating (Section 5.4).

The purpose of an air-conditioning system is to maintain a satisfactory environment for the comfort of the occupants (usually under varying conditions) and for the production of a constant quality product. As heat is liberated within a space, it must be removed or the temperature will rise. The cooling system removes unwanted heat or, better yet, transports it to where it could be more useful. Moisture, dust, and odors should also be held to certain limits. In the absence of natural ventilation (openable windows), fresh air has to be filtered, heated, cooled, humidified or dehumidified, and delivered in a regulated quantity to the point of need. Many of the HVAC systems installed during the cheap energy era have losses that can be reduced.

Energy Use for Cooling

The energy used for refrigeration and air conditioning on a national basis accounts for almost 5 percent of total energy use. This is broken down as follows: for refrigeration, 1.1 percent in each of the residential and commercial sectors and for air conditioning, 0.7 percent in the residential and 1.8 percent in the commercial sector. In the summer, according to the Department of Commerce, cooling systems represent 42 percent of the energy needs for the operation of buildings. (1)

Buildings with forced ventilation and air-conditioning systems tend to use two to three times as much energy as buildings with central heating systems alone. And ventilation becomes more important as the ratio of the volume to the perimeter increases. This is

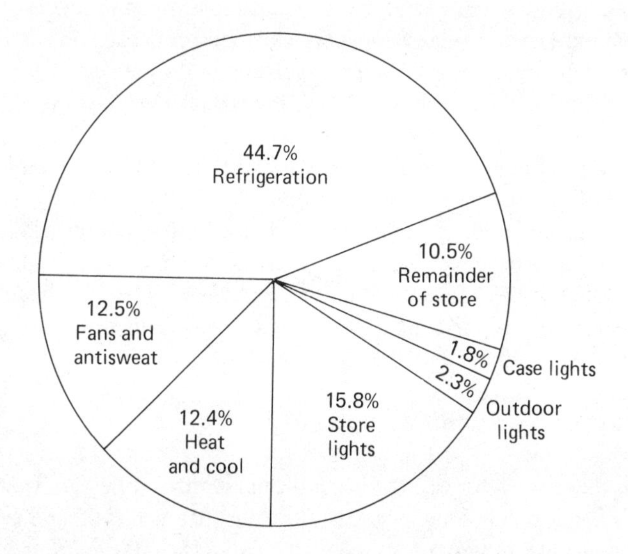

Figure 5.15. Energy use in supermarkets. [U.S. Department of Energy (101)]

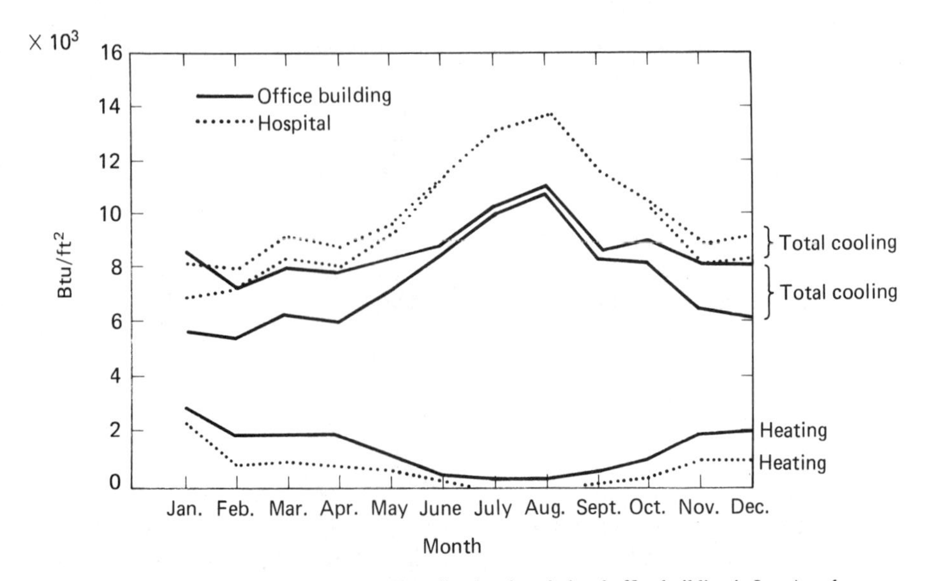

Figure 5.16. Energy use for heating and cooling in a hospital and office building in Los Angeles. For some buildings, cooling and heating occur simultaneously throughout the year. [NATO, *Technology of Efficient Energy Utilization* (102)]

why energy for artificial ventilation is such an important factor for large, modern buildings. The energy required, of course, depends on what goes on inside the buildings. Buildings in which noxious fumes and dusts require large amounts of exhaust and make-up air, or in which special temperature and humidity control is required, will use more energy. Kitchens in hotels and restaurants, for example, have special ventilation needs. More than one-third of the energy for a typical food service facility is used for HVAC. (19) Supermarkets, which use between 3 percent and 4 percent of the energy in the United States, use typically 70 percent of their energy for refrigeration and air conditioning (see Figure 5.15 [101]).

In a typical conventional office building, air conditioning may add between 15,000 to 35,000 Btu per gross square foot (gsf) (50 kWh to 110 kWh per gm^2) to the annual energy costs. In buildings with high ventilation rates, not only is the energy use higher, but it also tends to reflect variations in the local climate, as is illustrated in Figure 5.16 for a hospital and an office building in Los Angeles. (102) Computer buildings may have extraordinarily high air-conditioning costs, in the range of 400,000 to 600,000 Btu/ft^2 (1250 to 1900 kWh/m^2). (See Chapter 8.)

The Cost of Air Conditioning and Ventilation

The energy cost for cooling is equivalent to that for heating ventilation air for a given temperature and humidity change. However, there are other factors to consider. For example, while the energy released by the fans adds to the comfort heat, this heat becomes a parasitic load for the cooling system. Consequently, on a per unit heat basis, it is generally less expensive to add heat than it is to operate a mechanical system to remove heat. Furthermore, cooling equipment usually costs more to buy per unit of capacity than heating equipment.

Together with Figure 5.6 from which your cost of energy can be obtained, the following examples illustrate some of the operating costs for air-conditioning and ventilating systems:

1. Heating fresh air (without humidity treatment or the energy for fans, etc.) at the rate of 1,000 cfm uses approximately 40 MBtu per 1,000°F-days (45 MWh per m^3/sec for each 1,000°C-days).* The approximate costs per unit area for adding heat to outside air is shown as a function of the number of air changes per hour in Figure 5.17.
2. The annual cost of operating a refrigeration machine to remove a continuous heat load is 3000 kWh per kilowatt of heat removed.†

*Assuming continuous ventilation. The heat capacity of air is 1.08 Btuh per cfm°F (1.207 kW per °C m^3/sec).

†Assuming an annual COP of 2.9 (1.2 kW/ton). A ton of air conditioning is the cooling capacity to freeze one ton of ice per day. This corresponds to a heat removal rate of 3.52 kW or 12,000 Btuh.

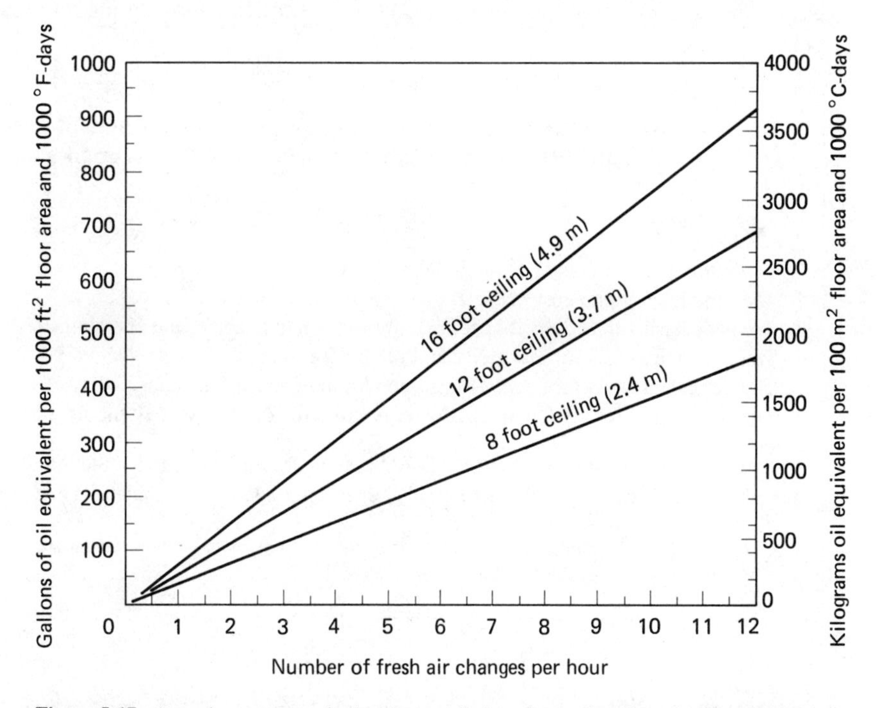

Figure 5.17. Annual energy cost for heating ventilation air as a function of ceiling height and outside air changes per hour, per 1000 degree-days and 1000 ft^2 ($100m^2$) floor area. The energy for humidification and for operation of fans is not included.‡ Heat recovery from the exhaust air can reduce the cost of heating ventilation air.

‡Assuming 65% production and distribution efficiency. Energy has been converted to kilograms of oil using the convention of 10,000 kcal/kg.

3. In the summer, a considerable part of the air-conditioning load is for dehumidification of outside air. For example, to cool 1,000 cfm of outside air at 90°F (32°C) and 70 percent relative humidity (RH) to 55°F (13°C) and 100 percent humidity, heat must be removed at the rate of 92,000 Btuh (27 kW). This can be thought of as 54,000 Btuh for removing water from the air (15.8 kW), and 38,000 Btuh (11.2 kW) for cooling.
4. For outside air at 90°F and 50 percent RH, the heat load is reduced by 30 percent over the example in 3. above, and for air at 80°F (27°C) but at 70 percent RH, the heat load would be 40 percent less than for the example in 3. The energy costs in the examples above do not include the requirements for the fans, pumps, cooling tower, etc.

With increasing energy prices, can air conditioning still be justified? Of course there are areas in which, from time to time, users will be asked to curtail HVAC use due to capacity limitations (brownouts, blackouts), or to fuel shortages. The use of air conditioning in most parts of the United States and in some parts of Europe may still be very worthwhile, both in terms of money and energy, for the improvement in working conditions and the resultant improvement in productivity. In a typical office building, the additional per capita operating cost for cooling and humidity control corresponds to the salary of an employee working for a few minutes a day. Some typical costs are placed in perspective below.

The Relationship of Some Typical Owning and Operating Costs of a Building and Its HVAC Equipment to the Salaries of Its Occupants

On a per-occupant basis and with respect to an occupant's annual salary:

1. The initial per capita cost of the building is 60 percent.
2. The initial per capita cost of the HVAC system is 10 percent.
3. The annual per capita costs for capital retirement, maintenance, and operation are 13 percent.
4. The per capita costs for electricity and oil are 2 percent.
5. The per capita costs for energy for air conditioning are 0.5 percent.
6. The per capita lifetime (equipment) energy costs for the HVAC system are 140 percent.[a]

Since the lifetime energy costs for the HVAC system are more than twice the initial cost of the entire building, it can pay to install initially, or to up-grade to, an efficient HVAC system.

[a]Annual costs for maintenance and renewal of the HVAC system of 0.56 percent are assumed over the 60-year life of the building.

Savings Potential

In the past, priority was placed on minimum first cost. The result is that too many air-conditioning systems have been designed and *operated* in such a way that they offer tremendous opportunities to reduce energy losses. Fortunately there is much that can be done at moderate cost, and there are operational changes that can be made at no cost at all. The potential for cost reduction for the owner is so large and so quickly accomplished that in many cases, improvements should receive high priority.

According to a study by the National Bureau of Standards, the energy requirements for cooling can be reduced by 30 percent with little sacrifice in comfort. (1) Energy costs for some businesses (e.g., supermarkets) may be as much as three times their net profits per unit floor area. (103) A 30-percent saving in air conditioning and refrigeration alone could result in doubling the net profit of such a supermarket. During the first year of an energy-saving program at MIT, involving 6 million square feet of floor area (600,000 m^2) in nearly 100 buildings, the use of chilled water was cut 48 percent by operational methods alone. (80)

Summary of Methods for Saving Energy

The most important methods for saving energy in air-conditioned buildings fall into the following categories:

1. Adjust winter and summer thermostat and humidity control to economy levels.*
2. Operate the HVAC systems only when necessary.
3. Minimize the heating and cooling of outside air by controlling infiltration and setting outside air rates to economy levels.
4. Set HVAC system controls to adapt to varying daily and seasonal occupancy and load conditions.
5. Minimize the energy from lighting and other equipment, which must be removed by the air-conditioning system.
6. Reduce the fan and pump energy use.
7. Recover energy from exhaust air and from other sources.

Why Do HVAC Systems Waste Energy?

An example from past practice shows why air-conditioning systems waste energy:

First Phase. The owner decides to air-condition his building. Only the best is good enough. The system desired must handle all possible load conditions, even if they will occur very rarely (for example, full occupancy of an open-plan office with large internal sources of heat from office and other equipment).

Second Phase. The architect decides to have clear (untinted) glass window walls, as large as possible; no external venetian blinds (he does not like their appearance); a very high lighting level; and high outside air rates.

Third Phase. The engineer designs a system handling all (maximum possible) internal and external loads, and adds her/his margin of safety. This avoids liability problems in the event of some unusual situation in which the guaranteed conditions could not be achieved.

*Each Fahrenheit degree rise in temperature can result in approximately a 3-percent energy saving (5 percent per centigrade degree). Lowering heating temperatures from their usual settings would result in a similar saving for each degree change. (104)

Fourth Phase. The equipment suppliers add their own safety margin for the same reason, and use undersized air handling units and ducting to save on manufacturing cost. The owner will have to pay for the increased operating and investment cost.

Fifth Phase. In the first year of operation there are complaints from the occupants, and the system is adjusted in such a way that it can handle any extreme situation. No attempt is made to readjust the system to reduce energy costs as operating experience is gained.

Last Phase. This is continuing in effect. The owner pays at least 30 percent too much each year. The system has an unnecessarily high air flow rate, and pressure drop. Therefore, the electrical energy used is a factor of 2 to 3 higher than what it needs to be.

In our example, the following questions should have been asked:

Is air conditioning necessary at all? If yes, where and when?

What is the optimum building design (floor plan, walls, windows) for minimum energy use? What are the marginal yearly costs for a deviation from these design conditions compared with the marginal benefits?

What are adequate and realistic requirements for human comfort, lighting, and internal loads? Can a departure from design temperature and humidity, which occurs infrequently and only for short periods, be tolerated (compare with the yearly energy saving by allowing this)?

When purchasing equipment, ask the supplier to demonstrate minimum life cycle costs (including investment, maintenance, and operating costs). Can you afford a higher initial investment, and how much would be saved yearly by doing this?

Is maximum system flexibility required, and how often would it be utilized?

Factors Influencing the HVAC Load

Heat losses and gains occur simultaneously in a building, although usually one far outweighs the other. This depends both on the activities inside the building and on the influences of the weather. The function of the air-conditioning system is to compensate for deficiencies. When the net heat flow from all of the factors is balanced, the temperature is constant. The heat flows during the cooling and heating season are diagrammed in Figures 5.18 and 5.19. The factors which are important are:

Infiltration. The exchange of air between the inside and the outside of the building through windows, doors, vents, and cracks. Infiltration depends upon the pressure difference between the inside and outside and the temperature which causes warmer air to rise inside tall buildings, drawing more air in through openings at the bottom, and upon wind effects. Not only must infiltrated air be heated or cooled, but it leads to problems where dust or humidity levels must be closely controlled.

Ventilation. Ventilation can be mechanical or natural. It can simply be outside air from open windows or doors, although in most new buildings ventilation is usually mechanical. The greater the flow of outside air, and the greater the temperature difference, the more energy is required to compensate for the heat gain or the heat loss involved.

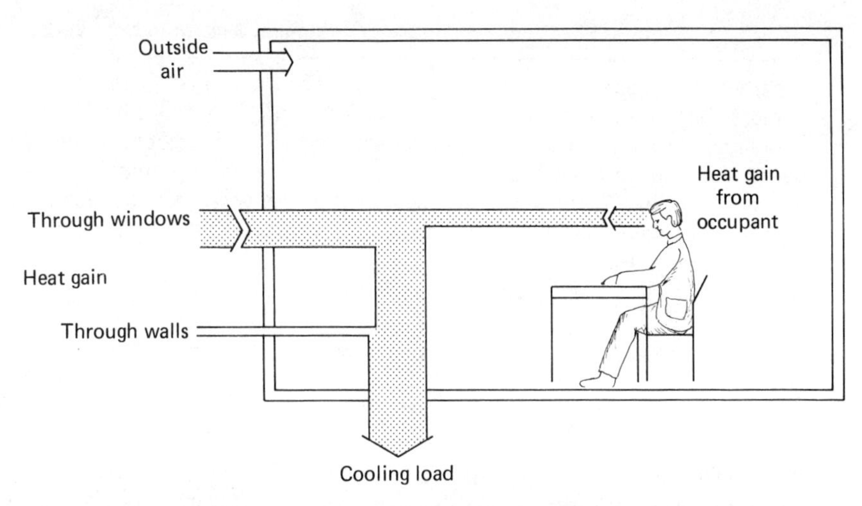

Figure 5.18. Heat flow for an office, summer conditions with lighting turned off.* Note that external blinds can eliminate 90% of the solar radiation through the window. Interior blinds hold back at the most only 50%, see Section 5.2. [After (105)]

*Proportions shown are for an outside temperature of 90°F (32°C) and an inside temperature of 79°F (26°C).

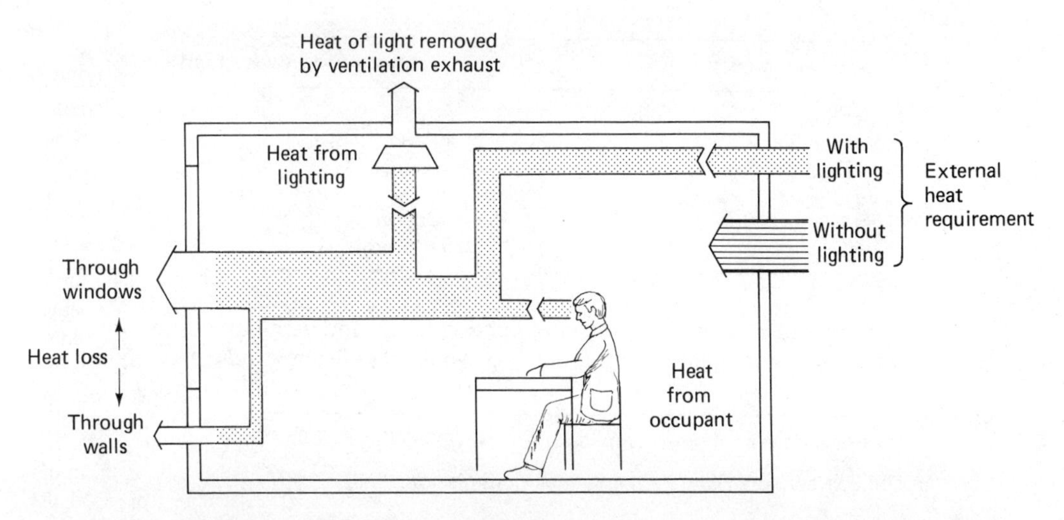

Figure 5.19. Heat flow for an office, winter conditions. Proportions shown are for an outside temperature of 12°F (−11°C) and a room temperature of 72°F (22°C). [After (105)]

Table 5.25. Summer Heat Loads on an Air Conditioning System in New York

Characteristics of the building: A 10-story office building was assumed with a 98,000-square-foot roof and a wall area of 125,000 square feet, half of it glass. In Case 1, the shading coefficient of the glass was assumed to be .55, with a U-value of 1.1; walls were assumed to have a U-value of .2, and one inch of insulation was assumed for the roof. In Case 2, the shading coefficient of the glass was assumed to be .23 with .50 U-value, walls were assumed to have a U-value of .098, and two inches of insulation were assumed for the roof.

The comparison of the reduced air-conditioning load is shown below.

	Case 1 (Btuh)	*Case 2 (Btuh)*	*Difference (Btuh)*	*Savings in refrigeration (tons)*
Solar gain	3,150,125	1,317,326	1,832,799	153
Other glass gain	1,306,250	593,750	712,500	59
Walls	246,500	120,785	125,715	11
Roof	2,003,270	1,163,381	839,889	70
Total	6,706,145	3,195,242	3,510,903	293 tons (293 kW)

The largest source of heat in most office buildings is lighting, followed usually by air brought into the building to satisfy ventilation requirements. For example, in the 10-story office building, assuming one ton of refrigeration equals 1 kW of electrical demand, the air-conditioning load is shown below for the Case 2 insulation and window treatment.

Source of heat	*Air-conditioning load (kW)*	
Lighting	1,633	60.0%
Ventilation	586 (minimum)[a]	21.5%
Glass	159 (as reduced)	5.8%
People	245	9.0%
Roof	97 (as reduced)	3.5%
Walls	10 (as reduced)	0.4%

These data suggest that reducing lighting levels and ventilation rates, in addition to decreasing heat gain from the outside, would significantly reduce air-conditioning demand.

[a]Ventilation in minimum compliance with New York City Building Code.

SOURCE: New York State Public Service Commission. (108)

Transmission. This is the heat flow through the walls, roof, and windows, the roof and the floor of the envelope, as discussed in Section 5.2.

Radiation. The sun's direct radiation through windows can be a major loading of the cooling system in summer (see Section 5.2), and often of benefit in winter.

Lighting. Lighting contributes to the cooling load in direct proportion to the wattage of the lamps in use. This can be the single most important factor loading the cooling system (see Table 5.25 opposite).

Occupants. The number of people in a building create a significant heat gain; in addition moisture is released through perspiration and respiration. This effect is noticed in theaters or large meeting rooms for which the air conditioning is inadequate. People sitting at rest give off moisture and heat equivalent to a cooling system load of 100 watts. For moderately active office work this load is 130 watts, and for very strenuous activity, 400 watts.

Dehumidification. Taking water out of the air is usually done by cooling it to a low temperature where the dew point corresponds to the desired humidity when it is mixed with the room air. When cooled to the dew point, water condenses out. Usually this is at a temperature too low for comfort, so that the air must be warmed—either by compensating for an unwanted room heat source, or by reheating it. To warm the air to comfortable temperatures requires additional energy for which the condenser heat from the chiller is a good source. For each pound of water removed, approximately 1,047 Btu of energy must be removed by the chiller. This corresponds to 0.68 kWh per liter or about 2.6 kWh per gallon of water removed. (To put moisture into the air during the winter, the same amount of energy is required.) Air may also be dehumidified by exposing it to a liquid or solid desiccant. The moisture-laden desiccant is dried by exposing it to heat from steam, a gas heater, or solar energy. (106,107)

Equipment. All powered equipment, fan motors, ovens, photocopying machines, etc., generates heat. If this heat is not removed, either by water cooling or by ducting it directly outside, the heat which may be welcome in the winter becomes a burden on the cooling system in the summer.

For an office building in New York City, two different wall and window conditions were analyzed by the staff of the New York Public Service Commission. (108) Examples of some heat loads on the cooling system are shown in Table 5.25. In this table, the resulting improved insulation and solar shading is shown to reduce the required cooling capacity by over 50 percent. It also illustrates the significance of lighting on cooling, even with the low-effective heat from a lighting load of 1.7 W per gross square foot (18 W/m^2).

Reducing the Cooling System Load

Many of the above factors which contribute negatively to the building's heat gain can be modified to some extent. One potential for saving is in reducing the amount of outside air that needs heating, cooling, or humidity control. On the other hand, outside air should be mixed into the building if its temperature or humidity reduces total air handling energy, as may be possible, for example, by using cool morning outside air during the summer.* Block the direct radiation from the sun. External blinds are the most effective, but venetian blinds, drapes, and shades can reduce by the sun's radiation through windows by 33 percent to 50 percent.

Table 5.25 suggests that reducing lighting would significantly reduce cooling load. This is confirmed by a Rand Corporation analysis of a high-rise office building in New York City, with 40 percent glass walls. Reducing lighting by 50 percent from 4.3 W/ft^2 (46 W/m^2) reduced annual cooling energy requirements by 40 percent, in addition to the lighting energy saved (see Section 5.6). (109) Whenever business machines, photocopiers, ovens, and other equipment can be operated for a fewer number of hours or vented directly outside, the annual cooling load will be reduced. Once the internal load is reduced, reduction of the air flow will result in a four-fold saving:

Reduced lighting and equipment kWh

Reduced lighting and equipment load on cooling

Reduced fan motor kWh

Reduced fan motor load on cooling

Consider setting room thermostats to higher temperatures during the cooling season (except where reheat is used for temperature control). Allow still higher temperatures in less critical spaces—halls, entryways, and storage areas—or simply turn off air conditioning to stairwells, etc.

Human comfort depends primarily on

Activity level

Thermal resistance of clothing

Air temperature

Radiant wall temperature

Relative air velocity

Relative humidity

As temperatures increase, evaporation of moisture from the body plays a larger role in maintaining comfort. (110) Maintaining lower room humidity may be a more cost-effective method for obtaining comfortable conditions than lowering room temperature, particularly if sources of moisture (infiltration of outside air, production processes, cooking, etc.) can be controlled. The federal government encourages the adoption of the temperature range of 78–80°F (26–27°C) for general comfort cooling, recognizing that certain activities or equipment may require special consideration. (111) From a health aspect, the inside

*This condition occurs when the outdoor wet bulb temperature is between the room dew point temperature and the room wet bulb temperature.

temperature should be governed by the outside air temperature: the temperature difference to the outside should be no greater than 10°F (5°C).*

HVAC Systems: How Do They Work?

Raising thermostat settings or turning off lights may not save energy with some air conditioning systems. It is important to understand how the system operates in order to determine both the effect of reducing the load and if the overall system is being correctly operated. In Figure 5.20, the elements of a basic HVAC system are shown. The method used for delivering cooling plays a greater role in defining energy use than is the case for heating. After measures to reduce the load have been implemented, there may be significant opportunities to improve the efficiency of distribution.

Much of the potential high efficiency of central cooling machines (chillers) is fre-

*In the past, people have accepted cooling in the summer to low temperatures about which they would complain during the heating season. Some stores cool to levels that are uncomfortable for shoppers in summer clothing. The shock encountered upon entering the store and again upon leaving is also unhealthy for business and the store's operating expenses.

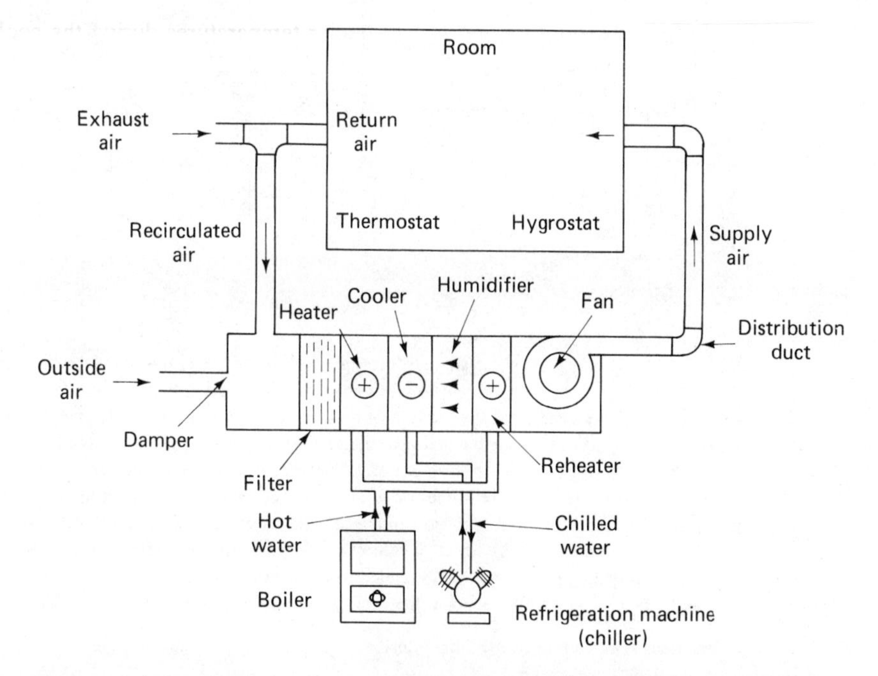

Figure 5.20. The elements of a central air-conditioning system. Not shown is the cooling tower circuit for cooling the condenser. The air-handling unit takes a mixture of outside air and recirculated air, filters it, and then either heats or cools it. The air is sprayed with water (or steam) to humidify it, then reheated and delivered by the fan through the distribution ducts to the room.

quently lost in the distribution and preparation of chilled air. The cold effect may be transported by chilled air, chilled water, and distribution of the compressed refrigerant, which is then expanded at the point where the refrigeration effect is desired (called "direct expansion"). Direct expansion is also commonly used for food freezing and refrigeration and industrial cooling where lower temperatures are required. Per unit weight, refrigerants and water have higher cooling capacities than air, so that the distribution system may be smaller and also require less pumping energy per unit of cooling. The system coefficient of performance is significantly lowered when large amounts of air must be transported. However, air distribution systems may provide more benefits in filtration and humidity control and are required with at least sufficient capacity to supply outside air to areas without openable windows.

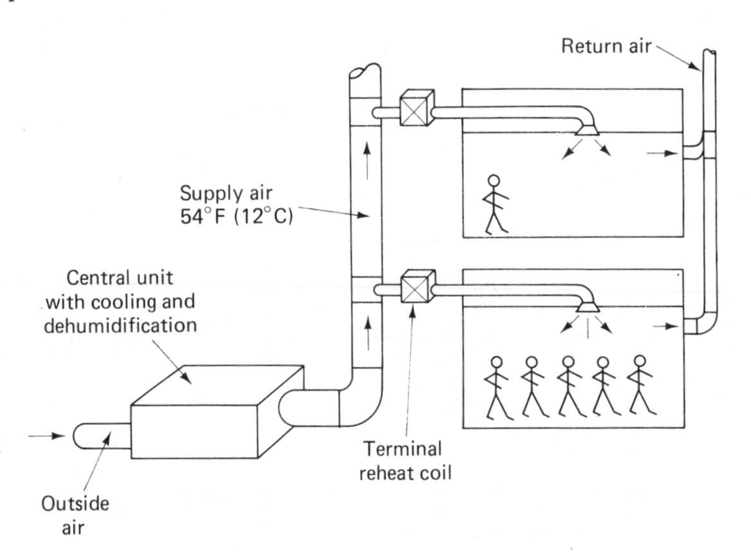

Figure 5.21. Terminal reheat system. With this system, the area or zone in the building with the most stringent heat removal and low humidity requirements determines the temperature of the supply air from a central supply unit. Just before the air is introduced into each room served by the system, it passes through a heating coil controlled by a thermostat located in each room. The thermostat causes the air to be reheated as necessary to maintain the desired temperature setting. Raising the temperature in the room during the cooling season by setting the thermostat higher simply causes more heat to be supplied to the air supply. Similarly, with the terminal reheat system, turning out the lights or removing lamps from their fixtures may also result in the equivalent quantity of energy being added to the supply air.

Methods to consider for reducing the energy used by this system are

1. Raise the temperature of the supply air.
2. Shut off the reheat coils.
3. Reduce the air volume.
4. If close temperature and humidity control must be maintained for certain equipment, consider lowering hot water temperature and reducing flow to reheat coils. This will still permit some control, but over-cooling may result as outside temperatures drop.
5. If close temperature and humidity control are not required, perhaps the system can be converted to variable volume by adding variable volume outlets and eliminating terminal heaters.

HVAC Air Handling Systems

Next we discuss some typical air handling systems. These are the units which may be known by the occupants as "air conditioners." Some different distribution systems are shown in Figures 5.21, 5.22, 5.23, and 5.24. From the descriptions given with these figures, it will be seen that some can have considerable losses. Better adjustment and modification of this equipment according to the methods given in the figure captions can achieve substantial savings.

Each of these systems supplies warm and/or cooled air during the heating and air-conditioning seasons. Air from each room is exhausted through a return air system. By

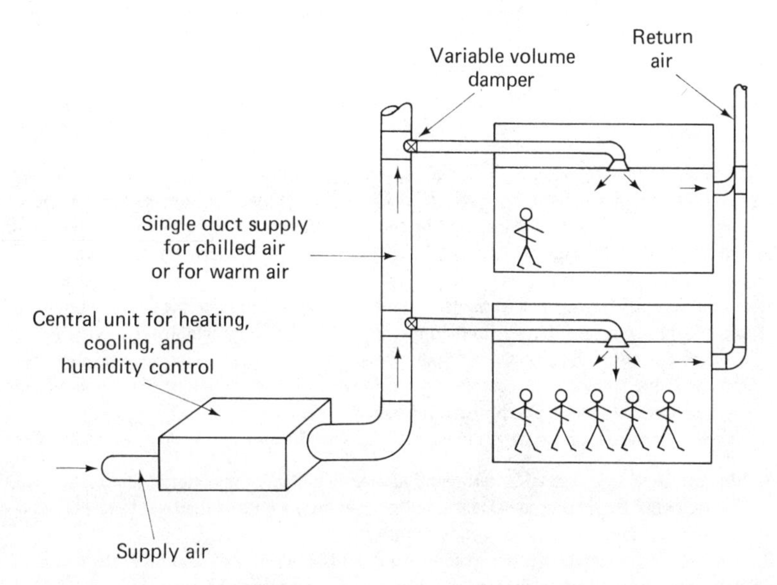

Figure 5.22. The variable volume system. A variable volume system provides heated or chilled air, depending on the season, through a single duct to each room or zone only in sufficient quantity necessary to maintain the temperature set by the thermostat. As the load conditions in the room change, the air flow damper adjusts the quantity of air supplied. The outlets for rooms with windows, or where there are large variable loads, must be capable of throttling the air flow without a significant change in the distributin pattern. Some variable volume systems are deceptive in that they merely divert the excess air from the room into a ceiling plenum. True variable volume systems can offer good operating economy at reduced loads.

Methods for saving energy with this system include:

1. Reduce the volume of air handled by the system to the minimum possible.
2. Raise the temperature of the air supply to that point which will result in the control valve being fully open to the room with the most extreme cooling load. (During the heating season lower the hot water temperature and reduce the air flow so that the air control valve is fully open for the greatest heating load.)
3. Consider installing static pressure controls for more effective regulation of pressure bypass (inlet) dampers.
4. Consider installing fan inlet damper control systems if none now exist.

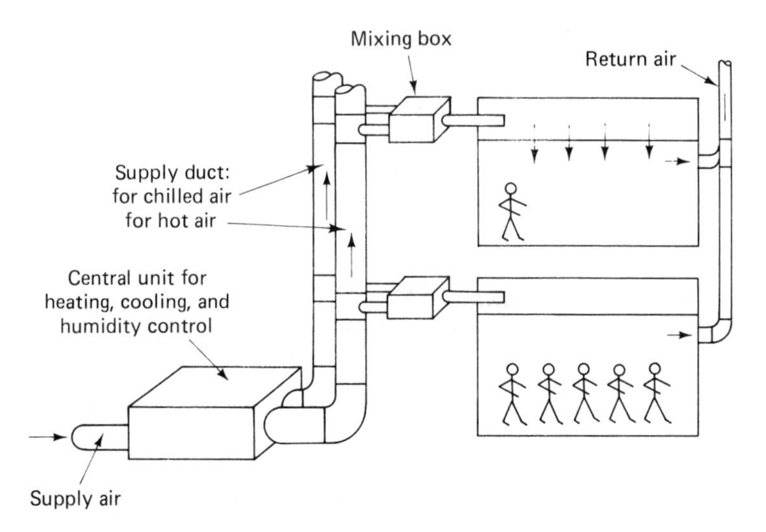

Figure 5.23. Dual-duct air-conditioning system. A dual-duct central HVAC system supplies hot and cold air through two separate sets of duct work. The two air streams are combined in a mixing box in a ratio that is controlled by the room thermostat. The two supply temperatures may be typically 53°F and between 90°F and 120°F (12°C and between 35°C and 50°C). The mixing of hot and cold air streams results in a wasteful energy loss. A more efficient arrangement is to convert this system to a variable volume by supplying either heating or cooling as the season requires, with the damper to the other duct completely closed. If no cooling is required at any time during the heating season, and vice-versa during the cooling season, then operating both ducts in the single-duct, variable-volume mode will result in a further reduction in the fan energy requirements as the result of the lower back pressure.

Other methods for improving the energy characteristics of this system would include:

1. Shut off the source of heating during the summer months. Temperature control is still possible for each area by mixing unheated air from one stream with chilled air from the other duct according to the setting of the room thermostat.
2. Increase the temperature of the cold air supply to between 60°F and 68°F (15°C and 20°C).
3. Reduce air flow to the supply ducts to the minimum acceptable level.
4. When no cooling loads are present, close off cold ducts and shut down the cooling system.

Figure 5.24. A room induction unit for an HVAC system. A more sophisticated form of the reheat system employs an induction unit(s) in each room. Supply air (1) is also prechilled to meet the requirements of the room with the greatest cooling load, which may be a room on the western side of the building receiving the afternoon sun. Primary air is discharged through nozzles (2) to induce an amount of secondary room air (3) to mix with it, thereby reducing the size of the primary air supply ducts. In the most common of these units the induced air is heated in coil (4) before exiting through the grill usually located along the outside walls under the windows. In some variations of the induction unit, coil 4 may be connected by three-way valves to either the chilled water lines, so that additional cooling may be obtained for heavy loads, or to the hot water lines for reheating. The advantages of the induction unit over the single-duct, reheat system are several; first, less air need be circulated throughout the building, resulting in smaller ductwork, refrigeration and air handling units. Second, if local cooling coils are provided,

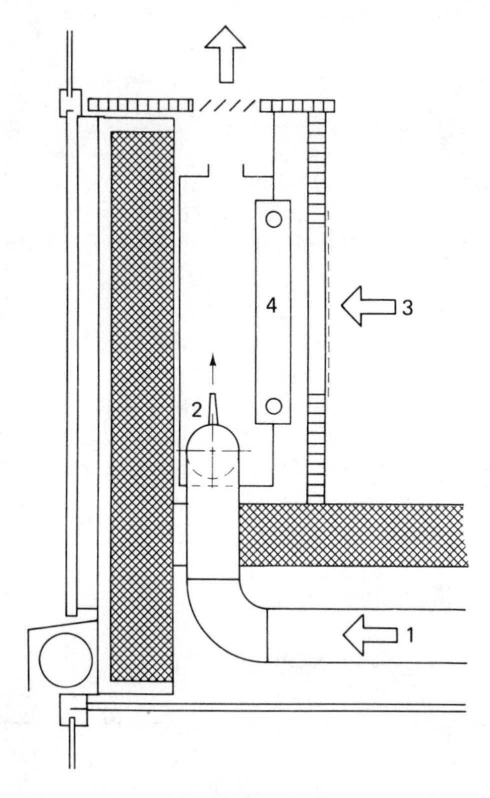

Figure 5.24

energy is saved since not all rooms need to be supplied with air chilled to the lowest temperature required in the building. However, under simple thermostatic control, at each change-over from cooling to heating, chilled water is flushed into the return hot water line. On the next cycle of the thermostat, hot water is flushed into the chilled return-water line. The resulting mixing can lead to considerable energy waste. Four-pipe induction units with separate heating and cooling coils avoid the mixing waste, but suffer from greater resistance to air flow which must now travel through two coils. The high supply pressure used to achieve adequate induced room air mixing already is the cause of additional energy loss in the distribution system. Although less wasteful than the simple reheat system, induction units cannot be used for deep rooms (over 25 ft or 7m).

Several methods for making the system more efficient include:

1. Increase primary air supply temperature to maximum possible value to minimize reheating.
2. If close temperature control is not required, shut off heat to secondary coil during cooling, vice-versa during heating season.
3. Inspect nozzles. If metal nozzles, common on most older models, are installed, determine if the orifices have become enlarged from years of cleaning. If so, chances are that the volume/pressure relationship of the system has been altered. As a result, the present volume of primary air and the appropriate nozzle pressure required must be determined. Once done, rebalance the primary air system to the new nozzle pressures and adjust individual induction units to maintain air flow temperature. Also, inspect nozzles for cleanliness. Clogged nozzles provide higher resistance to air flow, thus wasting energy.
4. Reduce secondary water temperatures during the heating season.
5. Reduce secondary water flow during maximum heating and cooling periods by pump throttling or, for dual-pump systems, by operating one pump only.
6. Consider manual setting of primary air temperature for heating, instead of automatic reset by outdoor solar controllers.

the use of a damper control, this air can either be exhausted from the building, or some portion of it can be returned to the central supply system. Typically the proportion of outside air (vs. recirculated air) can be varied from a maximum of 100 percent to 25 percent or lower. One hundred percent outside air may be used when the outside temperatures are between 54°F and 77°F (12°C and 25°C) to minimize the amount of cooling required in the central unit. At temperatures above this range, recirculating inside air usually requires less cooling energy than warmer outside air.

Below this temperature range, air is recirculated to reduce the quantity of air that is heated. To the extent that cooling in the winter is required, outside air below this range can be used. In making this decision, the humidity required should be taken into consideration and attention should be directed at avoiding any possible freeze-up of pipes or heat exchangers, etc., during cold weather.

Operation under Changing Load

The operation of the systems shown in Figures 5.21–5.24 can, in principle, be set to operate efficiently for a constant cooling load. If the load varies, due, for example, to the shifting position of the sun onto the western face of the building, the room thermostat will call for additional cooling. This is accomplished in a different way for each system.

Reheat systems (Figures 5.21 and 5.24) reduce reheating power. The dual-duct system (Figure 5.23) increases the volume of chilled air and reduces the volume from the other duct. The variable air volume system increases the flow of chilled air. Depending on the ratio of the areas with constant load to the areas with variable loading from any heat source, the efficiency and cost of operation of these systems will vary. Figure 5.25 shows the relative operating cost of the four systems in relation to the percentage of space with variable loading (exterior rooms with the influence of the sun and the weather). The variable volume system can adapt to changing loads with the lowest operating cost. However, systems with constant flow can be adapted to lower cooling needs by shutting off some of the units, or by intermittent operation. Because of heat storage in the room, the cooling can be on–off modulated to supply the cooling required by the area thermostat without reheating at the outlet terminals. Figure 5.25 shows that terminal reheat systems use considerably more energy.

Furthermore, energy-saving efforts can be frustrated; turning out or removing lights and turning off equipment can result in the equivalent energy being added in reheat, and setting the thermostat to a higher temperature results in still more wasteful reheating. If your system uses reheat, you could also be fooled into thinking that you are saving energy by turning off the lights! However, in most cases energy costs will be reduced because electricity costs are higher than oil or gas.

Check the operation of your system. Check the temperatures and the control system. Adjust the supply temperature upward after the cooling load has been reduced. Avoid simultaneous cooling and heating if at all possible. Turn off the reheat coils. This will sacrifice independent control of temperature in different areas, and may cause some loss of humidity control; however, experience has shown that it is not too difficult to find

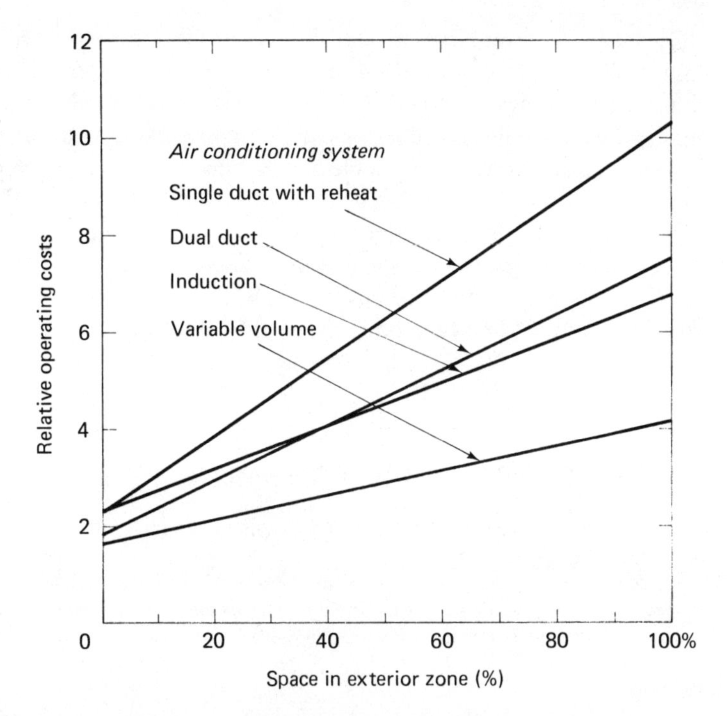

Figure 5.25. Comparison of the relative operating costs of several different air-conditioning systems. (105) This comparison includes fan and pump energy and considers the effect of the variable cooling load of the sun and weather. As the percentage of space subject to these varying influences increases, so does their cost of operation.

suitable adjustments that will provide satisfactory comfort. Some buildings may require a manual adjustment once or twice a day; however, those with less of a load variation will require adjustment less often. On days with high humidity, the supply temperature can be lowered, and reheat turned back on if additional control is required.

Reduce Air Flow Rates

Can recirculation rates or the exhaust of conditioned air be reduced? The air flow rate of an HVAC system selected for the maximum load is sized for conditions which occur only a few days during the year, and then only for a few hours during the day. If the flow rate cannot be reduced when the load is lower, two to four times more energy may be spent for air transport than is really required. Consider, for example, the internal space of an open plan office: the reduction of lighting wattage permits the reduction of air recirculation from six to eight air exchanges per hour (as was typical earlier) to three to four changes. This results in considerably less energy for cooling and transport of air.

Since fan horsepower increases with the cube of the rotational speed, while the volume of air flow varies directly with RPM, a small reduction in speed will reduce energy use much more than air flow. For example, motor horsepower is reduced by half if the speed of rotation (and air flow) is reduced by only 20 percent. A graph of this relationship is shown below, together with an example of the saving from changing the fan motor pulley. (112)

Where code specifies hourly air changes, the volume of air can be reduced by lowering the ceiling height. A new hung ceiling, incorporating luminaries where the light is actually required, would reduce the heat flow through the ceiling and the energy required for HVAC, while at the same time bringing an older building up to contemporary design.

Example of Saving Achieved by Reducing the Ventilation in a Warehouse.

Energy savings may be realized by reducing forced ventilation in buildings to a lesser but still adequate amount required to provide safe conditions.

The air flow from a centrifugal fan varies directly as its rotational speed. Thus, the amount of ventilation can be reduced by decreasing fan speed. The figure shows the air flow from a centrifugal fan versus the power required to drive it, both as percent of full rating.

EXAMPLE

A new 150,000 cubic foot warehouse was constructed with provision for five air changes per hour. This required a 10 hp motor (with a fan load of 9.83 hp) driving a 24 inch centrifugal fan at 915 rpm to deliver air at the rate of 12,500 cfm. Later information showed that only four changes per hour would be adequate, or 80% of the original design. Pulley changes were made, therefore, to reduce the fan speed to 915 × 0.80, or 732 rpm.

Reference to the figure shows that with the fan speed reduced to 80% of full rating, the power required to drive it is only 50% of full load. Assuming a motor efficiency of 80% at full load,

$$\begin{aligned}\text{Power at full load} &= 9.83 \text{ hp} \times 0.746 \text{ kW/hp} \times 1/0.80 \\ &= 9.166 \text{ kW}\end{aligned}$$

Assuming a drop in motor efficiency to 77%,

$$\begin{aligned}\text{Power at 50\% load} &= 9.83 \text{ hp} \times 0.50 \times 0.746 \text{ kW/hp} \times 1/0.77 \\ &= 4.762 \text{ kw} \\ \text{Electrical power saving} &= (9.166 - 4.762)\text{kW} \times 8760 \text{ h/yr} \\ &= 38{,}600 \text{ kWh/yr}\end{aligned}$$

If the utility consumes 10,000 Btu of fuel/kWh generated,

$$\begin{aligned}\text{Annual energy savings} &= 38{,}600 \text{ kWh/yr} \times 10{,}000 \text{ Btu/kWh} \\ &= 386 \text{ MBtu per year}\end{aligned}$$

If the cost of electric power is $0.02 per kWh,

$$\begin{aligned}\text{Annual cost saving} &= 38{,}600 \text{ kWh/yr} \times 0.02 \text{ \$k/Wh} \\ &= \$770 \text{ per year}\end{aligned}$$

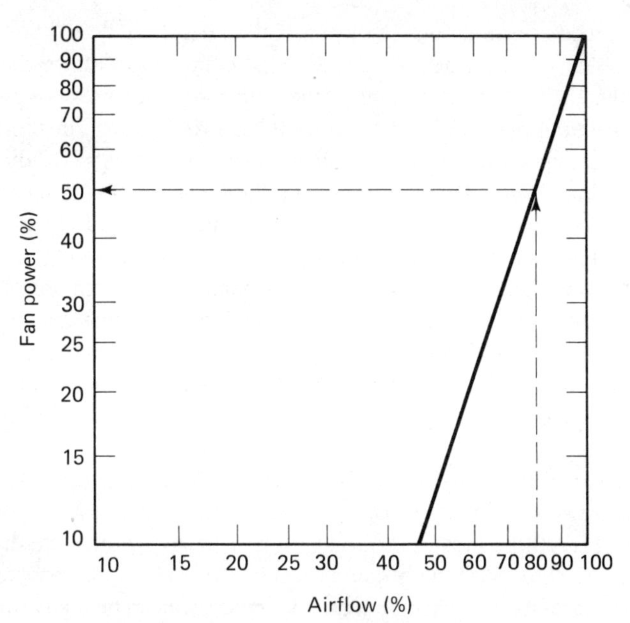

Decrease in horsepower accomplished by reducing fan speed (based on laws of fan performance).

SUGGESTED ACTION

Determine whether the number of air changes provided by your ventilation system can be reduced and still maintain safe conditions.

Fan speed can be reduced, and energy saved, merely by changing pulleys. If the motor operates at less than 50% of its rated load, however, its efficiency may be very poor and its power factor unduly high. In such cases, consult the motor manufacturer to determine electric efficiencies and power factors at low loads. In some cases a smaller motor rated for the job will produce greater savings.

(Note: Reducing ventilation may also reduce the energy requirements for heating and cooling.)

SOURCE: National Bureau of Standards (112).

For facilities in which the ventilation rate is set by the actual occupancy, various schemes are used to control the air exchange by the number of people in the building. Some supermarkets and department stores use photoelectric doorway sensors to count shoppers in the building. A computer operated control system regulates the outside air intake based on a people count. (113)

Generally, areas with high exhaust rates—kitchens, manufacturing areas, laboratories—with exhaust hoods are considered wasteful when there is much loss of conditioned air. Where high exhaust rates are required, some form of heat recovery should be considered as a supplement to, or in place of, the make-up air system. (Make-up air systems are designed to precondition fresh outside air during conditions of extreme outside weather or for process installations requiring high exhaust air flow.) Sometimes by reexamining the problem—excess heat or contaminant control—an alternative can be found. Where possible, eliminate contaminants at the source—use lower vapor pressure or dry lubricants, avoid painting parts which will burn off, control oil, etc., in coolants. Enclose or restrict contaminating sources to reduce air volume requirements. (Redirect or reduce forced air from ceiling fans and general room ventilation around contaminant emission points. This will reduce the capture velocity, and therefore the required exhaust volume.) Trap heat or contaminants over stoves and fryers, welding and plating areas, paint spray booths, etc., by ducting untempered air to and around the problem area so that flow vectors feed into exhaust hoods. Some examples of savings with heat recovery systems are given in References 114, 115, 116, 117, and in Chapters 6, 7, and 8.

The volume of outside air can be reduced by circulating ventilation air through charcoal filters or precipitators, air washers, etc. At the Minnesota Mining and Manufacturing plant in St. Paul, a saving of 33,000 gallons of oil per year and 1,700 ton-days of cooling was achieved by filtering and recirculating 17,000 cfm (30,000 cubic meters per hour) of exhaust air through charcoal filters. For a one-time investment of $12,000, the company saved $11,700 the first year of operation. The savings include the maintenance costs for the filters. (86) With such commercially available pollution control systems, solvents used as reactants or carriers in industrial processes can be recovered for reuse or combined with spent lubricating oil, etc., and used as boiler fuel. For example, the Gillette Company in Boston invested $70,000 in a carbon-absorption system to recover 500 lbs/hr (225 kg/hr) of trichloroethylene. The system recovers eleven times the cost of the system in reclaimed solvent each year, in addition to savings in reduced HVAC operating costs. (118)

Increase Production Efficiency

The production of refrigeration or cooling with a mechanical compressor will be discussed next. (Cooling can also be achieved with evaporative coolers in dry climates, and with the use of heat from steam or solar energy in an absorption chiller [see Section 8.3].) (119) The principle of mechanical cooling involves raising the pressure of a refrigerant gas in a compressor, passing the refrigerant through a condenser, and then passing it through an evaporator before it is returned to the compressor again.

In the condenser, the hot, compressed refrigerant gas is cooled to remove the heat of compression. It then condenses to a liquid. The heat of compression may be transferred to air in an air-cooled heat exchanger similar to that of an automobile radiator, or it may be transferred to a water loop in a water-cooled heat exchanger. The water loop, which may also be an ethylene glycol mixture, conducts the heat of compression to a cooling tower, or to an external body of water.

To produce the refrigeration effect, the liquid refrigerant is evaporated in the evaporator. The heat needed to evaporate the refrigerant is derived from cooling the heat load. This may be transferred to the refrigerant in direct expansion coils in the HVAC duct system, or in a heat exchanger with the chilled water system.

Refrigeration machines are heat pumps that transfer energy from the cooling heat load to some other place, often the external environment. Those refrigeration machines designed for dual purpose heating and chilling may be outfitted with "double bundle" heat exchangers. This is a second set of tubes on the condenser heat exchanger that allows the heat to be transferred to the building heating system in the winter rather than rejecting it outdoors. The second bundle in the evaporator is connected to a source of heat—an external body of water, or perhaps a source of low-grade waste heat.

The coefficient of performance (COP) of the chiller machine alone is defined as the ratio of the heat absorbed in the evaporator to the energy to drive the compressor. The heat absorbed in the evaporator is usually three to five times the energy for the compressor. As we mentioned above, the COP of the cooling system may be much lower than this value.* The COP depends strongly, and in a nonlinear way, upon the difference in temperature between the condenser and the evaporator, as shown in Figure 5.26. Typical temperatures are 38°F and 95°F (3°C and 35°C) respectively.

*The COP for the cooling system is the ratio between the heat load removed and the energy required to operate the cooling system. Thus the heat load excludes heat added by the fans and pumps of the cooling system. The energy input to the system includes that which operates the fans, pumps, compressor, and cooling tower. Because the distribution system may require considerable energy to deliver the cooling to where it is needed, it may suffer unwanted heat losses, and the COP of the overall cooling system can be quite low, less than even poor window air conditioners!

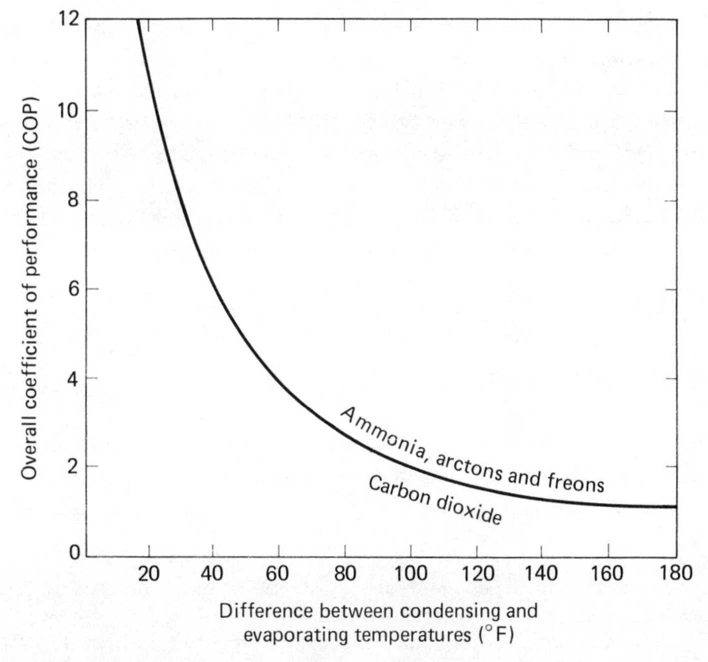

Figure 5.26. The coefficient of performance of refrigeration machines as a function of the difference between the condenser and the evaporator temperature. (120)

The performance of the chiller can be improved by decreasing this difference. Since many systems were overdesigned for rarely occurring conditions, and since they will have an additional surplus capacity after the energy-saving program, there may be more than adequate cooling capacity with the decreased temperature difference. This can be achieved by raising the temperature in the chilled water supply system to, say, 50°F (10°C) or higher. It may be possible to lower the condenser temperature simply by cleaning the heat exchanger surfaces. Since the condenser temperature is strongly dependent upon outdoor conditions, and since these may be poor when the cooling need is the greatest, the following operating mode may result in lower average condenser temperatures: overcool early in the morning when outdoor temperatures are lower. With cooling stored in the building's heat capacity, the cooling load (and consequently the condensing temperature) will be lower later in the day.

A lightly loaded chiller does not have a very good COP. If multiple units are operated in parallel, shut one or more down and valve off their chilled water lines to enable the remaining machine to operate near its peak COP.

Cooling Tower*

The cooling tower lowers the temperature of the condenser water with the aid of direct evaporation of water to the outdoor air. Each quart of water evaporated requires approximately 2000 Btu (0.63 kWh per liter). The rate of evaporation depends upon the wet bulb temperature of the ambient air, and the rate of air flow through the tower. Higher flow rates can produce lower temperatures. However the increased fan consumption should be balanced against the benefit gained from lower compressor operating power. (121) Shading will reduce the formation of algae growth and will reduce the heat load that must be carried by the tower. Solids left behind by the evaporated water can reduce water flow, and inhibit atomization from the spray nozzles. Nozzles should be kept free of scaling. Make-up water should be treated to prevent scaling both in the tower and in the condenser of the refrigeration machine. The generally lower performance coefficient of systems with air-cooled condensers can be improved somewhat by shading the condenser.†

CONCLUSIONS

The energy losses in our existing buildings with air-conditioning systems can be summarized as follows: the demand is too high in comparison with adequate comfort requirements, resulting in about 10 percent too much energy use. With air handling and

*Your local health department should be consulted for specifications on the germicidal treatment of the water in the cooling tower of your HVAC system.

†The annual average COP of direct expansion units may be as high as 4 (0.8 kW/ton) with water-cooled condensers, while if they are air-cooled, the COP may be as low as 2.5 (1.4 kW/ton).

distribution, another 10 percent can be saved, and in production of heat and coolants (including heat recovery) 10 percent to 20 percent. As much as 40 percent can be set as an average goal for reducing energy use for air conditioning a building, if the building was planned according to previous energy-wasteful design that focused on minimum first costs and not on minimum overall costs. Some examples of improvements to air-conditioning systems and the savings achieved by these investments are shown in Table 5.26.

Examples Showing Use of the Checklist

Our example is the air-conditioned office building modeled after the one shown in Figure 5.27. (11) This building has approximately 107,600 ft^2 (10,000 m^2) and 800 occupants. A schematic floor plan is shown in Figure 5.28. The energy use was calculated before EIP with a computer program. Its specific energy use is summarized in the following equation in which electricity is converted to its raw source energy equivalent using a factor of 3.*

$$E = E_{heating} + u(E_{motors} + E_{lighting}) = 145 + 3(75 + 85)$$

$$= 200{,}000 \text{ Btu/ft}^2 \text{ (625 kWh/m}^2\text{)}$$

*The specific energy use of this building as measured at the envelope was 95,000 Btu/ft^2 (300 kWh/m^2).

Figure 5.27. Electrowatt-House. This office building was constructed according to energy-saving design before the oil price shock. Even so, with the increased importance of energy and of minimizing operating costs, an EIP was able to reduce energy use further.

Table 5.26. Examples of Costs and Savings from Improvements to the Air Conditioning System

Energy saving measure	*Cost ($)*	*Savings ($/year)*	*Energy saved, MBtu (kWh)*	*SCP,[a] $/BPD ($/kW)*	*Comment*	*Reference*
Charcoal filters	12,000	11,700	6800 (2×10^6)	3900 (55)	At 3M's St. Paul plant, 33,048 gallons of oil and 1,700 ton-days of cooling were saved by filtering and recycling 17,000 cfm (30,000 m^3/hr) of previously exhausted air.	(118)
Computer control of HVAC	647,000	195,000	47,800 (1.4×10^7)	28,700 (405)	Saving achieved from load shedding and better matching HVAC system to actual conditions. Cost includes computer, controls, and installation at IBM Plaza, Chicago.	(122)
Thermocycle on chiller	65,000	50,000	5800 (1.7×10^6)	7790 (110)	Thermocycle system installed on 2,000-ton centrifugal refrigeration machine allows 2,000-hp motor to be shut down during periods of light cooling. A 10-hp pump circulates refrigerant between evaporator and condenser to effect heat transfer 4 months/year.	(45)
Demand controls on exhaust fans	600,000	180,000	40,000 (1.15×10^7)	35,600 (460)	Switches on exhaust fans in Chemical building at the Brookhaven National Laboratory are coupled with auto dampers in make up air supply. Part of saving is due to lowering temperature to 65°F (20°C) and allowing the temperature to fall back to 55°F (12°C) during non-occupied periods; 20% of annual average saving is elect.	(123)

Energy saving measure	*Cost ($)*	*Savings ($/year)*	*Energy saved, MBtu (kWh)*	*SCP,[a] $/BPD ($/kW)*	*Comment*	*Reference*
Intermittent ventilation, air ducts	8,800	6,000	1470 (4.3×10^5)	12,750 (180)	A variable volume ventilation of mechanical room was achieved by on-off cycling the ventilation. The on ratio is controlled to meet demand. Exhaust air from storage areas was used to ventilate garage. Costs include new ducting.	(124)
Enthalpy and Hygrometric controls	14,000	2,800	680 (2×10^5)	43,500 (615)	Mixing ratio of outside and recirculated air controlled by enthalpy sensor to minimize energy use. Humidity controlled by direct measurement (hygrometric) rather than by cooling all air to dew point and reheating.	(124)

[a]The specific capital productivity (SCP) is the cost of reducing the annual rate of energy use by one barrel of oil equivalent per day or by one kilowatt.

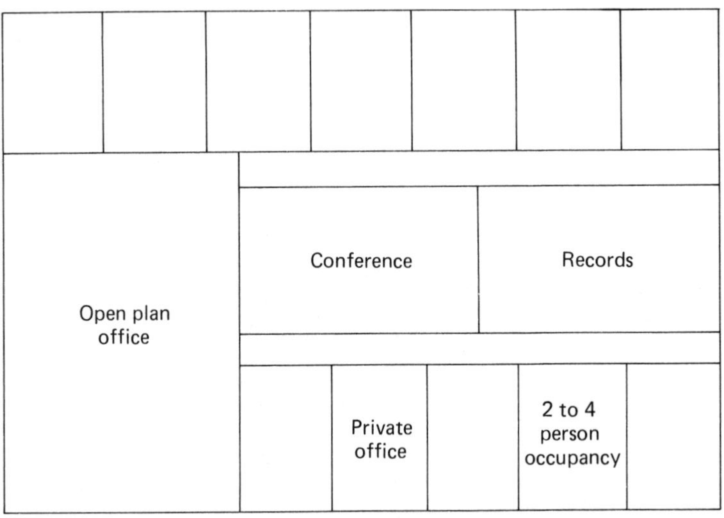

Figure 5.28. Schematic plan and data used for computer modeling of an office building in Zurich, Switzerland. (12)

Window area: 30% of wall area; clear glass, double glazed windows with venetian blinds.

Conductivity of walls: U = 0.18 Btuh°Fft^2 (K = 1 W/m^2 °C).

Air cooled lighting: 3.3 W/ft^2 (35 W/m^2), used 1500 h/yr. (i.e., 50% of the working and cleaning hours—7 A.M. to 7 P.M., five days a week).

Dual duct air-conditioning system (variable outside air mixing rate), oil-fired heating system. There are 6588°F-days (3660°C-days).

This building is located in a climate where cooling of the exterior offices would not be necessary. Nevertheless, since an HVAC system was required for the interior space, cooling and humidity control were provided for the exterior offices as well, to assure optimum comfort and productivity and protection from a noisy environment.

The following energy-saving measures were considered.

1. ***Cooling.*** Turn off cooling since temperatures exceeding 86°F (30°C) occur only 240 hrs/yr; saving: 8 percent.
2. ***Humidity.*** No humidity control* and no cooling. This means that humidity can drop below 30 percent or rise above 65 percent (few hours a year); saving: 10 percent.
3. ***Start-Up.*** Operating time reduced by one hour a day (start-up and shut-down time modified); saving: 7 percent.

*No humidification in winter, no dehumidification in summer.

4. ***Outside Air.*** Minimum outside air reduced from 65 cfm (110 m^3/h) per person to 35 cfm (60 m^3/h). Smoking is still permitted. Saving occurs in winter and summer at times of extreme outside temperatures when the minimum outside airflow rate is used; saving: 9 percent.
5. ***Blinds.*** Moveable outside venetian blinds are actually used only 30 percent of the sunny days. If consistent use could be achieved, a saving would occur because of reduced summer load of 3 percent.
6. ***Maintenance.*** System is adjusted four times a year so that all components, control, and regulation work properly. Estimates used for energy-saving calculation; leaks reduced, motor performance increased, air-conditioning supply temperatures optimized (hot and cold supply temperature difference reduced), ducting heat losses reduced by insulation; saving: 2 percent to 5 percent.
7. ***Lighting and Ventilation.*** Internal heat loads (lighting) reduced from 3.3 W/ft^2 (35 W/m^2) to 2.3 W/ft^2 (25 W/m^2). Rate of air circulation lowered. Saving: 16 percent.
8. ***Operational.*** Cooling circulation shut down when not needed: 0.2 percent saving.
9. ***Operational.*** Heating circulation shut down in the summertime; saving: 0.4 percent.
10. ***Heat Recovery.*** Heat recovery used between exhaust air and incoming outside air; saving: 5 percent.
11. ***Comfort Requirements.*** Instead of 72°F to 79°F (22–26°C) and 35 percent to 55 percent humidity, new limits of 68°F to 82°F (20–28°C) and 30 percent to 65 percent humidity are used; saving: 4 percent.
12. ***Special Startup.*** Outside air dampers are closed early in the morning while building is brought into proper temperature before occupancy; saving: 2 percent.
13. ***Enthalpy Controller.*** Volume of outside air used is regulated according to its humidity and temperature for minimum energy use; saving: 2 percent.
14. ***Night Set-Back.*** Temperatures lowered to 59°F (15°C) instead of 64°F (18°C). Air conditioning is off at night; saving: 2 percent.
15. ***Use Cold Night Air.*** Outside night air is used for cooling, so that room temperatures are lower when air conditioning system starts operation in the morning. Saving is small.
16. ***Lighting Reduced.*** Photocell controls turn lights off where and when daylight is sufficient. Lighting levels reduced by one-third during cleaning; saving: 5 percent.

The estimated savings for the office space are shown in Table 5.27. Don't forget: this is a model building for the purpose of demonstration; in an actual case one might expect as much or more saving in entrances, storage rooms, restaurants, and computer centers as in the office spaces!

CONCLUSIONS

Conservative estimates of savings as implemented add up to 30 percent using the measures summarized in the following table.

Table 5.27. Estimated Savings for Model Office Building with Air Conditioning (12)

No.		Heating, Btu/ft² (kWh/m²)	Cooling,[a] Btu/ft² (kWh/m²)	Air transport,[a] Btu/ft² (kWh/m²)	Lighting,[a] Btu/ft² (kWh/m²)	Total, Btu/ft² (kWh/m²)	Saving (%)
0.	Base case[1)]	12.78 (145)	4.58 (52)	15.52 (176)	22.57 (256)	55.46 (625)	—
1.	No cooling	12.78 (145)	—	15.52 (176)	22.57 (256)	50.88 (577)	8.3
2.	No cooling + no humidity control	11.46 (130)	—	15.52 (176)	22.57 (256)	49.55 (562)	10.6
3.	Optimum start up	10.58 (120)	4.23 (48)	14.28 (162)	22.57 (256)	51.67 (586)	6.9
4.	Reduce outside air	7.76 (88)	4.58 (52)	15.52 (176)	22.57 (256)	50.43 (572)	9
5.	Outside blinds are not used as intended (30% of time with sunshine windows are unprotected)	2.78 (145)	6.26[2)] (71)	15.52 (176)	22.57 (256)	57.13 (648)	−3
6.	Maintenance	[3)]	[3)]	[3)]	[3)]	[3)]	2–5 (estimation)
7.	Reduce lighting wattage	46 (145)	17 (43)	56 (176)	51 (162)	167 (526)	16
8.	Shut down cooling system circulation if unneeded	46 (145)	16 (50)	56 (176)	81 (256)	199 (627)	3

No.		Heating, Btu/ft² (kWh/m²)	Cooling,[a] Btu/ft² (kWh/m²)	Air transport,[a] Btu/ft² (kWh/m²)	Lighting,[a] Btu/ft² (kWh/m²)	Total, Btu/ft² (kWh/m²)	Saving (%)
9.	Shut down heating system circulation if unneeded	45 (143)	16.5 (52)	56 (176)	81 (256)	199 (627)	3
10.	Heat recovery	34[4)] (106)	16.5 (52)	59[4)] (185)	81 (256)	190 (600)	4.8
11.	Change comfort requirements	38 (120)	16 (50)	56 (176)	81 (256)	191 (602)	3.8
12.	Special start-up	43 (135)	16 (51)	56 (176)	81 (256)	196 (618)	1.6
13.	Enthalpy controlled outside air ratio	42 (133)	16 (50)	56 (176)	81 (256)	195 (615)	2
14.	Reduce base load heating (nights)	43 (135)	16.5 (52)	56 (176)	81 (256)	196 (619)	1.7
15.	Flush with cool outside air before start up	43 (135)	15[4)] (48)	57[4)] (180)	81 (256)	200 (629)	—
16.	Reduce lighting	46 (145)	16.5 (52)	56 (176)	72 (228)	191 (601)	4.4

Notes:

[1)]With air-conditioning, including cooling and humidity control.

[2)]Cooling need increased.

[3)]Heating need is reduced (recovery of heat with heat exchanger in exhaust air). Due to increased pressure drop (at heat exchanger) work needed for air transport has increased slightly.

[4)]Saving in cooling is offset by extra work for air transport.

[a]Electrical energy is converted to primary fuel equivalents using a multiplier factor of 3.

		Calculated (%)	Realistic for summing up savings (%)
3.	Start up time	7	10[a] (items 3 and 4 together)
4.	Reduce outside air	9	
5.	Use blinds as intended	3	1[a]
6.	Maintenance	5	2[a]
7.	Reduce lighting wattage	16	8[a]
10.	Heat recovery	5	5
13.	Enthalpy controlled outside air mixing	2	2
14.	Night set back	2	2
16.	Shut down unneeded lights	5	2[a]
			30%

Note: Approximately 20 percent additional savings would be possible at the expense of occupant comfort with measures 1, 2, 11.

[a]Assumption: only partly realized.

Saving Energy in a Shopping Center

The next examples we discuss are for energy use and some energy saving measures for shopping centers. The tenants usually have different requirements and different energy uses per unit area, as is illustrated in Table 5.28. (125) All the stores shown are in the same building in Allentown, Pennsylvania. They have the same hours of use, and identical HVAC systems that are physically located within the premises, individually metered, and under the control of each tenant. That the metered energy use shows almost a ten-to-one variation illustrates the importance of the activity within the space in determining energy use.

Our next example is an efficiency improvement program for a shopping center. (12) The average energy use (with electricity counted at its raw source value) is 365,000 Btu/ft^2 (1150 kWh/m^2). The saving measures evaluated are listed below, followed by a list of the resultant savings.

Vary the use of one, two, or all three air-handling units depending on occupancy.

Base the minimum rate of outside air used upon actual need as set by occupancy.

Schedule shut-down and start-up to avoid unneeded use of air-conditioning systems.

Shut down air-conditioning systems twice a day during time of minimum occupancy.

Reduce lighting, readjust air-flow rate.

Use a heat recovery system between exhaust air and outside air.

Table 5.28. Energy Use in a Shopping Mall

Shop	*Btu/sq ft/ year*
Auto center	74,000
Department store	114,000
Department store	102,000
Variety store	100,000
Restaurant	409,000
Bank	131,000
Drug store	129,000
Food market	205,000
Dry cleaner	688,000
Book store	104,000
Doughnut store	326,000

Note: The raw source energy for each of the all electric HVAC systems listed above would be three times the values shown.

Schedule cooling unit operation to avoid low load operation.

Shut off warm water system in summer.

Shut down humidifying earlier in the spring.

CONCLUSIONS

This example of an air-conditioned shopping center presented many opportunities for energy savings. In particular, the lighting levels need to be reduced as much as sales permit, with special attention given to reasonable use of accent lighting. Air flow rates should also be readjusted according to the level of customer occupancy, rather than maintaining a continuously high level of operation; this also gives significant saving.

The following energy-saving measures were considered for a shopping center:

1. In normal operation the energy budget is E = 365,000 Btu/ft^2 (1150 kWh/m^2). This includes yearly energy used for air conditioning, lighting, heating converted to primary fuels per unit net floor area. Lighting wattage: 5.6 W/ft^2 (60 W/m^2). Data computed for net sales area of 750,000 ft^2 (70,000 m^2).
2. Reduced lighting level by 20 percent to 4.6 W/ft^2 (50 W/m^2); saving: 10 percent.

3. Reduced air flow rates because of reduced lighting; saving: 15 percent.
4. Further reduction of lighting level by 40 percent to 3.2 W/ft^2 (35 W/m^2) level, adjusting flow rates; saving: 40 percent.
5. Modified start-up (reduction of operating time by one-half); saving: 2 percent.
6. Intermittent operation of air conditioning (shut down one-half to one hour during times of low occupancy); saving: 10 percent.

Conservative assumption of realistic savings (3, 5, 6): 20 percent to 30 percent. With retail oil and net energy prices at this time of $0.45/gallon and $16/1000 kWh respectively, the annual energy cost was $18/m^2 ($1.70/ft^2).* Hence the energy saving corresponds to $240,000 as the result of operational changes alone.

Refrigeration

Energy use for refrigeration equipment such as refrigerated display cases and walk-coolers is determined to a large extent by its design. Consequently to achieve refrigeration savings it is necessary to take issue with the original design of the equipment and make changes that the manufacturer could have made in the first pace if greater emphasis had been placed on minimizing energy use.

Refrigeration equipment often uses energy for defrost heaters and anti-condensate heaters, in addition to energy for compressors, condenser and evaporator fans that is common to air conditioning systems. Electrical energy for refrigeration may be reduced by reducing the heat load into the refrigerated space and by improving refrigeration efficiency. For example, display case loads can be reduced by improved loading measures, by avoiding unnecessary lighting, defrost cycling and mullion heating, and by keeping the environment around open cases cool and dry, and covering them after store hours.

The efficiency of the refrigeration equipment can be improved by minimizing the pressure difference through which the compressor must pump the refrigerant vapor, avoiding heat gains to the suction gas returning to the compressor, and by measures to cool the liquid reaching the expansion valve as much as possible. Temperatures in the refrigerated space should be no lower than necessary.

Some refrigeration compressors have adjustable head pressure control valves. In cold weather, the head pressure can be reduced. Set the head pressure to the lowest setting that will provide an adequate supply of refrigerant to the expansion valves. Check with your refrigeration mechanic about the advisability of installing manual by-pass valves around the refrigerant receivers. Savings between 10 and 20 percent may be possible in cold weather when cold refrigerant from the condenser could go directly to the expansion valve without first being warmed by hot gas in the receiver.

A number of systems are available to control the frequency of cycling of refrigeration equipment. These controls save energy by controlling the cycling of the compressors and improving the control of the temperature in the chilled space. One should consider a

*The $16/1000 kWh is the net cost of both oil and electricity at its raw source value (1,150 kWh/m^2 × $16/1000 kWh = $18.40/m^2). There are 6,588°F-days (3,660°C-days).

determination by actual trial, if necessary, of the possible savings from forcing the compressors to cycle on and off on a pre-determined schedule. As is discussed in the next section, such measures should be coordinated with measures to reduce total electric demand charges. The objective is to produce a temperature no lower than actually required, while minimizing excessively short and unnecessarily long compressor cycles.

Condenser fans may also be operating unnecessarily during night time and cool weather temperatures. Consider lower power motors, or thermostat actuation of the condenser fans. Increased air flow across the condenser coils lowers the refrigerant temperature and increases the compressor efficiency. However, this is achieved at the expense of increased energy for the condenser fans. Consequently, some experimentation may be necessary for the temperature to set the fan thermostat to achieve the lowest overall energy use. There is little, if any saving to be gained by sub-cooling the refrigerant, however, if it is subsequently warmed by flowing back into the receiver. Evaporator fans typically run continuously, and therefore account for a sizeable cost. This may be reduced with a device from NASA for improving the power factor and efficiency of fractional horsepower motors.

Mullion anti-condensate heaters may have been connected by the manufacturer for conditions to keep the surfaces free of moisture in the most humid weather. Rather than operate continuously, these can be wired for manual switching and turned on as necessary for the conditions in a particular location in the store. These heaters can also be connected to humidity sensors that are commercially available for this purpose. Anti-sweat heaters should be turned off after hours.

Powerful defrost heaters may be connected for periodic operation, rather than for actual need. When the defrost heaters operate when there is no longer any frost on the coils, electricity is wasted, additional refrigeration is required to remove their heat, and the product is warmed unnecessarily. Commercial controls can be installed to initiate the defrost cycle only when needed and to terminate the cycle when the ice has been removed.

The operation of the mullion heaters and the defrost heaters can be minimized by controlling moisture sources and covering the cases at night. Consider turning off the display lighting in refrigerated cases after hours.

We conclude this section on cooling and ventilation with a list of methods for saving in a supermarket, as compiled for the Industrial Workshop series sponsored by the United States Department of Energy.(101) This list includes many ways for reducing the cost of energy used for air conditioning and for refrigeration.

Ways to Save Energy in a Supermarket

Seal windows with caulking

Weatherstrip doors.

Use lower wattage or fewer bulbs.

Turn off lights when area is not in use.

Use window lights sparingly.

Turn off or dim lights in store parking lot.

During day, open curtains or shades.

At night, close curtains.

Keep windows clear of obstructions for maximum sunlight.

Restock during the day, since night hours require more heat and light.

Have store delivery of pickup trucks observe 55 mph rules.

Keep delivery doors closed when not in use.

Turn off heat in storerooms and other areas when not occupied.

Clean refrigeration coils to boost efficiency.

Maintain equipment in peak running order.

Provide regular maintenance for heating system and keep filters and heat transfer surfaces clean.

Add insulation on heating pipes passing through unused areas.

Keep wiring in good condition to minimize loss of power.

Consider not venting heat and moisture away from food service station or kitchen if feasible.

Apply light-colored finishes to walls, ceilings, and floors; dark surfaces absorb light and may require as much as 15 percent more wattage.

Open only one package of frozen food at a time for price marking, and then place food packages in cases immediately to prevent unneeded temperature gain.

Clean lamps and lighting fixtures to increase efficiency by as much as 40 percent.

Insulate ceilings and floors.

Use more efficient fluorescent, mercury, or sodium lights, rather than incandescent where possible.

Fix leaking faucets.

Use paper plates in food service to reduce hot water consumption for dishwashing.

Fully stock frozen food cabinets to load-line limits in order to maintain lowest temperature possible.

Use heat reclamation systems to heat the store environment.

Install demand load controls.

Install demand defrost controls.

If demand defrost controls are not used, stagger set defrost times.

Turn off one or two exterior lights if possible.

Change filters regularly on furnace, air conditioning, and meat preparation coils.

Set heating thermostat at 68°F (20°C).

Set cooling thermostat at 78°F (26°C).

Make sure air conditioning and heating are not on at the same time.

Vacuum all cooling coils and keep them clean.

Make certain that personnel reduce all water use to a minimum.

Keep reach-in case doors and walk-in cooler doors closed as much as possible. Use reminder signs.

Do not stack cases over return air grills or over refrigeration load lines.

Adjust case temperatures to the highest temperature that still allows for food preservation.

Keep supply diffusers and return-air grills free of stock, trash, and dirt. Restricted air flow causes equipment to work harder.

Train all personnel in energy-efficient operation of the store.

Institute a group relamping program.

Replace lamps with new ones designed for energy efficiency.

When removing fluorescent lamps, have an electrician disconnect primary side of the ballast.

For doors with air curtains, adjust the velocity and direction of air flow to prevent outside air from entering the store.

Adjust the timing of automatic doors so that they remain open only the necessary amount of time.

Use night covers on cases (if recommended by the manufacturer) strictly according to directions.

Position advertising or display cards where they will not interfere with air flow.

Preheat ovens in bakeries minimum amount of time.

Turn off ovens when not in use.

Do not load fryers beyond recommended capacity.

Buy display cases and coolers that have maximum life and are energy efficient.

Control anti-sweat heaters according to environmental conditions, rather than keeping them on high settings at all times.

Investigate feasibility of changing environmental set points. Remember to make sure this does not have an adverse effect on your refrigeration equipment.

When purchasing small units, look for Btu-per-watt or energy efficiency ratio.

Adjust outside air intake so as to only bring in minimum required by codes.

Keep temperature and humidity set points at optimum for refrigerator case and walk-in cooler operation.

Use reduced wattage case and valance lights that are designed for high output of light.

Make certain openings inside and outside for conveyor belt will close securely. Turn conveyor off when not in use.

CHECKLIST	OPERATING CHANGES
Turn Off AC Where Not Needed	*HVAC: Eliminate Unnecessary Demand*
Shut off AC in selected areas.	Shut off ventilation completely if possible (or only cooling and/or humidity control) after reducing loads (lighting, window area, solar shading). Shut off ventilation in unoccupied areas, infrequently used rooms (stairwells, entrances, utility rooms, basements, garages) and in areas with sufficient natural ventilation. Vent moist air from driers and hot air from photocopying machines directly to outside; turn off unneeded exhaust fans and kitchen exhausts. Isolate particular sections with higher requirements (or use local units) within larger areas with no requirement on HVAC.
Avoid artificial cooling where not needed.	Increase air flow or use cool outside air (e.g., morning) to handle internal loads without cooling; check if there is a saving by doing this because of additional energy for air transport.
Localize air flow.	Use drapes or canopies to minimize exhaust needed for welding, ironing units, process heaters, etc.; draw air from occupied areas.
Eliminate unnecessary exhaust hoods and roof ventilators	or reduce their size.
Turn Off AC When Not Needed	
Work out an occupancy schedule for working and also for non-working hours to achieve minimum operating time for HVAC.	Adjust working hours of personnel. Avoid occupancy peaks; organize office cleaning during normal working hours (or with reduced ventilation); concentrate night work in one section of a building.
No artificial ventilation in winter, temperate season.	Open windows, eliminate artificial ventilation where air quality, noise level, internal loads, and heating system permit.

Turn Off AC Where Not Needed (cont.)	*HVAC: Eliminate Unnecessary Demand (cont.)*
Shut off system at night, weekends.	Shut off ventilation (also warm and cooling water circulation) completely; arrange working schedules (cleaning, night work) so that a maximum of the system can be shut down.
Minimize operating time in areas with varying occupancy.	Avoid unneeded ventilation in assembly rooms, lecture halls, kitchens, restaurants, toilets. Shut down ventilation if not needed.
Discontinue humidity control (except working hours in winter).	Discontinue humidification, dehumidification: nights, weekends, temperate season and summer (if possible)
Minimize operating time of cooling	by reducing cooling in late afternoon; by flushing with cold outside air at night (e.g., increase air flow to avoid cooling when extra fan energy is less than for chiller). In infrequently used areas (auditoriums, etc.), turn off cooling whenever possible, or set thermostat at highest possible level during non-occupancy.
Optimize start-up/shut-down procedure.	Close outside air dampers during start-up. Vary start-up/shut-down time depending on minimum energy consumption, as a function of the outside and inside conditions (i.e., occupancy, production schedule, etc.). Work out operation schedule for ventilation. For each system, start-up and shut-down time must be determined.
Check possibilities for intermittent use of HVAC system.	In some office buildings intermittent use (e.g., 10 minutes off, 20 minutes on) is practiced with success!
Check building for negative pressure	to avoid infiltration.
Review Comfort Requirements	
Maintain required temperatures.	Encourage use of heavier clothing. Readjust thermostats to necessary level. Minimize drafts for comfort. Summer maximum: 80°F (27°C); encourage use of lightweight clothing.

Review Comfort Requirements (cont.)	*HVAC: Eliminate Unnecessary Demand (cont.)*
Allow adequate humidity conditions.	Allow wider variation. Winter: 30 percent to 40 percent maximum (check effects on electrostatical loads of rugs, on wooden furniture, and on health). Summer: if feasible, switch off dehumidification (or set thermostat to the highest allowable level).
Reduce minimum outside air supply.	Minimum outside air supply of 12–25 cfm (20–40 m^3/h) may be adequate during extreme outside conditions. Minimum outside air supply is used at particularly low or high outside temperatures. Is it possible to restrict smoking to areas vented directly outside and having no reinfiltration? This would allow a drastic reduction of minimum outside air supply. Identify problem sources of contaminant and odor (local control, filters, absorption, equipment, vent outside). Reduce outside air supply: morning, lunchtime (with reduced occupancy). Close outside air dampers during start-up. Adjust exhaust fan systems (kitchen, toilets, etc.) to minimum required exhaust air rate. Reduce the quantity of exhaust air; use local exhaust. Use low-volume, high-velocity exhaust systems for hoods, etc.
Review special environmental conditions.	Review for continued justification. Check areas with complaints (draft, temperatures). These areas usually have excessive energy consumption also, due to design or operating errors.
Lower requirements during non-working hours.	Is air conditioning needed at all or is only heating to 60°F (15°C) sufficient?
Achieve Energy Conscious Occupant Behavior	
Use sunshades optimally.	Close venetian blinds to the sun in the summer.
Do not open windows.	Opening windows in ventilated spaces will greatly increase energy use. Minimize infiltration by closing doors, unused chimneys, and vents.
Shut off lighting.	Shut off lighting if unneeded, reduce lighting level (also during cleaning), use natural light as long as possible. In many companies, a "lighting guard" will periodically turn off lighting in unoccupied spaces, which is a particularly large saving in combination with a variable air volume system that adapts to load.

Achieve Energy Conscious Occupant Behavior (cont.)	*HVAC: Eliminate Unnecessary Demand (cont.)*
Shut off heat producing machines if unneeded.	Is it possible also to adjust air flow rates? Minimize use of heat-producing devices at least during cooling and/or peak power consumption period.
Leave thermostats as set.	Leave thermostats at predetermined level where uninformed misadjustment would upset proper regulation of system.
Bill energy use.	Bill departments, tenants according to energy used, to encourage energy-saving measures.
Reduce refrigeration load.	Reduce display lighting in refrigerated cases. Turn off at night. Set thermostats in refrigerated spaces to higher temperature, if possible. Load freezers immediately after receipt of product, before unnecessary warming. Control moisture sources to reduce defrost cycling. Switch off anticondensation heaters after hours. Coordinate traffic into refrigerated areas to minimize door openings and infiltration.

Achieve Energy Conscious Maintenance	*HVAC: Reduce Air Distribution and Handling Losses*
Clean components.	Clean filters, heat exchangers, louvers, ducts. Check leaks (valves, steam lines, air ducts, duct work joints, doors to fan rooms, check tight sealing of louvers).
Check efficiencies.	Check component and system efficiencies at full and low load, analyze energy consumption in detail and compare with measured valves. Identify excessive demand and components with low efficiency.
Check air inlets.	Check air inlets whether closed or misadjusted by user of a room. In many cases this was done because of improper functioning of the HVAC system. The correction made by the occupants may cause difficulties in other areas.
Readjust system periodically.	Periodically check control and regulation, flowrates, recalibrate sensors. (Use specialists for this work.) Readjust HVAC system 4 times a year for optimum summer, fall, winter, springtime operation!
Check summer heat consumption.	Check and analyze summer heat production. In many cases it is surprisingly high; one reason for this may be simultaneous heating/cooling.

Increase System Efficiency	*HVAC: Reduce Air Distribution and Handling Losses (cont.)*
Avoid simultaneous cooling/heating.	Induction units: Check malfunction of three-way valves for intermixing hot and chilled water.
	Dual duct system: Select hot/cold check temperature as low/high as possible.
	Shut off reheat coils in summer.
Multiple use of heat.	Examine reason for high energy consumption; is multiple use of heat or air possible?
Readjust louvers.	Direct air flow to area needed.
Adjust to actual load.	Use high speed and all units only at time of full load. At other times, use lower speed or shut off one of several units.
Check air flow pattern.	Check adequate induced air flow with induction units (also part load), obstructions, air flow pattern. Avoid air stratification, such as cooling the ceiling, warming the floor; direct air where it is needed, check for air short circuits (where the inlet air is directly flowing to exhaust opening).
Reduce air change.	By reducing external and internal loads, can the air change rate be reduced? Reduce as far as possible without serious complaints (change pulleys).
	HVAC: Increase Production Efficiency
Increase cooling system efficiency. Optimize part load operation.	If the central controlled system is using the cooling units inefficiently, use manual start-up of cooling units, reduce time of part load operation.
Shut off cooling medium circulation after working hours.	Check whether start-up procedure has to be modified in this case.
Adjust cooling tower operation at lower outside temperature.	Shut off some of the fans to reduce air flow rate. The same applies to air cooled condensers.
Clean cooling tower condensers regularly.	
Increase chilled water temperature to highest possible level.	
Minimize operating time of absorbers	if compressor units with higher efficiency are available.
Check seals for oil leaks.	Oil contamination of the refrigerant can result in as much as a 25% reduction in the coefficient of performance for a centrifugal chiller.

Increase System Efficiency (cont.)	*HVAC: Increase Production Efficiency (cont.)*
Reduce head pressure.	Set pressure controls on compressor to lowest possible setting that provides adequate refrigerant supply to expansion valve. Check pressure setting seasonally. Reduce pressure in cold weather.

CHECKLIST	SHORT-RETURN CHANGES
Avoid Use When Not Needed	*HVAC: Eliminate Unnecessary Demand*
Exhaust fans.	Use automatic controls to operate exhaust fans as needed.
Time switches.	Install time switches for shut-down, start-up, part-load operation.
Install start-up system.	Optimize start-up/shut-off according to outside and inside conditions.
Reduce Requirements	
Select optimal position for thermostats.	Check optimum position of thermostats.
Achieve energy conscious occupant behavior.	
Install individual heat/coolant meters	and bill according to energy used to produce an incentive for saving.
Reduce Load	
Reduce load of areas with high load.	Ventilate attics, spaces under roofs during hot weather, to reduce temperature in occupied spaces below, etc. Reducing external or internal load of critical areas will allow new system balancing, which has to be made according to the zone with the highest load.
Reduce heat generated by lighting.	Install more efficient lighting system (fixtures at best location, deactivate certain fixtures, use lower wattage elements).
Cover hot and cold liquid tank surfaces.	Often this is done with plastic spheres.

Reduce Load (cont.)	*HVAC: Eliminate Unnecessary Demand (cont.)*
Improve controls on refrigeration system.	Install humidity sensing controls for anti-condensation heaters.
	Install defrost control to sense ice buildup and to turn defroster off when surfaces are ice free.
Achieve Energy Conscious Maintenance	*HVAC: Reduce Air Distribution and Handling Losses*
Repair faulty equipment and insulation.	Insulate and seal ductwork especially in non-conditioned spaces. Use optimum insulation thickness.
Increase System Efficiency	
Change air path to reduce air flow rates.	Connect in series instead of parallel if possible: Use exhaust air of suitable areas (no odor, low temperature) as inlet air of less critical areas (garages, corridors).
Use low quality heat.	Can low-quality heat (i.e., low temperature heat) from heat pumps or heat recovery be used for air pre-heating or re-heat?
Use enthalpy-controlled mixing of outside air.	Previously this was controlled only by temperature. Enthalpy control will include the effect of humidity also in determining optimum mixing of outside air and return air.
Increase Cooling System Efficiency	*HVAC: Increase Production Efficiency*
Install automatic blow-down valves on cooling tower.	
Install automatic shut-off for fans in cooling towers for lower outside temperatures or nights.	Reduce cooling tower operation in winter. Use water for cooling to evacuate heat load or recover this heat and transport to other areas. Avoid cooling tower operation during night; use outside air for cooling the rooms if ventilation is needed during this time.
Install freon alarm unit in compressor room	and start ventilation in compressor room only if freon alarm is present.
Install control and regulation to lower condenser temperatures of cooling units when not used as a heat pump.	
Check freon for oil contamination.	Assure purity of refrigerant to maintain high efficiency.

CHECKLIST	MEDIUM-RETURN CHANGES
	HVAC: Eliminate Unnecessary Demand
Reduce areas with high requirements.	Separate areas according to use: computers with special environmental conditions should not be in a room which is larger than necessary. Carefully seal off areas with different requirements.
Install solar shading devices.	Install remote controlled venetian blinds. Adjust venetian blinds according to sunshine.
Reduce glass	or install insulating glass, replace high conductive window frames.
Repair building to reduce infiltration.	Seal penetrations, joints, windows, door frames; install revolving doors, automatic door closures.
Replace air curtain doors.	Air curtain doors have a very high energy consumption due to the large amount of air which has to be circulated.
Reduce Internal Loads	
Use local suction for internal heat sources.	Computers, heat-producing equipment (e.g., escalators), also for cooling equipment; reduce electrical consumption, operating time, heat production; install local suction.
Install air-cooled lighting.	Remove a major part of the cooling load by air- or water-cooled lighting systems.
Install spot cooling.	Use separate spot or zone cooling units for spaces which are used when the majority of the building is unoccupied; turn off main, central cooling system.
Improve controls on refrigeration system.	Install thermostats in refrigerated spaces to control operation of compressor. Set to highest allowable temperature.
Increase Component Efficiency	*HVAC: Reduce Air Distribution and Handling Losses*
Replace motors and other equipment with low efficiency;	for example, oversized motors, undersized air handling units. Install high efficiency replacements. Check for minimum life cycle cost including operating costs.
Avoid dew point humidification.	Replace dew point humidification with humidistat controlled system
Reduce pressure drop.	If obsolete equipment is replaced, reduce pressure drop of ducts (elbows, discontinuities, cross section, shorter air path, air handling units, filters). Reduce fan speed at the same time.

Increase System Efficiency	*HVAC: Reduce Air Distribution and Handling Losses (cont.)*
Adjust outside air rate.	Adjust minimum outside air rate to occupancy (e.g., as a function of time in a shopping center). Make sure that building pressure is slightly positive, to minimize infiltration of outside cool air and dust.
Use two-speed fans.	Install two-speed motors, variable-speed drives, or several units for optimum part load operation of pumps, blowers, fans, compressors.
Use CO control in garages.	Use CO controlled start-up of exhaust fans.
Use variable volume system.	Replace obsolete systems with variable volume system; modify terminal reheat systems.
Install building automation system.	
Replace air-cooled equipment with more efficient systems.	Replace air-cooled equipment with water-cooled systems if possible; build insulated enclosures or use hoods and duct air outside from heat-producing units, computer modules.
Install additional zone controls or zone systems	to permit selective use of HVAC according to actual load.
Provide additional controls	on multi-zone or dual-duct systems to adjust hot and cold deck temperatures to a level which will just meet the actual demand.
Install free cooling mode.	Use outside air to lower cooling water temperature (reduced use of cooling machine).
Replace forced air heaters with infrared heaters.	Can infrared heaters be used to put heat where it is needed, rather than heat a larger space unnecessarily?
Reduce heat gain of refrigerant line from condenser.	Install valves to permit bypassing refrigerant receiver to allow refrigerant to go directly to expansion valve without being warmed by hot gas in the receiver.

Reduce Use of New Energy	*HVAC: Increase Production Efficiency*
Transfer heat from where it is in surplus to where it is needed	from sunny side to shaded side; from internal to external parts; from heat sources (computer rooms, utility rooms) to heat consumers with low temperature requirements (e.g., base load heating)
Make optimum use of outside air.	Install units with variable outside air mixing, maximum 100 percent! (Also for computer rooms.)

Reduce Use of New Energy (cont.)	*HVAC: Increase Production Efficiency (cont.)*
	Duct cool, outside air directly to fume hoods, critical hot areas.
Install heat recovery.*	Use optimum recycling of exhaust air, but avoid danger of contamination; increase outside air proportion during temperate season and warm winter, cool summer conditions.
	Install rotary wheel exchangers when exhaust and inlet air ducts can be brought together; where contamination is to be avoided use heat pipe or cross flow heat exchanger system (glass if there are chemical corrosion problems).
	Use heat exchanger loop to transfer heat from any number of widely separated exhaust ports to heat exchangers in fresh air ducts. Use low-quality recovered heat for preheat and reheat.
Use direct cooling.	Use lake or river water for direct cooling instead of cooling tower.
Use condensor heat of cooling units.	Use low quality heat 125°F (50°C) for heating during night, preheaters in ventilation units, floor heating systems.
Check possibilities for storage facilities in the cooling system.	In some cases (especially with cheap night electric energy), cold storage for a few hours may be cost-saving. Change start-up procedure to lower peak of cooling need at start-up.
Check possible use of a total energy system for cooling.	The combined on-site production of electricity, heat, and cooling energy may be economical if the price ratio between electricity and oil (or gas) is high, and if there are additional safety requirements for a standby energy supply.
Use solar energy cooling/heating system.	With high electrical energy and oil (gas) prices, such a system may be economical (summer: cooling water; temperate season: domestic hot water, heating).
Replace air cooled condensers with cooling towers.	
Increase the number of cooling tower sections	to lower condenser temperatures.
Install automatic valves on cooling tower bleed.	Save water costs using minimum flow required.
Install valves on compressor.	Install adjustable head pressure control to permit setting to minimum pressure providing continuous and adequate supply of refrigerant to expansion valves.

*Listed in order of decreasing efficiency.

5.6 LIGHTING

Lighting uses approximately 20 percent of the electrical energy generated in the United States, or about 5 to 6 percent of the total national energy budget.* Commercial, public, and industrial buildings account for 70 percent of this use. (126) The use of lighting for various categories is shown in Table 5.29. The component for office lighting is 10 percent of the total, and hence it is 2 percent of the total electrical generation or .5 percent of the total energy use in the United States.

Overpromotion, design economies, and other factors have encouraged lighting design favoring constantly increasing illumination to superfluous levels, while previously decreasing utility rates encouraged wasteful customs, with the following results:

1. Many areas are still overilluminated despite recent delamping programs and other reductions.
2. Intense, uniform illumination has led to a poor optical quality of our surroundings and to problems with glare and poor contrast.
3. High intensity lighting is supplied with uniformity to areas where it is of little benefit.
4. Adequate switching is lacking in some cases, to turn lighting off when it is unneeded.
5. Opportunities for more efficient lighting production and delivery have been missed because of the previous abundance of cheap electricity.

*The total energy used for all categories of lighting is greater than that given in the Stanford Research Institute study. (127). According to more extensive analysis, approximately 400 billion kWh of the 2 trillion kWh of electricity generated in 1976, was used for lighting; this amounts to about one-twentieth of the total energy use in the United States. (126,128)

Table 5.29. Distribution of Lighting in the United States According to End Use

End use	*Lighting energy (%)*
Residential	20
Store	19
Industrial	19
Offices	10
Outdoor advertising, area lighting, sports lighting	8
Schools	7
Streets and highways	3
All other indoor (public bldgs., garages, hospitals, etc.)	14
	100

SOURCE: Ross and Baruzzini (126)

6. Dark color or dirty walls, ceilings, and luminaries show a lack of maintenance probably due to oversight rather than to cost-effective tradeoff of the increasingly expensive lumens being absorbed.

Lighting practice in modern European buildings has quite often followed similar practices, so that there are many opportunities to reduce costs and similar needs to improve the optical comfort of working environments. It is estimated that as much as 43 percent of lighting, or about 2 percent of the total United States energy use, can be saved by improved lighting practice. (126)

A quantitative description of current practice, provided by information from two surveys of office lighting use, is shown in Table 5.30. On the average, lighting represents 26 percent of the energy use of an office building. (10) In a 307-building survey, this ranged from a minimum of 7 percent to a maximum of 75 percent. The average illumination

Table 5.30. Lighting Data for Buildings Built before the Energy Price Shock

	Office space lighting data for 19 buildings[a]			*Building lighting data for 307 buildings*[b]		
	Average	*Min.*	*Max.*	*Average*	*Min.*	*Max.*
Illumination level, footcandles (lux)	101 (1090)	63 (680)	175 (1900)	88 (950)	30 (325)	200 (2150)
Lighting load (peak) W/ft^2 (W/m^2)	3.9 (42)	2.4 (26)	8 (86)	3.2 (34)	0.9 (9.7)	11.22 (121)
Illumination level per connected lighting load $\frac{\text{footcandles}}{\text{W/ft}^2}\left(\frac{\text{lux}}{\text{W/m}^2}\right)$	25.9	21.9	26.3	27.5	17.8	34.5
Measured for 9 buildings	30	16	47	—	—	—
Annual hours of office occupancy	2610[c]	—	—	3035	1403	5211
Annual lighting energy kWh/ft^2 (kW/m^2)	7.3 (78.6)	5.2 (56)	10.7 (115)	9.8 (105)	1.8 (19.4)	25.8 (278)
Lighting energy as a fraction of total building energy (%)	—	—	—	26	7	75
Annual cost for lighting at 5¢/kWh, $/1000 ft^2 ($/100m^2)	365 (393)	260 (280)	535 (575)	490 (525)	90 (97)	1290 (1400)
Annual cost of lighting per occupant at 5¢/kWh ($/person)	—	—	—	110	14.83	404

[a]Source: Ross and Baruzzini (126)
[b]Source: Building Owners and Managers Association, International (10)
[c]Estimated annual hours of office use.

in these buildings was 88 footcandles (950 Lux) as provided by lighting installed at a density of 3.2 W/ft^2 (34 W/m^2)*. The lighting was on 3,035 hours per year and used 9.8 kWh/ft^2 (105 kWh/m^2). From data shown in Table 5.30, the average illumination obtained at the time of these surveys was from 27 to 30 footcandles (Lux) per installed watt per unit area. Example cost data per unit area and per occupant is also shown in Table 5.30.

Some other typical annual lighting costs computed at 10¢/kWh:

Each duplex fluorescent (2 × 40W) fixture costs $20 a year to operate just during working hours.

A 150-watt security lamp continuously illuminated costs $130 a year.

The penalty for leaving the office lights on overnight on a 20,000 ft^2 (1860 m^2) floor is $100, or $500 if they are left on over the weekend.

Details and other examples are shown in the box below. These examples show that the cost for unneeded lighting is often much greater than the cost of photosensitive and interval switches, or adding new master floor and local switches.

*One footcandle = 10.76 Lux. This is frequently approximated as 1 Lux = 0.1 fc.

Some Examples of the Costs for Lighting

Costs below evaluated at 10¢/kWh.

1.	A 2-lamp fluorescent fixture, on only during working hours (2 × 40 watts + 16 watt ballast transformer, 96 watts × 2080 hours/year × 5¢/kWh)	$20.00
2.	A 2-lamp fluorescent fixture, operating during 3,035 hours/year[a]	$30.00
3.	The annual cost for each 60-watt lamp burning continuously in a lavatory	$52.00
4.	The annual cost for 150-watt lamp burning continuously for security purposes	$130.00
5.	Leaving the lights on overnight on a 20,000 ft^2 floor (3.9 W/ft^{2a} × 20,000 ft^2 × 13 hours × 10¢/kWh)	$100.00
6.	Leaving the lights on over the weekend (3.9W/ft^2 × 20,000 ft^2 × 64 hours × 10¢/kWh)	$500.00
7.	The annual costs for elevator lighting (4 × 40 watts + 32 watts ballast × 8,760 hours 10¢/kWh)	$170.00
8.	The annual cost for each kilowatt of continuous connected lighting	$880.00
9.	The annual cost to power the ballast transformers in 100 hallway lighting fixtures from which the tubes have been removed.	$1400.00

10.	The annual cost if the lamps in no. 9 have not been removed.	$8400.00
11.	The lifetime operating cost of a 40-watt fluorescent lamp . . . (18,000 hours average lifetime burning 3 hours per start to 33,000 hour lifetime if on continuously.)	36 to 65 times purchase cost
12.	The lifetime operating cost of a 100-watt incandescent bulb (lifetime approximately 850 hours.)	$5.00

[a]See data in Table 5.30.

Examples of six different lighting arrangements are shown in Table 5.31, together with a comparison of the connected load, the source illumination, and the cost of operation for 1,530 hours per year. (40) Each of these systems is common, and all provide more than adequate illumination, but the connected wattage varies from 9 W/ft^2 to 1.7 W/ft^2 (100 W/m^2 to 18 W/m^2). Consequently, the cost of operation for the incandescent lighting system with the higher load is a factor of 4 greater than the most efficient system shown.

The large variation among these systems is explained by differences in the efficiencies of the lamps, the fixtures, and over-illumination. Incandescent lights require 55 to 70 watts to produce 1,000 lumens of light. Fluorescent bulbs use 12 to 20 watts to produce the same amount of light. System E draws 3.9 W/ft^2 (43 W/m^2) and produces more light than needed to meet an already generous code requirement. System B needs only 1.9 W/ft^2 (21 W/m^2). Furthermore, inefficiencies in the diffusers used to provide glare protection result in low utilization of the light emitted by the bulbs. Systems D and E are examples of low-efficiency diffusers. Note that sufficient lamps are installed to produce 242 and 271 footcandles (2,600 Lux and 2,915 Lux), respectively, inside the fixtures.

In addition to the direct cost of lighting, lighting is the major factor contributing to the expense of operating air conditioning in most office buildings. In the example of the ten-story New York office building shown in Table 5.25 lighting was 60 percent of the air-conditioning load. As a rule, when air-conditioning equipment is operating, each 100 watts of lighting causes an additional expenditure of about 30 to 50 watts in the air-conditioning system. It will be argued by some that, for all electric buildings at least, a decrease in lighting must be matched by an increase in heating. This argument is uninformed and false. First, the heat-of-light is concentrated in the ceiling, where it has an adverse effect on the operation and lifetime of the lighting equipment, rather than where it is required for comfort purposes. Second, even if the heat is partially reclaimed from lighting, the remainder must be removed from the core areas of the building at a net additional cost.

The effect of reducing lighting levels* is illustrated by two studies shown in Table 5.32. The first study shown, conducted by the Rand Corporation of a typical high-rise

*There may be no saving for some air-conditioning systems with reheat. (See Section 5.5).

Table 5.31. Typical Annual Costs for Different Lighting Arrangements Used in Some Massachusetts Schoolrooms

Fixture	*Ceiling plan*	*Total lamp wattage/room*	*Connected lighting*[b]	*Light*[a] *per watt*	*Annual energy cost*[c]
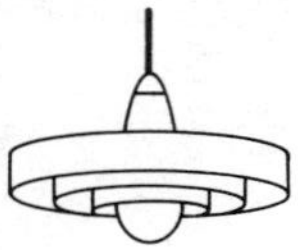	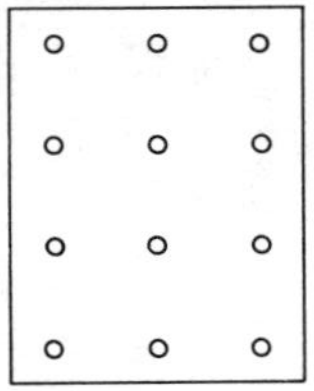	500 watts/lamp 1 lamp fixture 12 fixtures/rm. 6,000 watts/rm.	6,000 watts 190 fc (2,050 Lux)	7.7 fc/W (83 Lux/W)	$880.00
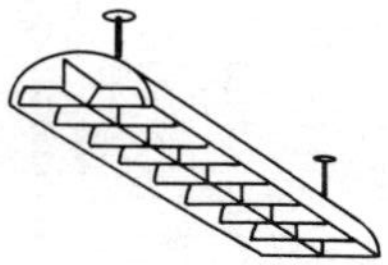	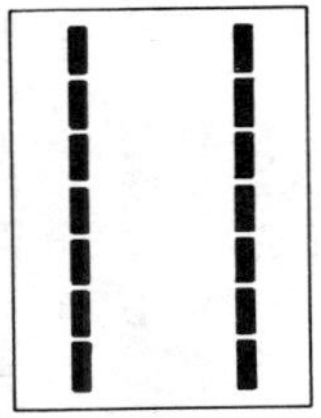	40 watts/lamp 2 lamps/fixture 14 fixtures/rm. 1,120 watts/rm.	1,316 watts 135 fc (1,450 Lux)	35 fc/W (380 Lux/W)	$200.00
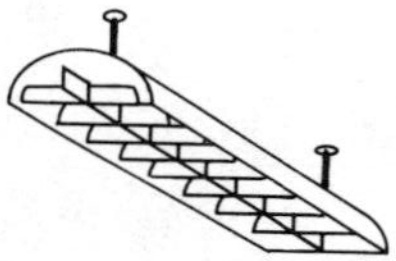	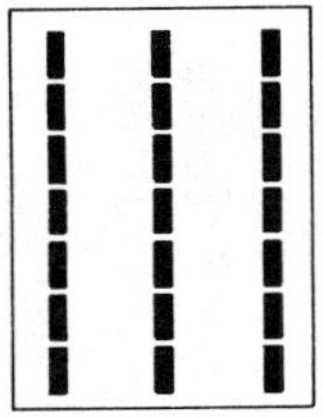	40 watts/lamp 2 lamps/fixture 21 fixtures/rm. 1,680 watts/rm.	1,974 watts 205 fc (2,185 Lux)	23 fc/W (250 Lux/W)	$300.00

Fixture	Ceiling plan	Total lamp wattage/room	Connected lighting[b]	Light[a] per watt	Annual energy cost[c]
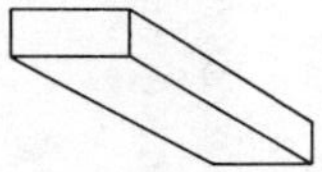		40 watts/lamp 2 lamps/fixture 25 fixtures/rm. 2,000 watts/rm.	2,350 watts 242 fc (2,600 Lux)	20 fc/W (210 Lux/W)	$360.00
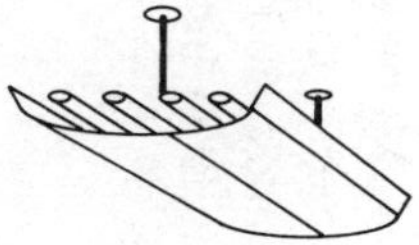	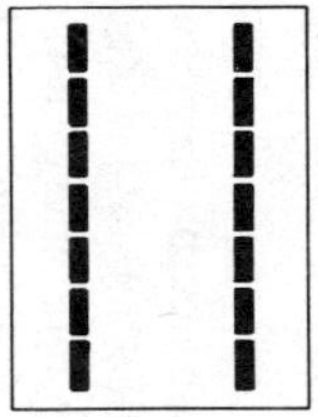	40 watts/lamp 4 lamps/fixture 14 fixtures/rm. 2,240 watts/rm.	2,632 watts 271 fc (2,915 Lux)	18 fc/W (190 Lux/W)	$400.00
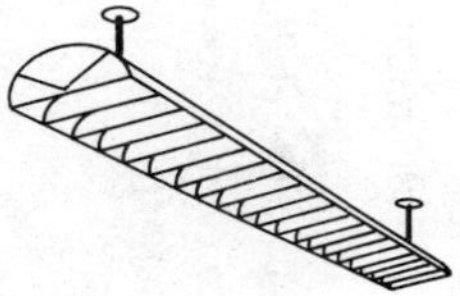	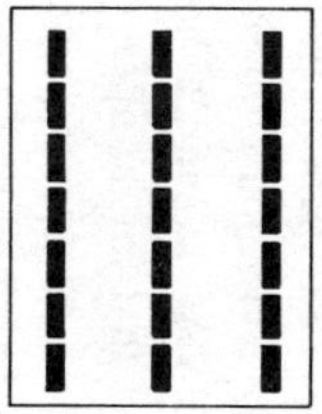	75 watts/lamp 2 lamps/fixture 9 fixtures/rm. 1,350 watts/rm.	1,602 watts 169 fc (1,820 Lux)	29 fc/W (310 Lux/W)	$240.00

[a]To supply 70 footcandles (700 Lux).
[b]Includes power consumed by ballasts.
[c]Based upon annual usage of 1530 hours; cost of electricity @ $0.10 per kWh; room area 660 square feet (60 m^2).

After Massachusetts Department of Community Affairs (40)

Table 5.32. The Effect of Reduced Lighting on Annual Heating and Cooling Requirements

		Annual			
Example	*Connected lighting load, W/ft² (W/m²)*	*Lighting, KBtu/ft² (kWh/m²)*	*Heating, KBtu/ft³ (kWh/m²)*	*Cooling and air transport, KBtu/ft² (kWh/m²)*	*Sum, KBtu/ft² (kWh/m²)*
High-rise office building with 40% glass in New York. (129)	4.3 (46) 2.1 (23) 50% reduction	56 (177) 28 (88) 50% reduction	14.4 (45.4) 17.9 (56.4) 24% increase	71.5 (225) 40.6 (128) 43% reduction	142 (447) 86.5 (273) 39% reduction
Interior zones of an open plan office building in Zurich. (130)	4 (42) 2.3 (25) 40% reduction	33 (105) 20 (62) 40% reduction	46.3 (146) 48 (151) 3% increase	16.5 (52) 14.3 (45) 14% reduction	96 (303) 82 (258) 15% reduction[a]

Note: On the average, for each 1 kilowatt-hour of reduced lighting, there will be an additional ¼ kilowatt-hour saving in reduced cooling and air transport. This saving will be realized, however, only if the air flow rates are reduced along with the lighting.

[a]There is a 24 percent reduction in raw source energy, of which 20 percent is due to reduced lighting and 4 percent to cooling and air transport.

office building in New York City, finds that a 50 percent reduction in lighting demand leads to a 40 percent reduction in overall energy demand for heating, cooling, and lighting. (129) The impact on the HVAC energy use depends also on climate and the details of the HVAC system. This is illustrated by the second study shown in Table 5.30, which found that for an open plan office building in Zurich, a 40 percent reduction in lighting led to a 15 percent reduction in total energy use. The results from a computer study of lighting, done on a national scale and accounting for the effect of different climates, also confirms the conclusion that lighting is a poor way to heat a building. Although for each kilowatt of decreased lighting power, 140 watts of additional heat energy were required on an annual average, this was almost completely offset by a decrease of 130 watts required by the chiller. (126) This study shows that there is net saving of 4 kWh of raw resource energy for each kilowatt-hour of lighting reduction, even when the additional fuel for heat is included.

Improving Lighting Efficiency

As with other energy conversion systems we have discussed, savings can be achieved with a concerted program focusing on reducing demand, improving the delivery of lighting, and increasing the efficiency with which lighting is produced. We will also include some guidelines for planning new lighting with reduced glare for a modernization program or for new construction.

New codes have been enacted to limit lighting in many areas. It is possible to achieve 70 footcandles (750 Lux) at desk level with a connected load of less than 1 W/ft^2 (10 W/m^2), as shown on page 239. This is a factor of 2 to 3 below the *average* specified by new codes. The measures discussed in this section should help meet any new code, provide more comfortable lighting with reduced glare, and help save money.

Turning off unnecessary indoor and outdoor lighting is one of the easiest and most obvious ways of cutting lighting costs. A practical example of such a saving is described below. Turning off fluorescent lamps when leaving the room not only saves energy; it also means that, contrary to popular belief, the lamps will need to be replaced less frequently than if left burning continuously.*

*Fluorescent lamps burning continuously 12 hours per day, 6 days a week, would last about 7 years. If operated at an average of 3 hours per day (as needed), they would last 19 years. Of greater significance is the fact that @ 10¢/kWh, the operating cost is reduced from $18 to $4.50 per year for each 40-watt tube. Futhermore, if turned off during a 15-minute absence, the electricity saved is worth 5 to 10 times more than the cost of the shortened life of the lamp.

An Example of Reducing Costs by Eliminating Nonessential Lighting

Energy may be conserved by reducing non-essential lighting. Such an opportunity to reduce costs by almost $7800 per year was identified by a fabric slip-cover plant that operates one eight-hour shift per day, five days per week. The plant consists of a 4800 sq ft office building attached to a 100,000 sq ft manufacturing building. Both buildings are lighted by a total of 2,200 eight-foot-long fluorescent lamps (110 watt plus 10 watt ballast per lamp). The normal practice of this plant was to leave all lights burning Monday through Friday, 24 hours per day, 50 weeks per year.

A lighting plan was instituted for non-operating hours which provided illumination only for required maintenance and security. The elimination of non-essential lighting resulted in a lighting plan equivalent to burning all lights only 10 hours per day, 250 days per year.

An annual power saving of $7090 was achieved, plus a reduction in lamp replacement cost of $710, for a total saving of $7800 per year.

The electrical energy saving by the new plan is calculated as follows:

$$\text{Annual Power Saving} = (24 - 10)\text{ h/d} \times 250\text{ d/yr} \times 120\text{ w/lamp} \times 2200\text{ lamps}$$

$$= 924{,}000\text{ kWh per year}$$

At a cost for electricity of $0.007677 kWh

$$\text{Annual Power Cost Saving} = 924{,}000 \text{ kWh/yr} \times 0.007677 \text{ \$/kWh}$$
$$= \$7090 \text{ per year}$$

If the utility consumes 10,000 Btu of fuel per kWh,

$$\text{Annual Energy Saving} = 924{,}000 \text{ kWh/yr} \times 10{,}000 \text{ Btu/kWh}$$
$$= 9240 \text{ MBtu/yr}$$

In addition to power costs for the two different schedules, one must consider lamp replacement costs. This depends, among other things, on the actual operating cycle. The annual replacement cost for fluorescent lamps may be estimated as follows:

$$\text{Annual Replacement Cost} = (P + h) \times (C\ d/L) \times n$$

Where P = price of replacement lamp
h = labor cost of replacing lamps
C = lamp life hours consumed by each operating period (hours per start) as estimated from the figure
d = number of operating periods per year
L = average lamp life at 3 hours per start
n = number of lamps

For this case the labor cost (h) is $2.00, and the price ($P$) is $1.42 per lamp with average life (L) of 12,000 hours. As stated previously, there are 2200 lamps (n).

For the old plan, there were $5 \times 24 = 120$ hours per operating period; hence from the figure, C = approx. 60 h. There were 50 operating periods (d) per year.

$$\text{Replacement cost (old)} = (2.00 + 1.42) \text{ \$/lamp} \times 60 \text{ h/period} \times 50 \text{ periods} \times 2200 \text{ lamps/12000 h}$$
$$= \$1880 \text{ per year}$$

For the new plan there are 250 periods (d) of 10 hours each; C from the figure is 7.5 h.

$$\text{Replacement cost (new)} = (2.00 + 1.42) \text{ \$/lamp} \times 7.5 \text{ h/period} \times 250 \text{ periods} \times 2200 \text{ lamps/12000 h}$$
$$= \$1170 \text{ per year}$$
$$\text{Replacement saving (net)} = \$1880 - \$1170$$
$$= \$710 \text{ per year}$$
$$\text{Total annual saving} = \$7090 + \$710$$
$$= \$7800 \text{ per year}$$

Electrical costs higher than rate used here will obviously result in higher savings.

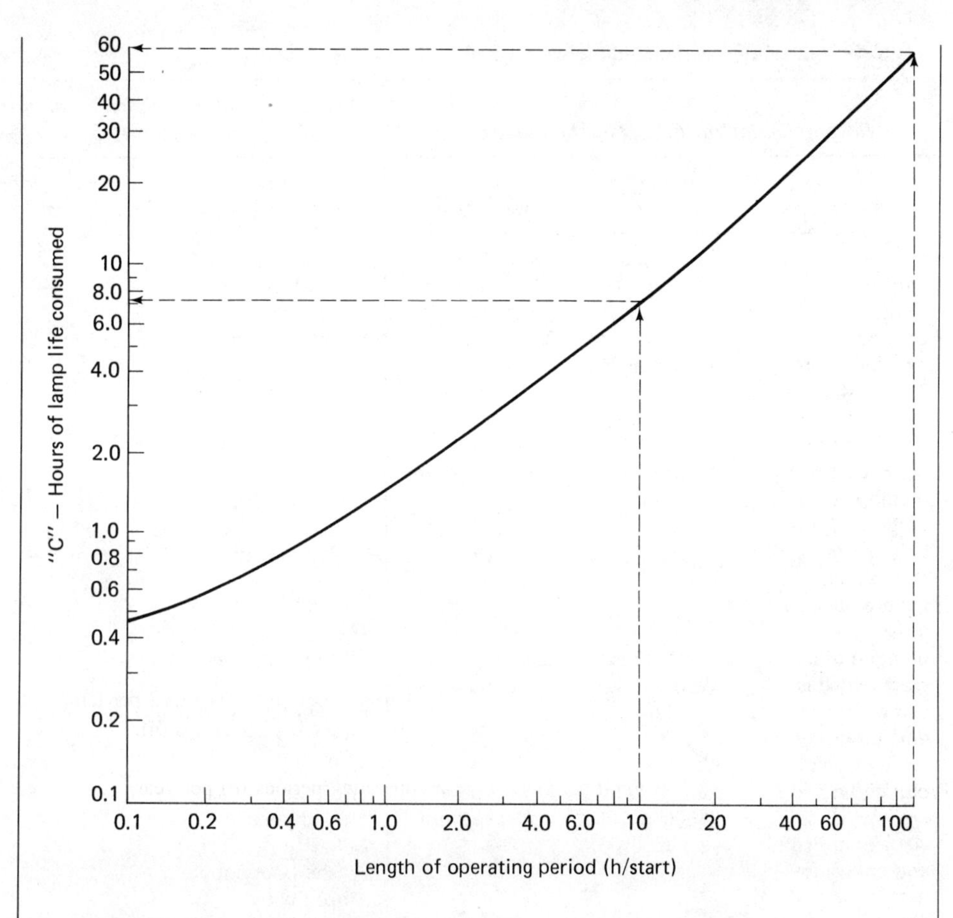

Typical lamp life as a function of the length of the burning period for 40 watt rapid start fluorescent lamps. [National Bureau of Standards (131)]

SUGGESTED ACTION

Review your lighting practices and eliminate all nonessential lighting during working hours. During nonworking hours, operate only those lights that are required for plant security and the performance of janitorial services. If possible, schedule all janitorial services to be accomplished simultaneously by area to minimize the use of lights throughout the plant for long periods of time. Install more switches to increase control of lighting. Consult your utility company, lighting consultant and manufacturer for recommendations.

SOURCE: National Bureau of Standards. (131)

Table 5.33. Typical Recommended Lighting Levels

Recommended maximum lighting levels			*Relative visual task difficulty for common office tasks*	
Task or area	*Illumination level*	*How measured*	*Task description*	*Visual difficulty rating*
Hallways or corridors	10 ± 5 fc (100 ± 50 Lux)	Measured average, min. 1 footcandle	Large black object on white background	1
Work and circulation areas surrounding work stations	30 ± 5 fc (300 ± 50 Lux)	Measured average	Book or magazine, printed matter, 8 point type and larger	2
Normal office work, such as reading & writing (on task only) store shelves, and general display areas	50 ± 10 fc (500 ± 10 Lux)	Measured at work station	Typed original	2
			Ink writing (script)	3
			Newspaper text	4
Prolonged office work which is somewhat difficult visually (on task only)	75 ± 15 fc (750 ± 150 Lux)	Measured at work station	Shorthand notes, ink	4
			Handwriting (script) in no. 2 pencil	5
Prolonged office work which is visually difficult and critical in nature (on task only)	100 ± 20 fc (1000 ± 200 Lux)	Measured at work station	Shorthand notes, in no. 3 pencil	6
			Washed-out copy from copy machine	7
Industrial tasks	[a]	As maximum	Bookkeeping	8
			Drafting	8
			Telephone directory	12
			Typed carbon, fifth copy	15

Note: The level of specific task illumination should be increased by 50 percent for someone 50 years or older, or for someone with uncorrectable eyesight problems. The level of illumination for a combination of different office tasks can be determined using the visual difficulty factors given above. Multiply the difficulty factor by the time spent at each task during an average day, and add the products together.[b] If the sum is greater than 40, the task should be illuminated with 80 footcandles (800 Lux). If the sum is greater than 60, the task should be illuminated with 100 footcandles (1,000 Lux). The use of the visual difficulty rating system is illustrated with the following examples:

Example A: An office worker spends 3.75 hours typing (visual difficulty 2) and 4.5 hours handwriting in number 2 pencil (visual difficulty 5). The sum of the products of time and visual difficulty are $(3.75 \times 2) + (4.5 \times 5) = 30$. Since this is less than 40, this person's work area should be illuminated with 50 footcandles (500 Lux).

Example B: If the person in the example above were older than 50 years, then 50 percent more illumi-

If people are to turn lighting off when it is not needed, switches must be convenient and clearly labeled. It is best to provide automatic control together with manual switching to override auto-control when local conditions change.

Costs can be reduced by installing timing mechanisms, dimmers, or automatic photocell devices. Timers can be installed easily on outdoor lighting fixtures—in parking lots, for example—or indoors in rooms that are used only briefly, such as storage and equipment rooms. Dimmers make it possible to vary the lighting level in a room; however, if that variability in illumination is not needed, eliminating some lamps or replacing lamps with those having a lower illumination is considerably more energy efficient.*

Automatic devices that turn lights on (when a door is opened, for example), are efficient for rooms not constantly in use, such as storage closets. Elevator and phone booth lighting can be controlled to turn on when the door is opened or a person is inside. Photocell switches can be installed to turn on lighting automatically when natural lighting dims at dusk or on a rainy day. These switches are useful both indoors and outdoors, and in entrances, offices, and other areas that use natural light for part of their illumination. Motion sensors, which will turn lights off after the last person has left the room, are good for lecture halls, etc., where responsibility for lighting control is absent.

It is important to recognize that tasks which require recognition of fine detail or tasks with low-contrast subject matter may require higher lighting levels *at the task,* than jobs with lower visual difficulty. Younger people, and people with no visual handicaps generally, require lower levels than, for example, older persons. If it is possible to do so without impairing the visual quality for neighboring people, it would be preferable in some cases to allow people to use individual task lights to optimize the intensity and orientation of their illumination. It would then be necessary to provide only lower-level, general illumination. In this way eye fatigue can be reduced if the illumination can be reduced for less demanding tasks.

Guidelines for lighting levels as recommended by the General Services Administration and the Department of Energy are given in Table 5.33, along with some examples of their application.† (15,48,126) In the description of office task difficulty, it is recognized

*Dimmers on incandescent lamp circuits are very inefficient. Lighting must be reduced by a factor of five in order to obtain a factor of two reduction in electricity consumption.

†There is a difference between these guidelines and lighting which can be derived from the ASHRAE 90-75 standard. (4) Lighting loads calculated by the 90-75 method yield essentially the same lighting power density as for designs commonly used before saving energy became an issue. Many people feel that the lighting standards as expressed in DIN 5035 in Germany are also too high.

---

nation, or 75 footcandles, should be provided. This lighting could be provided with an adjustable lamp at this person's desk.

Example C: A draftsman spends $7^1/_2$ hours a day at his work. The product of the visual difficulty (8), and the time (7.5 hours) is 60, so that 100 fc (1,000 Lux) should be provided. This level of illumination should be provided with an adjustable-arm lamp at the drafting table.

[a]Levels for industrial work are from the American National Standards Institute All.1–1973, Practice for Industrial Lighting.

[b]Time is to be computed in decimal hours, that is 3 hours 15 minutes = 3.25 hours.

SOURCE: Department of Energy.

that the amount of lighting required is a function of the optical properties of the work and the length of time at the task. Although with reduced veiling reflections someone with good vision could perform these tasks with less than one-third of the guideline illumination, it is felt that the levels recommended represent a practical compromise between comfort under typical conditions and more Spartan energy-saving.

To summarize Table 5.33, during working hours in office buildings, administrative spaces, retail establishments, schools, and warehouses, recommended illumination levels are 50 footcandles (500 Lux) at occupied work stations, 30 footcandles (300 Lux) in general work areas, and less than 10 footcandles (100 Lux) in areas that are seldom occupied or which have minimal visual requirements, such as hallways and corridors. Where needed, because of exceptional individual requirements or because of the difficult nature of a particular task, specific supplemental lighting should be provided for the duration of the task according to the levels indicated on page 222. Conveniently located individual switches should be provided to permit maximum control over both standard and supplemental lighting when not needed.

With present lighting systems, the 50/30/10 footcandles (500/300/100 Lux) levels of illumination should be achievable at connected lighting loads of 1.5 W/ft^2 (16 W/m^2) for 50 fc (500 Lux) in open plan offices, and 2 W/ft^2 (22 W/m^2) for 50 fc (500 Lux) in individual offices. Because the utilization factor for individual offices is lower than for larger rooms, 30 percent increased density may be required with the same lighting system to achieve the same lighting levels. These levels may be compared with 3.2 to 3.9 W/ft^2 (34 to 42 W/m^2) for the two averages from the lighting surveys of previous design discussed in Table 5.30. For comparison, the recommended lighting levels were specified for the Demonstration Federal Office Building in Manchester, New Hampshire, at a connected lighting load of 2 W/ft^2 (22 W/m^2). (132) At the end of this section, a new lighting design is given in which the recommended lighting levels would be obtained with a connected load of 1 W/ft^2 (11 W/m^2). (126)

The light levels referred to in Table 5.33 can be determined with portable illumination meters such as a photovoltiac cell connected to a meter calibrated in footcandles. These guidelines refer to average maintained horizontal footcandles at the task or in a horizontal place 30 inches above the floor. Measurements of work areas and nonwork areas should be made at representative points between fixtures in halls, corridors, and circulation areas. Daylight should be included but attention should be given to determination of levels when the lighting is used without available daylight.

Many existing buildings have been designed to provide a consistently high level of illumination in all areas. The designer's intention was to provide maximum flexibility in the use of rooms, by providing all areas with enough light to meet the most difficult visual tasks. The advantage of this flexibility is offset by the high costs of energy required to operate the system.

In many areas, lamps may simply be removed from fixtures to reduce light levels to the levels in Table 5.33. This is particularly true with rows of ceiling lights. Removing lamps from behind work areas will help keep tasks free of shadows. Check that lighting remaining in front of work areas does not contribute to glare.

If you are removing fluorescent lamps, have an electrician disconnect the primary side of the ballast transformer, which continues to draw some energy even after the lamp is removed. As one ballast is often shared by two or more fluorescent lights, removing one lamp will cause the others wired in series with the same ballast transformer to go out. Replace one of the two tubes with a "phantom tube" to maintain the operation of the other lamp at two-thirds of its original illumination. With the phantom tube in place, the electricity use will be approximately two-fifths (possibly even less) of that drawn previously by the two fluorescent tube fixture.

An important point to remember in lamp removal is that the level of room lighting is not proportional to the number of lamps in the room. For example, if half of the lamps are removed, the illumination may be decreased by as little as 20 percent. This will depend upon actual characteristics of the room and the lighting installation. The savings from removing lamps are easy to achieve. The dollar value will depend on the number and wattage of the lamps removed. Following the example given on page 214, the annual savings per lamp removed can be estimated.

Lower wattage lamps may be substituted in existing fixtures in situations where the removal of a lamp would be impractical. Check with your supplier for a current list of the variety of cost-saving incandescent, fluorescent and discharge lamp substitutes. Some of these new lamps will deliver more light from some luminaires, even while drawing fewer watts.

Improve the Delivery of Lighting

Perhaps optimizing delivery is one of the most neglected areas of lighting. Glare, dark walls, shadows, dusty fixtures, and dirty diffusers and ceilings waste light. Better positioning of fewer fixtures to direct their light onto the task can result in better illumination and productivity with much less cost and energy waste.

The design and arrangement of the lighting fixtures in a room and the colors and reflectivity of the walls, floor, and ceiling determine how much of the light produced by the lamps actually reaches the task area. A coefficient of utilization (CU) of .50 means that only half of the light supplied by the lamps is useful in performing a visual task; the remainder is absorbed by the fixture, the walls, and the ceiling before it can be used. For fixtures mounted on or recessed into the ceiling, only 45 percent to 65 percent of the light may actually reach the task area. For suspended fixtures, the utilization is about the same if the ceiling remains highly reflective. If the ceiling is only 50 percent reflective, only 35 percent to 45 percent of the lamp output reaches the task level.

Painting a dirty ceiling white will provide additional illumination for less cost than extra lamps burning in their sockets. Morgan Christensen, an industrial lighting specialist with the General Electric Company, cites a number of examples of improvement of between 15 to 40 footcandles (160 to 430 Lux) just from repainting the ceilings. (133) As the distance between the lamp and the task is reduced by half, the required lamp output is reduced by four times. In other words, a 20-watt lamp two feet away from a task provides as much light as an 80-watt lamp four feet away. Bringing fixtures closer to task areas in high-ceilinged rooms can lower the wattage for the illumination required. For example, a supermarket illuminated at 150 footcandles (1,500 Lux) from fixtures attached to a ceiling 18-ft (5.5 m) high can substantially cut its lighting cost by using far fewer fixtures; more than enough light is provided for customers to read the labels of goods on lower shelves by lowering the mounting height to 13 ft (4 m).

The placement of lamps in relation to task areas is difficult to anticipate when a building is designed. As a result, many offices, factories, stores, and schools are lit so that a uniformly high level of light falls on all areas. **Savings can be achieved by reducing the general light levels and providing supplemental light sources closer to the task surfaces.** Studies of task lighting by a consulting engineer in two school libraries and a town hall showed that between 44 percent and 64 percent of the existing lighting energy could be saved by reducing overhead lighting by half and using small table lamps. (40) If the general illumination level in stores were reduced, highlighting and other special lighting effects would become much more attractive. These effects should be taken into account when a lighting system is converted.

Improved lighting maintenance is also an inexpensive way to maintain the required illumination with fewer connected lamps. Several surveys have shown that the average lighting system in factory buildings delivers less than half of its original illumination. (133) Christensen tells of one maintenance contractor who would offer to charge his clients nothing if he could not double their amount of light simply by washing their fixtures and replacing their lamps.

Glare

Eliminating glare is important in improving the comfort of surroundings. When glare originates from artificial lighting, it represents misapplied energy, or energy waste in one of its worst forms. "Discomfort glare" is caused by a bright source of direct or reflected illumination in the field of view. It may originate from sunlight, direct light from an adjacent luminaire, or from a glass surface which reflects light into the eyes. A more subtle type of glare is called "veiling reflection." This type of glare causes a washing-out of the task contrast. It has been estimated that these two types of glare can cause a change of nearly 10-to-1 in effectiveness between two different types of common lighting systems. (126) The importance of eliminating glare as part of energy saving can be appreciated with the following example. Between two lighting systems, each providing 50 footcandles (500 Lux) at the task, the poorer system, because of glare, may be no more effective than the glare-free system at 5 footcandles (50 Lux).

The angular placement of the fixture, together with the distribution of light from the fixture, influences glare. Overhead and forward lighting cause high veiling reflections, while lighting from the side reduces veiling reflections. This can mean that one lighting fixture between two desks may be more effective than one fixture over each. High-intensity lights should be far away from a task, while conventional lights may be closer. It should also be noted in considering the efficiency of lighting, that the optical properties of the task greatly affect the amount of light required. Large-size, high-contrast objects on a dull mat background will cause fewer problems with glare and require less lighting for comfort.

The performance of the light fixture is important. Light is absorbed by poorly designed or aging diffusers and by reflecting surfaces that have deteriorated. Since the distribution of light in avoiding glare can make a 10-to-1 difference in the visual effectiveness of an illumination level, the light control characteristics of a luminaire deserve attention. Light directed at shallow angles causes veiling reflections. Light directed at steep angles begins to contribute to discomfort glare. As a result, a number of light control lenses have been designed to provide light distribution within a range of 40° to 60° from the vertical—the so-called "bat-wing" distribution. Some of these luminaires have as many as three separate prism types to redirect some of the light that would otherwise be distributed improperly into the preferred 40° to 50° vertical angular range. As illustrated on page 214, the operating costs of lighting soon exceed the first costs of the lamp and its fixture.

With the absence of glare and the presence of comfortable optical surroundings such an important part of productivity, good delivery of lighting is worth care, the time required for its design by trained professionals, and a greater initial cost for the system. Money for energy can be saved and productivity improved by replacing ineffective and obsolete fixtures with good luminaires during a modernization program.

Figure 5.29 illustrates an example of an air-cooled lighting fixture system with parabolic optical control elements providing better glare control, and reduced loading on the

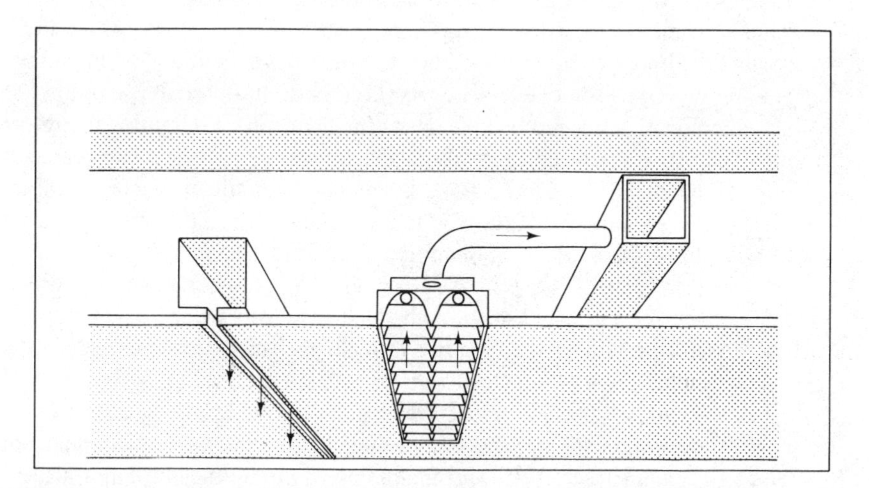

Figure 5.29. Air-cooled, controlled light distribution fixture integrated with the HVAC system. By controlling the distribution of light within the range of 40° to 60° from the vertical, annoying glare can be reduced. Without glare, tasks can be accomplished with less eye fatigue at lower light levels. Room air is exhausted through these luminaires to keep them cooler for higher light output and longer life. In winter this heat can be recirculated; in summer it can be exhausted from the building so that it does not become part of the chiller load. (12)

air-conditioning system. Part of the heat produced as a byproduct of every lighting system is extracted through the fixture, rather than being given off as cooling load. The air-cooled luminaire operates at lower temperatures, and therefore lasts longer and gives more lumens per watt. Because the heat from lighting is collected at a higher temperature than if allowed to dissipate into the general ceiling area, both heat rejection or recovering this heat is more efficient. In some systems the light fixtures are water-cooled and the heat of light is picked up by a piping network that also serves as the sprinkler system in case of fire. The water temperature from some lighting systems is high enough to be rejected directly by the cooling tower, when it is not needed elsewhere, so that the heat of light does not become a chiller load.

More Efficient Production of Lighting

Money can also be saved by using more efficient lamps. The comparison of some of the different light sources in Table 5.34 shows that incandescent lamps, which use 55 percent of the electricity for lighting in the United States, are very inefficient. As a consequence, most actual illumination is produced by fluorescent and discharge lamps which are 4 to 5 times more efficient than incandescent bulbs.

Incandescent lamps convert only 5 percent of the energy they use to light, and depending on the efficiency of their fixture, only as little as 50 percent of this may reach the task. Incandescent lamps are useful for supplementary lighting on tasks where mounting flexibility is desirable. "Long life" incandescents are the most inefficient type in comparison with any other lamp of the same wattage. Except where a very low-wattage lamp is required, a special mercury lamp with a self contained ballast and screw-in base would be a more cost-effective choice where color rendition is important. See Figure 5.30 for a comparison of efficiency vs. lifetime for different types of light sources. (126)

Fluorescent lamps provide the most efficient choice for lighting in low-wattage lamps, provided that aesthetic concerns for shape are not critical. New lamps are available which offer excellent color rendition and a lumen per watt efficiency as high as any of the older versions.* Furthermore, these new tubes maintain this high output for a greater number of hours and over a wider ambient temperature range.

There are several types of self-contained fluorescent lamps for use as a screw-in replacement for an incandescent bulb. One type uses a lightweight, self-contained transistor oscillator in place of a ballast to make an efficient fluorescent light source that is a direct replacement for the 100-watt incandescent lamp. (127)

In the design of new lighting systems and in the improvement of existing ones, the most efficient light sources that can provide the illumination required should be selected. The cost savings that result from upgrading to more efficient lamps are shown in Table 5.35. The strategy for replacing incandescent, general-area lighting with fluorescent is

*The new generation "warm white deluxe" has a color temperature of 3000 K, the "white deluxe" 4000 K, and the daylight deluxe 6000 K, respectively, at an efficiency of 80 lumens per watt.

Table 5.34. Types of Lighting Compared

Type & wattage	*Lumens[a] per watt (L/W)*	*Lifetime*	*Lumen efficiency*	*Equipment cost*	*Operating cost*	*Color characteristics*	*Recommended uses*	*Remarks*
					INCANDESCENT			
Standard incandescent								
15	8	750–1000 h shortest of all lamps	80% prior to failure	Low	High	Nearest to natural daylight. Skin tones heightened. Gives "warm" atmosphere where used	Task lighting, supplementary lighting, emergency lighting	
25	9							
40	12							
60	14							
75	18							
100	18							
250	20							
500	21							
Long-life								
100	13	2.3, or 5 yr		High	High		Not recommended	Produces less light than regular 75W incandescent. Replace with, e.g., screw-in mercury.
PAR	10.2		Reduced to 70% after 1,000 h				As narrow beam floodlights	PAR types have higher beam candle power than regular reflector lamps. Because they are filled with krypton gas, they have longer life.
75								
100								
150	11.6							
250	18.4							
Tungsten/halogen								
45	13	4,000 h min. for high-voltage lamps	90% after 3,000 h	Low	Low	Good color rendition: bright, white	Where strong light and good color are desired. General lighting for large rooms, production areas. In cornices and niches.	Low wattages available for single-purpose lamps. Not as flexible as standard incandescent.
100	18							
150	18							
200	19							
250 Spot	13							
500 Flood	14							
1,000 Flood	17							
Open gas flame	0.2						Listed for reference	
Gas mantle	1–2							

Table 5.34. Types of Lighting Compared *(cont.)*

Type & wattage	*Lumens[a] per watt (L/W)*	*Lifetime*	*Lumen efficiency*	*Equipment cost*	*Operating cost*	*Color characteristics*	*Recommended uses*	*Remarks*
					FLUORESCENT			
20–30 40 60 75 110	60 66 68 73 72	20,000 h	70% at 12,000–15,000 h	High Higher than incan-descent	Low Lower than incan-descent	Cool white. Warm white has the poorest color rendition; cool and deluxe warm white are better. Deluxe cool white most closely approximates natural daylight.	Replacement for 100W incandescent lamp. In production areas, offices, as display lighting. Deluxe cool white, deluxe warm, and white can be usually used in place of incandescent bulbs.	"Lite K" has screw base and self-contained ballast in familiar incandescent bulb shape. No separate ballast is required. Color-corrected lamps are available with the high efficiency of cool white. Efficiency affected by ambient temperature. Cool and warm white have high outputs. New, more expensive color-corrected lamps have high efficiency and are less temperature-sensitive.
High-efficiency fluorescent								
35 60	80 85							
					HIGH-INTENSITY DISCHARGE			
Mercury								
40 100 175 250 400	29 41 42 46 51	24,000 h	75% after 16,000 h	Low	Medium	Available in clear, white, color-corrected, and deluxe white. Deluxe white has best color rendition. Deluxe white interchangeable with cool white fluorescent	Replacement for long life incandescent. Indoors to light large spaces, manufacturing & production areas. Outdoors in parking areas and for merchandising or decorative lighting	Self ballasted types available with screw-in base. Cannot be dimmed; voltage requirements are precise. Not sensitive to frequent start-ups. Available with screw-in, self ballasted base.

Type & wattage	Lumens[a] per watt (L/W)	Lifetime	Lumen efficiency	Equipment cost	Operating cost	Color characteristics	Recommended uses	Remarks
Special mercury								
40 75 100	18 36 36	24,000 h	75% after 16,000 h	Low	Medium	Excellent color; preferred alternative to cool white fluorescent. Second best color choice for "warm" atmosphere.	Can replace incandescent lamps in interior fixtures.	Limited number of sizes; strictly for interior fixtures. Higher wattages and longer life than standard mercury.
Metal halide								
175 250 400	70 64 80	7,500–15,000 h	60% after 11,000 h	Medium	Medium	Better color than mercury; not as good as special mercury. Color-coated bulb has good, warm color; clear bulb less satisfactory. Best color rendition for outdoor lighting.	Parking areas, large work spaces, interior spaces, lighted from above.	Ballast required. Higher lumen output, lower lifetime than mercury.
High-pressure sodium								
150 250 400	89 to 97 90 to 100 106 to 115	12,000 h 15,000 h 20,000 h	80% at end of lifetime	High	Low	Poor color rendition; grays colors of red and blue objects. Similar to warm white fluorescent.	Outdoor, where color is unimportant: in parking spaces and security uses, or if illumination of building is enhanced by yellow light.	The most efficient lamp currently on the market.

Notes: Neon lights have not been included because they are commonly used only as decorative lighting.

[a]Lumen efficiencies and numbers of lumens per watt are approximations, and include ballast losses for discharge and fluorescent lamps.

SOURCE: DOE (101)

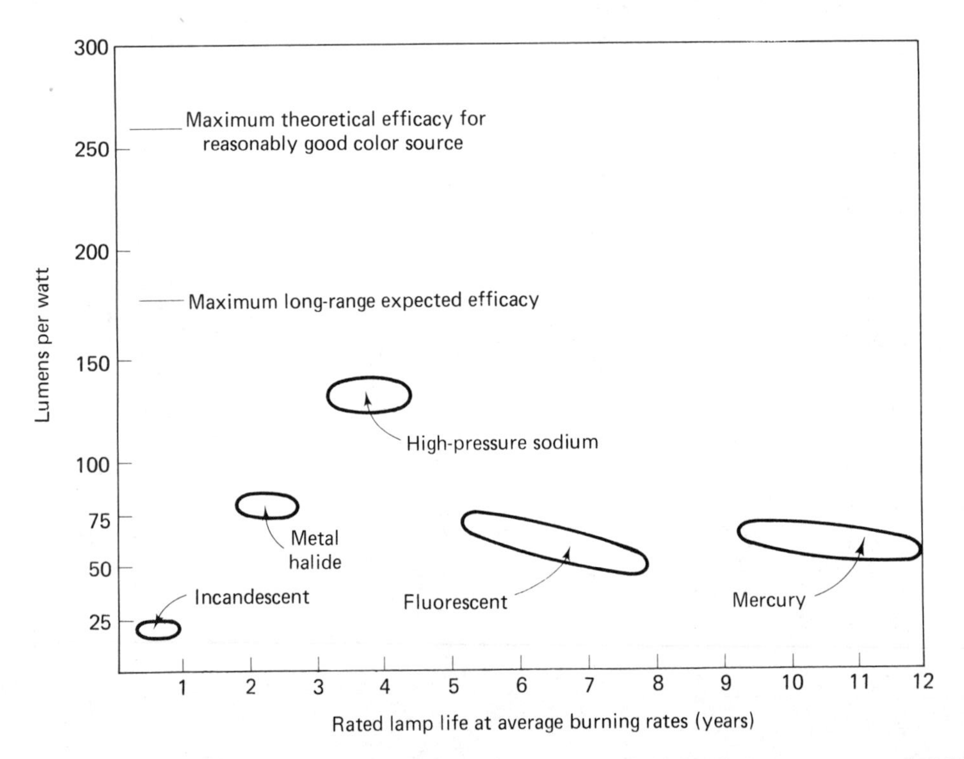

Figure 5.30. Comparison of efficiency (efficacy) and lifetime for some present practical light sources. Incandescent lighting uses 55% of the electricity for lighting, or 2.8% of the total United States energy use. Replacement of incandescent with fluorescent, which produces three times the light with the same amount of electricity, could reduce this amount by 70%. More efficient lamps can be expected in the future. Look for new fluorescent lamps producing 130 lumens/watt. More efficient incandescent lamps using selective optical filters to pass light and trap heat are also being developed. The maximum efficiency (efficacy) for a light source having acceptable color rendition is about 260 lumens/watt. The practical efficacy attainable in the foreseeable future may be as high as 180 lumens/watt. [DOE (126)]

also discussed, and an example of the quick payback from converting shop lighting to fluorescent is given.

Daylight can, of course, be the most efficient of all lighting, in addition to having great aesthetic value. Furthermore, illumination from a side wall or window can be more effective than lighting supplied by many overhead systems. (126) To be effective, glare, strong shadowing, and the unwanted flow of heat through windows must be overcome with proper screening. The best orientation for right-handed writers is with the window light coming from the left-hand side, or somewhat to the rear of the line of sight. Reorient the desk 180° for the left-handed writer.

Table 5.35. Comparison of Annual Savings from Relamping to More Efficient Lamps[a]

Change office lamps (2,700 hours per year) from	to	to save annually
1 300-watt incandescent	1 100-watt mercury vapor	$48.60 (486 kWh)
2 100-watt incandescent	1 40-watt fluorescent	$40.00 (400 kWh)
7 150-watt incandescent	1 150-watt sodium vapor	$236.00 (2,360 kWh)
Change industrial lamps (3,000 hours per year) from	to	to save annually
1 300-watt incandescent	2 40-watt fluorescent	$62.30 (623 kWh)
1 1000-watt incandescent	2 215-watt fluorescent	$161.70 (1,617 kWh)
3 300-watt incandescent	1 250-watt sodium vapor	$180.60 (1,806 kWh)
Change store lamps (3,300 hours per year) from	to	to save annually
1 300-watt incandescent	2 40-watt fluorescent	$68.50 (685 kWh)
1 200-watt incandescent	1 100-watt mercury vapor	$26.40 (264 kWh)
2 200-watt incandescent	1 175-watt mercury vapor	$67.00 (670 kWh)

[a] Costs have been computed at 10¢ per kWh and include ballast loss.

Two Apparently Contradictory Strategies: Replace Incandescent Light with Fluorescent Lighting, or Use Incandescent Task Lighting Instead of General Fluorescent Lighting.

1. ***Replace Incandescent Lighting.*** With incandescent lighting, only 5 percent of the electrical energy is used for production of light. The efficiencies of other systems are significantly higher:

Overall efficiency	*Lumen/ Watt*	*System*
15–20%	to 130	High pressure sodium
10–15%	35–80	Fluorescent lighting
3– 5%	15	Incandescent lamps

Comments:

(a) The efficiency of fluorescent lighting is reduced if plastic diffusers are used to reduce glare. If these diffusers, as is often the case, are dirty and yellow, there may be as much as 30 percent overall reduction of efficiency compared, for example, with the luminaire with a high-efficiency mirror reflector shown in Figure 5.29.

(b) Lately new types of fluorescent lamps have been developed which have a good color rendition and which produce as much light per watt as non-color-corrected lamps.

(c) At high power levels, existing high-pressure sodium lamps have a higher efficiency than fluorescent lamps (80–130 lumen/watt). They can be used in entry areas and in areas with large ceiling heights (40 ft [12 m] and greater). For smaller areas fluorescent lighting is, with respect to energy efficiency, about the same as high-pressure sodium or metal halide, because its efficiency is reduced at lower wattage. Some types of high-pressure sodium lamps have high efficiency up to 130 lumen/watt, but their color rendition is poor.

2. ***Task Lighting.*** General lighting can normally be reduced by 50 to 67 percent if task lighting is used. This will also improve comfort, since individual task lighting can be used according to the user's need. An example shows that the reduction of electrical power for lighting can be significant. In one office, for example, there are positions for eight fluorescent bulbs, for 570 watts of installed lighting in an area of 160 ft^2 (15 m^2). With half the bulbs removed and with a 60-watt desk lamp, the lighting load is only 2 W/ft^2 (23 W/m^2) and the illumination is more than adequate. In this example a fluorescent fixture could also be used as task lighting. With air-cooled lighting, assuming that 40 percent of the heat is entering the room, the heat load on the cooling system with the old lighting is 1.4 W/ft^2 (15.2 W/m^2). With the new lighting, it is only 1.1 W/ft^2 (11.6 W/m^2). If one assumes that lighting is two-thirds of the heat load of the room, then the air flow of the air conditioning can be reduced by 20 percent.

Replacement of Incandescent with Fluorescent Lighting in Retail Store

Incandescent lighting had originally been installed in this store, with a total connected load of 10 kW. These lamps burned 17 hrs/day, 5 days/wk, 52 wks/yr. The incandescent lamps were replaced by 37 fluorescent lamps with a power of 3.6 kW. The yearly energy use was reduced from 44,000 kWh to 16,000 kWh. With the price of electricity $0.043/kWh, the initial saving was $1,200/yr. Including the lower charge for demand and reduced replacement costs, the total saving came to $1,900/year. With the installed cost of the fluorescent fixture at $73, the investment costs were $2,700. This investment was returned in less than two years. The specific capital productivity of this efficiency improvement is $693/kW.

Comment:

Where the period of operation is shorter (for example, schools), the rate of return can be significantly longer (see, for example, Table 5.36). The specific capital productivity would remain unchanged (assuming the same installation costs).

For many lighting improvements, some cost is involved. Because of possible wiring changes as well as the cost of new fixtures, some changes can be expensive. However, because the cost of electricity has increased more than the cost of labor and materials, the payback for the conversion to more efficient lighting systems can, in many cases, be quite attractive. Table 5.36 shows a comparison of costs and savings for improvements in lighting. Also shown in this table are some examples with rather small costs which would rightly be charged to operation and maintenance—for example, the extra costs for energy-saving ballasts or phantom tubes. Some of the improvements shown are attractive not as financial investments, but rather for the better lighting environment they offer. Seven of the examples shown, however, have specific capital productivities lower than $370/kW, which was far less than the cost of installing new electric generation capacity.

New Lighting Systems

Guidelines for planning new lighting systems for use in modernization programs or new construction, are outlined below. In the course of renovating or building a new facility, there are excellent opportunities to provide energy-efficient as well as aesthetically pleasant and productive lighting environments. A guide for the non-uniform layout of office lighting with direct luminaires is provided below. Based on the survey of 307 buildings described in Figure 5.28, the typical lighting system in use was capable of delivering approximately 30 footcandles (300 Lux) per average watt of installed lighting. On page 239 the analysis for an arrangement is shown which is capable of supplying the illumination required to meet the standards given in Table 5.33, with a connected power of only 1 W/ft^2 (11 W/m^2).(126)

Guidelines for Planning New Buildings with Lighting Systems

Use high efficiency luminaires.

Optimize room conditions (for example, use light colors on walls and ceilings).

Table 5.36. Comparison of Costs and Savings for Improvements in Lighting

Energy-saving measure	*Cost ($)*	*Savings ($/year)*	*Simple payback period*	*SCP[a] ($/kW)*	*Comments*	*Reference*
Turn off unused lights	52	1,500	Immediate	15.20	Saving from turning off 1,000 120-watt lamps during a one-hour period when area is vacant. Cost is due to increased replacement labor and lamp purchases resulting from shorter lamp lifetime as the result of on–off switching. Saving computed at 5¢ kWh. Replacement labor at $2.00/lamp, lamp purchase at $2.95/lamp.	(134)
Energy-saving ballasts	1.88	1.05	1.8 years	270	Purchase of more expensive ($9.59) 9-watt ballast per 2 fluorescent 40-watt tubes, vs. $7.71 for 16w ballasts, saves $1.05 per 3000 hours operating time. Replacement of defective 16-watt ballasts with 9-watt ballasts returns investment in 1.8 years. 9-watt ballasts run cooler, and have a longer life.	(135)
Low level ballasts	0.77	3.00	3 months	38.50	In overilluminated areas, replacement of defective ballasts with low level ballasts (75% of light output costs $0.77 for more expensive ballasts $8.48 to save $3.00/ year.)	(135)
Central control of lighting	325	1,165	14 weeks	110	The burning time of 22.7 kW of lighting in two lecture halls is reduced 1,140 hours/ year by connecting existing relays to central computer used to control HVAC system.	(136)
Replace incandescent lighting with fluorescent	12,700	915	15 years	1060	Carrol School, Fall River, Mass. Initial illumination was 17 footcandles at desk top. After replacing 96 300-watt incandescent lamps with 213 fluorescent fixtures, the illumination is 27 footcandles. The installed demand is 12kW lower. The annual saving is greater than 18,300kWh.	(40)

Reduce area lighting, install task lighting	2,400	230	10.4 years	462	Concord-Carlisle School library. Initial illumination 80 footcandles as provided by 70 fluorescent fixtures. Replacement of 35 fixtures with 4 40-watt table lamps on 8 tables costs $75 for each lamp including $60 per lamp for the new wiring. If the lighting is used 10 hours each day, the annual saving would be 5,150 kWh.	(40)
Task lighting	$600	$70	8.6 years	$3,425/kW	Library in Town Hall, Dunstable, Mass. Replace half of fluorescent fixtures (2 kW) with task incandescent lighting at library tables. Annual saving 1,535 kWh. The lamps cost $75 each of which $40 each is required for new wiring.	(40)
Task lighting	$1,000	$110	9 years	$3,475/kW	Hill Roberts School, Attleboro, Mass. Replace half of 32 existing fluorescent fixtures with incandescent table lamps. Each table lamp cost $100 to install. The annual energy saving is 2,520 kWh.	(40)
Replace incandescent with fluorescent bulb	$2.65	$31.70	1 month	$33.75/kW	Continuously burning, 100 W incandescant ($1.66 for long-life 3,000 hour bulb) is replaced with Litek bulbs at $7.50 giving same output (1,290 lumens) with only 21.5 W.	(127)
Phantom tubes	$6.80	$8.11	10 months	$367/kW	Phantom tube permits operation of single tube in 2-tube luminary without rewiring at 70% of its rated energy, or 28 W. Phantom tube itself does not light up, costs $6.80 when purchased in lots of 30 tubes or more. Saving shown does not include that from added lamp life and improved power factor.	(137)
Replace incandescent with fluorescent	$2,700	$1,900	17 months	$235/kW	Forty 250 W incandescant fixtures replaced with 37 2-lamp fluorescent luminaires reducing load from 10 kW to 3.6 kW. Savings computed at $0.043/kWh and $2.80 kW demand charge, include reduced labor for incandescent bulb replacement.	(136)

[a] The specific capital productivity is the cost of reducing electric demand by one kilowatt.

Strategically place luminaires to minimize glare.

Locate clearly labeled switches for convenience in turning off unneeded light.

Use automatic switching (time interval, light sensitive, clock controlled, etc.).

Computer control and zone switching of lighting should have flexibility to enable adaptation to local requirements.

Make maximum use of daylight. Bounce window light from the ceiling to reach deeper into interior space. Use lightwells and skylights.

Recycle heat-of-light with exhaust air fixtures.

Install high-pressure lamps (for example, sodium lamps) in factories, or other buildings with high ceilings.

Guide for Nonuniform Office Lighting Layouts with Direct Luminaires

1. *Locate work surfaces and determine lighting needs for tasks.* Desks, tables, credenzas, files, etc., will have tasks. Determine their difficulty, specularity, and the plane on which they lie.

2. *Within the limits of the luminaire supporting system, locate the luminaires as close to directly over the task as possible without creating excessive veiling reflections.* With highly specular tasks—shiny paper, pencil writing—try to avoid the trapezoidal ceiling area defined by (for a 9′-0″ ceiling height): the working edge of the desk projected vertically to the ceiling forming a line 4′ long centered on the worker; a line parallel to and 8′ forward of this line that is 12′ long; the sides of the trapezoid connecting the ends of these two base lines. See figure opposite.

If luminaires are kept out of this zone, the LEF[a] should be 1.0 or better. If part of a luminaire projects onto the edge of this area, only minor visibility losses will occur, producing an LEF on the order of .8 to .9. If the bulk of a luminaire or parts of several luminaires project onto this zone, particularly near the desk, LEF can be .4 to .7. Higher LEF will also result with more light coming to the task from well outside this offending zone. An interior boundary is shown to define the most sensitive part of the offending zone.

For less specular tasks—typing, ink writing, and printing on matte paper—there is less concern with veiling reflections and LEF's closer to 1.0 will result. An effort should still be made to avoid the offending zone, however, particularly the center.

3. The result of 1. and 2., above, is to *locate luminaires just beyond the ends and working edge of the desk,* but avoiding as much as possible the offending zone for adjacent work positions.

[a]LEF—(Lighting Effectiveness Factor): The ratio of equivalent sphere illumination to ordinary measured or calculated illumination.

SOURCE: GSA, (48)

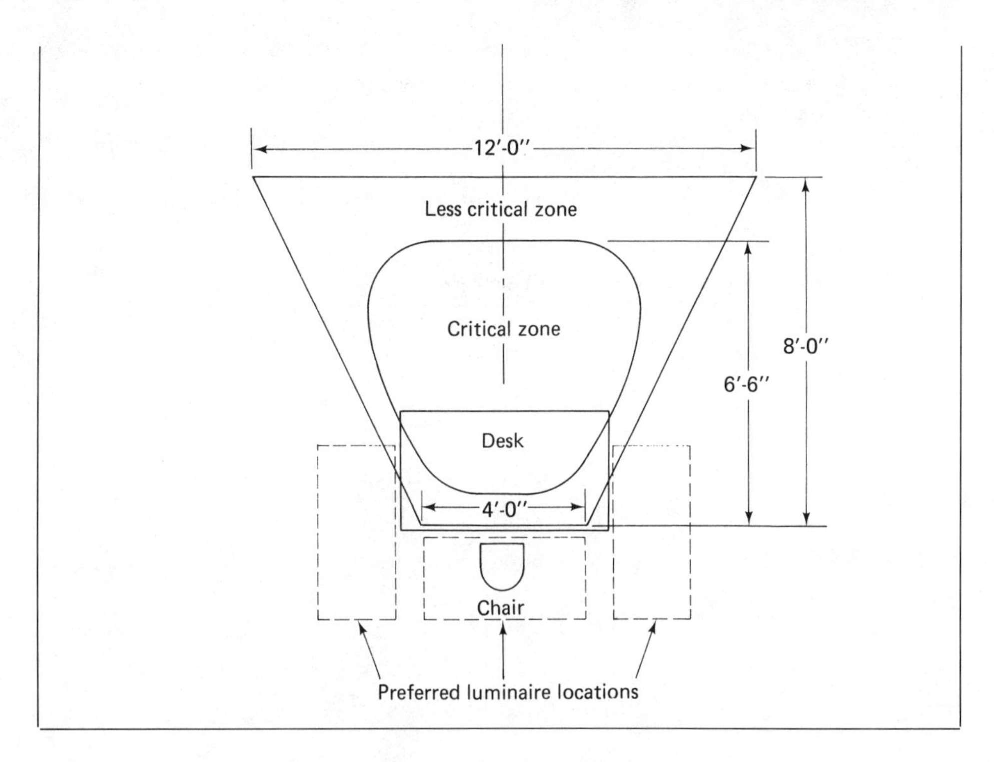

Calculation of Idealized Watts/Sq Ft Required for Lighting for Unit Space of 160 Sq Ft per Worker

1. Assume 25 sq ft for task areas with 70 percent of the tasks requiring 50 FC, 20 percent requiring 70 FC, and 10 percent requiring 100 FC.
2. Assume 110 sq ft at 30 FC.
3. Assume 25 sq ft at 10 FC.
4. Assume that the 30 FC and 10 FC are area lighting in their respective areas (part of a larger space) and that the task lighting is achieved by *supplementing* the 30 FC area lighting with an added 20 FC (70 percent of the time), 40 FC (20 percent of the time), and with 70 FC (10 percent of the time), respectively. The task lighting comes from specialized directional sources providing a lighting efficiency factor of unity.
5. Assume source with efficacy equivalent to the fluorescent lamp and with proportional ballast losses.

$$3{,}200 \text{ lumens for } 40 \text{ watts} + 6.5 \text{ watts in ballast}$$

$$3{,}200/46.5 = 68.8 \text{ lumens/watt}$$

6. Use coefficient of utilization for area lighting of .6 and use a maintenance factor (MF) for office space of .8. For task lighting, use an overall average efficiency of .6 rather than the

maintenance factor and coefficient of utilization (CU). These factors are selected as being the reasonably best obtainable values for the non-uniform lighting system envisaged, in clean office space.

Then, using the relationship:

$$FC = \frac{\text{Lumens} \times CU \times MF}{\text{Area}} \text{ for each incremental area,}$$

solving for lumens and converting to watts;

$$\text{Lumens} = \frac{FC \times \text{Area}}{CU \times MF}$$

$$\text{For 30 FC area; } \frac{30 \times 135}{.6 \times .8} = 8438 \text{ lumens}$$

$$\text{10 FC area; } \frac{10 \times 25}{.6 \times .8} = 521 \text{ lumens}$$

$$\text{50 FC area; } \frac{25 \times 20 \times .7}{.6} = 583 \text{ lumens}$$

$$\text{70 FC area; } \frac{25 \times 40 \times .2}{.6} = 333 \text{ lumens}$$

$$\text{100 FC area; } \frac{25 \times 70 \times .1}{.6} = 292 \text{ lumens}$$

$$\sum = 10{,}167 \text{ lumens/160 sq ft} = 63.5 \text{ lumens/sq ft}$$

$$63.5\text{/lumens sq ft/68.8 lumens/watt} = .92 \text{ watts/sq ft}$$

This is the minimum feasible average lighting power input per worker, with the technology of today.

The assumption is made in the above analysis that suitable light sources are available in the necessary increments and proper efficacy to provide only the lighting necessary.

SOURCE: Ross and Baruzzini. (126)

CHECKLIST	LIGHTING: OPERATING CHANGES
Turn off unneeded lights.	When no one is using the lights, switch them off. Reduce lighting after working hours, over weekends, holidays, lunch periods, etc. Alternate, where possible, the lights shut down to balance tube and ballast life.
Shut off lighting during periods of sufficient daylight.	Where can lights be turned off during daylight hours? In peripheries of offices from 10 to 13 ft (3 m–4 m) in depth, lighting can be reduced 1,500–2,500 h/yr almost completely.

	For office zones with a greater depth, lighting can be shut off in stages of 1/3, 2/3. Orient desks, machines to make best use of window light, minimize glare and shadow. Venetian blinds should be adjusted to deflect light to the ceiling or to minimize need for artificial lighting. Check all exterior rooms, corridors, restrooms, eating areas for possible reduction of artificial lighting.
Reduce lighting in corridors, storage rooms, garages, etc.	Turn off or substitute with lamps of lower wattage. Shut off most of the lighting in garages and parking lots, leaving only necessary corridors lit. But keep safety in mind.
Turn off outside lighting during daylight.	Reduce use at night. Minimize unnecessary display, decorative, and advertising lighting. Shut off sign, advertising, decorative lighting after 10 P.M.
Develop a lighting use program.	Establish an effective lighting usage program to control lighting according to occupancy and conditions of need. Determine the amount of lighting needed for safety and security purposes.
Put someone in charge of lighting.	Advise responsible personnel to control illumination level according to available daylight and occupancy. Give adequate instruction to the responsible personnel to assure understanding and compliance with the lighting use program. Provide detailed instructions for system operation by means of time charts and color coding of switches. Color code those switches to be left on at night. Security and cleaning personnel should be properly advised. Periodically inspect lighting for adequacy, glare, maintenance, waste. Survey out-of-the-way places for forgotten lighting waste. Survey exterior lighting also.
Orient occupants.	Post a small sign or chart near each switch which identifies which lights are controlled by the switch. Campaign for better utilization by using posters, memos, and personal contact to encourage occupants to use lighting only when it is needed, to use only the amount of lighting required, and to turn off lights whenever they are not being used. Put "help, last leaving lights out" ("HLLLO") notices in conference, storage, utility, supply, closets, etc.
Note.	Turn off all unnecessary fluorescent and/or incandescent lighting, even if they will be unused only for a few minutes.
Reduce lighting level during cleaning.	If possible, cleaning work should be performed during normal working hours. If custodial work must be performed

outside regular working hours, the light intensity should be reduced and limited to the area where the work is being performed, or in an area no greater than can be cleaned in one hour.

Disconnect unneeded lighting.

Unoccupied or low occupancy areas, such as lobbies, hallways, passageways, and storerooms, need not be brightly lit.

Remove or disconnect some of the lamps from fixtures within 10 to 12 ft (3 to 4 m) of window walls.

Reduce lighting level.

Eliminate unneeded lamps, use lighting with lower wattage. Uniform lighting in any facility often represents wasted energy. It is unnecessary to light non-task areas at the same level required for task areas.

Use task lighting with low background lighting. Use bench, machine, or desk lamps to supplement light in specific work zones. The requirements of the work should dictate the illumination level.

Can exterior lighting be reduced?

Reduce display lighting.

Turn off or reduce the wattage of unneeded case and shelf lights.

Optimize contrast and accent lighting in place of uniform, high-level lighting which contributes to glare and washes out color.

Turn off internal refrigerator lights to reduce the amount of energy used both for lighting and for the additional refrigeration required to compensate for the heat it gives off.

Can the lights in vending machines be removed?

Remove unneeded lights.

Reduce light levels by substituting smaller bulbs and reducing the number of bulbs left burning.

Light levels are almost uniformly too high today. It is not uncommon to find many areas with five to six times brighter lighting than is actually necessary, not only in office and shop areas, but also in support areas such as corridors, storage rooms, etc.

Check optimum placement of remaining fixtures to minimize glare.

An area which once required high-level illumination should not continue to be as brightly lit if it becomes a storage area or some other inactive operation.

When removing tubes from four-bulb fluorescent fixtures, remove two tubes in each fixture rather than all the tubes in alternate fixtures to maintain even lighting levels.

After reducing lighting save in the HVAC system.	By cutting the illumination level in half, the energy requirement is cut by more than 50%. Further energy will be saved through reduced heat load in air-conditioned buildings. For still further savings, reduce air-conditioning air flow rate after lighting power reduction. Save energy with thermostatic valves installed on radiators or individual heating units to take advantage of heat of lighting. Check that energy saved in the lighting system is not being lost by extra reheat supplied in the air-conditioning system. Do not raise the heating thermostat or lower the air-conditioning thermostat when you reduce your lighting levels.
Optimize task lighting.	Orient desks, other work surfaces to take advantage of existing fixture positions, with minimum shadow and glare. Group together activities requiring the same level of high illumination. Reduce lighting in other areas. Locate work stations requiring the highest illumination levels nearest the windows. Arrange work surfaces so that window lighting crosses the task perpendicular to the line of vision and from the left-hand side for right-handed people. Arrange lighting from broad source, ceiling luminaires to come either from somewhat behind the head, or from over the left or right shoulder. Arrange desks with ceiling fixtures between them, rather than in front of the writing surface.
Remove diffusers.	Remove light diffusers to gain additional lighting. In areas where the diffuser is not required for glare control or lamp protection, for example, in garages, bus stations, stairwells, corridors, light levels can sometimes be increased several-fold.

CHECKLIST	LIGHTING: MAINTENANCE
Schedule periodic lighting maintenance.	Luminaire efficiency can be maintained by properly cleaning the reflecting surfaces, diffusers, shields, etc. Replace lens shielding that has yellowed or become hazy with a clear acrylic lens with good non-yellowing properties. For some applications, a clear glass lens can be considered if it is compatible with the luminaire and does not present a safety hazard. If fixtures are yellow or rusty, repaint them with high reflectance white enamel.

	Lamps should be wiped clean at regular intervals to assure maximum efficiency. Lamps which are exposed to an atmosphere with substantial amounts of dirt, dust, grease, or other contaminants should be cleaned more frequently than lamps in a relatively clean atmosphere.
Keep ceiling and walls clean.	Clean ceilings, walls, and floors frequently to improve reflective qualities. Keep windows, skylights clean to provide more satisfactory utilization of daylight.
Schedule group relamping.	Lamp efficiency deteriorates over the life of a lamp. When the light output of a group of lamps has fallen to approximately 70 percent of the original light output, relamp all fixtures in the group at the same time. For most commonly used fluorescent lamps and incandescent bulbs, this occurs at about two-thirds of their rated life. This is also a good time to *check whether more efficient or lower wattage lamps are suitable*. Regular replacement of lamps increases the overall efficiency of the lighting system and may permit the use of fewer fixtures, or a reduced wattage retrofit lamp.
Check exterior lighting.	Trim trees and bushes that may be obscuring lighting fixtures and creating unnecessary shadows.
Check up on the condition of your lighting system.	Conduct lighting surveys periodically (at least once a year) to check illumination levels and positioning of installed lighting, and need for maintenance or relamping.

CHECKLIST	LIGHTING: SHORT-RETURN CHANGES
Use automatic light switching.	Use clock operated light switching for automatic shut-off when not required. Automatic 7-day programmers may be used to handle weekends and photocells may be included for various length days or storm conditions. Use photoelectric control to turn off periphery lights when sufficient daylight is available. Use photocell and/or time clock controls for outdoor lighting whenever feasible. Parking areas, building exteriors, identification signs, etc., usually require lighting for only a part of the period of darkness. Such lighting should be turned off automatically during late evening and early morning hours except for security and safety lighting. Consider using motion sensors to turn lighting off after last person has left meeting rooms, restrooms, lounges, etc.

Control lighting with master switching.	Use master floor switches to switch off a number of separately controlled areas. Connect area lighting and circuits with large connected lighting wattage to building computer system.
Install local switches for small areas.	Consider small-area switching where an individual task area may be only intermittently used. This prevents wasted light over large unused areas and can provide reductions in energy both in lighting power and air-conditioning power. Install manual switches for the outlying parking lot fixtures to ensure operation only during peak nighttime traffic hours. All rest rooms should be equipped with individual "on-off" light switches, so that the last person leaving the room can turn off the lights, or use interval timers. Install chain pull switches on infrequently needed, overhead fluorescent and incandescent fixtures.
Install interval switches in infrequently used areas.	Use time controls for those areas of a building which are used infrequently and only for brief periods. These controls turn off lights automatically after being activated for a set period of time. Use interval switches in storage rooms, basements, garages, supply rooms, etc.
Use interval switches where responsibility is ambiguous.	For classrooms, conference rooms, coffee rooms, and other community areas, interval timers (for periods of 15 minutes, half-hour, 1 hour, 2 hours, etc.) may be used where the responsibility of the last person leaving the room to turn out the lights may be ambiguous.
Switch to lower levels.	Use alternate switching or dimmer controls when spaces are used for multiple purposes and require different amounts of illumination for the various activities. It is possible to provide multiple levels by providing switching for alternate fixtures, alternate sets of lamps in fluorescent fixtures, etc. Remember that dimmer controls on incandescent circuits reduce lighting levels far more than they reduce energy costs.
Eliminate double illumination.	Reduce store lighting where lights on shelving gondolas and refrigerated displays already provide adequate lighting for sales purposes.
Use lighter colors.	When remodeling, use light finishes on ceilings, walls, floors, furnishing. Use reflective wall and ceiling paints to take best advantage of minimum lighting. The upper acceptable limits for reflectance are dependent on glare effects.

Choose more efficient lights.	Use one large lamp rather than two or more smaller ones. For example, one 100-watt incandescent lamp produces more light than two 60-watt lamps. Eliminate indirect lighting where possible; it can waste as much as 75 percent of the light output from a lamp. Replace incandescent lamps with fluorescent lamps wherever possible. For example, use the new generation fluorescent lamps with good color rendition (30 percent higher efficacy than previously available), color-corrected fluorescent lamps with much better aging characteristics and improved light output at high ambient temperature. Do not use "long life" incandescent lamps. Choose fluorescent lamps with screw-in base and self-contained ballast. Use screw-in mercury vapor lamps to replace high wattage incandescent lamps, for longer life and lower operating cost.
Use phantom tubes.	Replace one tube in two-tube fluorescent fixtures with phantom tubes for more even light distribution when removing tubes.
Replace defective ballasts with efficient types.	Energy saving ballasts cut ballast loss 44 percent yet cost only 24 percent more. Replace defective ballasts with low-level ballasts to reduce power input from 96 to 75 watts; low-level types cost only 10 percent more than standard replacement ballasts.
Install task lighting.	Disconnect 50 percent to 80 percent of general area lighting and install desk lamps. Use fluorescent desk lamps for writing and drafting. Use incandescent on machinery to avoid stroboscopic disturbances, or use 2-lamp lead-lag ballast. In computer installations, provide on-task lighting for changing tapes, disks, key punching, repairs, etc. Reduce area lighting to 25 footcandles to enhance CRT visual display, and decrease cooling load.
Switch from incandescent to fluorescent lamps.	Especially in areas with high operating hours, e.g., stores, shopping centers, etc. In rooms with lower operating hours, e.g., restaurants, offices, schools, the rate of return may be longer. Replace outdated or damaged luminaires with modern luminaires which have good cleaning capabilities and which use lamps with higher efficiencies and good lumen maintenance characteristics.

	Use relocatable light fixtures for flexibility vs. high general area lighting. Consider installing lighting around the perimeter of interior walls. Perimeter lighting can make any area or a store appear brighter with fewer fixtures.
Install lighting systems with low heat production.	Electrical consumption and heat load of lighting system is very important in areas with air conditioning, where the heat load has to be removed. With high efficacy discharge lamps, good reflector configuration (e.g., parabolic reflectors), and proper placing of luminaires, the heat load at rational lighting levels can be reduced by a very large factor compared with previous design.
Reduce height of fixtures.	Consider lowering luminaires so they will provide recommended illumination levels on the task area at reduced wattage. With lighting closer to where it is required, the same illumination level can be achieved with a lower connected lighting power. Consider reducing the ceiling height and using fixtures mounted in the dropped ceiling.
Improve glare control.	Less shielding than is used today may be sufficient to reduce glare. Check glare protection and replace it if necessary. Where appropriate, consider installation of lenses which provide special light distribution patterns to increase lighting effectiveness. As examples, linear batwing, radial batwing, parabolic louvers or polarizing lenses may provide better visibility with the same or even reduced wattage. Relocate luminaires to provide light on task areas at an angle outside the zone which causes veiling reflection if relocation of work station is impractical.
Buy furniture, fixtures, display gondolas with integrated task lighting.	Desks with integrated lighting give better glare control with ample light for typing or writing with far fewer luminaires. With light close to merchandise, fewer watts are needed for customer appeal. Use light colors with mat finishes for desk tops, filing cabinets, and other furnishings within the field of vision.
Reduce heat-of-light in refrigerators and freezers.	Either keep all refrigerated product lighting in front of displays or install remote ballasts to reduce the amount of heat in or near freezers and refrigerated display cases.
Better utilize skylight.	Install or utilize skylights wherever feasible to reduce daytime lighting requirements. Keep skylights clean. Improve distribution of skylighting by keeping adjacent surfaces highly reflective. Repaint with white enamel if necessary.

	Provide external solar shielding. Use external or internal reflectors or bounce panels to better distribute skylight. Adjust venetian blinds for optimum distribution of daylight and rejection of air-conditioning load.

CHECKLIST	LIGHTING: MEDIUM-RETURN CHANGES
Use cooled luminaires.	Use air-cooled luminaries (water-cooled for high intensity lighting in theaters, TV and photo studios, operating rooms). If necessary for heating, this heat can be used with heat recovery or recirculated for heating in the winter. In the cooling season 60 to 70 percent of heat can be ducted directly outside.
Provide selective switching.	Extend switching to areas with varying occupancy and lighting requirements. Initial cost economies and lack of knowledge about final space subdivision often lead to the use of central panel-boards as the only means of controlling large blocks of lighting. This precludes the potential for turning on only the amount of lighting that is actually needed after the use of the space has been determined by the occupant. Localized switches can be provided near doorways. Remotely controlled switches can be located near panelboards to control groups of lights, low voltage control circuits can be used to provide local control of switches situated in remote locations (these relays usually are relatively inexpensive). When properly used, localized switching usually will save enough energy to provide a payback on the investment within a moderate period of time.
Monitor lighting with pilot lights.	Outside refrigerators, closets, transformer and utility rooms, install pilot light to indicate when room lights are on. Use neon or LED indicators for long life and low operating cost. Lighting use in remote areas can be monitored by providing neon indicator lights at central stations. Personnel will be alerted to investigate and turn off lights not being used.
Install multiple level light switching.	In spaces which require more than one level of illumination to compensate for variable daylight availability, or different use conditions (for examples, after-hours cleaning, or during non-selling hours in stores), wire alternate fixtures or rows of fixtures to separate switches. Arrange for $^1/_2$, $^1/_4$ reductions, in addition to security lighting.

	Consider installing fluorescent luminaires with 2-level or multiple level ballasts in multiple-purpose spaces which require more than one level of illumination.
Reduce heat from lighting in chilled display cases.	Since refrigeration display lighting generates significant amounts of heat, move lighting from the chilled space to the exterior. Remove ballasts from remaining fixtures to non-refrigerated location.
Get expert help.	Competent technical advice should be obtained to help plan a comfortable, minimum glare optical environment. Insist that energy-saving guidelines be used rather than previous, superfluously high, and uniform lighting standards. You can combine high efficiency and a pleasant, productive lighting environment.

5.7 ELECTRICAL SYSTEMS

Electricity accounts for close to 30 percent of the United States energy use. (138) Because of production losses, only 32 percent of this energy is converted to electricity, and another 10 percent is lost in transmission and distribution to final consumers.

The Recent Price History of Electricity

Between the mid 1960s and 1980, the cost of new central station plants rose more than a factor of 30, far outpacing the cost of labor, the wholesale price index, and even the cost of oil. Today the facilities used for the generation and transmission of this high-quality energy form represent some of the most capital-intensive equipment in the energy system. The recent history of the average cost of electricity in the United States (including demand charges) is shown for commercial and industrial users in Figure 5.31. Following 1973 these costs began to climb rapidly—at 13 percent annually for commercial users, and 17 percent annually for industrial users. Nevertheless, current electricity prices are still far below the cost for energy from current design, central station power plants.

Users should be aware of two likely consequences which recommend that stringent measures be taken to conserve electricity: First, future prices will climb to much higher levels than those known today. (139) Electricity savers will be rewarded handsomely. Second, despite urgent arguments for the desirability of substituting electricity for oil, etc. the construction of central station power plants will be constrained due to severe capital shortages. Unless there are stringent conservation practices, the surplus generating capacity with which the United States began the 1980's will disappear, and the reliability of the electric system will deteriorate. Consequently, an increasing number of utilities

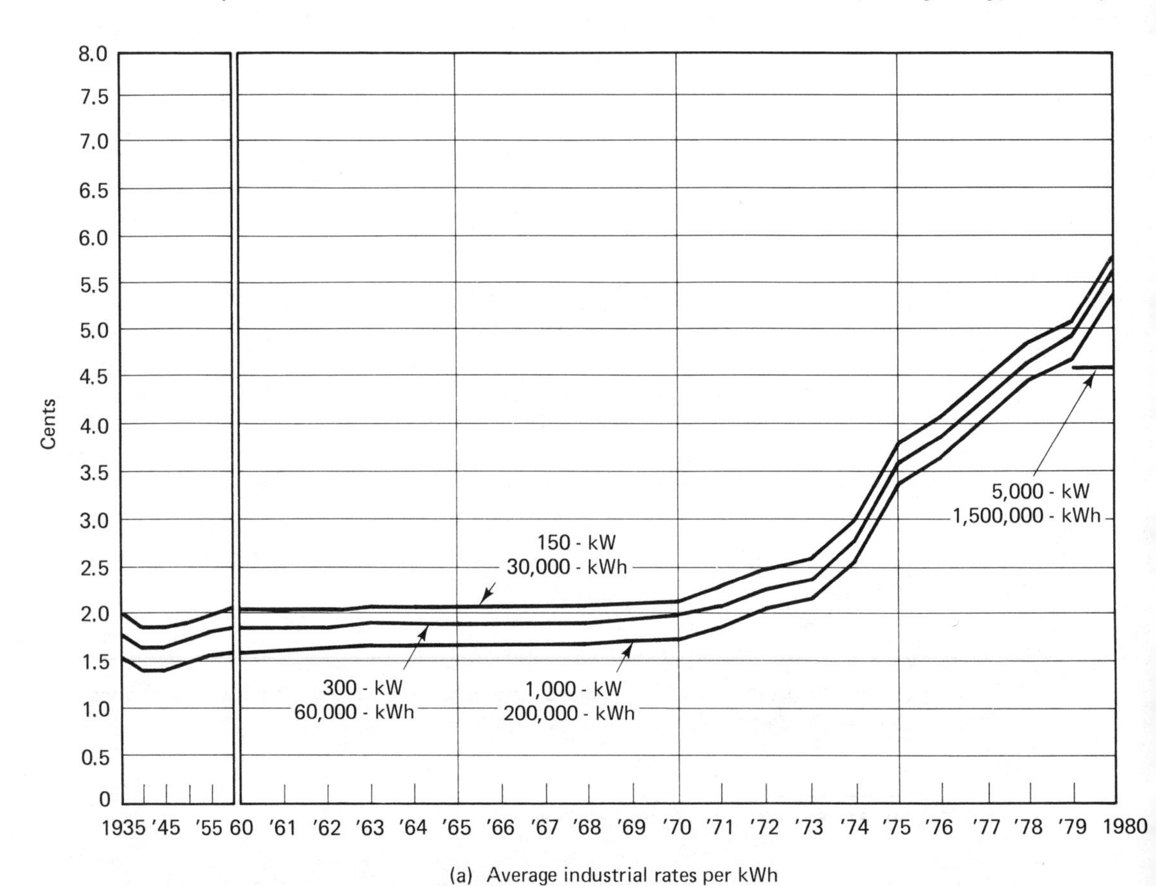

Figure 5.31. Average rate for electricity for industrial (a) and commercial (b) users (including demand charges) in $/kWh. [U.S. Department of Energy]

are discovering that it is far less expensive to provide the capacity for new users by investing in conservation than by building expensive plants to serve present waste.

How to Save*

The kWh of electrical energy used depends on three factors: the size of the connected load (lighting, ventilation, data processing units, manufacturing equipment, etc.), the length of time it is drawing power, and characteristics of the load such as its power factor, and in many cases, peak demand. The characteristics of the distribution system and the type of electric service also affect the efficiency of energy use.

As a result of having implemented savings described in the previous sections, the

*In addition to sources specifically cited, references 140 and 141 are supplementary references for this section.

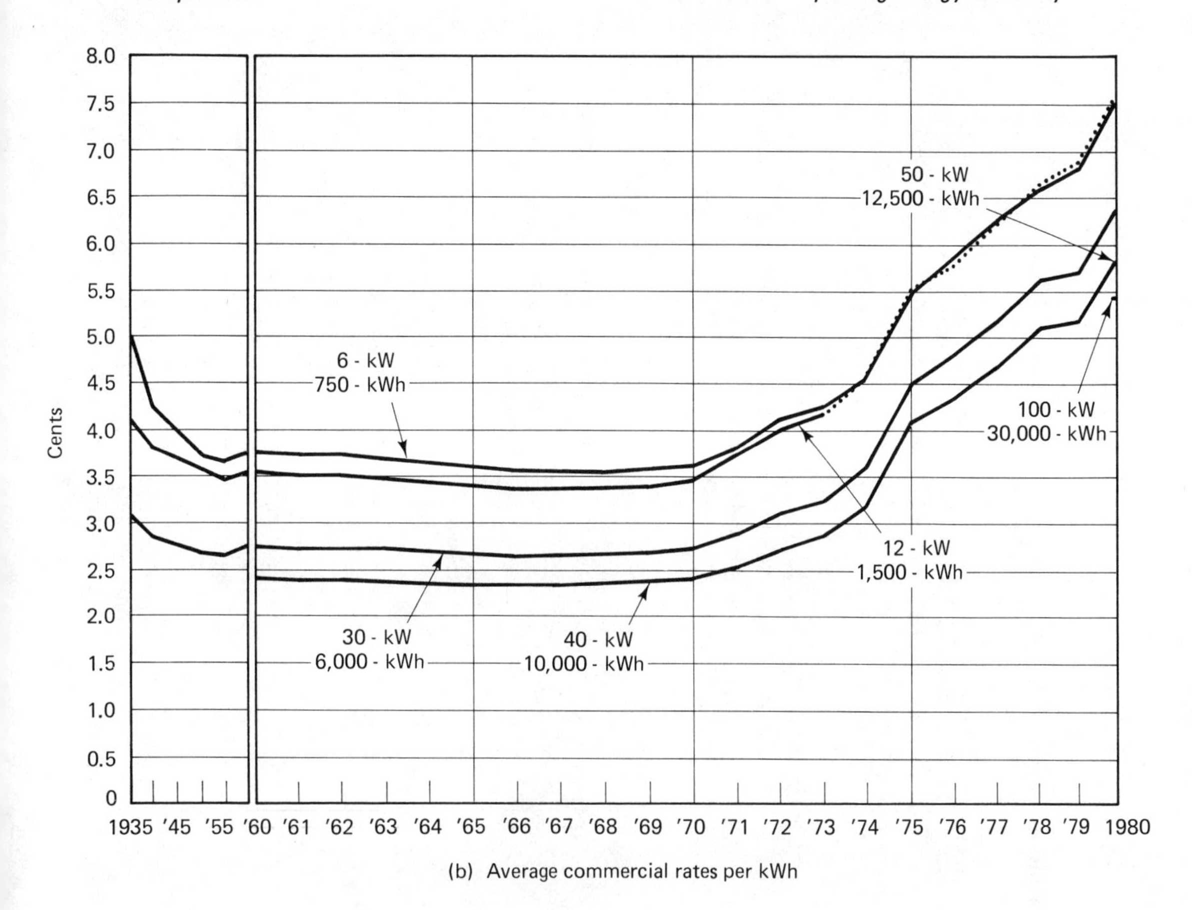

(b) Average commercial rates per kWh

electrical system itself may be ripe for additional savings. For example, as a consequence of reducing air flow rates, the fan motors are now oversized. It may be possible to move motors from some smaller units to replace larger, oversized units. After following the sequence of reducing and leveling demand, and then improving power factor and the distribution system, it may be possible to eliminate one or more superfluous transformers. Elimination of parasitic losses in the electrical system will not only improve its reliability and safety, but will also cut costs where no productive benefit is received. Table 5.37 shows a comparison of some improvements.

If the structure of your electrical tariff includes a charge based on peak power demand, substantial cost savings may result with the lower peak demand following an EIP, and from shifting loads away from peak load periods. As an initial step, a survey of the electrical system should be made. The load survey will indicate any problem areas. The power demand of the system should be recorded and the operation of the equipment analyzed to determine what units need to be operating when. The object of the analysis

Table 5.37. Comparison of Capital Improvements in the Electric System

Energy-saving measure	*Cost ($)*	*Savings ($/yr)*	*Pay-back period*	*SCP[a] ($/kW)*	*Comments*	*Reference*
Select more efficient kitchen equipment	130	315	5 months	44	Choice of better insulated deep fat fryer with 25% lower power rating saved 7,500kWh/yr and helped to reduce demand charge. Demand was reduced an average of 3kW during 210 hrs/month of use.	(136)
Shutting down milling machines	1,517	1,673	1 year	175	Installation of thermostatically controlled heaters enabled ready-to-use conditions to be maintained in five numerically controlled milling machines. Savings resulted from shutting down hydraulic power units when machines are not in use, in addition to reduced wear.	(142)
Improved motor efficiency	44	49.20	11 months	107	Polyphase 10 hp motor of improved efficiency (85.2%) is used in place of standard motor with 81.8% efficiency, at $224 first cost vs. $180. Saving is computed at 3¢/kWh for 4,000 hrs annual use.	(143)
More efficient motor design	72	72	1 year	120	Increasing quality and amount of steel in motor core raises efficiency to 86.8%, increases cost of 10 hp motor to $252.	(143)
High efficiency motor design	99	82.80	$1\frac{1}{2}$ years	143.50	Increased copper, quality of motor core, and closer tolerances raise efficiency by 107.6% over standard motor (81.6%) and increase its cost by 155% over $180 for conventional motor. Pay-back is computed at 3¢ kWh over 4,000 hrs annual use.	(143)
Correct low power factor	2,900	2,150	16 months	23	Effective capacity of electricity generation and distribution system is improved by raising power factor of 350kW from 0.65 to 0.85. Annual cost saving results from reduced demand charge.	(144)

Incandescent lighting replaced with mercury vapor	9,090	11,357	10 months	117	In an electrical equipment manufacturing plant, 187 750-watt incandescent lamps were replaced with 300-watt, screw-in replacement mercury vapor lamps. Labor costs for replacement and cleaning fixtures were $748 and capacitors to correct power factor cost $1,422. Demand charge was reduced $3,010; savings during 2,500 hours computed at average cost of 4.3¢/kWh.	(136)
On–off demand control for cooling tower motor	4,000	6,220	8 months	169	Evaporative cooling provides sufficient temperature reduction to permit shutting down one to three of four cooling tower fans depending on weather conditions in Houston, Texas. A temperature sensor in the outlet keeps temperature below 95°F (35°C) in this 100 MBtu/hr (30MW) capacity cooling tower.	(121)
Load matching with dual speed motors	4,800	6,890	8 months	183	Installation of two dual speed motors in the 100MBtu/hr (30MW) cooling tower of the previous example allows for closer matching of air flow to ambient weather conditions at an increased cost of $800 and an additional annual saving of 22,420kW.	(121)
Variable pitch fans	6,500	8,379	9 months	204	In the cooling tower above, use of continuously variable air flow fans controlled by the outlet temperature enables air flow to be adjusted to hold outlet temperature at 95°F (35°C). Cost for four auto-variable pitch fan hubs is $4,000, other controls $2,500.	(121)
Reducing HVAC fan speed by 25%	385	1,306	4 months	28	Air flow is reduced 25% by replacing motor sheave at a cost of $100. Motor hp is reduced from 30 to 13 hp. Power factor correcting capacitors costing $285 installed are required to bring power factor to 85%. Savings are computed for 2,100 hours operation at $4^{1}/_{2}$¢/kWh.	(136)

[a]The specific capital productivity is the cost to reduce annual average power demand by one kilowatt.

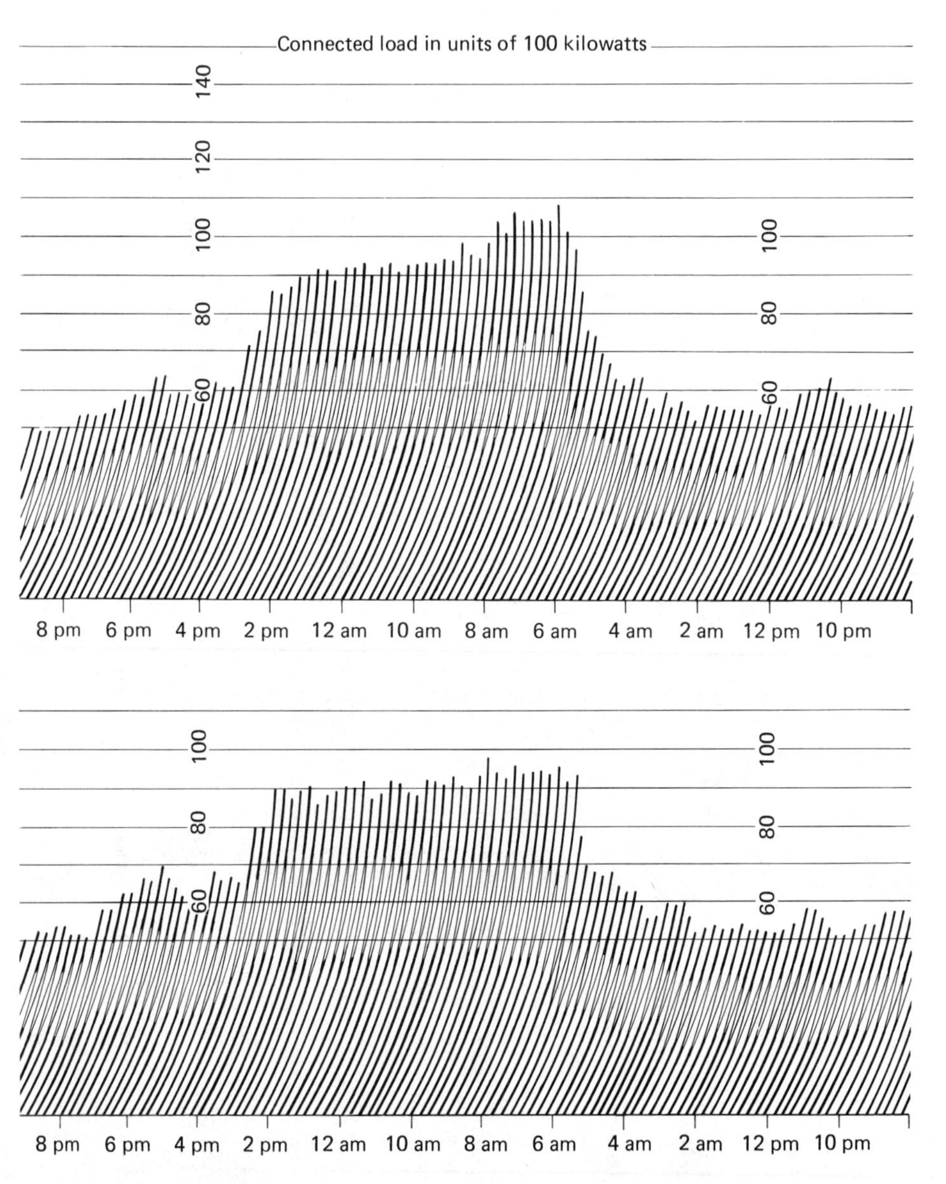

Figure 5.32. Demand recordings. Each arc represents the average load during a 15-min period. In the first example, a peak load of 10.9 MW (10,900 kW) was recorded shortly before 6:00 a.m. In the lower chart, the peaks were below 9.7 MW due to reduced load and to the staggering of loads during morning start-up. The starting-up of all equipment simultaneously can cause large peaks due, for example, to the surge current in equipment until it comes up to speed. This can overload circuits and will cause peaks in the demand load resulting in penalties that could be avoided with improved start-up management.

is to identify obvious waste and deficiencies and to compare expected with actual use. Attention should first be directed to operational changes that reduce the load, and then to maintenance and any necessary improvements in the switching and distribution system. Unneeded equipment should be de-energized. Your electrical utility may be able to provide assistance. It may be able to determine the power factor of your system and provide a record of load demand similar to the demand chart shown in Figure 5.32. Electrical measurements and alterations should be done by properly trained people in conformance with the requirements of local codes.

Maintenance and Voltage Checks

A maintenance program can help reduce losses in the electrical system, increase the serviceable lifetime of equipment, and—most important—maintain necessary safety conditions. Because much of the electrical system is out of sight and may operate without obvious problem, it may receive inadequate maintenance, and in fact hazardous conditions may have developed.

Check for hot contacts and tighten loose connections which can cause voltage and energy loss. Undersize wiring also causes voltage and energy loss. New, high capacity lines may be needed to avoid line voltage drops due to equipment additions.

Low voltage on motors results in poor starting torque and can cause higher current loading which further increases the voltage drop problem. For many types of motors, speed remains almost constant if the line voltage drops. Current load increases so that power output remains nearly constant. However, ohmic winding losses increase with the square of the currents, causing motor efficiency to decrease. For this reason reducing peak demand may improve voltage regulation and system efficiency.

Critical installations may use constant voltage systems with flywheel-motor generators, or electronic regulators. Since constant voltage supplies are not only expensive, but also cause a significant efficiency loss, only the minimum, absolutely necessary equipment should be operated from constant voltage circuits.

Reducing Demand

There are often loads which are out of sight (or out of mind) that operate longer than necessary, some of them 24 hours a day, 365 days a year; these include elevator lighting, elevator and other control systems, ventilators, pumps, photocopying machines, computer systems, water coolers, vending machines, transformers, escalators, electric heaters, and furnaces. An example is shown below. (144) The basic plan is to identify these loads and to find methods for disconnecting them when they are not needed. The cost of unneeded operation must be compared against the cost of an automatic controller, or against the effort of cajoling users to save. Where there are multiple units in simultaneous operation, perhaps better scheduling or identification of slack periods will enable elevators, process equipment, ovens to be shut down.

Use automatic timers to turn off equipment after working hours, or need-of-use sensing controllers to de-energize unneeded equipment. Although the largest individual savings come from turning off large motors and high-powered heaters, in the aggregate de-energizing a number of smaller motors and lights could quickly exceed the savings from a single larger unit. There is an optimum point in the number of switches used to control widely dispersed loads that depends on your method of management. A single switch makes it easy for supervisory personnel to turn off equipment and lighting after hours, but this is very inefficient under conditions of partial use, unless local switching is also provided. Switch panels should be clearly identified.

Energy Conservation by Efficient Use of Water Coolers and Vending Machines

For the convenience of their employees, most industrial and office installations have various types of vending machines, water coolers, etc. These items are generally taken for granted and therefore are seldom considered in a program of energy conservation. Yet, these convenience appliances waste relatively significant amounts of electricity if left operating during non-working hours. Electric water coolers can easily use over 1000 kilowatt hours per year each and vending machine display lighting another 300 to 400 kilowatt hours per year. These seem like small amounts, but translated into dollars, they amount to $12 to $48 per machine or more per year.

EXAMPLE

As a sample of energy savings in this area, consider a company which has eleven water coolers and eight vending machines. The water coolers use 48.4 kWh per day or 17,700 kWh per year and the vending machine lighting uses 8.64 kWh per day or 3160 kWh per year.

Potential savings from operating only 250 days per year.

Present energy use = 17,700 kWh/yr + 3160 kWh/yr

= 20,860 kWh/yr

Reduced energy use for 250 days operation = 20,860 kWh/yr × 250d/365d

= 14,300 kWh/yr

Energy saving = 6,560 kWh per year

Cost saving at $0.03/kWh = 6,560 kWh/yr × 0.03 $/kWh

= $197 per year

SUGGESTED ACTION

Determine whether the number of electric water coolers and vending machines in your plant exceeds actual requirements. Perhaps the number could be reduced by relocation. Remove excess coolers and vending machines and consider shutting down the remainder during extended off-work periods. Consult vending machine service representative about shutting down machines during non-working periods.

Some Representative Appliances

Type	Energy use	Per day use	Per yr use (365 days)	$ per yr (approx.)[a]
Small water cooler	Cooling water	2.0 kWh (20 min/h operation)	730 kWh	22
Large water cooler	Cooling water	4.4 kWh	1600 kWh	48
Coffee vending machine	Display and button lighting	0.48 kWh	175 kWh	5
Candy vending	Display lighting	1.08 kWh	394 kWh	12

[a]Based on cost of 0.03 per kilowatt hour.

SOURCE: National Bureau of Standards. (145)

A color-coded time chart relating to color-coded switches will help cleaners, guards, and substitute personnel identify what should be on and when. As a matter of safety, at least, all equipment to be left in operation after hours should be clearly designated with the responsible person's home telephone number in the event of a mishap, power outage, water leak, etc.

Improved operation of elevators and escalators, which can account for up to 10 percent of the electrical energy of large office buildings and department stores, may be able to help cut electricity costs. (27) Escalator energy use depends on their capacity and they use electricity whether they are carrying passengers or not.

Load can be reduced by switching off escalators and de-energizing one or more units in banks of parallel operating elevators during periods of light use. Where satisfactory stairways exist, elevator service can be limited to alternate floors, and to providing service only above the third or fourth levels. Information can be obtained from the manufacturer which can help optimize the cost of operation against the cost of less frequent passenger service. Older, DC elevators, supplied by motor generator sets, draw power whether or not the elevator is in operation. These should be shut down after hours or during periods of infrequent traffic with clock timers and demand-sensing controllers. In some cases it may be cost-effective to supply the DC for these elevators with new, SCR controllers. When replacing elevators, consider that an elevator with its load partially offset by counterweight is more efficient. Slow-speed hydraulic lifters are generally less efficient than geared elevators.

Finally, do not overlook the lighting in elevators. Elevators may have 300 watts of lighting burning 24 hours a day, 7 days a week. The cost of operating elevator lighting may even be more than the energy cost for the hoisting mechanism.* Between 20 and

*A one-ton capacity elevator making 600 stops per day (4 car-miles) uses typically 2,100 to 2,600 kWh annually.

40 watts of fluorescent lighting would be adequate. With newer models, this lighting will automatically switch on after the elevator has been called, and after a short delay it will switch off when the elevator is not occupied.

Demand Management

Large demand for electric power during peak loading periods can incur expensive demand charges, or even penalties for exceeding contractual limits. Most utilities include a demand charge that is based on the highest value of power (in kW) drawn by a facility during a 15-minute to 30-minute interval. The consequence may amount to one-third of the total cost of electricity, and is based on the extra cost of providing standby generating facilities just to supply the peak rate of energy use. Since usually older and less efficient equipment is called upon to meet peak demand, reducing peaks can also save fuel.

Critical periods in meeting peak demand often occur during hot summer days. Consequently, air-conditioning systems are a fundamental part of this problem. The single unit that draws the most power in the air-conditioning system is the compressor. If a commercial building or factory has a large cooling system with an electric motor-driven water chiller, a substantial power saving can be realized by shutting the compressor down one hour before the normal end of the day. If the chilled water pump and the air-conditioning fans are allowed to continue operation, no objectionable increase in space temperature may be noted. This same procedure can be applied during any peak period. The actual power saving will vary with the size of the chiller. However, for each ton of capacity, at least .75 kW can be removed.

Another power saving during the peak load period for air conditioning (11:00 A.M. to 5:00 P.M.) can be effected by starting the refrigeration equipment an hour or two earlier than normal and pulling the building down a few degrees below normal. By pulling the temperature down lower than usual and then resetting the water chiller temperature regulator higher than normal, the power demand during the peak load period will be less. If the equipment has a demand limiter, further reductions in peak power demand can be realized by resetting the limiter to 80 percent during the working day.

The coolant (chilled water or refrigerant) supply temperature should be as high as possible to satisfy design conditions. Raising coolant supply temperatures allows higher compressor suction temperature. As suction temperature rises, the compressor delivers more cooling capacity per kW. Raising suction temperatures by 5°F (9°C) can result in a 7 percent to 9 percent reduction in compressor load.

A significant power saving can be realized if the air-conditioning system does not use reheat for control. The main reason for using reheat control is to avoid overcooling of the air. For example, the load on the refrigeration plant with a terminal reheat system could be 20 percent to 30 percent higher than necessary for other systems, such as variable volume. As explained in Section 5.5, by raising the coolant temperature it is possible to reduce the amount of reheat, or even completely to eliminate the need for reheat. This would achieve a triple saving: saving from improved compressor efficiency, saving from reduced reheat, and saving from lower peak demand charges.

If your utility billing is the sum of a number of individual meters, each with its own demand recorder, there may be a saving in going to a single meter on the primary side of the transformer feeding these meters. (Retain individual meters for internal energy accounting purposes.) Demand would then be computed from the single meter also. Usually, though not always, the aggregate demand is lower since not all of the individual demand peaks occur at the same time. There may be other economies in primary side metering which your utility representative should be willing to explain for your particular situation.

Information obtained during the load survey, and as the result of recording the connected load during the day, should be helpful in reducing the costs of electric service. Every facility has some loads that can be considered non-essential (sheddable) for some periods during the day. Some services have an innate potential for storage within the system. Although storage of electricity itself to even out peaks and valleys can be done only with difficulty, this limitation may not exist for product flows, compressed air, pumped water, heating and cooling, etc. For example, instantaneous hot water heaters sized for peak demand will cause excessive costs if their use coincides with periods of peak demand. An example of cost reduction by improved demand scheduling is shown below.

Schedule Use of Electrical Equipment to Minimize Peak Demand

Reschedule the use of electrical equipment to lower the demand peak. This action will not reduce the amount of electrical energy used, assuming the same equipment is continued in operation, but will reduce the "surcharge" (demand charge) paid to the power company to provide, in effect, the standby equipment that must be maintained to meet your *peak* demand for power. Theoretically, if you and your community reduce the peak demand, this reduces the standby capacity required, which in turn may postpone the power company's need to install additional capacity to meet an increasing load on its systems. While the individual industrial power consumer's contribution to the utility-wide demand peak may be small, reducing his peak demand can be rewarding.

EXAMPLES

1. A plant operates a group of twelve 30 kW resistance heated furnaces. Each furnace draws its full load of 30 kW for two hours after being turned on and then falls back to a temperature holding rate of 10 kW. All furnaces go through one cycle of heating up, holding, and cooling every 24 hours. By scheduling their use so that no more than two furnaces are on heat-up simultaneously, the following savings in electrical demand charge can be achieved.

Where all 12 furnaces heat-up simultaneously:

$$\text{Peak demand} = 12 \text{ furnaces} \times 30 \text{ kW/furnace}$$
$$= 360 \text{ kW}$$

Where 2 furnaces heat-up simultaneously and the remaining 10 furnaces are on hold:

$$\text{Peak demand} = (2 \text{ furnaces} \times 30 \text{ kW/furnace}) + (10 \text{ furnaces} \times 10 \text{ kW/furnace})$$
$$= 160 \text{ kW}$$
$$\text{Peak demand} = 360 \text{ kW} - 160 \text{ kW}$$
$$\text{Reduction} = 200 \text{ kW}$$
$$\text{Annual demand charge cost saving} = 200 \text{ kW} \times 1.50 \text{ \$/kW mo} \times 12 \text{ mo/yr}$$
$$= \$3600 \text{ per year}$$

2. A small city utility uses an 800 hp pump for eight hours out of each 24. By operating the pump only at night, an off-peak rate reduction of $1.40/kW mo in the demand charge results in the following annual savings.

$$\text{Annual demand charge cost savings} = 800 \text{ hp} \times 0.746 \text{ kW/hp} \times 1.40 \text{ \$/kW mo} \times 12 \text{ mo/yr}$$
$$= \$10{,}000 \text{ per year}$$

SUGGESTED ACTION

A plot of demand versus time is helpful in evaluating the possibilities for savings. If one is not available, your local power company will usually cooperate in preparing such a plot. If the plot shows some high cyclical peaks, usually some savings are possible by altering equipment use or possibly scheduling the use of equipment during off-peak hours.

SOURCE: Bureau of Standards. (146)

The basic idea involves the minimization of the amount of simultaneously operating equipment. The time during which the peak occurs is determined from the demand chart, and equipment is identified which must be operated at this time, taking into consideration requirements of production. Lower-priority equipment is shifted to other times, so that their cumulative energy requirements are satisfied. The duration of the peak demand may last only 20 to 40 minutes. For the equipment to be turned off the quiescent period can be quite short, lasting anywhere from one minute to ten minutes.

A typical priority ranking for shut-down would be: First, heat- or cold-producing equipment such as ovens, driers, kitchen cooking appliances and refrigerators, electric heat and reheat in the HVAC system, and air compressors or water pumps with storage. **Second** would be electric motor-driven equipment, such as chillers, electric fan motors, process equipment (schedule a coffee break during this period). **Third** would be periphery lighting, one or more of a bank of elevators, etc. Care must be exercised in energizing equipment with high starting current, such as large motors, in order not to undo the saving. Likewise it is important to stagger start-up of equipment at the beginning or during the work day in order that these transient starting peaks do not coincide.

Load shedding can be controlled manually or with clocks, thermal switches, or computer systems. With the knowledge of those conditions (equipment testing, etc.) which are likely to cause a peak demand, or by watching the demand power meter, you can manually begin to switch off non-essential loads until the desired demand is achieved.

When the time during which certain loads can be shut off is predictable, (water heaters, for example), load shedding can be controlled at relatively low cost with a clock-operated switch. As an example, at the start of the workday it may be possible to shut down large air-conditioning units during equipment start-up, or when all of the elevators are in use. Peak loads which are not predictable can be limited with various types of controllers which range in complexity from simple thermal sensors to computers. Low-priority loads can be controlled by a thermal activated circuit breaker that switches off when the building load reaches a preset limit. Automatic load controllers combining clock-operated switches with the demand indicating, watt-hour meter are also common. For large installations, more sophisticated, multiple-priority level control can be obtained with computers. These are capable of shedding and restoring loads according to a variety of input conditions.

There are several control schemes used, some of which are only expensive devices to avoid the utilities' demand rate charge, rather than limit the *peak power* drawn. They are based on limiting the average energy used during the utility's demand interval (usually 15 minutes).* These devices cause large surges in power that are undesirable both for the user and for the utility.

Controllers which simply hold the instantaneous peak power below a preset level are a less expensive approach to demand control, and avoid added calculations. This method does not require synchronization with the utility's demand meter. It effectively limits peaks in electricity demand without causing unwanted, periodic surges.

For equipment that is frequently turned on and off, attention should be directed to the temperature rise caused by the surge starting current and to adequate maintenance of motor controllers. Normally a 40°C temperature rise above the ambient temperature is permissible. As long as you can place your hand on the motor frame, the temperature is not excessive. Some new equipment with higher temperature insulation materials may operate at still higher temperatures; this would be indicated on the name plate. For equipment that is available immediately after being energized (conveyors, fans, etc.), and as long as the temperature remains below the manufacturer's recommendation, it is usually more energy- and cost-effective to shut down equipment when it is unused for brief intervals. When this interval exceeds 10 minutes, usually longer lifetime and reduced maintenance are obtained.

*The "Predicting Principle," the "Ideal Rate Principle", and similar schemes start each period with all of the loads connected and start removing various units if it is predicted that the energy used during the period will exceed the demand limit. Some method of synchronizing the computer to the start of the utility's demand interval (a contact closure) may be required where this period does not start on the hour. Because this scheme produces large peaks on the utility's lines every 15 minutes, they often refuse to provide this contact closure. These computers, which are widely advertised by some of the largest computer companies, will have to be modified if this method of "demand control" is disallowed.

Efficient Electric Motors

Electric motors between 1 and 125 hp consume 26 percent of the electric energy in the United States. It is estimated that by 1990 potential savings of 5 percent of the total United States electric consumption could be achieved by the cost-effective incorporation of more efficient motors. Most of the potential improvement would be for motors between 1 and 50 hp, with the largest savings for 5–20 hp motors.*

The energy used by a motor depends on its mechanical load, on the type and efficiency of its construction, and, to an extent, on losses in the circuitry external to the motor by its power factor. The electrical power drawn by the motor, on the other hand, is increased directly by the reciprocal of the power factor. The power loading, energy demand, and operating costs for some different motor sizes are shown in the Table 5.38, for some typical examples of mechanical loading. It is interesting to compare the two motors supplying mechanical loads of 33 and 32 horsepower, at 110 percent and 64 percent respectively, of their rated capacity. The lightly loaded motor draws 120 percent more power and costs $550 more to operate than the more heavily loaded motor. We will discuss motor efficiency next, and power factors and their improvement later.

Efficiency specifications according to the International Electrotechnique Commission (IEC) are presently superior to the unnecessarily vague American Institute of Electrical and Electronic Engineers (IEEE) specifications. Use of the precise IEEE dynamometer method would be superior to both IEEE and IEC methods in current use, and some care should therefore be used in using present efficiency claims.

More efficient motors will have more copper to reduce the resistance in stator and rotor conductors (I^2R loss), closer tolerances to allow for decreased air gap (to reduce magnetizing current), lengthened rotor and stator cores to lower magnetic density, better

*See *Energy Efficiency and Electric Motors*, DOE (Aug. 1976). (143)

Table 5.38. Power Loading, Energy Demand, and Costs for Typical Motor Examples

Size (hp)	*Typical efficiency (%)*	*Assumed loading (hp)*	*Power factor (%)*	*Power load (kW)*	*Hourly energy use [a] (kWh)*	*Cost of operation [b] ($)*
1	73	.85	75	1.16	.87	227
5	80	2.1	57	3.44	1.96	510
10	85	9.2	78	10.35	8.07	2069
30	86	33	82	34.91	28.63	7181
50	86	32	66	42.06	27.76	7731

[a] Energy costs do not account for parasitic losses due to reactive circulating currents in the circuitry external to the motor.

[b] Annual cost of operation calculated for 4000 hours at 5¢/kWh and a monthly demand charge of $2.85 divided by the decimal power factor when lower than 0.85.

rotor and stator design for reduced leakage reactance, and higher quality, non-oriented silicon steel used in thinner core laminations.

Operating efficiencies can be improved as the result of closer specification of motor horsepower to actual connected mechanical load and correction of power factor, as well as the selection of more efficient motor designs. (143) Motors should be closely matched to the actual load required. Oversized motors cost more initially and they cost more to run. They do not necessarily last longer, nor are they more reliable. Figure 5.33 shows the efficiency vs. load curve for a typical 10-hp motor. It can be seen that the power factor drops substantially as the load is decreased. This leads to larger line losses from circulating currents and the liability to a low power factor surcharge on the kWh rate. The efficiency starts to fall off sharply as the load drops below 50 percent of full rating.

The different losses intrinsic to an induction motor are shown in Figure 5.34. This explains why the efficiency is very low at light loads and high near full load. If presently installed motors are greatly oversized, consideration should be given to replacing them with smaller, more efficient designs. An example of the return on investment with the

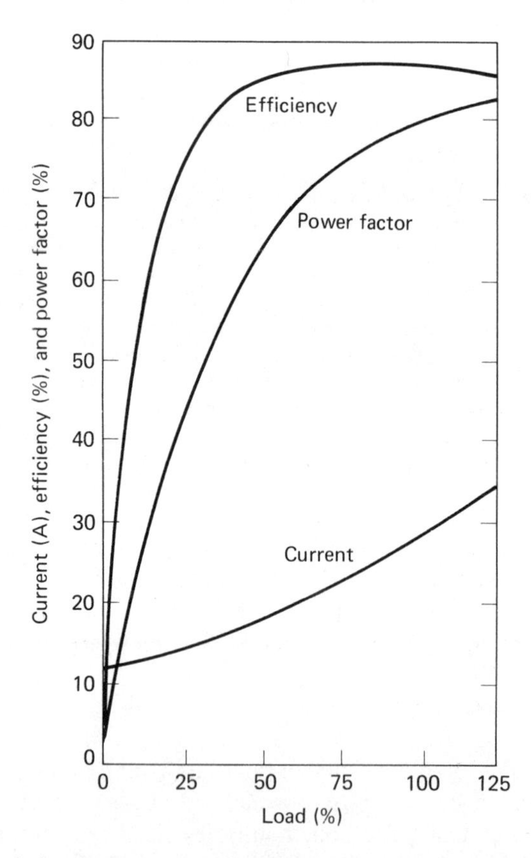

Figure 5.33. Typical performance curves for design B, 10-hp, 1800-rpm, 220-V, three-phase, 60-Hz induction motor. (143)

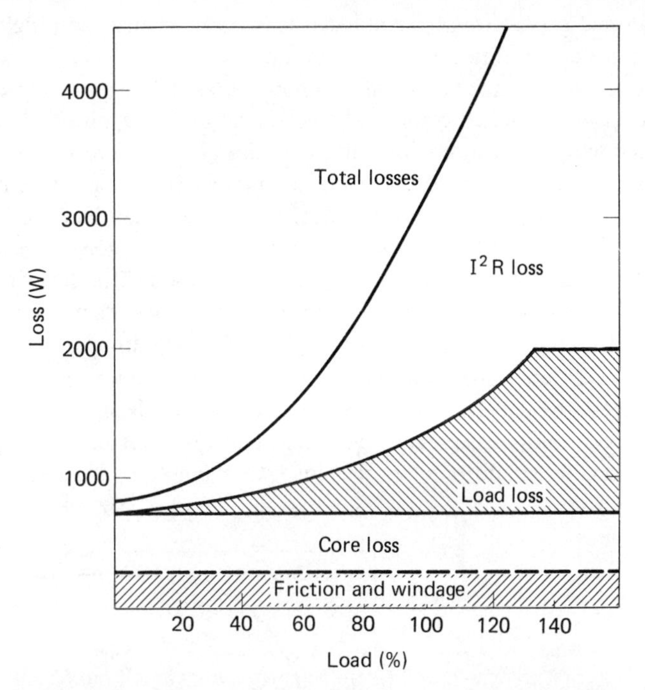

Figure 5.34. Typical load vs. loss curve for design B, 50-hp, 1800-rpm induction motor. (143)

Friction and Windage: This is the input power required to make up bearing and fan windage losses. Since speed varies so little from no load to full load, this loss is constant, unaffected by load.

Core Loss: Core loss is made up primarily of hysteresis losses in rotor and stator iron caused by the 60-Hz magnetization of the core. This loss is also independent of load.

Load Loss: Load loss also occurs in the rotor and stator iron. This loss is roughly proportional to load, or to current squared, and is induced by leakage fluxes caused by load currents.

I^2R Losses: These are heating losses in rotor and stator conductors caused by the current flowing through the conductor resistance. Because it varies as the square of the current, it is generally small at no load but of major proportion at full load.

selection of more efficient motors is shown in Table 5.39. Older motors may have more efficient cores than newer, lightweight designs. Rather than replacing old, heavy frame motors with newer, possibly less efficient models, consider rewinding and installing modern bearings. The result could be both cheaper and more energy-efficient.

Power Factor Improvement

Inductive loads from motors and discharge lighting circuits cause an out-of-phase component in the current. As mentioned above, oversized motors also result in large circulating currents. This condition is described by a term known as the "power factor." A low power factor increases losses in electrical distribution equipment such as wiring,

Table 5.39. Cost-Payback Comparison of More Efficient Motor Choices (Example of a 10-HP AC-Polyphase Induction Motor)

	Standard motor	*High-efficiency motors*		
		A	*B*	*C*
1. First cost	180	224	252	279
2. ÷ life = annual cost	22.50	28.00	31.50	34.88
3. Electricity required (kW)	8.93	8.52	8.33	8.24
4. H use/yr	4,000	4,000	4,000	4,000
5. Efficiency	81.8	85.2	86.8	87.4
6. kW-h/yr	35,720	34,080	33,320	32,960
7. Cost/kW-h ($) (Energy + demand)	.03	.03	.03	.03
8. Annual electric cost ($)	1,071.60	1,022.40	999.60	988.80
9. Difference in elec. cost	0	49.20	72.00	82.80
10. Total annual cost ($)	1,094.10	1,050.40	1,031.10	1,023.68
11. Payback—yrs	—	.89	1.0	1.2

Note: This is a comparison of four different 10-hp motors, with different first costs and different efficiencies: a standard motor and three high-efficiency motors with increased efficiencies of 4.5 percent, 6.7 percent, and 7.6 percent compared to the standard. The wholesale price is increased respectively by 24.4 percent, 40 percent, and 55 percent over that of the standard motor. For the conditions assumed, all three high-efficiency motors are cost-effective. The added costs over the standard motor are paid back in 0.89, 1.0, and 1.2 years respectively, for the three different models.

SOURCE: DOE (143)

switches, and transformers, and reduces the load-handling capability and voltage regulation of your electrical system. In low-power-factor circuits with long distribution lines, as for example to lightly loaded irrigation pump motors, the line losses may exceed the useful energy supplied. At unity power factor these losses are a minimum.

From the power company's point of view, a low power factor requires a larger percentage of the generator output to supply the same effective power. In addition, its transmission and switching systems must be larger to handle the increased currents. However, the larger circulating currents do not register on the watt-hour meter.* A power factor of 0.75 means that of the entering energy, only 75 percent is convertible to usable form. Yet the production equipment, transformers, lines, and switches have to be sized

*If your electricity billing is computed according to volt-ampere-hours, or kvah rather than kWh, your energy charge contains a penalty which is the reciprocal of the power factor. By correcting the power factor to unity, you can receive the same energy at a lower cost.

for 100 percent. Since the low power factors unnecessarily tie up poorly utilized capital equipment, and since the larger currents cause avoidable distribution losses, the utility levies a power factor surcharge when the power factor is below a given level. This is usually 80 percent or 85 percent.

The power factor can be determined by measurement, and the penalty you are paying for a low power factor can be determined from your electricity billing. Proper sizing of motors can improve power factors and the remaining power factor correction can be achieved with the installation of phase-correcting capacitors to completely eliminate the power factor surcharge.

Previous convention considered power factors above 85 percent acceptable. However if the charge for electricity is based on kvah rather than on actual power consumed, or if the power factor penalty in the demand charge results in essentially the equivalent rate, then the power factor for your system should be maintained above 95 percent. Additional savings could occur if this also resulted in the use of smaller transformers and switches. See below for the explanation of a typical example. (147)

Power factor correction can be accomplished with solid state devices, with synchronous motors, or with automatic switching capacitor banks. Synchronous motors can be cost-effective in some cases, but are usually more expensive than capacitors. Power factor correction should be located as close to the offending source as possible. The higher cost of electricity means that the use of central phase correctors, as opposed to power factor correction connected at the point of the offending load, should be reviewed. New solid state circuits developed as an offshoot of the United States Space Program can be built into or attached directly to motors, transformers, etc. Power-factor-corrected equipment, ballast transformers, etc., may cost more initially; however, their performance should be reviewed for attractive life-cycle cost advantages.

Transformers and Switching Equipment

Water, dirt, heat, and corrosive atmospheres adversely affect the efficiency and operation of electrical equipment. Consequently, a clean, dry, cool environment for electric equipment usually pays off in extended lifetime. Opportunities to improve the efficiency of the power system include replacing obsolete equipment with new, more efficient equipment such as SCR controllers for elevators, and transformers with low loss factors.

Correct Low Power Factors

The penalty charge for a low electrical power factor can easily be saved by installing capacitors. The installation will normally pay for itself in one or two years.

EXAMPLE

Consider the case of a plant with a maximum demand of 350 kW, and operating with a power factor of 0.65. Since the power contract has a penalty clause for power factors of less than 0.85, the monthly demand charge is

$$\text{Original demand cost} = 350 \text{ kW} \times 0.85 \text{ PF}/0.65 \text{ PF} \times 1.67 \text{ \$/kw}$$
$$= \$764 \text{ per month}$$

Reference to the figure shows that by installing capacitors rated at 0.55 kvar for each kilowatt of demand, the power factor could be improved to 0.85. Assuming an installed cost of $15 per kvar,

$$\text{Cost of capacitors} = 350 \text{ kW} \times 0.55 \text{ kvar/kW} \times 15 \text{ \$/kvar}$$
$$= \$2900$$

At the improved power factor of 0.85,

$$\text{New demand cost} = 350 \text{ kW} \times 0.85/0.85 \times 1.67 \text{ \$/kW}$$
$$= \$585 \text{ per month}$$
$$\text{Annual cost savings} = (764 - 585) \text{ \$/mo} \times 12 \text{ mo/yr}$$
$$= \$2150 \text{ per year}$$

BRIEF EXPLANATION

Every inductive device (e.g., electric motors, transformers, magnetic vibrators, solenoids, etc.) has one or more magnetic coils through which flow two different components of electric power.

One component, measured in kilowatts (kW), does the useful work and is the quantity recorded by a watt meter. It is approximately proportional to the amount of fuel burned by the electric utility.

The second component, reactive kilovolt-amperes (kvar), represents the current needed to produce the magnetic field for the operation of a motor, etc. This component does no useful work, is not registered on a watt meter, but does some heating of generators, transformers, and transmission lines. Thus it constitutes an energy loss.

The relative amount of the kvar component in an electrical system is designated by the power factor *(PF)*.

$$PF = \text{Useful power/Total power}$$
$$= \text{kw/(kW + kvar)}$$

A light bulb or an electric heater, both non-inductive devices, has a power factor of 1.0. A motor on the other hand, will typically have a power factor of 0.3 to 0.9 as shown in the figure.

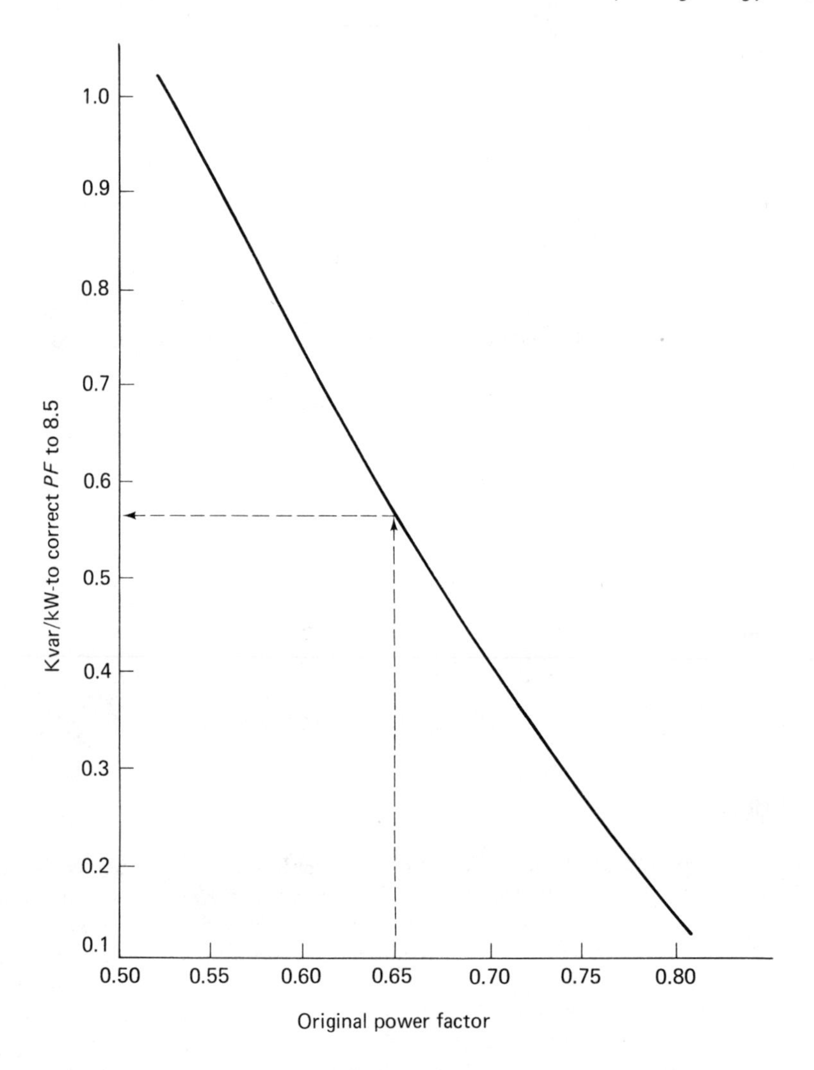

Electric utilities assume a power factor of 0.85 or more in their rate structure. If the overall power factor of a commercial customer is less than 0.85, they add a penalty in the demand charge. The usual formula,

Monthly demand billing = Maximum demand × 0.85/measured power factor × $/kW

The usual cure for a low power factor is to install a capacitor in parallel with the offending machine, or across the line feeding a group of such machines. If the capacitor is properly sized, the kvar currents will not flow between motor and power plant, but will only shuttle back and forth between motor and capacitor. The figure shows the correct size capacitor, in kvar, for different power factor conditions. It is important that the capacitor be as near the correct size as possible.

SUGGESTED ACTION

Determine whether or not your plant is paying a penalty charge for a lower power factor. If so, consider installing corrective capacitors. The advice of an electrical consultant or an engineer from your utility will be helpful in planning an installation.

SOURCE: National Bureau of Standards. (147)

Most transformers range from 93 percent to 98 percent efficient. The losses occur in the core and coils. New transformers have better steel but tend to have smaller cores and smaller wire. This results in higher iron and copper losses compared to older transformers, especially at higher load. Consequently, operating two parallel connected transformers at less than half load may be more efficient than disconnecting one of them because of ohmic losses which go as the square of the current. If you lack the necessary information to make this determination, check with your utility or the transformer manufacturer. One indication of internal losses is given by the temperature rating of the transformer at full load. **Lower temperature rise corresponds to lower internal loss. Transformers with lower operating temperature at full load will also have a longer life expectancy, thereby justifying the higher initial cost for the larger core and greater amount of copper.** For optimum efficiency, keep transformer heat exchangers clean and free of obstruction. In the summer, forced ventilation of the transformer vault may be cost-effective because of the greater efficiency at lower temperature. De-energize the primary circuit of transformers with no secondary load.

CHECKLIST	ELECTRICAL SYSTEM: OPERATIONAL IMPROVEMENTS
Reduce Demand	
Turn off equipment when not in use.	Through letters, memoranda, signs, personal contact, and other means, encourage all building personnel to turn off all electric equipment after working hours and to shut down unneeded systems during operating periods. Reduce lighting to levels actually required. Turn off lights in non-occupied areas.
	Turn off equipment and devices which will be unused for a period of time, for example, vending machines (in which spoilage would not be a problem), and drinking fountains which can be turned off for the weekend. (You may wish to consider time controls to do the job for you. It would be feasible, then, to also turn off such devices during the evening hours.)
	Turn off window displays and revolving signs at least at the end of closing hours, possibly during low traffic periods, and perhaps at all times.

Reduce Demand (cont.)

	Encourage employees to turn off electric typewriters or process equipment, etc., when not actually in use.
Check operating time of all electrical systems.	Is operation at the present level necessary? Can the operating time be shortened or is operation during periods of lower electrical tariffs possible?
Schedule work, processes to avoid idle operation.	Improve product flow, work load scheduling for employees. For high volumes, continuous processing at near peak capacity should be the most energy-efficient. For small volume, quick processes requiring little warm-up, batch processing may be more economical than continuous, if equipment can be shut down between batches, especially if operations can be scheduled to avoid demand peaks. For example, encourage chefs to preheat ovens no earlier than necessary and to forgo preheating completely except for baked goods. Turn off equipment when not in use—would timers be helpful on certain equipment? Interlock accessory equipment with primary production equipment so that it is automatically shut off.
Use electricity appropriately.	Reexamine all uses of electricity for heat and the cost of alternatives. Manual snow and ice removal, for example, is much less costly than electric-resistance snow melting systems. Disconnect electric heating where alternate source is available (delivered at the point of use) at lower cost. Check for wasteful use of radiant electric heaters in open areas. Electric heat tracing should be turned off when there is no fluid in pipes or when the outdoor temperature is above freezing.
Reduce peak loads.	Modify operation to reduce peak loads. Sequence motor start-ups so that current surges while coming up to speed do not coincide. Do likewise for lighting. Use intermittent operation of systems with large electrical consumption during peak hours—for example, refrigeration, electrical heating, air-conditioning systems. Use timers or building automation system.
Take advantage of low tariff periods.	Take advantage of low nighttime tariff to charge systems with sufficient storage to provide service when needed. Operate heat pumps if possible during low tariff periods and store energy in hot water, chilled water, etc. Fill water

Reduce Demand (cont.)

tanks, fuel storage tanks, etc., or take delivery of raw material requiring electric operated transport during this period.

Reduce fan and pump speeds.	Reduce fan and pump motor speeds where flows are in excess of requirements. Since fan or pumping power is a function of the cube of the speed, while flow is directly proportional to speed, operating a fan or pump at the lower flow sufficient to accomplish the task will result in an overproportionate energy saving. In some cases it may be satisfactory to reduce average flow with intermittent switching on and off.
Interchange underloaded motors.	Move lightly loaded motors to heavier duty to match motor's horsepower to load as closely as possible. After reducing air flow, for example, fan motors may be overrated by 33 percent to 66 percent. Improve efficiency and power factor by moving this motor to heavier service. Replace with a smaller horsepower motor. If a motor needs replacement, consider this as an opportunity to interchange motors.
Balance loading on polyphase circuits.	Balance the single phase loads on three-phase, four-wire systems (120/208v or 227/480v) so that no excessive power loss will concentrate on only one of the three phases.
Bill electrical use.	Bill tenants for electrical use. Bill operating departments for energy use.

Maintenance

Upgrade maintenance schedule.	Provide adequate preventive maintenance on power distribution equipment to assure minimum power loss from loose connections and/or contacts. Tighten cable connections to large rotating or vibrating equipment. Use proper compression connector lugs for aluminum cables to avoid high resistance and power loss at the terminals. Clean primary and secondary distribution equipment to avoid unnecessary leakage. High humidity and dirt build-up on electrical parts can cause inefficiencies and, in some cases, total breakdowns.
Provide proper maintenance and lubrication of motor-driven equipment.	Mechanical transmissions frequently are the source of considerable loss. Check clutch operation, proper adjustment. Lubricate motor and drives regularly to reduce friction. Replace worn bearings and seals.

Maintenance (cont.)

Check that drive belts on motors are not loose, worn, or slipping.	Keep belt drives adjusted to the proper tension. Tighten belts and pulleys at regular intervals to reduce losses due to slip. Check alignment between motor and driven equipment to reduce wear and excessive vibration.
Keep electrical equipment clean.	Keep electrical motors clean, and assure adequate ventilation. Hot motors are less efficient and fail more quickly. Additional forced ventilation should be considered for equipment in hot environments. Clean transformer heat exchanger so that they run cooler. Provide shade for transformers, and assure adequate air flow.
Elevators and Escalators	*Note: In considering the following measures which may cause employee waiting time to be increased, keep in perspective that the decrease in their productivity may outweigh the value of the energy saved.*
Optimize operation of vertical transportation.	Reduce the number of elevators in service during hours when majority of persons are not leaving or entering the building. Encourage employees to walk up and down one flight of stairs where staircases are safe rather than to use vertical transportation systems. Reduce the number of elevators operating by scheduling their use for stops at every other floor to eliminate single floor trips and enforce the use of staircases (if the staircases are safe).
Increase elevator load factor.	Consider lengthening floor dwell time to increase passenger load factor. If elevators have demand-type controls, adjust controls so that the least number of elevators travel the shortest distance that demand on the system allows. Reduce speed of elevators and escalators. This will increase the lifetime of this equipment.
Operate escalators only during peak traffic periods.	During light traffic, schedule elevators rather than escalators for vertical travel. For light traffic, elevators are more efficient than escalators, although they do cause indirect energy loss due to stack effects created by the shaft and infiltration around cabs. Where a building usage has high-density traffic flow between floors, escalators may be more energy-efficient than elevators. Check with your manufacturer for comparative operating costs.

Elevators and Escalators (cont.)

Set escalator to agree with maximum traffic flow.	Turn off all "down" escalators during periods of light traffic. Consider turning off "up" escalators on alternate floors during periods of light traffic. Escalators not running will still provide access to upper floors. Turn off escalators after hours and during cleaning.
Reduce elevator lighting.	Disconnect all but one fluorescent fixture per car. Disconnect all unnecessary ventilation fans in elevators where smoking is not permitted.
Attend to elevator drive motors.	Drive motors on older elevators should not be left running for extended periods of time when the elevator is not in use. Turn off the motor-generator set located in the elevator machine room when not in use—nights, weekends, holidays, and slack periods during the day. Do not permit elevators to time out and shut down too rapidly. They should idle long enough so that power consumption is equal to or just less than power consumed in motor generator starting.
Avoid excessive transportation capability during electrical demand peaks.	Connect elevators to the automatic load-shedding or demand-limiting system to enable automatic shut-down of one or more to limit peak demand.

Reduce Demand Peaks

Schedule to minimize electrical demand charge.	Investigate the possibility for scheduling use of power and reducing demand. Adjust start times on large equipment to stagger loads and prevent high peak demands. This is also applicable to heavy shift start times, lunch, and rest periods. Start equipment in an unloaded condition to reduce starting torque requirements. (For example, start pumps against closed valves.) Utilize the heat storage capacity of a building to shut down large fan motors, chillers. Plan storage of hot or chilled water to avoid running equipment during demand peaks, or in an inefficient mode. Turn battery chargers, melting pots, ovens, etc., off for short intervals, during which energy stored will continue to provide service required.
Take advantage of off-peak-load pricing.	Charge batteries for transport equipment at night or during low tariff periods. Observe safe charging rate.

Improve Transformer Management	
De-energize transformers whenever possible.	De-energize excess transformer capacity. Transformer losses range from 0.2 percent to 0.4 percent at no load to 0.9 percent at full load.
	Disconnect primary power from transformers for unused equipment, unoccupied sections of the building, heating equipment transformers during the cooling season, transformers for cooling equipment when cooling system is shut down, etc.
	Use caution in possibly de-energizing needed housekeeping circuits.
	Where there is a bank of two or more transformers, operate transformers at the most efficient loading point.
	Confine clocks, fire alarms, heating control circuits, emergency lighting, etc., and needed process equipment to branches supplied by a single transformer.
Keep transformers cool.	Ventilate transformer banks to keep them as cool as possible. Clean transformer heat exchanger surfaces.
	Shade exterior transformer banks from solar radiation to prevent heat build-up and resultant losses.

CHECKLIST	ELECTRICAL SYSTEMS: SHORT- RETURN IMPROVEMENTS
Improve switching.	Install switches on individual equipment, on local area lighting.
	Install main floor switches so that supervisory personnel can switch off all equipment, lighting after hours.
Automatic switching.	Install demand sensing switching, proximity sensors, etc., to operate lighting equipment when needed.
	Use time interval switches on pumps, conveyors, fans, process equipment to automatically turn off units whose operating duration relates to a given length of time.
Time of day switches.	Provide automatic timers with remote control switches to de-energize equipment which is not required at night and during weekends.

	Operate vending machines, water coolers, office equipment, etc., from a circuit(s) controlled by a time of day switch. Use clock timers to operate equipment during off-peak, low tariff periods.
Meter electrical use.	Increase metering points to better monitor electrical use. Increase maintenance on instruments, meters, and controls where justified by need of accurate measurements and close control to operate at peak efficiency. Install energy and demand meters for operating departments and ouside tenants. Bill according to use. Consider primary side metering (review with utility).
Add load shedding controls.	Use low-cost thermal controls to limit demand peaks. Install manual or automatic controls to disconnect loads when they are not required. To reduce demand charge, size water heaters as small as possible to satisfy average load. Likewise for compressed air and other systems with sufficient storage to meet peak demand. Inexpensive timers can be used for avoiding regularly occurring demand peaks.
Consider energy efficiency when purchasing new equipment.	Choose better insulated refrigeration and cooking equipment. Size electric motors for peak operating efficiency; use the most efficient type of electric motors. Do not oversize motors. Select the electric motors closely matched in function and size for the task. Replace oversize or inefficient motors with newer, more efficient designs. If a motor needs replacement, consider interchanging motors to achieve closer to full loading. Choose two-speed or variable-speed motors to better match changing load conditions. Improve power factor by replacing oversized units with more efficient, properly sized motors. Where it is impracticable to replace motors which have low load and power factors, use capacitors at motor terminals to correct the power factor to 90 percent.
Improve power factor.	The overall power factor should be 85 percent or higher (as close to unity as possible). High peak demand and low power factor both increase your cost. They also increase the electric generating plant size required to service a given job. Install capacitors or solid state devices on motors or inductive equipment that require power factor correction.

Size equipment for low power factor.	Change oversized motors to improve the system power factor and motor efficiency. Reduce power factor correction-line losses by installing capacitors at the source of the phase shift, rather than in a central location. This can increase the load capacity of the switching and distribution system also. For systems with varying power factor due to changing loads, use automatic switching capacitor banks rather than sizing for maximum correction. Note that improper use of capacitors can result in over-voltage conditions.
Get professional advice.	If power factor is low, have a specialist recommend changes.
Select high efficiency discharge lamp ballasts.	Use only high power factor replacement ballasts (90 percent or more) for fluorescent and HID lamps.
Minimize lifetime transformer costs.	When replacing transformers or adding new ones to the system, select those designed for a lower temperature rise but furnished with high temperature rise insulating materials. Consider power loss as well as initial loads and load growth in sizing transformers.

CHECKLIST	ELECTRICAL SYSTEM: MEDIUM-RETURN IMPROVEMENTS
Replace electric resistance heating.	Consider improving efficiency with heat pumps or switch to fossil fuel heat.
Replace oversized motors.	Select smaller motors rated slightly below peak load requirements and allow some overloading to occur. Use 440-volt or higher voltage motors in lieu of 220- or 208-volt motors wherever practical. Use direct drive whenever possible to eliminate drive train losses. Where could large synchronous motors be used for power factor correction? Use two or more motors, or variable speed systems, for equipment with large load variations.
Control demand peaks.	Install automatic load shedding devices to reduce the peak loads. Consider diesel generators to supply peak loads. Equip system with heat recovery. Use standby emergency generators for this purpose.
Recover transformer heat.	Store heated/cooled water for use during peak demand periods.

	Use liquid cooled transformers and capture waste heat for beneficial use in other systems.
Install solid state elevator controls.	Use solid state controls for elevators to shut down motors during periods of light loads, and after hours.
	Consider installation of demand-type elevator controls on elevators having through-trip or collective-type controls so elevators may be adjusted to ensure that the fewest number of cabs travel the shortest distance that demand on the system allows.
	For conditions where an escalator is subject to low usage, it can be equipped with treadle switches to operate on demand. Minimize escalator operating time especially in areas where heat produced has to be evacuated by the air-conditioning system.
Use building exhaust air for equipment cooling.	For example, transformer vaults, elevator equipment rooms, and the like.
Replace electric drive.	Where steam is available, use turbine drive for large items of equipment. Electric motors may be replaced if steam is in surplus, or being vented. Electric motors can be replaced with back-pressure turbines to reduce steam down to process-pressure levels.
	Use steam pressure reduction to generate power.
Consider the use of a total energy system integrated with all other systems.	Use combined cycle gas turbine generator sets with water heat boilers connected to turbine exhaust.
	Or install diesel generators with exhaust heat recovery and cooling system heat recovery.
Replace vacuum tube equipment with solid state electronics.	Replace thermionic tube rectifiers, amplifiers, oscillators, with semiconductor devices available as plug-in replacements. This will avoid the energy use for heaters in the thermionic tubes, and reduce subsequent maintenance costs.
	Replace filament lamps with light-emitting diodes (LEDS).
	Replace thyrotron dimming and speed controls with SCR units.

5.8 *SAVING ENERGY WITH FRESH AND WASTE WATER, COMPRESSED AIR, AND WASTES* *

An establishment which aims at minimizing total inputs and costs must maintain a critical eye toward both *process* and *post-process* energy and material losses. The first category focuses on increasing production efficiency and eliminating energy waste. Post-

*The assistance of Susan Levene with this section is gratefully acknowledged.

process considerations are more qualitative: they require us to determine whether discards and discharges have further industrial, commercial, or domestic value. Both initial resource conservation and resource recovery techniques are involved.

REDUCING PROCESS LOSSES

Water

It is only recently that water has been recognized as a resource that must be conserved. In the past, most areas have taken for granted that water—like sunshine—exists in unlimited supply. Previously, using moderate quantities efficiently has frequently been less economical than excessive water consumption. Large amounts of high-quality water traditionally have been so easily and cheaply obtained, transported, and disposed of that severe and widespread water shortages seemed almost unimaginable.

The problem is not an overall physical scarcity of water. Rather it is having enough, at the right time, in the right place, at a reasonable cost, and of a suitable quality. Highly populated areas and other geographical concentrations of demand are depleting readily accessible water sources. Already 50 percent of the population uses groundwater for domestic purposes. Yet most underground reservoirs (aquifers) are below pumping depths that are economically feasible. The contamination of groundwater is particularly serious since it is virtually impossible to reverse. An aquifer damaged with leachate can be disqualified for decades as a source of drinking water. In addition, water experts foresee a time when the only remaining new sources may lie in basins so deep that the water will be too salty for human use. Some of the western states in particular are quite aware that the mere proximity of an oceanful of water is not enough to ward off the social and economic impact of a dry spell or drought. In the near future, increasing public attention will be drawn to the implications of water shortages. Not far behind (and perhaps a bit ahead) will be economic and legislative incentives and regulations to protect our "daily water."

Principles of Water Saving

Repair and Maintenance of Equipment. The losses and waste involved in the most minor equipment dysfunctions are rarely appreciated. A single slow leak—perhaps from a faulty valve—can result in a loss of 50,000 gallons a year. (148) To heat that water to 160°F would cost approximately 500 gals of oil equivalent energy, plus $100 for the water, in a city whose sewage and water costs are $2.00 per thousand gallons. The repair of six such leaks could save more than $35,000 over a ten-year period, based on 1980 costs.

Reducing Quantity of Water. Closing the gap between the amount of water that is *used* and the amount that is actually *needed* will yield considerable savings. Periodic assessment of water usage can suggest changes in the flow rate and operation time as well as in the total quantity applied to a given process or task. A southern industrial laundry, for example, lowered the water level in two washers by one inch. The minor change in the formula cards conserved

Table 5.40. Examples of Costs and Savings from Methods Used to Save Water

Energy-saving measure	*Cost ($)*	*Savings ($/year)*	*Water saved, gal (ℓ)*	*Comments*	*Reference*
Once-through cooling changed to mechanical cooling	6,000	10,800	6×10^6 (2.3×10^7)	Water-cooled equipment converted from once-through flow to closed system with mechanical refrigeration. Thinner ice used on hockey rink and flow controls installed. Clarkson College, Potsdam, New York.	(24)
Once-through cooling recovered for domestic hot water	20,000	60,000	4.8×10^7 (1.7×10^8)	The Sulzer iron foundry reclaimed once-through air compressor cooling water for use as domestic hot water. Saving is in addition to heat recovered (listed elsewhere).	(150)
Closed-cycle cooling for a casting foundry	160,000	20,000	1.6×10^7 (6×10^7)	Cooling water from the Sulzer casting foundry is now used in a closed-cycle, plant heating system. The investment is also recovered from the value of the heat supplied.	(150)
Cooling water recycled	160,000	80,000	7.4×10^7 (2.8×10)	An electrical equipment manufacturer in Zurich installed a roof-mounted cooling tower to convert once-through equipment cooling to a closed-cycle system.	(151)
Cooling tower make-up water		8,000	10×10^6 (3.8×10^7)	Value on make-up water line to the cooling tower for the ETH computer center was wide open, resulting in an excessive flow of more than 10 million gallons (40,000 m^3) of water per year. Water savings result from adjusting flow from the quantity permitted by the through-put of the piping, to actual requirements.	(152)

535,000 gallons of water, 7,258 therms of gas, and $1,500 a year for each washer. (149) Some examples of costs and savings from methods which save water are shown in Table 5.40.

Matching Water Quality with Task. Financial and environmental costs of developing and delivering new, high-quality water supplies undoubtedly will increase. Using lower-quality water where higher qualities are currently used would conserve available and less expensive resources. Alternate sources of water—such as rainwater—frequently can satisfy quality requirements, and would circumvent the standard supply-treatment system.

Compressed Air

Savings in compressed air systems can be achieved most readily by reducing distribution losses and re-evaluating the uses of compressed air. The first aspect incorporates the basic principles of water conservation: regular repair and maintenance of equipment, matching quantity and availability with task requirements (refer to checklist), and upgrading the efficiency of the system.

Detecting air leaks can be a bit tricky since they are odorless, colorless, and easily hidden by other plant noises. Corrective measures should include:

1. Regular Inspections: Replacing worn parts or packing, cleaning and unclogging air valves, crankcases, heat exchangers, suction filters, and all passages, and adjusting working parts are important in keeping air compressors operating at peak economy. Most manufacturers provide instruction books which describe operation, maintenance, lubrication, and repair.
2. Leak Repair:* Leaks in air pipelines are easily detected by swabbing the joints with soapy water. Any air that is escaping will produce air bubbles.
3. Monitoring Air System: Documenting industrial gas use can help determine standby losses. In one installation, 47 percent of the annual consumption of CO_2 was being lost from a system that appeared to be in good condition. Leakage was detected by graphing CO_2 consumption as a function of equipment operation. Recouping the $8,000 yearly losses mainly required greater care in packing valves. As with most minor water leaks, the cost of small air leaks is frequently undervalued. As is demonstrated below, three quarter-inch leaks in a 100 lb/in^2 line at a power cost of $.015 per kWh would waste 107,000,000 cubic feet of "free air," 2,920 MBtu/year of raw source energy, and $4,370. (153)

Delivering the appropriate discharge pressure is critical to optimum efficiency in pressurized air systems. On the one hand, it is common sense to operate a compressor at the lowest air pressure required. One plant, for example, was using two 2-stage reciprocating compressors driven by 250-hp electric motors to compress 2,500 cfm of free air to 110 psig. It was determined that lowering compressor discharge to 95 psig would not diminish performance, and would save 161,000 kWh and $4,830 per year. (154)

*Because gas viscosities are typically 100 times smaller than liquids (such as water, for example), leaks at valve packings, connections, etc. may be leaking much greater volumes of matter. One method for measuring the leak rate would be to measure the time interval for the pressure to drop to one-half under no load conditions and with the power to the compressors off. Or note the period between on/off cycling of the air compressor under "no load" conditions.

Eliminate Leaks in Compressed Air Lines

The cost of leaking compressed air is often considered insignificant. The following example illustrates that appreciable energy savings can be realized by repairing leaky air lines.

EXAMPLE

A complete inspection of a plant compressed air system was conducted at the start of a regular monthly leak detection program. Air compressor discharge pressure was 100 psig. At a power cost of $0.015 per kWh, the cost of the leaks found in the compressed air system were:

Number of leaks	*Estimated diameter (in.)*	*Free air wasted (cu ft/yr)*	*Fuel wasted (MBtu/yr)*	*Cost of power wasted ($/yr)*
3	1/4	107,000,000	2920	4370
7	1/8	62,200,000	1700	2550
12	1/16	26,600,000	727	1090
15	1/32	8,300,000	227	341
37		204,000,000	5570	8350

The energy and cost saving possible by fixing compressed air system leaks can be estimated from the following table:

Hole diameter (in.)	*Free air wasted (a) (cu ft per year) by a leak of air at*	*Fuel wasted (b) (MBtu/yr)*	*Cost of power wasted (c) ($/yr) at unit power cost of*		
	100 psig		*$0.010/kWh*	*$0.015/kWh*	*$0.020/kWh*
3/8	79,900,000	2190	2190.00	3280.00	4370.00
1/4	35,500,000	972	972.00	1460.00	1940.00
1/8	8,880,000	243	243.00	364.00	486.00
1/16	2,220,000	60.6	60.70	91.00	121.00
1/32	553,000	15.1	15.10	22.70	30.30
	70 psig		*$0.010/kWh*	*$0.015/kWh*	*$0.020/kWh*
3/8	59,100,000	1320	1320.00	1980.00	2650.00
1/4	26,200,000	587	587.00	881.00	1170.00
1/8	6,560,000	147	147.00	220.00	294.00
1/16	1,640,000	36.6	36.60	54.90	73.30
1/32	410,000	9.2	9.18	13.80	18.40

Assumptions used:
(a) Based on nozzle coefficient of 0.65,
(b) Based on 10,000 Btu fuel/kWh,
(c) Based on 22 brake horsepower per 100 cu ft free air per min for 100 psig air and 18 brake horsepower per 100 cu ft free air per min for 70 psig air.

SOURCE: National Bureau of Standards (153).

Lower Pressure of Compressed Air to Minimum Necessary Level

If the users of a compressed air system are surveyed and it is determined that the air pressure can be lowered to a certain amount without causing operating problems, the horsepower for the air compression can be decreased resulting in energy savings.

EXAMPLE

A total volume of 2500 cfm of free air was being compressed to 110 psig by 2 two-stage reciprocating compressors, each driven by a 250 hp electric motor fully loaded. The compressors operated 8000 hours per year. A survey revealed that all plant air users could operate satisfactorily with 15 psi lower air pressure. The approximate savings that could be realized by lowering compressor discharge pressure from 110 psig to 95 psig are determined as follows:

From Figure (b) for two-stage compressors, the decrease in horsepower would be about 5.4% from reducing the pressure to 95 psig.

$$\text{Annual savings in kWh} = 5.4\%/100\% \times 500 \text{ hp} \times 0.746 \text{ kW/hp} \times 8000 \text{ h/yr}$$
$$= 161{,}000 \text{ kWh per year}$$

If the utility consumes 10,000 Btu of fuel/kWh generated,

$$\text{Annual savings in Btu} = 161{,}000 \text{ kWh/yr} \times 10{,}000 \text{ Btu/kWh}$$
$$= 1610 \text{ MBtu per year}$$

If the cost of electric power is $0.09/kWh

$$\text{Annual savings in \$} = 161{,}000 \text{ kWh/yr} \times 0.09 \text{ \$/kWh}$$
$$= \$14{,}490 \text{ per year}$$

SUGGESTED ACTION

Canvass users of plant air to determine the practicality of lowering the air system pressure. To lower the discharge pressure on some air compressors requires a simple adjustment of the pressure control. On other compressors, modification may be necessary and the compressor *manufacturer should be consulted*. The manufacturer can also provide you with the performance data for your particular compressor and inform you of any limitation on lowering your compressor discharge pressure.

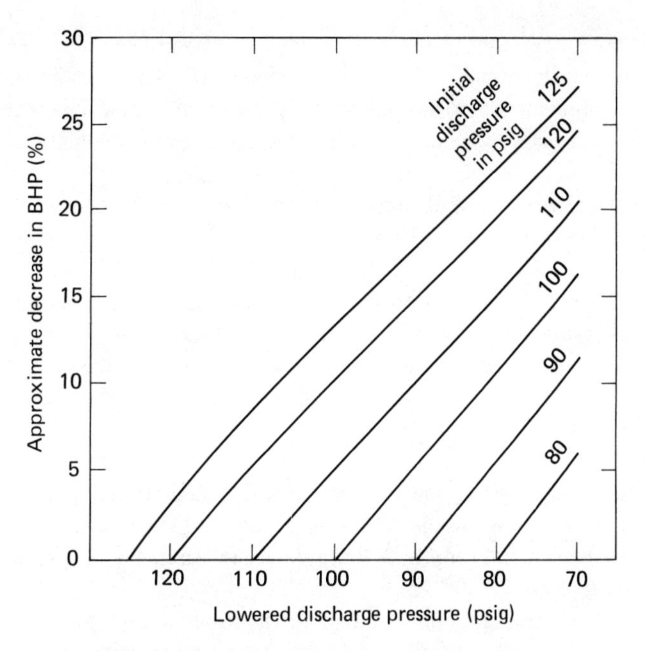

(a) Single-stage reciprocating and rotary screw compressors.

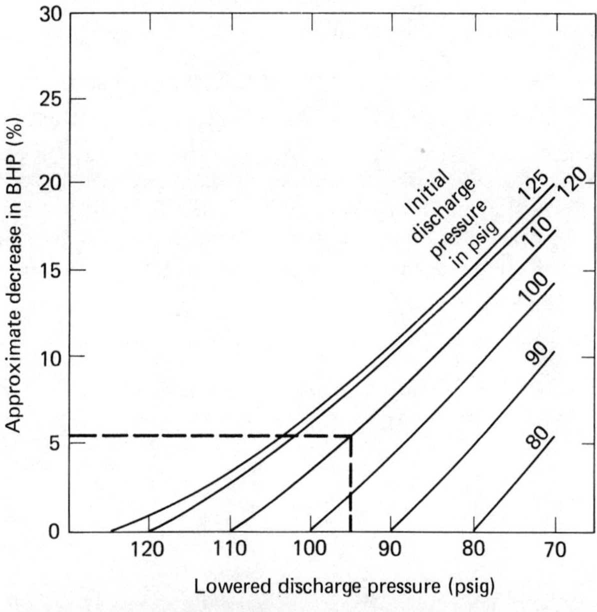

(b) Two-stage reciprocating and centrifugal compressors. [National Bureau of Standards (154)]

SOURCE: National Bureau of Standards (154)

On the other hand, insufficient discharge pressure can also cause energy waste. If air to a hand-held pneumatic tool is discharged at 80-psig pressure, rather than the required 90 psig, efficiency can be cut by 15 percent. Accurate calculations of the discharge pressure will need to account for the pressure drop in the distribution lines to the equipment.

Reducing the size and number of compressor units can result in significant energy savings. The power cost for operating a 500-cfm unit at full load is about $5.00/hour at 5¢/kWh.* A 2,500 cfm compressor at light load, however, would also cost about $5.00/hour. The use of one 100-hp compressor at full capacity during one shift, rather than two 100-hp units each at half capacity, could save over $2,500 a year at typical 1980 prices. It is worth noting that compressors at part load not only waste energy, but also have a poor power factor, which can result in utility penalty charges not included in these examples. (155)

Significant energy savings also can be achieved through a more selective use of pressurized air. Operations such as cleaning and drying parts, agitating liquids, and comfort conditioning should not be performed by such an energy-intensive method. The question, then, is not only how to upgrade the efficiency of a system, but whether to use that system at all.

The production of compressed air can be improved by lowering the intake temperature of the air. The saving possible by installing an intake pipe to draw in outside air, rather than warm air from the equipment room, is explained below.

*At 5¢/kWh including demand changes.

Install Compressor Air Intakes in the Coolest Locations

Wherever feasible, the intake duct for an air compressor should be run to the outside of the building, preferably on the north or coolest side. Since the average outdoor temperature is usually well below that in the compressor room, it normally pays to take the cool air from outdoors. The energy savings potential in lowering the air intake temperature is illustrated in the following table.

Temperature of air intake (°F)	*Intake volume required to deliver 1000 cu ft of free air at 70° F*	*% HP saving or increase relative to 70° F intake*
30	925	7.5 saving
40	943	5.7 saving
50	962	3.8 saving
60	981	1.9 saving
70	1000	0
80	1020	1.9 increase
90	1040	3.8 increase
100	1060	5.7 increase
110	1080	7.6 increase
120	1100	9.5 increase

Regardless of the outside temperature, the compressed air in a shop pipeline will closely approximate the temperature in the shop by the time it reaches the tools. Assume the indoor temperature to be 70 F. If the compressor takes in 1000 cu ft of free air from the shop, it will also deliver 1000 cu ft of free air at the tools because the initial and final temperatures are the same.

Suppose the outside air averages 50 F and the compressor is supplied with intake air from outdoors. Only 962 cu ft of free air will be required to deliver 1000 cu ft of free air at the indoor temperature of 70 F, a saving of 3.8% in the horsepower required.

EXAMPLE

A compressor takes its inlet air directly from the compressor room where the average temperature is 80 F. The compressed air at 100 psig is delivered to a shop building where the temperature is maintained at 70 F. The compressor delivers 750 cu ft per min of free air at 70 F for 8000 hours per year. The 150 hp electric motor drive operates at full load. The average outside air temperature is 50 F. The energy savings to be realized by taking compressor inlet air from outdoors are calculated as follows:

From the table:
For 80 F air at the intake, volume to deliver 1000 cu ft free air at 70 F = 1020 cu ft.
For 50 F air at the intake, volume to deliver 1000 cu ft free air at 70 F = 962 cu ft.
The saving on power from using the cooler intake,

$$\text{HP saving} = (1020 - 962) \div 1020 \times 100\%$$
$$= 5.69\%$$

$$\text{Annual power savings} = 5.69\%/100\% \times 150 \text{ hp} \times 0.746 \text{ kW/hp} \times 8000 \text{ h/yr}$$
$$= 50{,}900 \text{ kWh per year}$$

If the utility uses 10,000 Btu fuel per kWh,

$$\text{Annual energy savings} = 50{,}900 \text{ kWh/yr} \times 10{,}000 \text{ Btu/kWh}$$
$$= 509 \text{ MBtu per year}$$

If electric power cost is $0.06/kWh,

$$\text{Annual cost savings} = 50{,}900 \text{ kWh/yr} \times 0.06 \text{ \$/kWh}$$
$$= \$3{,}054 \text{ per year}$$

SOURCE: National Bureau of Standards (156)

Waste

In late 1973, the National League of Cities and the United States Conference of Mayors identified waste management as one of the most critical problems confronting cities. Two issues immediately targeted were "the sharp decline of available urban land for disposal sites" and "the sky-rocketing volume of solid waste." (157) The rapid increase of waste generation has intensified the concern over inadequate disposal techniques; threats to our air, water, and food; resource depletion; and escalating collection and disposal costs. In 1977, 145 million tons of municipal trash and garbage and 260 million tons of industrial waste were generated. Urban collection and disposal costs are currently $6 billion—and rising. By 1985, the collection and disposal of one ton of waste is expected to rise from $30 to $50. In 1979, federal guidelines stipulated that disposal sites throughout the nation must be shut down or upgraded by 1983. Substantial funds will have to be made available to inspect, improve, and/or properly close down the 12,000 land disposal sites (out of a total of 18,000) which do not meet current state standards.

The conclusion: Waste is inflationary. The more waste that is produced, the more it will cost to handle, store, treat, collect, transport, and dispose of. Effective waste management calls for a reduction in the quantity and volume of waste that is generated. Reducing waste will also reduce the energy it represents as well as the energy to collect it, to haul it away, and to dispose of it. (158)

A considerable amount of energy is represented in just the waste paper that is produced each day. Furthermore, consider the time and money spent in handling this paper each day before it is declared waste, as for example the time and money spent preparing a photocopied product. Consider the use of electronic mail and improved word-processing techniques as systems with an enormous potential to reduce the quantity of paper that flows through your establishment each week.

A controversial yet significant example that emphasizes the connection between waste management and source reduction is packaging. Back in 1971, packaging accounted for "approximately 47% of all paper production, 14% of aluminum production, 75% of glass production, more than 8% of steel production, and approximately 29% of plastic production." (159) The 40–50 million tons of solid waste that come from packaging activities clearly demonstrate the need for design practices which are less material- and energy-intensive and which support recycling efforts.

These source reduction principles are also applicable to atmospheric wastes. One facility examined the cost of operating exhaust systems designed to remove airborne contaminants from work spaces. At $.30/gallon for oil (and $.04/kWh), the heating and cooling of 10,000 cfm of make-up air cost $1,530 for fuel and $4,830 for electricity. The company subsequently installed environmental control units containing electrostatic precipitators which ionize and strip dirt and fume particles from a recirculating air stream. The $41,200 to buy and install an electrostatic precipitator that passes 1000 cfm was clearly an investment that cut costs, energy demand, air emissions, and make-up air requirements. (160)

The second aspect of waste management centers on volume reduction, primarily of solid wastes. It has been estimated that urban solid waste volumes have doubled in the past twenty years, and continue to grow five times faster than the population. Industry, business, and public institutions are confronted with overflowing storage facilities in building service areas where access for removal is limited and which contain contaminated or infectious materials. A number of volume reduction methods exist which permit more material to occupy a given space. Baling, crushing, and packing, for example, are useful in lowering long-haul transportation costs. Equipment such as shears, hammermills, rasp and disk mills, and wet pulpers shred solid waste into fairly homogeneous materials.

European successes using shredded waste processes range from reducing refuse fires to reducing litter and vector problems. The pulping process grinds and shears loose or bagged dry wastes in the presence of water, producing a slurried wet waste. Passing this waste through a dewatering press and junk remover yields sludge forms for subsequent landfilling or incineration.

POST-PROCESS CONSERVATION

The second avenue for efficient utilization of energy involves recapturing and reusing valuable and depletable resources from refuse. The basic argument in favor of resource recovery is that it makes economic and environmental sense to extract materials and energy from the waste stream and divert them back into the production stream. The importance of reclamation lies in its potential to influence conservation and waste management efforts. For example, 400,000 tons of municipal waste are generated daily. (161) EPA reports estimate that mixed waste from the larger urban areas could produce 830 trillion Btu's of energy, saving 400,000 barrels of oil per day.

The nation consumes in excess of 200 million tons of major metals, paper, glass, rubber, and textiles, of which 75 percent comes from virgin resources. (162) Potentially, of the United States total annual use, 19 percent of all tin, 14 percent of paper, 8 percent of aluminum, 5 percent of copper, and 3 percent of lead could be recovered from our waste stream. (163)

A system using secondary rather than virgin materials causes less water and air pollution, produces less solid waste, and consumes less energy. Consider, for example, that recycling scrap aluminum requires only 5 percent of the energy needed for processing the virgin ore. (164)

Essentially, waste is the damaged, residual, or defective materials that remain unused by a specific manufacturing process or consumer activity. Conventionally, our leftovers are discarded, removed, flushed away, buried, and forgotten. Resource recovery, however, is built around the idea that waste production—solid, liquid, and gaseous—is one phase in an overall resource *cycle*. Materials recovery focuses on reclaiming physical resources such as paper, ferrous metals, non-ferrous metals, glass, oil, and water. Energy recovery, in contrast, aims at harnessing the waste heat from manufacturing processes (see Box below and the examples cited in Chapter 6), and combustible wastes which cannot be reclaimed properly. (165)

Recovery of Heat from Hot Waste Water

EXAMPLE

An opportunity to save $34,000 per year while conserving 10 MBtu per hour was identified by a large textile company by recovering heat from hot waste water. A major source of hot water discharge in the textile industry is the griege preparation ranges which perform a continuous fabric washing operation.

For this particular application, the discharge of waste water from the griege preparation ranges is 360 gpm at 160 F. Rather than discharging this hot water to a drain, it was decided to preheat the 360 gpm of cold inlet water having a yearly average temperature of 63.3 F, by passing it through a counterflow heat exchanger with automatic back flushing to reduce fouling (characteristic of textile processes).

Based on a heat recovery factor of 58%, and operation 4440 hours per year,

$$\begin{aligned}\text{Annual energy saving} &= 360 \text{ gal/min} \times 8.34 \text{ lb/gal} \times (160 \text{ F} - 63.3 \text{ F}) \times 60 \text{ min/h} \\ &\quad \times 4440 \text{ h/yr} \times 1 \text{ Btu/lb F} \times 0.58 \\ &= 44900 \text{ MBtu}\end{aligned}$$

Using an energy cost of $0.90 per MBtu,

$$\begin{aligned}\text{Annual energy cost savings} &= 44{,}900 \text{ MBtu/yr} \times 0.90 \text{ \$/MBtu} \\ &= \$40{,}400 \text{ per year}\end{aligned}$$

Assuming a 5 year capital recovery period, the estimated annual amortization cost of the 10 MBtuh heat exchanger plus the installation cost is $6,000 per year.

$$\begin{aligned}\text{Net cost savings} &= 40{,}400 \text{ \$/yr} - 6{,}000 \text{ \$/yr} \\ &= \$34{,}400 \text{ per year}\end{aligned}$$

These savings will, of course, increase if energy costs become greater.

SUGGESTED ACTION

Evaluate the feasibility of heat recovery from all hot waste streams. Selection of proper counterflow heat exchange equipment for a particular system is very important. Recognize the potential of excessive maintenance due to fouling problems.

SOURCE: National Bureau of Standards (165)

Materials Recovery

Often, waste products retain a portion of their original value and can be recycled, i.e., upgraded and used again in the same kinds of processes. The single largest component of municipal solid waste, and our most important renewable resource, is paper, which constitutes 32 percent by weight and 50 percent by volume of all commercial and residential discards. Given present market resale value, tab cards and high-grade white paper, which compose between 30 percent and 80 percent of average office waste, are the most financially plausible materials to retrieve from office waste streams. The Environmental Protection Agency has quoted the following prices from 1976 for recycled paper: (164)

Grade	*Price range ($/ton)*
Manila tab cards	165–220
White ledger paper	70–100
Newsprint	20–40
Mixed Paper	5–20

To optimize recycling efficiencies, recoverable materials should be set aside wherever they are generated to avoid the task of separation later. For example, employees can be provided with separate wastebaskets or desk-top containers for recyclable paper. The custodial staff can then empty either the individual wastebaskets or centrally located containers into which employees have deposited their own waste paper. There are several advantages of paper recycling: it reduces the amount of waste, recovers resources and lowers disposal costs; it can save money for a product which otherwise costs money to remove; it does not require large equipment investments; collection is frequently included in recycled materials sales contracts; and it is easily incorporated into office routines.

Although still in a developmental stage, multi-material recovery subsystems have been designed which can retrieve ferrous and non-ferrous metals, glass, and paper. In a "typical" system, oversized and hazardous materials are screened out from the raw refuse, which is then fed to a shredder—most often a hammermill. The refuse is then air classified into a light fraction (primarily organic and combustible materials) and a heavy fraction (ferrous metals, aluminum, glass, zinc). Salable ferrous metals are easily and inexpensively recovered through magnetic separation, which is a commercially established technology. The nonmagnetic material is sent through a series of screening and classifying operations to produce a heavy fraction (copper and zinc) and a light fraction (aluminum and glass). The final non-color sorted glass, baled paper, and nugget-like mixture of aluminum, copper, and zinc-based alloys are suitable for recycling into existing markets.

The longest operating materials recovery system is located in Franklin, Ohio. Solid waste is ground in a hydropulper; fibrous material is screened, washed, and pumped in a slurry to a nearby roofing felt mill; and reject fiber and sewage sludge are burned in a fluidized bed incinerator. EPA implementation grants have encouraged a number of industries and municipalities to construct resource recovery systems. In addition, investment tax credit may be applicable for equipment which can recycle metals, paper, textiles, rubber, and other materials. It should be noted that this investment incentive is on top of an already existing credit for the rehabilitation of industrial and commercial buildings which has been in use for at least twenty years. Private industry can also further reclamation efforts by developing processes that use secondary materials and by designing products that are reusable, more durable, and easier to recycle.

Recover Heat from Waste By-Products

EXAMPLE

A chemical manufacturing firm will save $240,000 per year by burning a liquid chemical by-product to generate 64,000 lb/yr of 300 psig process steam, which is presently produced by burning purchased fuel oil. Use of the by-product as a fuel also eliminated a disposal problem for an environmentally objectionable material. The cost of the boiler, with special design features for corrosion and soot deposits, is approximately $400,000, and the cost of installation and accessories is approximately $300,000.

The by-product has an average heating value of 6000 Btu/lb and is completely oxidized at 1800 F. A total of 200,000 cfm of flue gas will be produced and quenched to 200 F before passing through a scrubber into the stack.

The flue gas flow rate corresponding to 200,000 cfm at 1800 F is 170,000 lb/h. The specific heat of the flue gas is 0.30 Btu/lb F and the saturation temperature of 300 psig steam is 422 F. The waste heat boiler was designed to have an exit flue gas temperature of 500 F. Therefore the energy available from the flue gas, is:

$$\text{Energy available} = 170{,}000 \text{ lb/h} \times 0.30 \text{ Btu/lb F} \times (1800 - 500)\text{F}$$
$$= 66.3 \text{ MBtuh}$$

If the feedwater is at 220 F, the heat required to produce saturated 300 psig steam is 1027 Btu/lb. Allowing for 1% heat loss,

$$\text{Steam production} = 66.3 \text{ MBtuh} \times 1 \text{ lb/1027 Btu} \times 0.99$$
$$= 63{,}900 \text{ lb/h}$$

The company was paying $1.25 per 1000 lb of steam when it was produced from fuel oil. Based on 3000 hours per year operation,

$$\text{Annual cost savings} = 63{,}900 \text{ lb steam/h} \times \$1.25/1000 \text{ lb steam} \times 3000 \text{ h/yr}$$
$$= \$240{,}000 \text{ per year}$$

SUGGESTED ACTION

Liquid by-products and solvents should be considered as fuel sources for steam generation, space heating, etc. If your fuel oil costs are higher than the approximate 18¢/gal in this example, annual savings will accrue at a faster rate. Consult manufacturers of boilers and waste heat recovery equipment for recommendations.

SOURCE: National Bureau of Standards (167)

Although materials recovery has been oriented chiefly to the reclamation of solid wastes, liquid recycling also warrants consideration. It is particularly significant in the case of petroleum products, since they are derived from nonrenewable raw materials. It has been estimated that crankcase oil consumed nationally is equivalent to about 10 billion barrels of oil, and that waste machinery oil translates into approximately 5.25 million barrels. The collection and burning of 75 percent of all waste oil would save in excess of 11 million barrels of residual oil. (166)

Another example, explained below, illustrates how process heat can be produced from a waste byproduct from a chemical manufacturing process. (167) Four examples are summarized in Table 5.41.

The re-refining of waste oil into used lubricating products is the most energy-conserving method for waste oil utilization. The procedure involves, first, the removal of contaminants, spent additives, and compounds formed during initial use; second, distillation into light fuel and heavy lube oil fractions; and third, final purification prior to blending. (168) The end result is a high value product, which requires less energy and releases lower levels of air and water emissions than initial petroleum production. Currently, waste oil sales are made principally to processors who, at best, remove many of the contaminants before marketing it as a fuel, and to suppliers of road oil, which is used in asphalt manufacture and as a dust suppressant. As petroleum prices continue to climb, in-house purification and secondary use will become more attractive. The University of Rhode Island, for example, has begun purchasing 2,500 gallons of used crankcase oil from Narrangansett Improvement Company in order to supplement the No. 6 oil in its heating boilers. The 20¢/gallon spent for waste oil—compared with 34.6¢/gallon for fresh No. 6 oil—saved URI more than $7,000 a month in its heating bill. (169)

Any materials recovery program should examine the feasibility of water reclamation. For the most part, industrial water usage is well suited to recovery and renovation processes. In the first place, most industrial operations that require water are non-consumption processes, so that the water remains physically accessible. Actual consumption occurs principally through evaporation in heat exchanger systems. Second, almost all water must be treated prior to industrial use. It would seem reasonable for an industry to pay at least as much money to recondition waste water as it would pay to treat its incoming fresh water and polluted discharge water. Third, waste water technology is sufficiently developed to produce industrial process waters of almost any required quality. Fourth, the large volumes of water used in industrial processes generally do not need to comply with drinking water standards bacteriologically or chemically. And fifth, most industrial water demands are fairly constant, and waste water flows are usually available continuously.

In many cases it is not appropriate to talk about recycling water in the strict sense of restoring it for its original purposes. The term "reuse" more aptly conveys the idea of successive, lower grade applications, in which the resource requirement for one task is met by using the water effluent of another. For example, potable water (sanitary, domestic, or drinking water) can conveniently be reconditioned and reused for parts washing, toilet flushing, and particularly recirculating systems. Cooling processes, which require minimal if any treatment, account for the largest proportion of industrial water usage. It is both economical and energy wise to recirculate cooling water in an essentially closed system. Only a small percentage of outside water is needed to restore losses due to evaporation, leakage, and contamination. Significant energy losses also occur when domestic hot water is allowed to drain from a building after single use. Clean water used for cooling should be reused in warm water systems; and condensates, for instance, can be used for boiler feedwater.

Table 5.41. Examples of Costs and Savings from Utilizing Waste Products from Industrial Processes as Sources of Energy

Energy-saving measure	*Cost ($)*	*Savings $/year*	*Energy-saving MBtu (kWh)*	*SCP[a] $/BPD ($/kW)*	*Comments*	*Reference*
Heat from liquid chemical wastes	700,000	240,000	200,000 (5.8×10^7)	2550 (36)	A chemical firm disposed of a liquid chemical waste with environmentally objectionable properties, by burning to produce 30 tons of steam per hour. The exhaust gas was quenched to 195°F (90°C) before passing through a scrubber into the stack. (See Table 5.40.)	(167)
Preheating with waste water	30,000	40,000	44,400 (1.3×10^7)	710 (10)	Hot waste water from fabric washing at 160°F (70°C) is sent through a counterflow heat exchanger with automatic back flushing to clear lint fouling. Inlet water at an annual average temperature of $63^1/_2$°F (17.5°C) is preheated, with a 58% heat recovery factor.	(165)
Waste heat boiler	37,000	58,200	60,000 (1.77×10^7)	850 (12)	Asphalt saturated air released at 1,400°F (760°C) is subsequently burned in a waste heat boiler at a cost of $37,000, to produce 20,000 lbs (9 metric tons) per hour steam at 150 psi (10 atmospheres) pressure.	(74)
Process heat from wastes	13×10^5	600,000	1.13 million (3.3×10^8)	2480 (35)	By increasing concentration of a waste gas stream to combustible levels and installing a special boiler, it was possible to recover process heat and reduce natural gas use at Union Carbide's Brownsville, Texas, plant by 3 million cf^3 (85,000 m^3/day).	(123)

[a]The specific capital productivity (SCP) is the cost of reducing the annual rate of energy use by one barrel of oil per day energy equivalent (BPD) or one kilowatt.

One final form of materials recovery deserves mention: waste swapping. It involves company A selling a waste product that it has generated to Company B, which can use the product in its own manufacturing processes. The passive approach consists of industry trade associations, regional chambers of commerce, or other groups serving as clearinghouses to match generators with users. A few small American and European companies have assumed the more active role of becoming material exchanges which handle, treat, and certify the characteristics of chemical materials. Wastes that may have transfer and reuse value include solvents, alkalis, concentrated acids, catalysts, wastes with a high concentration of recoverable metals, oils, and combustibles. The estimated 6 million metric tons/year (wet basis) that can be profitably exchanged certainly will not solve the waste problem. Waste swapping, however, can capture an important niche for particular chemicals. And the real significance of waste transfers is that it suggests the kind of creative and resourceful thinking that is needed in developing a comprehensive waste management program.

Solid waste, in addition to containing a wealth of recyclable and reusable materials, can be recovered as a source of energy. Most of the technologies being developed in this area involve waterwall combustion, pyrolysis to produce gaseous or liquid fuels, and the production of refuse-derived fuel. Waterwall combustion, attractive in its simplicity, pneumatically fires the light combustible waste fraction into existing boilers. Steam that is generated can be used for district heating/cooling, industrial processes, and electric generation. The Refuse Energy Systems Company in Saugus, Massachusetts, for example, processes 1,200 tons of solid waste per day from 16 communities north of Boston, and provides process heat equivalent to over 17 million gallons of oil a year to the General Electric Company. (170)

Pyrolysis, or destructive distillation, involves the chemical and physical decomposition of organic material through heat treatment. In one type of system, such as the "Landguard" unit in Baltimore, Maryland, shredded solid waste is fed into a pyrolytic incinerator which operates at very high temperatures and in the absence of oxygen. The organic fraction is gasified and subsequently burned in an afterchamber to reduce pollutants. In the second type of unit, such as the one designed by the Occidental Research Corporation, incoming waste is shredded and air classified, and inorganic materials are removed for recycling or landfill. The organic waste is then reshredded and fed to a pyrolytic reactor which first gasifies and then rapidly condenses the refuse into an oil-like liquid. The system designed by Occidental can produce a liquid which has an average heating value of approximately 12,000 Btu/lb., a low sulfur char with a heating value of about 9,000 Btu/lb., and a gas stream with a heating value of around 600 Btu/ft^3.

Refuse-derived fuel systems involve the use of shredded waste as a supplementary fuel in fossil-fueled electric generating plants. For instance, the RDF and materials recovery system for Monroe County, New York, handles 2,000 tons of solid waste per day. Shredded waste is air classified into light and heavy fractions. Approximately 65 percent of the input is reclaimed as refuse-derived fuel, which then is used by Rochester Gas and Electric as supplementary fuel for coal-fired boilers. A plant constructed in East Bridgewater, Massachusetts, by Combustion Equipment Associates pulverizes combustible waste into a fine, powdered substance. Known as Eco-Fuel II, each ton of the dust-like fuel is equivalent to 2.5 barrels of residual oil.

In a sense, the field of resource recovery is on trial. It is unclear which of the tried and untried technologies will ultimately provide relief for waste and energy problems. It is difficult to predict which factors—economic, institutional, or technological—will determine what constitutes maximum recovery. In part, the rate and extent of development will hinge on government incentives, assistance, and regulations. The Resource Conservation and Recovery Act passed in 1976 provides a framework for understanding and anticipating federal priorities. The EPA grant program to help finance demonstration projects in resource recovery and source separation, the technical assistance available to state and local governments around sound management practices, and the 10 percent recycling investment credit illustrate the federal approach to stimulating growth. Regulatory legislation (e.g., enforcing landfill standards and controlling hazardous waste disposal) offer insight into federal interpretations of key problem areas and issues.

CHECKLIST	WATER: OPERATIONAL CHANGES
Periodic controls of washrooms and toilets.	Check water taps for possible leakages. Close off unneeded water taps.
Shut off cooling water. Shut off water-consuming installations if installations are not in operation.	Instruct safety personnel on how to shut off, what are the consequences, who has to be notified.
Shut off circulation pumps for domestic hot water outside of operation.	If possible, install automatic shut-off. Circulation should be shut off in any case if domestic hot water is not needed for periods longer than 4 hours.
Reduction of water quality and quantity.	Is a change in process or installations possible to reduce water consumption without consequences on product? Adapt to temporary changed conditions—for example, shut off cooling water in winter.
Cover open tanks.	Reduce evaporation.
Adapt water flow rates to actual needs.	Normally water flow rates are regulated according to maximum amount.
Check and if necessary replace seals.	Repair leakages immediately on piping connection pieces, valves, water taps, etc. Apply additional insulation on warm water pipes, do not allow uninsulated water pipes with surface temperature over 30°C.
Minimize cooling water quantity.	Clean heat exchanger surfaces. With a reduced cooling water flow rate, the exit water temperature can be increased, allowing with this the use of heat recovery. The pressure drop can be reduced with shortening of pipes or the use of larger diameter.

Reduce water overflow quantities.	For example, toilets. In cooling towers, water quality should be selected in such a way that a minimum of fresh water demand results.
Periodically decalcify water lines.	Periodic cleaning, for example, of boilers.

CHECKLIST	WATER: SHORT-RETURN CHANGES
Limit flow rate.	Limit flow rate and operation time. Install push-button operation and spray nozzles. Regulate flashing valves in toilets.
Install water meter and billing of the consumers according to effective water consumption.	Notify consumers if water consumption is abnormally high.
Use cooling water in series.	If the exit temperature is low enough, cooling water can be used in cascades. In the same way, washing water can be used several times if quality is sufficient.

CHECKLIST	WATER: MEDIUM-RETURN CHANGES
Check process for reduced water consumption.	Discontinue use of water jets and of systems with large waste water production. Is the use of de-ionized or distilled water really necessary? Is drinking-water-quality necessary?
Compare water and air cooling.	Use closed water systems instead of air cooling (for example in air-conditioning units). Check possibilities for lower cost using direct cooling with outside air. Use air-cooled cooling towers, air-cooled condenser instead of direct fresh water cooling. For systems with a large heat production, heat removal should be with water, not with the air-conditioning system (with air).
Use thermostatic valves for the domestic hot water.	Use body-temperature water. With this mixing, losses are limited and operating time is shortened.
Recirculate clean water.	Recirculate cooling water, reuse condensate water, use local water boiler (if there are large heat losses in long lines).
Install rain water tank.	Use filtered rainwater, for example, for boiler make up.
Use untreated fresh water instead of drinking water.	Use your own source or use ground water. Use industrial water with industry quality only.
Use closed loop cooling.	For cooling purposes source water or lake water can also be used directly.

Recover cooling energy.	If cold water from sources, lakes, or ground water, or from drinking water, is used, check whether it can be used for cooling before consumption.
Water tanks.	Avoid peak demand, collect process water.

CHECKLIST	COMPRESSED AIR
Monitor air distribution system.	Check air consumption when all consumers are shut off. Graph air consumption as a function of equipment operation to determine standby losses.
Repair leaks.	Piping, valves, blow guns, hose connections. Check leaks with soap solution.
Perform regular inspection for optimum performance efficiency.	Clean or replace dirt-clogged filters; remove scale from air cooling fins on cylinders and intercoolers; replace defective work tools; remove obstructions from all ports and passages; check mechanical and electrical systems; correct abnormal intercooler pressure.
Investigate possibility of shutting down compressor when not in use.	Automatic controls are available to start and stop compressors according to air demands.
Avoid oil contamination and inadequate lubrication.	Before refilling, thoroughly clean crankcase. Check seals for oil leaks. Oil contamination of the refrigerant can result in as much as 25 percent reduction in the coefficient of performance for a centrifugal chiller.
Check purity and quantity of cooling water.	
Operate at lowest air pressure possible.	Survey users of plant air to determine feasibility of lowering the air system pressure. Consult manufacturer about performance data and modification procedures for your compressor. Insufficient discharge pressure can also cause energy waste.
Check air distribution system.	Drain water traps and inspect automatic water ejectors.
Start standby or load peaking compressors on demand.	Avoid allowing compressors to idle at part load.
Use coolest possible intake air for compressors.	Air compressor intake frequently is located in the boiler room. Whenever feasible, relocate air intake in a cooler location, e.g., at an outside (north) wall or on the roof.
Avoid improper use.	Use vacuum cleaners rather than compressed air for cleaning.

	Use low-pressure blowers or fans for parts drying and comfort conditioning.
	Avoid compressed air for agitating liquids and drying processes.
Determine effluent piping system.	Use adequate-size piping to limit pressure drop.
Match compressors to system demands.	To meet large intermittent load, a compressor smaller than one required for peak demand may suffice in conjunction with storage capacity.

CHECKLIST	WASTES: MEDIUM-RETURN CHANGES
Analyze waste.	What is its origin? Could waste amounts be reduced or waste be incinerated (with heat recovery)?
Separate and reuse materials.	With this the energy use for introduction of new materials will be lower.
Collect used oil.	Is incineration in your boiler or in a central oil and waste incineration unit more economical?
Install a local waste incineration unit.	Compare local production of hot water at a site remote from central plant as a method of reducing transport losses and also solving a refuse disposal problem.

5.9 MEASURES TO REDUCE TRANSPORTATION ENERGY*

Truck and automobile transportation is sensitive to the world price and availability of oil. One quarter of the energy used in the United States goes for transportation, and 96 percent of this is oil. Given the tendency for rising fuel prices, there is significant economic incentive to reduce its consumption and to improve the efficiency of publicly owned and business transportation fleets. This section describes measures to reduce transportation costs and energy. Reducing fuel for transportation can also have important benefits for environmental quality.

There are three categories for reducing transportation fuel consumption—demand reduction, operations, and equipment. First, reduce the number of miles vehicles travel and/or hours of operation through coordination of loading and routing. This would also include choice of other modes where this would result in overall economy, such as train

*The assistance provided by Stanley B. Manes, and by Ray Murphy of the Freightliner Corporation is gratefully acknowledged.

and ship vs. air freight and truck. The logistics of employee commuting deserves special attention. Second, greater vehicle economy can be obtained through improved driver training and better maintenance. Third, vehicles can be modified to achieve greater efficiency, and more efficient equipment can be purchased.

Reducing the Energy Used for Commuting

Consider the energy used to commute to work compared with the energy used while at work. An employee commuting 30 miles (48 km) per day, alone in an automobile achieving 20 miles per gallon (11.8 ℓ or 8.73 kg per 100 km), would use 47 MBtu/yr (13,840 kWh/yr). This is approximately that given in Table 5.1 as the average energy per occupant for 307 buildings surveyed. This comparison shows that during the 200 to 300 hours spent commuting each year, employees may use, collectively, as much energy as is required to operate the building they work in for the entire year, and that measures to reduce their fuel use could be of significance comparable with those possible for the building. Such measures could include modification of operating times to reduce traffic congestion, operating longer hours per day but fewer days per week, and promotion of carpooling and the operation of vanpools. Eventually, future planning should consider the coordination of working and living locations to permit less energy-intensive access to work.

Vanpooling, an Innovative and Energy-Saving Approach to Employee Transportation

Vanpooling is an innovative approach to employee commuting which utilizes employee-driven, 12-seat commuter vans. (171,172,173) A fuel-saving idea developed by private industry in 1973, vanpooling has become very successful in the United States. Groups of neighboring employees pay monthly fares that cover the costs of acquiring and operating the vans, so the program pays for itself. Based on the experience of 60 firms using 700 vans to transport over 8,000 employees, each 12-seat van saves about 5,000 gallons of gasoline annually and cuts commuting costs by an average of two-thirds. Since these vehicles travel with a payload nearly 100% of the time, vanpooling is more cost-effective than carpooling and is the most energy-efficient commuting vehicle in the United States after the bicycle. The advantages to the employer are that it improves the morale of employees, minimizes tardiness and absenteeism, broadens the labor market, reduces the need for parking spaces, brightens the corporate image, and results in favorable publicity.

For example, the 3-M Corporation was able to avoid the construction of a $2.5 million parking facility through the establishment of a self-liquidating vanpooling program. A second example shows the solution of a similar problem. In a Washington, D.C., suburb, the Government Employees Insurance Company (GEICO), with 3,800 employees, was faced with a parking shortage. With only 1,100 spaces, zoning approval was denied for the construction of additional parking facilities. As a result, GEICO pursued a positive approach by establishing priority parking spaces for carpools of three or more, and provided carpool matching service and a buspool program. Over 300 parking permits were issued, effectively raising the average auto occupancy for all parking spaces to about two persons per car. Eight free buspools, carrying 300 people, pick up complete loads at three scattered suburban shopping centers. Managers at the shopping centers have approved specific fringe parking areas for the bus riders' use. Another 200 employees ride subsidized buses from more distant origins of approximately 35 miles. The three-month trial of these programs has been successful and is now considered permanent. (173) Some employers have found that they have been able to avoid building new parking facilities for their employees that would have cost between 15 percent and 85 percent of the employees' annual salary, plus saving between 1 percent and 3 percent additional for maintenance and taxes.

Employees have an alternative to driving personal cars that is safer, more dependable, less tedious, more enjoyable, and more economical. The advantages for the community and for the nation are reduced pollution and traffic congestion, easing of the pressures for transportation-related facilities, and a potentially significant saving of fuel.

The feasibility of vanpooling should address itself to these questions:

What benefits does vanpooling have to offer the company, its employees, and the community?

What legal requirements must be satisfied by a vanpool program and how will it be administered?

What are the program's costs and revenues?

Is there a market? Are the geographic and workday characteristics of the employees compatible with the proposed vanpool service?

Should vans be leased or purchased?

Could those vans necessary for work-related activities also be used in the vanpool program?

The United States Department of Energy has developed material to help implement vanpooling which may be obtained from the Government Printing Office. A step-by-step handbook is available which uses examples from other firms to illustrate common approaches to program implementation. (173)

Given the obvious need to save energy, vanpooling is expected to play a growing role in employee transportation. It offers enormous potential for saving energy, and its low cost, speed of implementation, and multitude of benefits make vanpooling an increasingly attractive alternative to the private automobile.

Plan Routes and Loads to Reduce Transportation Demand

Where can unnecessary miles be eliminated? Coordinating urban vehicle travel times and routes, as well as the choice of vehicle used, can help to cut costs. Reduce operating hours by planning routes to avoid congested areas and rush-hour traffic. Could the use of a trailer solve routing problems?

Routes should be analyzed to maximize loads and minimize distances traveled. It has been estimated that in 1976 alone as much as 20.4 percent of the miles traveled on the

Table 5.42. Near Term Energy Intensity for Automobiles and Buses Compared with Operating Energy Intensity (174)

Vehicle type	Gross weight (1000 lb)	Trip length (statute miles)	Average trip hrs @ mph	Fuel type[a]	Vehicle statute miles/gal	Number of seats		Specific energy, stop/start			
								Seat-miles/gal		BTU's/seat-mile	
						Available (full load)	1972 Actual aver. oper.	Available (full load)	1972 Actual aver. oper.	Available (full load)	1972 Actual aver. oper.
Urban, subcompact auto	2.0–2.4	10.0	.24/25	Gas	24.0	4.0	1.6	96	38.4	1,302	3,255
Urban, compact auto	2.5–3.4	10.0	.24/25	Gas	18.0	5.0	1.6	90	28.8	1,389	4,340
Urban, standard auto	3.5–4.4	10.0	.24/25	Gas	14.4	6.0	1.6	86.4	23.0	1,447	5,435
Urban, luxury auto	4.5–6.0	10.0	.24/25	Gas	9.0	6.0	1.6	54	14.4	2,315	8,681
Urban, bus	18.5 (empty 20.3–26.0)	13.0	1.25/ 10.3	Diesel	3.6–4.0	50	12	180	48	771	2,891
Intercity, bus	28.7 (empty 45.0)	100.0	1.81/55	Diesel	6.0	46	19.4	276	116.4	503	1,192
Intercity, subcompact auto	2.0–2.4	100.0	1.81/55	Gas	30.0	4.0	2.0	120	60	1,042	2,083
Intercity, compact auto	2.5–3.4	100.0	1.81/55	Gas	22.5	5.0	2.2	112.5	49.5	1,111	2,525
Intercity, standard auto	3.5–4.4	100.0	1.81/55	Gas	18.0	6.0	2.6	108	46.8	1,157	2,671
Intercity, luxury auto	4.5–6.0	100.0	1.81/55	Gas	13.0	6.0	3.0	72	36	1,736	3,472

[a]Gasoline = 125 × 10^3 BTU/gal, Diesel = 138.8 × 10^3 BTU/gal.

SOURCE: U.S. Department of Transportation

Table 5.43. Energy Intensity for Trucks, 1974 to 1980 (174)

Use	Cargo density (lb/ft³)	Maximum payload (tons)	Trip length (statute miles)	Average trip time (h at mph)	Type of fuel	Vehicle statute miles/gal	Specific energy, stop/start cycle	
							Ton-miles/gal	Btu/ton-mile
Urban	20–100	9	10	0.4/25	Gas	8	64	1953
Urban	20–100	8	10	0.4/25	Diesel	12	96	1446
Urban	10–30	3.1	10	0.4/25	Gas	8	25	5040
Intercity	20–100	25	100	1.8/55	Diesel	5	125	1110
Intercity	15	14.3	100	1.8/55	Diesel	4.8	69	2023

SOURCE: U.S. Department of Transportation.

interstate highway system by trucks with three or more axles and by tractors without trailers were logged entirely empty. Owner-operated trucks generally had a lower percentage of empty truck miles than non-owner-operated trucks. While it is unlikely that, due to routing problems, empty trucks can be entirely eliminated from travel on the highways, there does exist potential for reducing the energy used by empty freight vehicles.

Dispatch equipment with the lowest operating cost that will do the job. Choice of transportation vehicle and mode should also be based, where scheduling permits, on the use of the least expensive mode of operation. The energy intensity of different vehicles is given in Tables 5.42, 5.43, and 5.44. (174,175) Intercity trains require between 330 and 550 Btu per ton-mile (0.066 to 0.11 kWh per met. ton-km), compared with 1,110 to 2,023 Btu per ton-mile for trucks (0.22 to 0.41 kWh per met. ton-km), and compared with 7,700 to 14,700 Btu per ton-mile for freight aircraft (1.5 to 2.9 kWh per met. ton-km). (176) Not only do subcompact automobiles cost less to buy, but as is shown in Table 5.37, they require considerably less energy to operate.

Table 5.44. Truck Operating Efficiency and Intensity by Weight Class (175)

		10,000 to 14,000	14,000 to 16,000	16,000 to 19,500	19,501 to 26,000	26,000 to 33,000	Over 33,000
Gas	Btu/ton-mile	7353	7764	6944	3079	2129	1701
	Ton-miles/gal	17	16.1	18	40.6	58.7	73.5
Diesel	Btu/ton-mile				2872	2102	1622
	Ton-miles/gal				48.3	66	85.5

SOURCE: A. D. Little

Reducing Operating Costs with Records, Maintenance, and Driver Education

Operating and maintenance cost may exceed the original cost of a vehicle by one-and-a-half to two times. Cut this cost with improved driver training and cost-effective maintenance.

This must be based on an accurate accounting system that logs the time, mileage, and use of a vehicle according to each operator and relates this data to fuel use and other operating expenses and maintenance data. Without a complete accounting system, the efficient use of vehicles and personnel can only be guessed at. The ability to ascertain the true costs of operation and maintenance is basic to cost reduction strategies that can reduce fuel costs by 50 percent to 70 percent in a typical fleet. (177)

Over 10 million automobiles—comprising over 10 percent of the entire automobile population—are owned by business and government. Caught in a financial squeeze since the end of the oil embargo, many firms have kept their fleet cars long after they should have been replaced. Compounding the problem, many owners have no method of examining fuel, repair, and operating costs of their vehicles. At the minimum, mileage, fuel use, maintenance labor, and cost of replacement parts should be recorded on a monthly basis.* Accurate records of these costs provide a fleet with the justification to commit capital funds to measures that improve the energy efficiency of the fleet.

One method of fuel monitoring is a system mounted directly on each fuel pump used by the fleet, which consists of an array of numbered keyholes and counters. Each vehicle using the pump is provided with a numbered key, which must be inserted in the appropriate keyhole to allow fuel to be pumped. The counter for that key records the quantity of fuel pumped and maintains a cumulative total of fuel use for that vehicle.

It has been reported that using this fairly simple and inexpensive pump-activated mechanical monitoring system usually pays for itself over time through the improved accuracy of the cost data generated, reduced pilferage, and 24-hour accessibility by emergency vehicles. (177)

Preventive maintenance can improve fuel economy and reduce vehicle repair costs. Consider the dollar savings that would result from reducing the fuel consumption of a passenger vehicle by just one mile per gallon. Yet an improved, regular maintenance schedule is capable of raising the average fuel economy of some typical municipal fleets by one to three miles per gallon. (177) A program of maintaining correct tire pressure, regular tune-ups, wheel alignment, etc., may return its cost three times over. (177) The effect of tune-ups on fuel economy is illustrated in Table 5.45. (178) For small fleet owners without a maintenance facility, the possible advantages of a regular maintenance contract with a local shop should be investigated.

*The most accurate method for scheduling maintenance is according to a given amount of fuel used. Engine fuel consumption is preferred over time or mileage because it allows for variations in driver habits, load, road conditions and speed, and weather. For a diesel engine, a starting point for maintenance would be after 500 gallons consumed. Gasoline powered vehicles should be scheduled for maintenance according to a similar plan.

Table 5.45. The Effect of Tune-ups and Air Conditioning on Vehicle Fuel Efficiency

Tune-ups

Operating speed (miles per hour)	Miles per gallon		*Improvement after tuning*	
	Before tuning	*After tuning*	*Miles per gallon*	*Percent*
30	19.30	21.33	2.03	10.52
40	18.89	21.33	2.44	12.92
50	17.29	18.94	1.65	9.54
60	15.67	17.40	1.73	11.04
70	13.32	15.36	2.04	15.32

Air Conditioning

			Effect on fuel consumption caused by use of air conditioning	
Operating speed (miles per hour)	*Air conditioning in use (miles per gallon)*	*Air conditioning not in use (miles per (gallon)*	*Reduction (miles per gallon)*	*Percent saving with air conditioning "off"*
30	18.14	20.25	1.91	10.53
40	17.51	19.71	2.20	12.56
50	16.42	18.29	1.87	11.39
60	15.00	16.25	1.25	8.33
70	13.17	14.18	1.01	7.67

Note: All the cars included in this table were equipped with air conditioning in good operating condition

SOURCE: U.S. Department of Transportation. (178)

Driver Operating Procedures

Operating efficiency may be improved with improved driving habits, less idling time, enforcement of speed limits, better vehicle utilization, and slower engine speeds. It is the owner's responsibility to establish the protocol for vehicle use and operation, and procedures for its enforcement. To do this it is necessary, first, to establish what constitutes necessary vehicle usage, second, to determine norms for route mileage, elapsed time, etc., and third, to monitor vehicle usage and to measure fuel consumption to ensure compliance.

The cost of vehicle operation is very sensitive to poor or sloppy operator habits. Note, for example, that a good number of drivers apparently do not operate their vehicles in the correct gear for fuel economy. Experience with some vehicles has shown that because of lack of skill with a manual transmission, better fuel economy is achieved with a good automatic transmission. (177) Driver education courses should include as part of the course lessons in fuel conservation techniques. Experienced operators with good records and technique consistent with the need to conserve higher priced fuel could be used to help educate other drivers and to evaluate their performance. One company decreased fuel use by its 177,000 cars and trucks by 3 percent (while fleet size increased by 7 percent), in large part simply by teaching employees to use economical driving skills.

Sharing the dollar savings or other incentives could be helpful in improving fuel economy as well as in reducing unnecessary maintenance and repair. Consider operator ownership of vehicles.

Several vehicle operating procedures deserve comment. First, excessive engine idling can reduce vehicle economy and shorten engine life. Yet one finds too many idle vehicles with running motors—at truck stops, loading platforms, and at work sites, as well as in city, county and state operations. Consider using a line-operated battery charger during off-duty hours to maintain a full charge, together with a second battery, if necessary, to power radios and other equipment so that the engines can be shut off during stops, while working, and while loading and unloading, etc. It should be stressed that idling is an extremely inefficient method of maintaining battery charge. No one would couple such an enormous engine with a small automotive or truck alternator to charge batteries in the shop!

Second, slow acceleration and minimizing braking (either from engine compression or brake application) can improve economy. Installation of a manifold vacuum gauge will provide a visual indication of the accelerator pressure conducive of good economy. The vacuum level drops close to zero when the accelerator is fully depressed. In general, the higher the vacuum that can be maintained during acceleration, the better the fuel economy.

Third, unnecessary operation of powered accessory equipment (lighting, radio, etc.) will waste fuel. The saving from turning off air conditioning when it is unneeded is shown in Table 5.45. (178) This data shows that cars equipped with air conditioning save approximately 10 percent with it off.

Fourth, the effect of speed on fuel economy is shown in Table 5.46. (178,179) This table shows that fuel economy decreases as speed increases. For example, increasing truck speed by 15 mph to 65 mph (24 km/hr to 105 km/hr) decreases fuel economy by 23.3 percent. This table also shows that operation in a lower gear than necessary, with correspondingly higher than necessary engine speed, decreases fuel economy.

More Efficient Equipment

It may be possible to make technical modifications in existing equipment to improve its fuel economy. Improved aerodynamics, demand-activated fans, and low rolling resistance tires help improve fuel economy. Consider, also, that a trailer added to an existing

Table 5.46. The Effect of Speed on the Fuel Consumption Rates of Cars and Trucks (178,179)

	Miles per gallon at selected speeds					*Percent increase in gasoline consumption caused by increase in speed*				
	30	*40*	*50*	*60*	*70*	*30 to 40*	*40 to 50*	*50 to 60*	*60 to 70*	*50 to 70*
Cars without air conditioning	21.05	21.07	19.49	17.51	14.93	0.00	8.11	11.31	17.28	30.53
Cars with air conditioning	18.14	17.51	16.42	15.00	12.17					
Percent improvement in MPG after tuning	10.52	12.92	9.54	11.04	15.32					
	50	*55*	*60*	*65*		*50 to 55*	*55 to 60*	*60 to 65*	*50 to 60*	*50 to 65*
Heavy-duty highway trucks	5.11	4.81	4.54	4.09[a]		6.20	6.66	5.50[a]	12.48	23.30[a]
Percent increase caused by "wound-up" driving[b]	14.4	8.0	7.6							

[a]Only three of the six test vehicles were operated, as the governor setting that controls fuel injection did not permit the other three vehicles to be operated at 65 miles per hour.

[b]"Wound-up" driving is driving a diesel truck at or near its maximum horsepower. This practice results in an increase in miles per hour but a decrease in miles per gallon, especially on downgrades.

vehicle may be able to provide additional load capacity at far less fuel, labor, and capital cost than use of a second vehicle.*

A list of measures to improve truck fuel economy is given in Table 5.47, together with the potential percentage of improvement for different size trucks. (180) Use of diesel engines and continuously variable transmissions are measures with the greatest potential improvements. Diesel engines in light trucks and in medium-duty vehicles that have been conventionally equipped with gasoline engines, as well as diesel-equipped passenger vehicles, offer significant potential for reduced fuel use, considerably reduced fuel cost, higher reliability, and lower maintenance. Trucks equipped with high torque engines and 5 speed transmissions will operate with better fuel economy and longer diesel engine life.

The effects of four measures on the vehicles typically found in a municipal fleet are

*A solo truck is pushing its load by the rear drive axle. A semi-trailer, or a two- or four-wheel trailer, is being pulled. A vehicle is capable of pulling more weight than it will push with the same energy consumption. Truck/trailer tandem doubles and triples have proven greater fuel economy in states where these "trains" are permitted on the highways.

Table 5.47. Possible Improvements in Truck Equipment to Increase Fuel Economy

Measure	Fuel economy improvement (%) — Truck class: I and II	VI	VIII
Substitution of a diesel for a spark-ignited engine	20–25	55–60	
Lean-burn engine	10–15		
Turbocharging	5–15		2–5
Stratified-charge engine	15–25		
Derating of diesel engine			1–6
Reduction of parasitic loads (especially cooling fan)	2–3	3–4	3–5
Tag axle			2
4-speed automatic transmission with lockup	7–15	0–10	
Continuous variable-ratio transmission	12–30		10–15
Radial tires	2–3	2	6–9.5
Aerodynamic improvement			
10% reduction of drag	1–2	2	1–5.5
Streamlining of trailer			12
Full streamlining of tractor-trailer combination			30

Note: Due to the variety of different styles of trucks and operating modes, the improvements listed above may be expected to vary. The lower percentages given would apply to local service, the higher values would apply to long-distance hauling. (180)

SOURCE: U.S. Department of Transportation (180)

shown in Table 5.48. (181) Each major vehicle type is classified by two driving modes: urban (local), which consists primarily of stop-and-go driving in an urban setting, and suburban (rural), which consists of a mix of urban, suburban, and open-road driving under 200 miles per day. Depending on the type of vehicle and the type of service, the measures shown in Table 5.43 have the potential to improve efficiency by at least 20 percent, and in some cases well over 50 percent.

The weight of a vehicle is an important factor in its fuel economy. Table 5.49 shows how the fuel economy decreases as the weight of the vehicle increases. (182) This data clearly shows the importance of reducing inertial weight. The lighter-weight automobiles not only cost less to buy, they may cost less than half as much to operate. The following example indicates that lighter automobiles can provide savings in more than fuel economy. (177)

Table 5.48. Comparison of Different Measures for Their Percentage of Fuel Economy Improvement for Municipal Trucks and Buses (181)

		Diesel engine instead of gas (%)	*Temperature actuated cooling fan (%)*	*Radial tires (%)*	*Auto. trans. or CVT (%)*	*Total (%)*
Light pick up (6,000 lb)	(urban)	25–30	4–7	3–6	0	32–45
	(suburban)	25–30	6–9	4–5	0	35–44
Heavy pick up	(urban)	20–40	4–7	3–6	0	27–53
(10,000–14,000)	(suburban)	10–20	6–9	4–9	0	20–38
Dump truck (19,500–26,000)	(urban)	20–40	4–7	3–8	0	27–55
	(suburban)	10–20	4–7	5–10	10–30	29–67
Heavy dump/garbage	(urban)	20–40	4–7	3–8	0	27–55
(26,000–33,000 lb)	(suburban)	10–20	4–7	5–10	6–15	25–52
School bus		20–40	6–9	0–5	10–15	36–69
Transit bus	(urban)	20–40	4–5	3–8	10–20	37–73

SOURCE: U.S. Department of Transportation.

In 1974, the Seattle Police Department initiated a policy of substituting smaller (3,700 lb or 1,680 kg) Dodge Dart cruisers with 318 V-8 engines for the existing stock of full-size (4,700 lb or 2,130 kg) Plymouth Furies. In 1974, 113 compact vehicles (Darts) were purchased. (177)

According to the Seattle Police Department, the use of compact cruisers reduced fuel consumption by about 50 percent and maintenance costs by 30 percent (attributed to the superior maneuverability of the compact cruisers). In comparing the operating and maintenance costs of the compact and full-size cruisers, the Seattle Police Department noted that:

1. Average gas mileage for the compact cruisers was 8.9 mpg (26.5 l per 100 km) versus 5.8 mpg (40.6 l per 100 km) for the full-size;
2. Maintenance costs averaged 3.6¢ per mile for the compact vs. 5.2¢ per mile for the full-size cruisers;
3. The vast majority of officers testing the compact voted to approve the compact as a replacement for the full-size cruisers.

The great body of satisfactory experience in Europe with lighter (ca. 3,000 lb or 1,375 kg) and safer vehicles such as the Volvo indicate that there may be several advantages to using lighter vehicles.

Further economy results when power accessories such as power steering, air conditioning, and automatic transmissions—frequent adjuncts to passenger cars—are not included in fleet vehicles. Data in Table 5.41 shows that cars with air conditioning suffer a 14 percent fuel penalty at 30 mph (48 km/hr), rising to 16 percent at 50 mph (80 km/hr). (178) Vehicles which are safer and more economical to operate frequently cost more

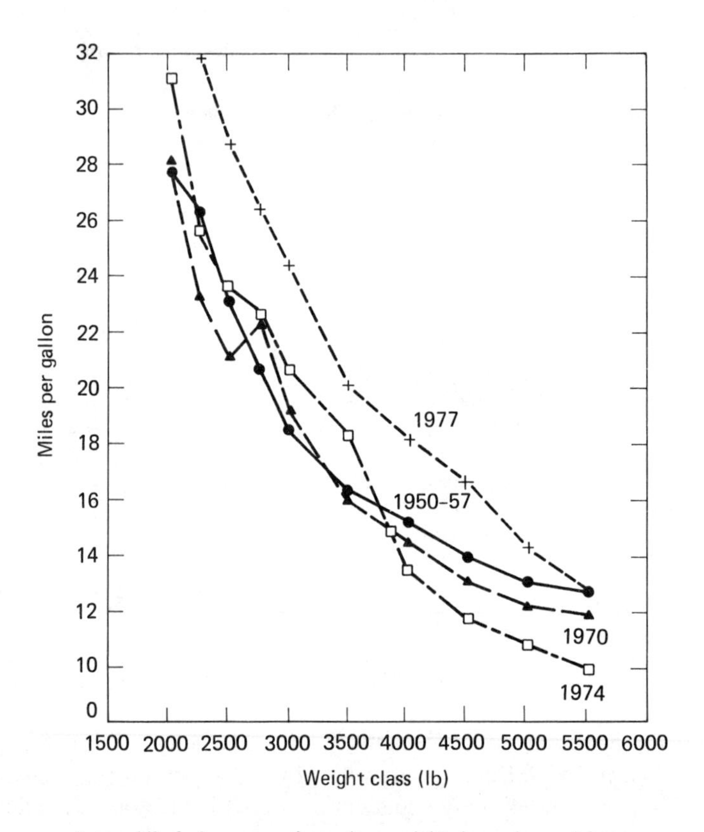

Automobile fuel economy for various weight classes by model year.

Table 5.49. City/Highway Combined Fuel Economy by Model Year and Weight Class (miles per gallon)

	Inertia weight class									
Model year	*2000*	*2250*	*2500*	*2750*	*3000*	*3500*	*4000*	*4500*	*5000*	*5500*
'57–'67 av.	27.8	26.3	23.1	20.7	18.5	16.3	15.2	14.0	13.1	12.7
1968	23.3	24.7	22.3	23.8	18.8	16.0	14.5	13.6	11.2	10.7
1969	26.9	24.5	22.7	20.3	18.6	16.0	14.4	13.6	11.0	13.0
1970	28.2	23.3	21.1	22.3	19.2	16.0	14.5	13.1	12.2	11.9
1971	27.3	25.8	23.3	22.1	17.8	14.7	14.1	12.9	11.6	13.1
1972	27.7	26.4	23.6	24.1	17.4	16.0	13.4	12.9	11.6	11.2
1973	28.7	26.4	23.8	21.1	18.8	16.8	13.0	12.2	11.2	10.4
1974	31.2	25.7	23.6	22.5	20.6	18.3	13.5	11.8	10.8	9.9
1975	31.3	28.1	24.5	22.4	21.6	17.6	15.5	14.6	12.8	12.0
1976	32.1	29.1	25.9	24.4	23.4	19.1	17.4	15.6	14.6	13.3
1977		31.8	28.7	26.4	24.4	20.1	18.2	16.6	14.3	12.7

Note: The inertia weight of a vehicle is an important factor in determining its fuel economy. The graph and the table show the EPA estimated fuel economy (in miles per gallon) by inertia weight class for the model years 1957 to 1977. The difference in fuel economy from the low average mpg to the high average mpg ranges from a 100 percent to a 200 percent increase for most of the model years.

SOURCE: J. D. Murrell, et al. (182)

to buy. Automobiles and trucks with diesel engines cost more than vehicles with gasoline engines; however, the initial extra cost may be returned many times over during the life of the vehicle.

Bidding techniques for fleet vehicles should be based not on the lowest initial price of the vehicle, but on the lowest *life-cycle cost*. The operating and maintenance costs of a vehicle throughout its life in the fleet may exceed its purchase cost by one-and-a-half to two times. Yet these costs are often not considered in the evaluation of bids at the time of acquisition. As a result, an inefficient vehicle may be purchased because it costs less initially than an alternative vehicle that has lower total ownership costs. Purchasing specifications written to include operating maintenance and safety components can help avoid the large cost penalty associated with low-first-cost equipment. The life-cycle cost of passenger cars for fleet purchases can be developed from the following criteria:

1. The bid price of the vehicle
2. The price of gasoline
3. The EPA mileage rating of the vehicle
4. The expected annual mileage
5. The number of years the vehicle is expected to remain in the fleet
6. The anticipated resale value of the vehicle when it leaves the fleet
7. The expected annual maintenance cost
8. The current interest rate of money
9. The cost of insurance

Table 5.50 illustrates typical cost benefit data for vehicles purchased for a municipal fleet.

Table 5.50. Life-Cycle Fuel Cost Savings versus Higher Purchase Costs of Diesel Trucks and Buses (177)

Vehicle	*Driving mode*	*Incremental purchase cost[a] ($)*	*Cumulative[b] fuel savings at 100,000 mi ($)*	*Cumulative fuel savings at 150,000 mi ($)*	*Mileage when fuel saved = incremental purchase price*
Light pickup	Urban	1,000	2,000–2,200	3,000–3,300	45,000–50,000
	Suburban	1,000	1,900–2,000	2,850–3,000	50,000–52,000
Heavy pickup	Urban	2,500	2,500–3,000	3,750–4,500	83,000–100,000
	Suburban	2,500	2,000–2,500	3,000–3,750	100,000–125,000
Dump truck	Urban	2,500	3,500–4,500	5,250–6,750	13,300–71,000
	Suburban	2,500	3,000–3,500	4,500–5,250	71,400–83,000
Heavy-duty dump/ garbage	Urban	3,600	4,000–4,500	6,000–6,750	66,700–75,000
	Suburban	3,000	3,500–4,000	5,250–6,000	75,000–85,000
School Bus	Suburban	3,000	3,000–3,500	4,500–5,250	85,700–100,000
Transit Bus	Urban	3,000	4,500–5,500	6,750–8,250	54,500–66,000

[a]The incremental costs shown in the table are intended as illustrative examples only and may vary significantly from actual incremental costs.

[b]The fuel savings and break-even mileages shown in the table are based on undiscounted 1975 fuel prices; gasoline at $.50/gallon, diesel fuel at $.35/gallon.

SOURCE: Massachusetts Department of Community Affairs.

(177) The specific capital productivity of investments in transportation equipment to save fuel is generally attractive.

In summary, management has the opportunity to reduce energy costs by eliminating unnecessary demand and operating hours, by improvements in operator training and maintenance, and through cost-effective acquisition of more efficient equipment. Good planning requires data upon which to predict the outcome of decisions. Management decisions can be improved by increasing the emphasis and money necessary to acquire this knowledge. Consider that an operating protocol is essential to efficient, economical operation, but that flexibility should be encouraged to accommodate unforeseen circumstances. The importance of management leadership in nurturing employee morale and willingness to save cannot be too highly stressed.

CHECKLIST	MEASURES TO REDUCE TRANSPORTATION DEMAND
Help reduce commuting energy.	Encourage use of public transportation.
	Consider matching operating times to local public transportation service.
	Consult with public transportation officials to obtain routes and time-of-service better suited to employee needs.
	Reduce parking problems by subsidizing employees' use of public transportation.
	Provide shuttle bus service between establishment and public transportation terminal, railroad station, etc.
Encourage carpools.	Encourage employee carpools by assigning reserved parking spaces for those people participating.
	Include data collection matching riders and routes, public information calling program to employees' attention, and incentives for successful continuation of the program.
Organize a vanpool program.	Coordinate purchase of company vehicles with those which could be used in vanpool program.
	Identify employees with particularly difficult access to public transportation in order to initiate a limited vanpool program.
Reduce internal transportation requirements.	Provide van or shuttle bus service for employees between division facility locations to eliminate single personal vehicle travel.
	Coordinate transportation among division facility locations to reduce empty truck and van miles on vehicles.
	Eliminate the use of trucks and other energy-intensive vehicles for personal transportation, message delivery, coffee breaks, etc.

Consider the use of lightweight vehicles, mopeds, bicycles, etc., for single-person transportation.

Consider two-way radio communication to aid dispatching.

Consolidate freight shipments and/or deliveries.

Can you make do with fewer vehicles?

Increase vehicle utilization.

Maintain tight schedules of shipping and supply trucks so they carry full loads through the shortest route as much as possible.

Consider the use of trailers which could be detached in mid-route after they are empty, etc.

Strive for better planning to minimize rush deliveries.

Coordinate with firms supplying your establishment to have your own trucks pick up your orders when they are making deliveries in the suppliers' vicinity, to avoid suppliers' shipping charges and to reduce empty truck miles.

One longer trip combining three tasks or deliveries is better than two or three short ones.

Reduce operating hours of vehicles.

Arrange routes to minimize overlaps, repetition.

Choose least-congested routes.

Plan trips to avoid peak hour traffic.

Consider the cost-effectiveness of very early morning deliveries.

Use public carriers, bus and taxi parcel service, etc., when this would save energy and make more effective use of employees.

Reduce unnecessary employee travel.

Use the telephone, conference calls, electronic mail, etc., when this could help avoid personal travel.

Dispatch the lightest, most economical to operate vehicles for single-person transportation.

Coordinate personal transportation to reduce the number of vehicles.

Eliminate unauthorized use of fleet vehicles.

Establish fuel and mileage record system.

Assign specific vehicles to individual drivers.

Improve customer deliveries with fewer trips.

Urge customers to take merchandise with them.

Make certain that a supply of inexpensive car-top carriers is available for purchase.

Ship on-order merchandise and parts directly to the customer whenever possible.

Increase the density of stops by restricting the number of days when home service to a particular area is offered.

	Establish a procedure to ensure that the customer will be home for service calls and merchandise delivery. Pre-edit daily route sheets to ensure minimum mileage for the stops required.
Establish route mileage norms.	Establish a norm for route mileage and operating hours. Establish norms for per-mile costs. Review trip and route reports regularly to check for abnormalities. Abnormal fuel use may be a signal for needed maintenance. Discourage off-route mileage for lunch and coffee breaks.
Improve data base record mileage and fuel consumption for each vehicle.	Record mileage per trip, or at least when fuel tank is filled. Record fuel use and driver/operator assignments for each piece of equipment, by trip. Incorporate data logging with operator reports such as Operational Safety and Health Act (OSHA) reports.

CHECKLIST	MEASURES TO IMPROVE OPERATION AND MAINTENANCE
Record vehicle maintenance.	Record monthly and annual direct maintenance labor hours and cost of replacement parts for each vehicle. Encourage drivers to report maintenance and repair needs before they develop into major problems.
Improve maintenance tune engines regularly.	Tune engine according to manufacturer's recommended specification. A poorly maintained vehicle (e.g., spark plugs misfiring, clogged air filter, improper carburetor adjustment) can increase gas consumption by 6–10 percent. Consult vehicle manufacturer for procedures to tune for maximum economy according to the altitude conditions, seasonal weather, and local characteristics.
Care for tires.	Inflate tires to manufacturer's specifications. Underinflation increases friction, thus increasing fuel consumption. Ensure proper wheel alignment; improper alignment reduces engine efficiency and increases fuel consumption.
Develop a program of preventive maintenance.	Have operators and service personnel read manufacturers' service manuals. Establish regular mileage and time intervals for service and maintenance. Vehicles subject to heavy usage and vehicles with high mileage should be checked more frequently.

Coordinate listed measures with procedures in service manual for each piece of equipment.

1. Check air cleaner.
2. Spark plugs: check and replace regularly.
3. Lubricating oil: purchase and use oil type that matches the local ambient temperature and driving conditions.
4. Tune-ups, points, and timing: replace worn points, ignition wires.
5. Air pollution equipment: keep it free of sludge, gum, and other foreign material.
6. Carburetor: clean and adjust periodically.
7. Choke and throttle: keep fuel-feeding mechanism clean and well lubricated.
8. Filters: keep clean and replace regularly.
9. Wheel alignment: check regularly.
10. Belts: maintain correct belt tension.
11. Radiator: keep free of clogging material.
12. Exhaust system: check for and eliminate any blockage.
13. Battery: check water level frequently; use distilled water only.
14. Hydraulic lines on trucks and other equipment: check for leaks, proper pressure relief setting; examine controls, linkage, pump, and fittings.
15. Moveable components on equipment: check for signs of damage, binding, or misalignment.
16. Electrical wiring: note condition and check for exposure to damage from road salt or ice build up.
17. Tire pressure: maintain at recommended levels.

    Ten percent overinflation of tires can increase gas mileage 5 percent. Check with your dealer about the overall economy with regard to any adverse effects on tires or suspension.

Tire Underinflation = Poor Mileage

Percent underinflated	*Percent loss in gasoline mileage*
10	5
20	16
30	33
40	57

Service and maintenance for economy.

Avoid fuel waste caused by overfilling tank. Establish the practice of removing the hose the *first* time the automatic valve closes.

Avoid fuel waste (from fuel expansion) by filling tank no more than 90 percent of capacity in warm weather.

	Fill gas tanks at end of day. This minimizes condensation, thereby improving gasoline mileage.
	Check to see that all vehicles have a gas cap (either conventional or locked) to reduce fuel evaporation.
	Indoor automobile and truck parking in cold weather will reduce fuel required for start-ups.
	Replace antifreeze with water, and consider removing thermostat for cooler engine operation in hot weather and for better economy. (Water has a higher heat transfer capacity than most antifreeze fluids.)
	Keep radiator surfaces clean and free of insects.
	Clean engines for cooler operation.
	Keep air filter clean to avoid restricting air flow. An engine requires 7,500 gallons of air for every gallon of gasoline it burns. A dirty filter changes this ratio and can reduce mileage by 10 percent.
	Charge lift truck batteries and other batteries during off-peak, low tariff periods (at night and during the weekend).
Improve vehicle operation and driver training.	Implement special instruction lectures for drivers covering related fuel economy recommendations described within this checklist. Review your existing fleet service and maintenance procedures with supervisors to optimize vehicle performance.
	Periodically review operating procedures with each operator. Discuss overall performance for his suggestions for possible improvement.
Reduce idling.	Shut off engines during loading and unloading.
	Shut off engines when stopped for longer than one minute. Engine idling leads to sludge build-up and abrasion of tappet faces, rings, and valve stems.
	According to manufacturers' instructions, "warm-up" for most equipment is unnecessary and wastes fuel.
	Wait to start engine until vehicle is ready to move.
Check engine speed.	Keep engine revs down—it's more efficient.
	Engine speed recorders provide a cost-effective method of controlling avoidable engine wear. These recorders also can provide an effective record of vehicle and driver activity, excessive lunch breaks, etc.
	Install vacuum gauge. Use manifold vacuum level as an indicator of engine operating economy.
	Controlled acceleration and shifting reduces fuel consumption. High gear allows realization of greatest fuel economy. For

	vehicles with 3-speed transmissions, first gear consumes 55 percent more fuel, and second gear consumes 20 percent more fuel than high gear, respectively.
Avoid fast starts.	Fast starts burn up to 60 percent more fuel than smooth, easy starts.
Avoid unnecessary braking.	Decelerate gradually when approaching a stop, using engine compression when necessary to assist in braking action.
	Avoid excessive lane changing to reduce unnecessary acceleration or braking.
	Think ahead to minimize braking. Remember that braking in any form represents energy waste.
Maintain safe driving distances.	Safe driving distances (one car or truck length for every 10 mph) should be maintained between vehicles to avoid unnecessary stops and starts. Fewer stops can mean 10–25 percent better gasoline mileage.
Keep speed down.	Consider the advantages of fewer accidents and reduced fuel use with lower speed protocol.
	Consider establishing a speed limit of 50 miles per hour for delivery and service vehicles. (On average, 50 mph requires 11 percent less gasoline than 60 mph.)
Reduce unnecessary friction.	Keep weight to a minimum.
	Every extra pound to be moved requires energy.
Reduce wind resistance.	Over-the-road transports should keep trailer aligned with cab to reduce wind resistance. Wind resistance can increase fuel consumption by as much as 6 percent.

CHECKLIST	MEASURES FOR MORE EFFICIENT EQUIPMENT
Retrofit More Efficient Components	
Use radial tires.	Radials cost approximately 20 percent more but outlast conventional tires by 40 percent or more and reduce fuel consumption by 5–7 percent. Radial tires maintain a firmer grip on the road (important on ice and snow) and thus have less rolling resistance. Many drivers report that traction with steel-belted radials is equivalent to that from conventional snow tires.
Reduce wind resistance.	Install wind deflectors to reduce wind resistance. The cab-mounted wind deflector prevents the wind from rushing directly against the higher front of the trailer and swirling

Retrofit More Efficient Components (cont.)	into the gap between the trailer and cab. Fuel savings estimated in the range of 5–7 percent.
	Install vortex stabilizers on trailers to reduce the effect of crosswinds flowing through the gap between the cab and trailer, causing the trailer to sway and reducing engine efficiency. Fuel savings estimated at 5–6 percent.
Install diesel engines.	Dieselize trucks and buses using gasoline engines.
Install a demand fan.	A clutched, or temperature-actuated fan saves motor horsepower.
Purchase more efficient and durable vehicles.	Replace old, fuel-inefficient vehicles.
	Do not purchase vehicles larger than required. Size trucks to the job.
	Consider buying automobiles, trucks and buses with diesel engines.
Obtain energy-saving optional equipment.	Install temperature-actuated cooling fans, radial tires, and 4 or 5 speed automatic or continuously variable transmissions. Consider the use of air foils to reduce wind drag.
Purchase vehicles of lower weight and with fewer power accessories.	Switch to smaller and/or lighter-weight vehicles without options such as power steering, power brakes, and air conditioning. Purchase passenger cars with manual transmissions.
Purchase vehicle on basis of life cycle cost.	Include analysis of bid price of vehicle, price of fuel, EPA mileage rating, expected annual mileage, number of years vehicle expected to remain in fleet, resale value of vehicle, annual maintenance costs, and interest rate of money.

REFERENCES

(1) Office of the Chief Engineer, *Guidelines for Energy Conservation for Immediate Implementation: Small Business and Light Industries,* 2nd ed. (Washington, D.C.: Department of Energy, February 1974).

(2) Richard T. Salter, *Energy Conservation in Non-Residential Buildings* (Santa Monica, Cal.: Rand Corporation, June 1974).

(3) Arthur D. Little Co., *Energy Conservation in New Building Design,* An Impact Assessment of ASHRAE Standard 90-75, FEA Conservation Paper 43 B (Washington, D.C.: Department of Energy, Dec. 1975).

(4) ASHRAE Standard 90–75, *Energy Conservation in Building Design,* (New York: American Society of Heating, Refrigeration and Air Conditioning Engineers, 1975).

(5) DOE, "Building Energy Performance Standards (BEPS)," (Washington, D.C.: The United States Department of Energy, 1980).

(6) Roger W. Sant, "Energy Conservation," Address given at the "Energy Awareness Symposium," sponsored by the Office of Energy Conservation and Environment, United States Department of Energy, Knoxville, Tennessee, February 27, 1976.
(7) Jeffrey Cohen and Ronald White, *Energy Conservation in Buildings: The New York Metropolitan Region,* (Washington, D.C.: Environmental Law Institute, July 1975), Chapter 3.
(8) R. G. Stein, "A Matter of Design," *Environment,* 14, no. 3 (October 1972), 16–29.
(9) Lee Schipper, "Towards More Productive Energy Utilization," Lawrence Berkeley Laboratory report LBL-3299, (Berkeley, Cal.: University of California, October, 1975), p. 70.
(10) DOE, "Building Owners' and Managers' Association International Lighting Energy Survey of 307 Buildings," *Lighting and Thermal Operations,* (Washington, D.C.: Department of Energy, 18 April 1975).
(11) Thola Theilhaber, "Energy Bulletin No. 10," from the office of the Corporate Energy Conservation Engineer, (Lexington, Mass.: Raytheon Co., 1975).
(12) M. Kiss, H. Mahon, and H. Leimer, *Energiesparen Jetzt! Working Methods and Checklists to Reduce Costs in Existing Buildings and Industrial Installations* (Wiesbaden, West Germany: Bauverlag, 1980).
(13) NBS/GSA, "Energy Conservation in Public Buildings, $E = mc^2$," a report from the workshop held by the National Bureau of Standards and the General Services Administration (Gaithersburg, Md.: 23–24 May 1972).
(14) NBS, *Technical Options for Energy Conservation in Buildings,* The National Bureau of Standards Center for Building Technology Publication NBS-TN-789, (Washington, D.C.: U.S. Department of Commerce, 1973).
(15) GSA, "Energy Conservation Guidelines for Existing Office Buildings," Public Buildings Service, (Washington, D.C.: General Services Administration, January 1974).
(16) Lewis Allen Felton, *Energy Conservation: A Case Study for a Large Manufacturing Plant* (thesis for the Master of Science degree in the Department of Ocean Engineering, Massachusetts Institute of Technology, June 1974).
(17) DOE, *Guidelines for Saving Energy in Existing Office Buildings: Building Owners and Operators Manual ECM 1,* Conservation Paper number 20, prepared by Dubin, Mindel and Bloome, (Washington, D.C.: Department of Energy, 1975).
(18) DOE, *Guidelines for Saving Energy in Existing Buildings, Engineers, Architects and Operators Manual ECM 2,* Conservation Paper number 21, prepared by Dublin, Mindel and Bloome, (Washington, D.C.: Department of Energy, 1975).
(19) DOE, *Guide to Energy Conservation for Food Service,* prepared by the FEA Food Industry Advisory Committee, (Washington, D.C.: Department of Energy, October, 1975).
(20) T. Nejat Vezirojlu, ed., *Energy Conservation: A National Forum,* proceedings of the Forum sponsored by ERDA, School of Engineering and Environmental Design, (Coral Gables, Fla.: Clean Energy Research Institute, University of Miami, 1–3 December 1975).
(21) Walter Carnahan, Barry Casper, Kenneth Ford, et al., *The Efficient Use of Energy: A Physics Perspective,* The Study Group on Technical Aspects of Efficient Energy Utilization, (New York: The American Physical Society, 1975).
(22) IFI, *Energy Cost Reduction in The Fabricare Industry,* (Joliet, Ill.: International Fabricare Institute, 1976).
(23) DOE, "Total Energy Management," A practical handbook on energy conservation and management developed by the National Electrical Manufacturers Association and the National Electrical Contractors Association, (Washington, D.C.: Department of Energy, 1976).

(24) DOE, *Energy Conservation on Campus,* prepared by the Energy Task Force of the Physical Plant Administrators of Universities and Colleges, *Volume I: Guidelines, Volume II: Case Studies* (Washington, D.C.: Department of Energy, December 1976).
(25) Thola Theilhaber, "Energy Bulletins," from the office of the Corporate Energy Conservation Engineer, (Lexington, Mass.: Raytheon Co., 1975).
(26) G. S. Springer and G. E. Smith, *The Energy-Saving Guidebook,* (Westport, Ct.: Technomic Publishing Co., Inc., 1974).
(27) Robert R. Gatts, Robert G. Massey, and John C. Robertson, *Energy Conservation Program Guide for Industry and Commerce,* National Bureau of Standards Handbook 115, (Washington, D.C.: U.S. Department of Commerce, September 1974; Supplement, December 1975).
(28) S. Sterner, *Energie Sichern—Energie Sparen,* (Düsseldorf, West Germany: ECON Verlag GmbH., 1975).
(29) GDI, *Wie Spare Ich Energie im Betrieb?* Proceedings of the Symposium (Rüschlikon/Zurich, Switzerland: Gottlieb Duttweiler-Institut, 12–13 March 1976).
(30) GEK, *Zwischenbericht,* (Bern, Switzerland: Eidgenössische Kommission fur die Gesamtenergiekonzeption, May 1976).
(31) AIA, *Energy Conservation in Building Design,* The AIA Research Corporation, (Washington, D.C.: The American Institute of Architects, May 1974).
(32) Proceedings of the Seminar "Bauliche Massnahmen zum Energiesparen in der Gemeinde," (Rüschlikon/Zurich, Switzerland: Gottlieb Duttweiler-Institut, 24–25 June 1976).
(33) M. Kiss, H. Mahon, and H. Leimer, *Energiesparen Jetzt!* (Berlin, Germany: Bauverlag, 1980), Chapter 6.
(34) Kneeland A. Godfrey, Jr., "Energy Conservation in Existing Buildings," *Civil Engineering—ASCE,* (September 1975), pp. 79–84.
(35) Jack E. Snell, Paul R. Achenbach, Stephen R. Petersen, "Energy Conservation in New Housing Design," *Science,* vol. 192 (June 25, 1976), pp. 1305–11.
(36) M. Kiss, H. Mahon, and H. Leimer, *Energiesparen Jetzt!* (Berlin, Germany: Bauverlag, 1980), 85–91.
(37) Flach and Kurtz, Consulting Engineers, Exhibits 172 and 243, in the New York Public Service Commission Hearing as reported in J. C. Cohen and R. H. White, *Energy Conservation in Buildings in the New York Metropolitan Region,* (Washington, D.C.: Environmental Law Institute, July 1975), p. 27.
(38) Alfred Greenberg, Exhib 241, PSC case 26292, ibid.
(39) L. G. Spielvogel, "More Insulation Can Increase Energy Consumption," *ASHRAE Journal,* vol. 16, no. 1, (1974), 61.
(40) Daniel A. Harkins, et al., *Energy Management in Municipal Buildings,* a report of the Energy Conservation Project, (Boston: Massachusetts Department of Community Affairs, 1977).
(41) See reference 27, pp. 3–44.
(42) Ibid., pp. 3–60.
(43) Ibid., pp. 3–60A.
(44) See reference 29, pp. 47–53.
(45) R. E. Doerr, "Six Ways to Keep Score on Energy Savings," *Oil and Gas Journal* (May 17, 1976), pp. 130–45.
(46) John M. Fox, "Energy Consumption for Residential Space Heating, a Case Study," Report no. 4 of the Princeton Center for Environmental Studies (Princeton, N.J.: 1 September 1973).
(47) SIA, "Winterlicher Warmeschutz im Hochbau," SIA Empfehlung 180/1, (Zurich, Switzerland: Schweizerischer Ingenieur- und Architektenverein, 1977).

(48) GSA, *Energy Conservation Guideline for New Office Buildings,* Public Buildings Service, (Washington, D.C.: General Service Administration, July 1975).
(49) Jeffrey Cohen and Ronald White, *Energy Conservation in Buildings: The New York Metropolitan Region,* (Washington, D.C.: Environmental Law Institute, July 1975), Chapter 3.
(50) William J. Jones and James W. Meyer, *Solar Energy and Conservation at St. Mark's School,* Energy Laboratory Report No. MIT-EL-77-001, (Cambridge: Massachusetts Institute of Technology, February 1977), Appendix III.
(51) M. David Egan, *Concepts in Thermal Comfort,* (Englewood Cliffs, N.J.: Prentice-Hall, Inc., 1975), p. 203
(52) B. Y. Kinzey, Jr. and H. M. Sharp, *Environmental Technologies in Architecture,* (Englewood Cliffs, N.J.: Prentice-Hall, Inc., 1963).
(53) J. R. Burrows, "The Technical Aspects of the Conservation of Energy for Industrial Processes," Federal Power Commission National Power Survey Technical Advisory Committee on Conservation of Energy, Position Paper No. 17, (Midland Mich.: Dow Chemical Company, May 1973), 48 pp.
(54) AIPE, "Better Ideas for Conserving Energy," Energy Conservation Committee, (Cincinnati: American Institute of Plant Engineers, August 1974).
(55) DOE, Managing the Energy Dilemma: Increasing Energy Efficiency, Technical Reference Manual for Industrial Workshops (Washington, D.C.: United States Department of Energy, 1977).
(56) DOE, *Economic Thickness for Industrial Insulation,* Conservation Paper no. 46, (Washington, D.C.: Department of Energy, August 1976).
(57) Reference 27, pp. 3–21.
(58) Thola Theilhaber, "Process and Plant Heating with High Temperature Fluids," *Modern Power and Engineering* (January 1970), pp. 38–40.
(59) J. M. Jesionowski, W. E. Danekind, and W. R. Rager, "Steam-Leak Evaluation Technique Quick, Easy to Use," *The Oil and Gas Journal* (January 12, 1976), pp. 100–2.
(60) M. Kiss, H. Mahon, and H. Leimer, *Energiesparen Jetzt!* (Berlin, West Germany: Bauverlag, 1980), 100–110.
(61) Thomas F. Widmer and George N. Hatsopoulos, "Summary Assessment of Electricity Cogeneration in Industry," (Waltham, Mass.: Thermo Electron Corporation, 1977).
(62) "Total Energy Handbook," (Peoria, Ill.: Caterpillar Tractor Co., 1969).
(63) Reference 50, pp. 38–47.
(64) R. S. Spencer and G. L. Decker, "Energy and Capital Conservation Through Exploitation of the Industrial Stream Base," Dow Chemical Co. Report for the NSF "Energy Industrial Center Study," (Washington, D.C.: National Science Foundation, 1972), pp. 343–362.
(65) *A Study of Inplant Electric Power Generation in the Chemical, Petroleum Refining and Paper and Pulp Industries,* a report for the Department of Energy (Waltham, Mass.: Thermo Electron Corporation, 1976).
(66) E. Gyftopoulos, Lazaros Lazaridis, and Thomas Widmer, *Potential Fuel Effectiveness in Industry,* (Cambridge, Mass.: Ballinger Publishing Company, 1975).
(67) R. M. E. Diamant, *Total Energy,* (New York: Pergamon Press, 1970).
(68) Reference 17, pp. 330–335.
(69) HUD, "Economic Evaluation of Total Energy: Guidelines," (Washington, D.C.: U.S. Department of Housing and Urban Development, 1973).
(70) HUD, *Economic Evaluation of Total Energy; Technical Report,* (Washington, D.C.: U.S. Department of Housing and Urban Development, 1973).

(71) J. P. Zanyk, "A Case History on the Application and Operation of Gas Turbine and Waste Heat Boilers in a Total Energy Cycle for a Major Chemical Manufacturing Complex," Contributed by the Gas Turbine Division of the American Society of Mechanical Engineers at the Gas Turbine Conference and Products Show, Zurich, Switzerland, 30 March–4 April 1974.
(72) W. H. Day, p. 363 in CERI, Reference 20.
(73) Reference 27, pp. 3–42G.
(74) Ibid., pp. 3–42.
(75) Ibid., pp. 3–39.
(76) Ibid., pp. 3–47.
(77) Ibid., pp. 3–50.
(78) T. Kusuda, J. Hill, et. al., *Pre-Design Analysis of Energy Conservation Options for a Multi-Story Demonstration Office Building,* Institute for Applied Technology, National Bureau of Standards, (Washington, D.C.: U.S. Government Printing Office, 1975).
(79) M. P. Zabinski and J. Y. Parlange, "The Thermostat as a Source of Energy Savings," *ASHRAE Journal,* (January 1978), pp. 72–5.
(80) MIT, "Energy Conservation at MIT," (Cambridge, Mass.: Department of Physical Plant, Massachusetts Institute of Technology, 1975).
(81) "Medical Association Recommends Lower Temperatures in Homes," *New York Times,* (November 27, 1977), p. 20.
(82) John C. Moyers, "The Value of Thermal Insulation in Residential Construction: Economics and the Conservation of Energy," ORNL-NSF report no. EP-9, (Oak Ridge, Tenn.: Oak Ridge National Laboratory, December 1971).
(83) Stephen R. Petersen, "Retrofitting Existing Housing for Energy Conservation: An Economic Analysis," (Washington, D.C.: Center for Building Technology, National Bureau of Standards, December 1974).
(84) Fred S. Dubin, "More Efficient Utilization of Energy in Buildings," pp. 493–525 in reference 20.
(85) J. Batey, Vincent Gazerro, F. Salzano, and A. Berlad, "Energy Management in Residential and Small Commercial Buildings," Brookhaven National Laboratory report BNL 50576, (Upton, New York: July 1976).
(86) Thola Theilhaber, "Improving Combustion Efficiency of Oil-Fired Boilers," *Plant Engineering,* Part 1, (May 26, 1977), pp. 117–20, Part 2 (June 9, 1977) pp. 134–5.
(87) Reference 27, pp. 3–42E.
(88) Ibid., pp. 3–37.
(89) Thola Theilhaber, Raytheon Corporate Energy Conservation Engineer, Private Communication, 1978.
(90) M. Kiss, H. Mahon, and H. Leimer, *Energiesparen Jetzt!* (Berlin, West Germany: Bauverlag, 1980), Chapter 7.
(91) Reference 27, pp. 3–51.
(92) Ibid., pp. 3–52.
(93) Ibid., pp. 3–42 D.
(94) Ibid., pp. 3–45.
(95) Ibid., pp. 3–49.
(96) *Steam—Its Generation and Use,* 38th ed., (New York: Babcock & Wilcox, 1972).
(97) M. Stadelmann, "Neue Energiesparende Gasheizungssysteme," *Neue Zürcher Zeitung,* no. 441, (23 September 1974), p. 37.

(98) EPA, "Guidelines for Burner Adjustments of Commercial and Oil Fired Boilers," Industrial Environmental Research Laboratory report EPA-600/2-76-088, (Research Triangle Park, N.C.: U.S. Environmental Protection Agency, March 1976).
(99) David W. Locklin, et al., "Guidelines for Residential Oil-Burner Adjustments," Battelle Columbus Laboratory Report PB-248-292 (Research Triangle Park, N.C.: U.S. Environmental Protection Agency, October 1975).
(100) ASHRAE, *Fundamentals* (New York: American Society of Heating, Refrigerating, and Air Conditioning Engineers, 1972), Chapter 13.
(101) DOE, *Energy Management Workshop: Supermarkets,* Handbook for the Regional Workshops sponsored by the FEA (Washington, D.C.: U.S. Department of Energy, 1977), Section 1.
(102) NATO, *Technology of Efficient Energy Utilization,* ed. E. G. Kovach, report of a NATO Science Committee Conference held at Les Ares, France, 8–12 October 1973 (Brussels: Scientific Affairs Division, NATO, 1973), p. 48.
(103) Reference 101, pp. 1–3.
(104) DOE, "Lighting and Thermal Operations: Guidelines," Conservation Paper number 3, (Washington, D.C.: Department of Energy, 1974).
(105) See Reference 12 and "Klimaanlagen, Energieverbrauch," (Winterthur, Switzerland: Gebruder Sulzer, Aktiengesellschaft, January 1975).
(106) W. F. Rush, "Statement Relative to HR 10952" (Chicago, Ill.: Institute of Gas Technology, 14 November 1973).
(107) James W. Meyer, "Solar Energy Dehumidification Experiment on the Citicorp Center Building," Energy Laboratory NSF report no. MIT-EL-77-005, (Cambridge, Mass.: Massachusetts Institute of Technology, June 1977).
(108) R. Gerald Irvine, Testimony in PSC Case 26292, Transcript P.P. 1803-1816, (1973), in Reference 7.
(109) Deane N. Morris, *Evaluation of Measures to Conserve Energy,* (Santa Monica, Cal.: The Rand Corporation, February 1974).
(110) Frederick H. Rohles, Jr., "Humidity, Human Factors and the Energy Shortage," *ASHRAE Journal* (April 1975), pp. 38–40.
(111) DOE, "Lighting and Thermal Operations: Guidelines," Conservation Paper number 3, (Washington, D.C.: Department of Energy, 1974).
(112) Reference 27, pp. 3–14.
(113) Jean Dietz, "Stores Beaming in to Save on Energy," *Boston Globe,* (28 July 1977), p. 28.
(114) Reference 86, p. 508.
(115) M. Kiss, H. Mahon, and H. Leimer, *Energiesparen Jetzt!* (Berlin, West Germany: Bauverlag, 1980), Sections 7.12 and 9.6.
(116) Kennard L. Bowlen, "Energy Recovery from Exhaust Air," *ASHRAE Journal* (April 1974) pp. 49–56.
(117) NBS/DOE, *Waste Heat Management Guidebook,* NBS Handbook 121, (Washington, D.C.: U.S. Dept. of Commerce/National Bureau of Standards, and Department of Energy, February 1977) 200 pp.
(118) Energy Research Reports (Newton Mass.: Dec. 1975), p. 6.
(119) Reference 12, pp. 270–278.
(120) Reference 56, page 30.
(121) DOE, "A Study of the Energy Saving Possible by Automatic Control of Mechanical Draft Cooling Tower Fans," Conservation Paper No. 41 (Washington, D.C.: U.S. Department of Energy, Nov. 15, 1975).

(122) R. J. Moss, "Computer Control System for One IBM Plaza, Chicago," IBM Tagung-Energie Sparmassnahmen (Zurich, Switzerland: January 29, 1976).
(123) Energy Users Report, No. 204 (Washington, D.C.: Bureau of National Affairs, 7 July 1977), p. 18.
(124) M. Kiss, H. Mahon, and H. Leimer, *Energiesparen Jetzt!* (Berlin, West Germany: Bauverlag, 1980), pp. 270–278.
(125) L. G. Spielvogel, "Energy Conservation Methods for Buildings," in Reference 20, pp. 407–30.
(126) DOE, "Lighting and Thermal Operations," Office of Buildings Programs (Washington, D.C.: Department of Energy, 1975).
(127) SRI, *Patterns of Energy Consumption in the United States,* Stanford Research Institute for the Office of Science and Technology (Washington, D.C.: Executive Office of the President, January 1972).
(128) DOE, "The Electrodeless Fluorescent Lamp Fact Sheet " (Washington, D.C.: The U.S. Department of Energy, 14 April 1976).
(129) Deane N. Morris, *Evaluation of Measures to Conserve Energy* (Santa Monica, Cal.: The Rand Corporation, February 1974).
(130) M. Kiss, H. Mahon, and H. Leimer, *Energiesparen Jetzt!* (Berlin, West Germany: Bauverlag, 1980), Chapter 8.
(131) Reference 27, pp. 3–170.
(132) T. Kusuda, J. Hill, et al., *Pre-Design Analysis of Energy Conservation Options for a Multi-Story Demonstration Office Building,* Institute for Applied Technology, National Bureau of Standards (Washington, D.C.: U.S. Government Printing Office), 1975.
(133) Morgen Christensen, "The Wise Use of Electricity," *Lighting Design and Application* (July, 1974), pp. 29–35.
(134) Refer to pp. 219–221.
(135) Thola Theilhaber, "Energy Saving Ballasts for Fluorescent Lamps," Energy Bulletin No. 26, (Lexington, Mass.: Office of the Corporate Energy Conservation Engineer, Raytheon Manufacturing Company, March 1977).
(136) From the authors' files.
(137) Thola Theilhaber, "Phantom Tubes," Energy Bulletin No. 22, (Lexington, Mass.: Office of the Corporate Energy Conservation Engineer, Raytheon Manufacturing Co., Aug. 26, 1976).
(138) DOE, *Monthly Energy Review* (Washington, D.C.: Department of Energy, June 1977), parts 5 and 7.
(139) "Electricity Cost Projections," *The Oregonian,* (21 July 1977), p. D7.
(140) Fred S. Dubin, "Energy Conservation Through Building Design and a Wiser Use of Electricity," presented at the Annual Conference of the American Public Power Association, San Francisco, California (June 6, 1972).
(141) Craig B. Smith, *Efficient Electricity Use, A Practical Handbook for an Energy Constrained World* (New York: Pergamon Press Inc., 1976).
(142) Thola Theilhaber, "Energy Saving by Shutting Down Hydraulic Power Units of Numerically Controlled Machine Tools While Maintaining Oil Temperature by Electric Heaters," Raytheon Energy Bulletin no. 20 (Lexington, Mass.: Office of the Corporate Energy Conservation Engineer, May 1976).
(143) Robert E. Hunt, Frank Seabury, and Philip Valence, *Energy Efficiency and Electric Motors,* Conservation Paper Number 58 (Washington, D.C.: U.S. Department of Energy, August, 1976), 169 pp.
(144) Reference 27, pp. 3–18B.

(145) Ibid., pp. 3–16.
(146) Ibid., pp. 3–18.
(147) Ibid., pp. 3–18B.
(148) International Fabricare Institute, "Energy Cost Reduction in the Fabricare Industry " (Joliet, Ill.: International Fabricare Institute, 1976), p. 21.
(149) Ibid., p. 27.
(150) M. Kiss, H. Mahon, and H. Leimer, *Energiesparen Jetzt!* (Berlin, West Germany: Bauverlag, 1980), Section 7.13.
(151) Ibid., Section 7.11.
(152) "ETH Will Neues Energiesparkonzept Testen," *Tages Anzeiger* (Zurich, Switzerland: 13 September 1970), p. 15.
(153) Reference 27, pp. 3–31.
(154) Ibid., pp. 3–29.
(155) Ibid., Section 3.
(156) Reference 27, pp. 3–33.
(157) "A Report of the National League of Cities and the U.S. Conference of Mayors Solid Waste Management Task Force, Cities and the Nation's Disposal Crisis " (Washington, D.C.: EPA, 1973), p. 1.
(158) James W. Meyer, " 'Saved' Fuel As An Energy Resource," Energy Lab Working Paper #15 (Cambridge, Mass.: MIT, June 1974).
(159) Sheldon Meyers, "Status of Solid Waste Management in the U.S.," address delivered to the Second International Congress of the International Solid Waste Association (Washington, D.C.: EPA, 1976), p. 14.
(160) Thola Theilhaber, "Saving Energy by Recirculating Air," from the Corporate Office of Manufacturing, (Lexington, Mass.: Raytheon Company, 1976), p. 5.
(161) Steffen Plehn, "Resource Conservation: How Industry Might Help " (Washington, D.C.: EPA, 1978), p. 1.
(162) Reference 159, p. 7.
(163) Ibid., p. 9.
(164) Environmental Action Foundation, Garbage Guide, (Washington, D.C.: EPA, 1977), p. 1.
(165) Reference 27, pp. 3–42F.
(166) Reference 158, p. 18.
(167) Laurence McEwan, Jr., "Re-Refining of Waste Lubrication Oil: Federal Perspective " (Washington, D.C.: EPA, 1976).
(168) *Boston Herald American* (January 8, 1979), p. 8.
(169) EPA, Waste Clearinghouses and Exchanges: A Summary. New Ways for Identifying and Transferring Reusable Industrial Process Wastes (Washington, D.C.: EPA, 1977), p. 4.
(170) "Utility Will Buy Power from Refuse Plant," *Electrical World* (1 August 1975), p. 80.
(171) DOE, "Vanpool Executive Summary," Prepared for the Executive Conferences and Workshops—"Managing the Energy Dilemma" (Washington, D.C.: Department of Energy, February 1977).
(172) DOT, "Car and Bus Pool Matching Guide," Third Edition (Washington, D.C.: Department of Transportation, November 1973).
(173) DOE, "Vanpool Implementation Handbook " (Washington, D.C.: Department of Energy, February 1977).
(174) W. F. Gay, *Energy Statistics* (Washington, D.C.: Department of Transportation, 1975), p. 137.

(175) D. A. Hurter and W. D. Lee, "A Study of Technological Improvements to Optimize Truck Configurations for Fuel Economy," (Washington, D.C.: Arthur D. Little Inc., September 1975), pp. 3-4–3-6
(176) D. B. Shonka, A. S. Loeble, P. D. Patterson, *Transportation Energy Conservation Data Book: Edition 2,* (Oak Ridge, Tenn.: Oak Ridge National Laboratory, October, 1977), pp. 195, 200.
(177) Daniel A. Harkins, John Haynes, "Energy Management in Municipal Fleets," report of Energy Conservation Project, (Boston, Mass.: Massachusetts Department of Community Affairs, February 11, 1977), 69 pp.
(178) E. M. Cope, "The Effect of Speed on Automobile Consumption Rates," (Washington, D.C.: Department of Transportation, Federal Highway Administration, October 1973), pp. 6 and 7.
(179) E. M. Cope, "The Effect of Speed on Truck Fuel Consumption Rates," (Washington, D.C.: Department of Transportation, Federal Highway Administration, August 1974), p. 14.
(180) D. A. Hurter and W. D. Lee, "A Study of Technological Improvements to Optimize Truck Configurations for Fuel Economy, (Washington, D.C.: Department of Transportation, Transportation Systems Center, September 1975).
(181) DOT, "A Study of Potential Motor Vehicle Fuel Economy Improvement," (Washington, D.C.: Department of Transportation, January 10, 1975).
(182) J. D. Murrell et al., "Light Duty Automotive Fuel Economy—Trends Through 1977," SAE Paper 760795, presented at the Automobile Engineering Meeting, (Dearborn, Mich.: October 1976).

6

Heat Recovery and Cogeneration

Much of the energy that is presently wasted could be put to further use before being rejected into the environment. Examples would be the heat that could be recovered from electric power plants, office buildings, kitchens, schools, laundries, supermarkets, hospitals; from industrial processes, bakeries, brick kilns, metal heat treating furnaces; and from equipment such as computers, air and refrigeration compressors, air-conditioning chillers, boilers, and furnaces.

Heat recovery from ventilation air, industrial processes, and cogeneration are discussed in this chapter. Other examples will be found in Chapters 7 and 8.

6.1 HEAT RECOVERY FROM VENTILATION AIR

The technology for heat recovery from ventilation air has been known for a long time. Most existing buildings, however, were constructed when energy was readily available and inexpensive, with little thought toward recovering energy from ventilation air. The oil price shock and the threat of shortages has led to intensive interest in getting more for one's energy dollar, and hence to interest in heat recovery installations and to the development of more efficient recovery devices. In the following pages, some different heat recovery systems will be described, and their advantages and disadvantages will be pointed out. References 1, 2, and 3 contain additional information.

Most building heat recovery systems have a dual purpose. In winter, they are used to transport heat from exhaust air to the (incoming) outside air. In summer,

the air flow is in the opposite direction, with outside air cooled and dehumidified (by exhaust air). Some heat recovery systems function only with low outside temperatures. Selections of the optimum system for a particular building would depend on several factors:

The temperature of the heat sources and heat sinks

The latent heat content of the heat source and sinks

The mean flow rate through the recovery device

The additional energy required to operate the recovery device

The physical constraints on the recovery system size and location

Some of the systems discussed here are shown in Figure 6.1. Their efficiency can be defined as shown in Figure 6.2. With a recuperative plate heat exchanger, the exchange

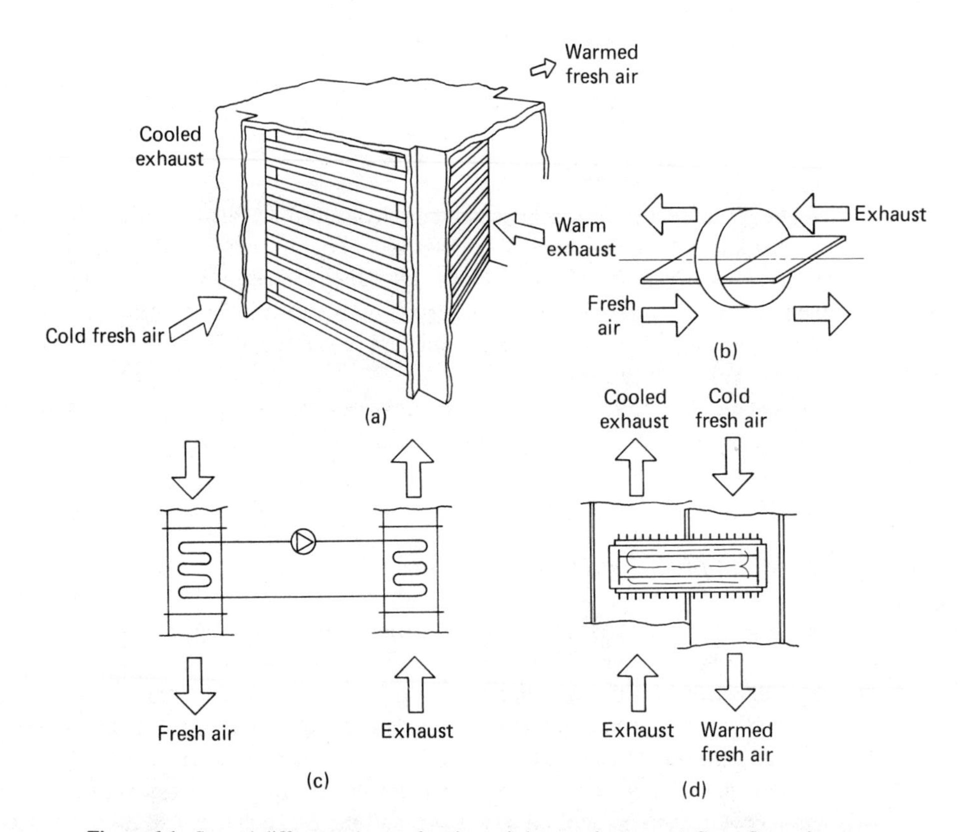

Figure 6.1. Several different schemes for air-to-air heat exchanges: (a) Cross-flow, plate type, (b) Regenerative heat wheel, (c) Run-around, coils and pump, (d) Capillary wick, heat pipe. Other types not shown include twin tower with desiccant spray, gravity return heat pipes, and units optimized for energy recovery from grease-laden kitchen exhaust.

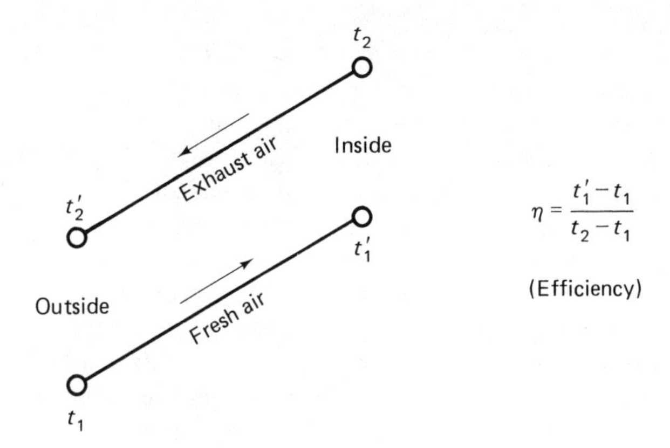

Figure 6.2. Definition of the efficiency for heat recovery. During the heating season, for example, the efficiency is the ratio of the energy added to the incoming air to the total energy difference between the inside and the outside. For systems that recover the latent component of heat, the temperature, t, in the above equation, should be replaced with the respective components of enthalpy, h.

of heat is across a plate constructed from aluminum, glass, etc. Normally, the heat exchange utilizes the cross-flow principle. Therefore, the mean temperature differences are smaller than with counterflow devices, which require larger temperature differentials between the two airstreams to make the economics justifiable. As a result of laminar flow, the heat exchange is not very efficient. It is used where special attention has to be given to leakage problems. To increase heat recovery efficiency, several units are often staged, providing even greater transfer surface area. Further, **the plate heat exchanger is used where no contamination between exhaust air and outside air can be allowed, for example, in chemical factories or hospitals.** A disadvantage of this system is that the exhaust air ducts and outside air ducts have to be brought together, thereby necessitating a large space requirement.

If the exhaust air is dry, the plate exchange can only transfer sensible heat. The sensible heat is transmitted with conduction, radiation, and convection. Whenever air passes over a dry surface that is at a temperature higher than the dry bulb temperature of the air, sensible heating occurs. During such a process, the specific humidity remains constant but the dry bulb temperature rises and approaches that of the surface. If air is passed over a surface or through a spray of water that is at a temperature less than the dew point temperature of the air, condensation of some of the water vapor in the air will occur simultaneously with the sensible cooling process. Latent heat is transmitted through evaporation or condensation. Exhaust air having a relative humidity of over 50 percent normally means that condensation will occur in the heat exchanger. In some cases, formation of ice may occur which must be avoided to prevent deformation of the plates and reduction of the flow area. Possible solutions would be to preheat the outside air or to bypass air using control dampers.

With a rotary heat exchanger (regenerative heat exchanger), a rotating mass periodically contacts the outside air and the exhaust air. Heat is absorbed from the exhaust air and transmitted thereafter to the outside air. There are two types of rotative heat exchangers: one transmits both latent and sensible heat, while the other transmits only sensible heat. The recovery of latent heat is obtained by the absorption of moisture from the exhaust air and its transferral to the outside air. A hygroscopic heat wheel composed of a material like lithium chloride is an example of a latent and sensible heat recovery device. In the summer the process is reversed, with the use of drier exhaust air to dehumidify incoming outside air.

With operation under 32°F (0°C), preheating is necessary to prevent formation of ice. **Regenerative heat exchangers are used in comfort and industrial air conditioning systems and have a high efficiency rate—on the order of 75–80 percent.** With this, the payback period can be short. One application of heat wheels is in space heating situations where unusually large quantities of ventilation air are required for safety reasons. As many as 20 or 30 air changes per hour may be used to remove toxic gases or to prevent the accumulation of explosive mixtures. Comfort heating for that quantity of ventilation air is frequently expensive enough to make the use of heat wheels economical.

One disadvantage is cross contamination, since the pores of heat wheels carry a small amount of gas from the exhausts to the intake duct. Where such contamination is undesirable, the carryover of exhaust gas can be partially eliminated by the addition of a purge section, where a small amount of clean air is blown through the wheel and then exhausted to the atmosphere. **Another disadvantage is the large space requirement and complicated routing of exhaust and outside air.**

A list of typical applications follows:

Heat and moisture recovery from building heating and ventilation systems

Heat and moisture recovery from moist rooms and swimming pools

Reduction of building air-conditioning loads

Recovery of heat and water from wet industrial processes

Heat recovery from steam boiler exhaust gases

Heat recovery from gas and vapor incinerators

Heat recovery from baking, drying, and curing ovens

Heat recovery from gas turbine exhausts

Heat recovery from other gas-to-gas applications in the low through high temperature range

A heat exchanger system with a closed water loop consists of two or several heat exchangers, for example, in the exhaust and outside air. In the recirculation loop, a pump recirculates a mixture of water and glycol, etc., and the heat content of the warm exhaust air is transmitted to the cold outside air. In summer, the process is reversed. The efficiency of the system is usually 35–50 percent. The danger of ice formation has to be considered similarly to other heat recovery systems. The advantages of the system are that it requires little space, it can be installed using standard heat exchangers, and it needs no modification of the ducts for exhaust and outside air. It is possible to collect waste heat from several exhaust ducts and use the heat to preheat the outside air. These advantages, together with **relatively small investment costs, encourage the use of this system.** Some typical applications are domestic hot water heating, boiler feedwater heating, hot water space heating, absorption-type refrigeration or air conditioning, and process liquids heating. The specific applications depend on the type of heat exchanger used.

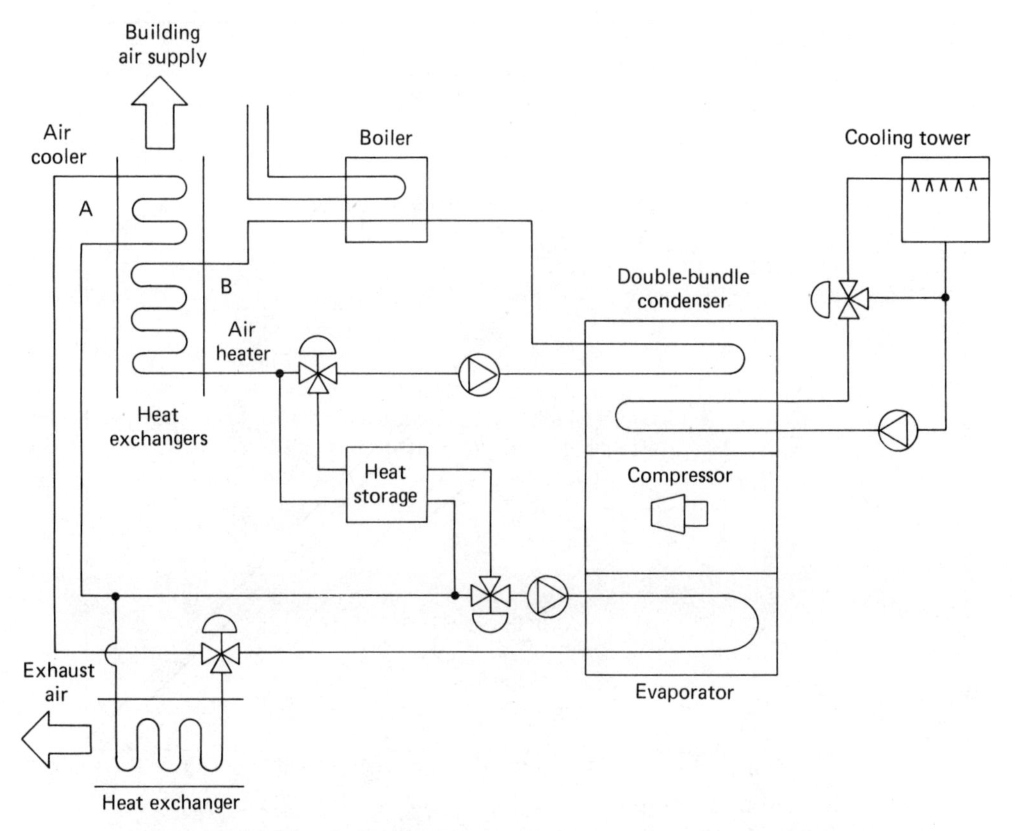

Figure 6.3. Explanation of the operation of a heat pump with a double bundle condenser. In summer the supply air is cooled and dehumidified by the heat exchanger (A), with the heat being transferred by the chilled water loop to the evaporator. The temperature of this heat is raised by the compressor and sent via the condenser to the cooling tower.

In winter heat from the condenser is used to preheat cold outside air at the supply air heater (B). Heat is recovered from the exhaust air via the heat exchanger, heat pump, and air preheater. If more heat is recovered from the inner core exhaust air than is required, it can be held for later use in the storage tank. Should there be insufficient heat for the operation of the system, the oil- or gas-fired boiler makes up the deficit. At night, with the fresh air supply off, the exhaust air is no longer a source of heat. Heat is obtained from the storage tank, which was charged during the day.

The heat pipe consists of a closed pipe, half of which is located in the exhaust duct, the other half in the outside air duct. The pipe consists, on the inner side, of a porous material which is filled with refrigeration medium, e.g., freon or alcohol. Circulation will occur in the pipe, with one end of the pipe acting as a condenser (where the cold outside air is heated and the refrigerant medium is condensed). The liquid flows to the evaporator side which is in contact with the warm exhaust air. The fluid evaporates in

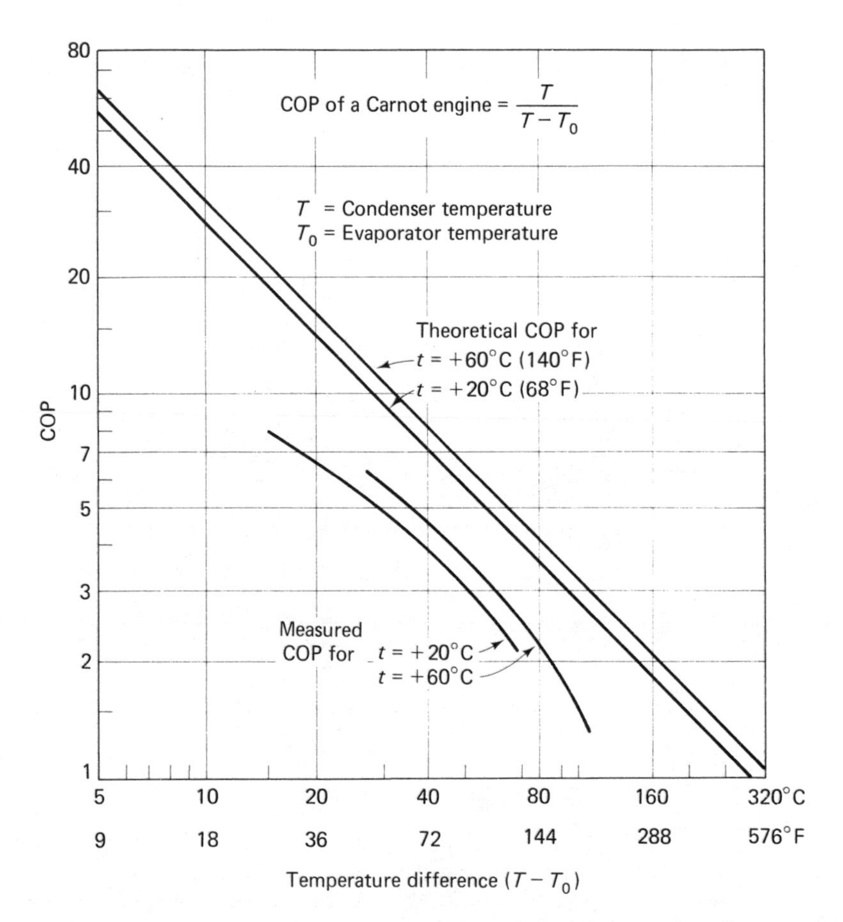

Figure 6.4. Theoretical and typical measured coefficients of performance for heat pumps as a function of the condenser-evaporator temperature difference. The COP is the ratio of heat output to the work required to operate the compresser. High COP corresponds to good economy of operation. The COP increases as the temperature difference between condenser and evaporator is reduced. Theoretical values for COP are computed from the Carnot expression in which the temperature must be expressed in absolute units.*

*To convert to absolute temperature, add 273 to °C temperature, or add 427 to °F temperature.

this part and the heat evaporation is removed from the exhaust air. **The efficiency of this system is between 55 percent and 70 percent.**

Another system for heat recovery in air conditioning is the heat pump. In the United States, the refrigeration units of larger installations have been used for heating as well since 1965. With the increased use of open-space offices, pumps are used to transport the heat surplus of the inner zone to the outer zones, where it is required. As with other heat recovery systems, the heating energy source is exhaust air. As shown in Figure 6.3, the heat pump increases energy quality, which means that the heat recovered is transformed

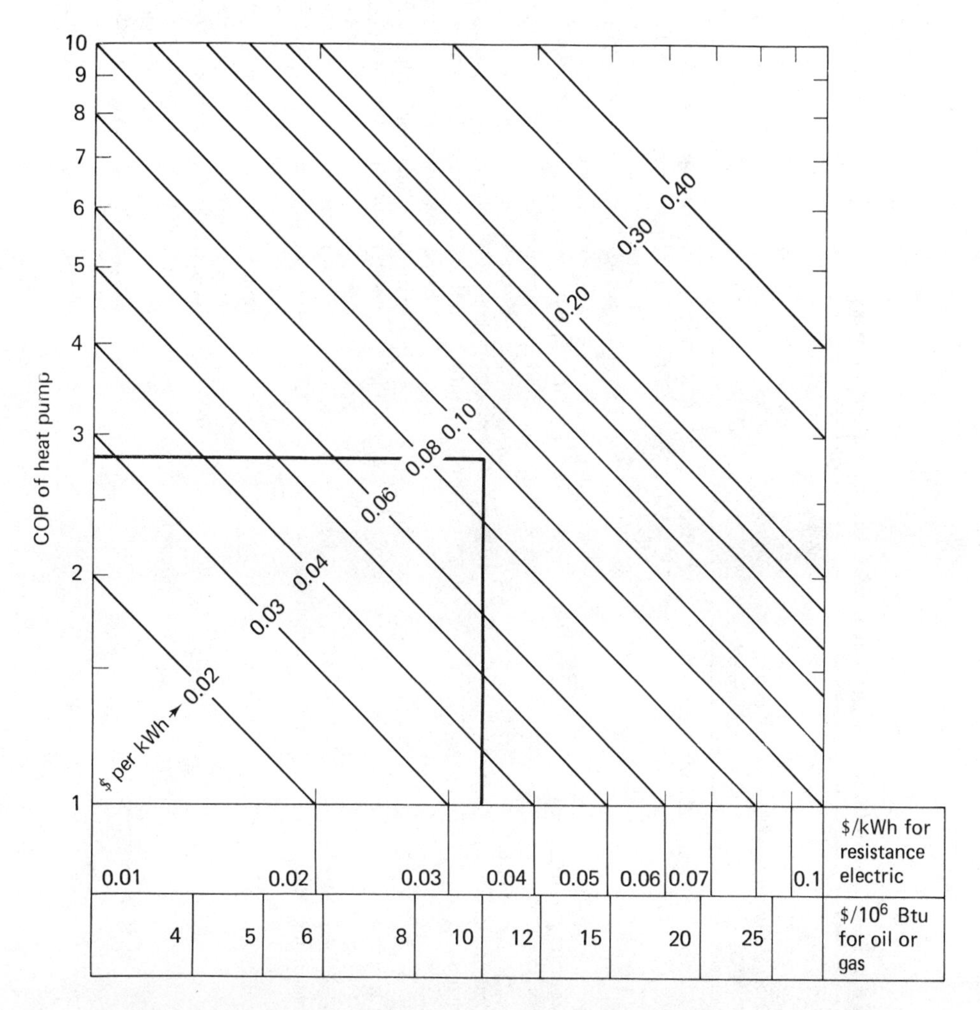

Figure 6.5. Comparison of the operating cost for a heat pump with the cost of conventional resistance electric or oil- or gas-fired heating. For example, if oil heat costs $10/10^6$ Btu, then a heat pump with an annual average COP of 2.9 would be cheaper until electricity costs more than 10¢/kWh. Capital cost is not included in this comparison.

Table 6.1. Examples of Costs and Savings from Exhaust Air Heat Recovery

Energy-saving measure	*Cost ($)*	*Savings ($/year)*	*Energy saved, MBtu (kWh)*	*SCP[a], $/BPD ($/kW)*	*Comment*	*Reference*
Glycol loop	20,000	63,000	31,000 (9×10^6)	4250 (60)	Exhaust air heat recovered with a runaround glycol loop between fresh air and exhaust air heat exchangers at 3M plant in St. Paul. Saving computed at $2/MBtu ($7/1000 kWh), averaged over entire year.	(4)
Heat recovery from exhaust air	67,700	9,200	2,250 (6.6×10^5)	63,700 (900)	Heat exchange between intake and exhaust ventilation in headquarters building for Swiss engineering company.	(5)
Hospital, Zurich, Switzerland	53,400	43,200	11,250 (3.32×10^6)	10,000 (140)	Heatwheel with hygroscopic coating transfers enthalpy between exhaust and fresh air without any compromise in hygienic quality of air supply. Energy saved is net (including heatwheel power), referred to raw source.	(6)
Enthalpy recovered in Jacksonville, Fla.	18,000	3,860	7,500 (2.2×10^6)	5,300 (75)	Including heat recovery in building design reduces peak capacity of cooling and heating equipment required. (Heat recovery is most efficient when the temperature differences are largest.) Net reduction in first costs are $27,000 for cooling equipment and $2,500 for heating for a net total saving in construction of $11,500. Energy saving is reduced by 39,000 kWh for the drive motor to obtain raw source energy saving shown.	(2)

Energy-saving measure	Cost ($)	Savings ($/year)	Energy saved, MBtu (kWh)	SCP[a], $/BPD ($/kW)	Comment	Reference
Enthalpy recovered in Chicago, Ill.	18,000	4,960	6,000 (1.75×10^6)	6,370 (90)	Planning for heat recovery at construction saves $22,000 in reduced cooling equipment size and $3,900 in reduced heating size (i.e., construction costs $7,900 less with heat recovery). Less energy for drive motor gives the net raw source energy shown.	(2)
Electronics manufacturing plant	29,700	11,540	575,000 kWh	450	Ventilation operating continuously to remove fumes from plating and other manufacturing operations is diverted through glycol runaround coils. This reduces electric load by a yearly average of 65.8 kW.	(7)

[a]The specific capital productivity (SCP) is the cost of reducing the annual rate of energy use by the energy equivalent of one barrel of oil per day (BPD) or by one kilowatt.

Table 6.2. Comparison of Costs and Savings for Process Heat Recovery (8)

Example	*Cost ($)*	*Savings ($/yr)*	*Energy savings, MBtu/yr (MWh/yr)*	*SCP, $/BPD ($/kW)*	*Comments*
Baking lithographed steel sheet	100,000	20,000	40,000 (12,000)	3,540 (50)	Heat recovered from incinerator of solvent gas following lithographed printing of steel sheet for cans. Costs are in 1970 dollars for 6,000 h/yr operation.
Boiler feed water	12,092	30,000	8,300 (2,430)	2,600 (36.60)	Small packaged boiler in glass plant retrofitted with economizer to heat boiler feedwater. Costs are in 1972 dollars for 6,000 h/yr operation.
Heat reclamation from air-conditioning cycle	2,370	805	210 (62)	24,000 (340)	Hot water heating by the use of rejected heat from standard air-conditioning cycle. Costs in 1961 dollars for 8,760 h/yr operation.
Heat recovery in a glass plant	4,520	2,950	4,300 (1,259)	2,200 (31)	A glass company installed a stack recuperator to preheat combustion air for one of its day tanks. The unit can withstand temperatures of 2,100–2,300°F (1,150–1,260°C). Costs are in 1971 dollars.
Heat recovery from a rotary forge furnace	6,400	2,000	1,530 (448)	2,340 (33)	Metallic radiation recuperator was used to remove waste heat from flue gas. This was then used to preheat combustion air for the rotary forge furnace. An additional $1,690 is required every 5–7 years for inner shell replacement. Costs are in 1968 dollars for 3,800 h/yr operation.

Example	*Cost ($)*	*Savings ($/yr)*	*Energy savings, MBtu/yr (MWh/yr)*	*SCP, $/BPD ($/kW)*	*Comments*
Turbine driven compressor	41,000	15,187	20,250 (6,000)	1,415 (20)	Exhaust at 825°F (440°C) from gas turbine supplies 6.75 MBtuh to heat regeneration gas for dehydration. Costs are in 1970 dollars for 3,000 hr/yr operation.
Food product processing plant	780,000	150,000	1×10^5 (3×10^4)	16,000 (225)	Heat formerly rejected to stack from fired product dryer is recycled to molasses evaporator. Neighborhood air pollution problems were greatly relieved, in addition to estimated energy saving shown.
Fume incinerator	90,000	46,000	3.8×10^4 (1×10^4)	5,000 (70)	Noxious fume incinerator uses heat wheel to reclaim heat from incineration to preheat fumes from a strip oven. Heat recovery of ceramic wheel is 70 percent.
Radiation recuperator for an automatic transmission plant	4,950	1,075	860 (252)	12,000 (172)	High temperature, bent tube batch furnace equipped with exhaust gas recuperator.

to a higher temperature level. The definition of efficiency given in Figure 6.2 cannot be used here. It is common practice to use the coefficient of performance (COP) to characterize the merits of this system. This is defined in Figure 6.4.

Because of losses in the compressor, in the motor, and in the cooling recirculation system, the heat pump has a COP of about 50 percent of its theoretical value. In the past, the heat pump has not received much industrial application. However, several manufacturers are now redeveloping their domestic heat pump systems for industrial use. These will make it possible to use large quantities of low grade waste heat with relatively small expenditures of work.

In Figure 6.5, the heating cost of a conventional oil- or gas-fired heating system is compared with that of a heat pump. Figures 6.3, 6.4, and 6.5 are generally applicable to heat pumps, not only if they are used for heat recovery, but also if the source of heat is from well water or some other source.

With the use of the heat pump for heating, a common design practice is to use it to meet the average heating load. For colder outside temperatures a supplementary heating system using oil or natural gas is employed. Since the supplementary system is used a relatively small percentage of the heating season, the additional fuel use will not increase average energy costs significantly, although consideration should be given to possible excessive demand charges. Examples of the cost of some heat recovery systems and their savings are shown in Table 6.1, and additional information will be found in references 2, 4, 5, and 6. The application of these heat recovery devices to a variety of industrial processes is given in a publication of the National Bureau of Standards. (8) Many of these applications are summarized in Table 6.2. The examples listed in Tables 6.1 and 6.2 show that the specific capital productivity for most heat recovery applications is quite favorable.

6.2 COGENERATION OF HEAT AND ELECTRICITY

An important opportunity to save energy and lower annual costs in industry is afforded by the practice known as cogeneration. This is the combined production of electricity with industrial process steam (or other forms of process heat). Such systems are also known as "total energy systems." Similar advantages exist from putting total energy systems into commercial buildings to produce comfort heat, cooling, and domestic hot water together with electricity. As explained in Chapter 2, the opportunity for saving arises because most process and comfort heat is used at relatively low temperatures and hence does not make full use of the fuel's capability to produce high-temperature heat. The usual practice of burning fuel for low-temperature metallurgical processes or to raise low-pressure steam in a boiler is thermodynamically inefficient. More efficient use of the fuel can be made by first using it to produce high-quality work or electricity in a machine that exhausts heat at the appropriate temperature level required by the desired process heating application.

Electricity can be cogenerated with usable process heat using less than 50 percent of the fuel required in a conventional central station power plant. Furthermore, the capital cost can be as low as one-fifth that required to add the same capacity to a central station power system and a separate heating system. With the advent of *time-of-use rates* (TOUR), the dollar savings could be significant for those industries which cannot shut down during peak load periods. See the Report of the Governor's Commission on Cogeneration (9) for information on its economics and regional benefits, as well as references 10 through 21.

About 45 percent (13 $\times$ 10^{15} Btu/year) of all the fuel consumed by United States industry is used for steam generation, chiefly at low pressure. An additional 29 percent (8.3 $\times$ 10^{15} Btu/year) is consumed in fossil-fueled direct heating processes. The remaining 26 percent is consumed in the generation of electricity for electrical heating, motors, lighting, electrolysis, etc. Only about 15 percent of the electricity consumed by industry and 5.8 percent of that consumed nationwide is generated in-plant; the remainder is generated by utilities with condensing steam turbines or gas turbines without heat recovery. (10) In West Germany, industry produces 28 percent of that country's total electricity. (11)

The potential for fuel savings has not been tapped in the United States and in many other countries because of economic, regulatory, and institutional barriers. Opposing interests of the electric utilities have raised economic and legal barriers. Disadvantageous rates exist which utilities have used to discourage total energy systems. Because of the advantages to society at large from the reduced fuel and capital demands, utilities must now purchase cogenerated electricity from industrial and commercial plants. Grid connection can lead to cheaper and more dependable electricity for the cogenerator and the other utility customers. In most cases, it can provide new generating capacity far more cost-effectively than can construction of central station base load plants.

Cogenerating Systems

Figure 6.6 compares three systems for cogeneration of process heat and electricity with conventional practice. (11) In each of the cogeneration schemes a prime mover, i.e., a back-pressure steam turbine, a gas turbine, or a diesel engine, drives a generator which supplies electric power. Waste heat is reclaimed from the exhaust and/or jacket cooling water to produce steam or hot water. In addition to process heat applications, this steam or hot water can provide space heating, domestic hot water, or cooling with an absorption chiller. The exhaust gas from a diesel or gas turbine can sometimes be used directly for a variety of purposes such as annealing or drying. The engines employed in these systems are known as topping engines.

Table 6.3 is a summary of nominal incremental fuel rates based on the assumption that all of the heat cogenerated with electricity can be used. The object of cogeneration is to produce electricity more efficiently and to reduce annual costs. All of these systems can generate electricity *and* deliver it to the user at fuel rates considerably lower than the 12,000 Btu/kWh presently used to generate electricity in a central station power plant.

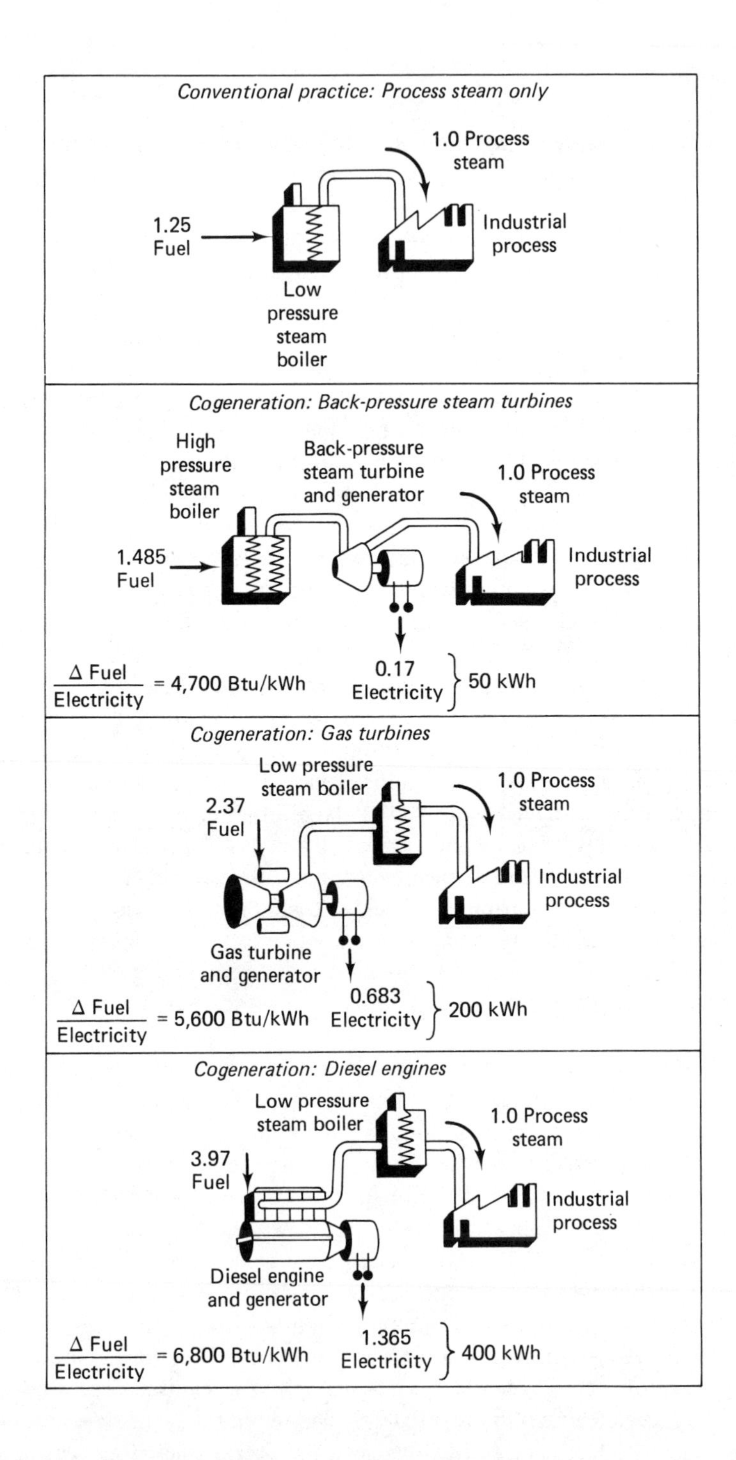

Figure 6.6. Energy balances for alternative cogeneration systems. The incremental fuel consumed in the diesel engine is 43,000 Btu/kWh if the steam and/or hot water produced in the engine water jacket is utilized for various industrial processes. (11)

Table 6.3. Comparison of Electricity per 10^6 Btu (0.3 MWh) of Process Heat and Incremental Fuel Rates for Cogenerating Systems

Type of cogeneration system used in topping cycle	*Applicable fuel types*	*Electricity generated per 1000 pounds of process steam (or per 10^6 Btu process steam energy) (kWh)*	*Incremental fuel consumed per kWh of electricity output (Btu)*	*Efficiency of electricity production (%)*	*Efficiency of fuel use (%)*
Back-pressure steam turbines	Coal, residual oil, or waste materials (i.e., boiler fuels)	50	4,700	72	79
Gas turbines with exhaust recovery boilers	Distillate oil, natural gas, or low Btu byproduct gas from certain processes in industry	200	5,600	61	71
Diesel engines with exhaust recovery boilers	Residual oil, distillate oil, natural gas	400	6,800	50	60
Above with water jacket recovery	As above	400	4,300	79	80

Note: The efficiency of electricity production is the ratio of the electricity produced to the additional fuel required by the system for its production. The efficiency of fuel use is the ratio of electricity and useful heat recovered to the fuel input.

There is considerable latitude in adjusting the ratio of electricity to process heat. Although the heat rate for steam topping is very favorable, the quantity of electricity that can be generated per unit of process steam is relatively low: between 20 and 70 kWh per million Btu (0.3 MWh). **The steam turbine is oriented for industrial processes where large quantities of heat are required.** The gas turbine is less suited for following a varying heat load. **Gas turbines have low intrinsic efficiency** (18–22 percent) and exhaust large quantities of high temperature gas. **For high-system efficiency, the available work in the exhaust gas from the turbine must be utilized for its high quality,** i.e., at its high temperature of 900 to 950°F (480–510°C). The diesel systems are best suited where there is a relatively large need for electricity compared to heat. They are thus also suited to achieve good economics for modern office buildings, shopping centers, and computer centers. Furthermore they can be adapted to a wide variation in the electricity required per unit heat output: between 300 kWh and 460 kWh per 10^6 Btu (0.3 MWh). It is for this reason that the classic "total energy system" of past years often consisted of diesel engine(s) and heat recovery boilers to generate steam and/or hot water for commercial buildings. In addition to generating electricity, the shaft horsepower from the steam or gas turbine or diesel engine can be used to drive a compressor directly for air conditioning, refrigeration, or compressed air, to drive a pump, or for any other normal engine load. The use of diesel driven heat pumps, for example, may make it possible to use a waste heat source.

Practicability

Maximum fuel savings and related benefits occur if all of the heat and all of the electricity can be used simultaneously. This is described by the plant load factor which, together with the relationship of fuel costs to the cost of purchased electricity, affects the economics of the total energy application. Considering these factors, the ideal application would have a high load factor for electric power and a favorable steam or heat utilization factor along with a relatively low fuel cost as compared to purchased power. The extreme favorability of one factor can sometimes compensate for the others being something less than favorable. An industrial plant working three shifts per day would generally represent a favorable load factor, while an office building operating five days a week, eight hours a day, might not present an especially feasible load factor. However, each could represent a favorable opportunity for a total energy plant. The unfavorable load factor, which would have recommended against the office building in the past, may, in fact, be the source of high peak demand charges with TOUR that makes the total energy system very attractive. The economics have to be considered in light of the details for each application; however, since additional personnel are not necessarily required, the major factor will be the cost of fuel compared to the cost of purchased electric power.

Completely unattended, automatic operation can be achieved in many instances. Maintenance and operation can be handled with in-plant personnel, or guaranteed maintenance and service contracts and provide the reliability and convenience associated with utility power together with efficiency and economy. Different arrangements can provide everything from complete maintenance and service, including all parts, supplies and labor, to contracts that provide only a guaranteed cost for engine rebuild. Fully automatic systems include load sensing to add or drop engines from the line. Any number of units can be operated automatically and unattended if they have adequate safety and status read-out devices to notify when other than routine servicing is required.

Where high standby charges are in effect, an increasing number of cogenerators are finding it economically beneficial to assemble a somewhat larger system that is completely independent of the utility, except for possible resale of surplus electricity. Such a system would be configured with a sufficient number of smaller engines so that all essential peak load can be covered with one engine off-line in case of failure, and a second engine down for regular maintenance.

An example of such a system was installed in midtown Manhattan in the 692,000 ft^2

office building at 11 West 32nd Street. (12) The building's new power plant, consisting of eight 700-kilowatt diesel generators, will permit it to operate completely independently of Consolidated Edison. The system actually requires only six of the generators; the other two are for backup duty. Although Consolidated Edison has been experiencing difficulties in increasing the capacity of its distribution network in downtown Manhattan, and in financing new power plant construction, its response to the opportunity for relief from this cogeneration system nevertheless follows the traditional line. Although conversion of 100 similar buildings could achieve a 5 percent to 6 percent reduction in its load, the utility would not remain "complacent" under such a development. (12)

The total energy system was included as part of a complete renovation of this 32-story building that was accomplished for less than $40 per square foot. About 30 percent of the fuel used will produce the building's electricity, and 45 percent will be recovered to supply its hot water and steam needs for heating and air conditioning. Ten percent is expended as radiant heat in the basement where the system is located; the remainder is rejected up the stack. According to the design engineers, about 1.25 million gallons of fuel per year is saved compared to a conventionally operated building. Operating costs are reduced by about 10 percent per year.

Lost Efficiency with Cogeneration

When evaluating a cogenerating system such as the one described above, it would be wiser to compare fuel saving against the most efficient system economically feasible, rather than against existing practice. For example, an absorption chiller produces the refrigeration effect with a very low coefficient of performance (COP). Today's mechanical compressors have seasonal COPs that are four to seven times greater than absorption chillers. Consequently, less primary fuel may be consumed by driving a mechanical chiller directly from an engine, or even with electric drive with power from the utility. In other words, caution must be exercised when using an absorption chiller with a total energy system, so as not to diminish greatly or even negate its potential fuel efficiency. The best operating economics with cogeneration are achieved by employing it *in conjunction with* all other possible efficiency measures such as heat storage, etc., rather than in lieu of them.

A second caution concerns small steam turbines; these are extremely inefficient. It is only in the larger sizes that efficient steam turbines are available at prices comparable to gas turbines or diesel systems. As a consequence of these economics, small steam-turbine-driven chillers operating with saturated steam, or with the moderate-pressure steam from district heating systems, have efficiencies as poor as the absorption chiller, or even worse. Now that fuel prices have increased, any equipment driven by a small condensing-steam turbine should be examined for a possibly more efficient alternative.

Heat Recovery with the Diesel Engine

To improve the thermal efficiency of an on-site energy system, at least part of the rejected heat must be recovered and put to useful purpose. Perhaps the most convenient source of heat is that from ebullient cooling of the diesel engine. Essentially 100 percent

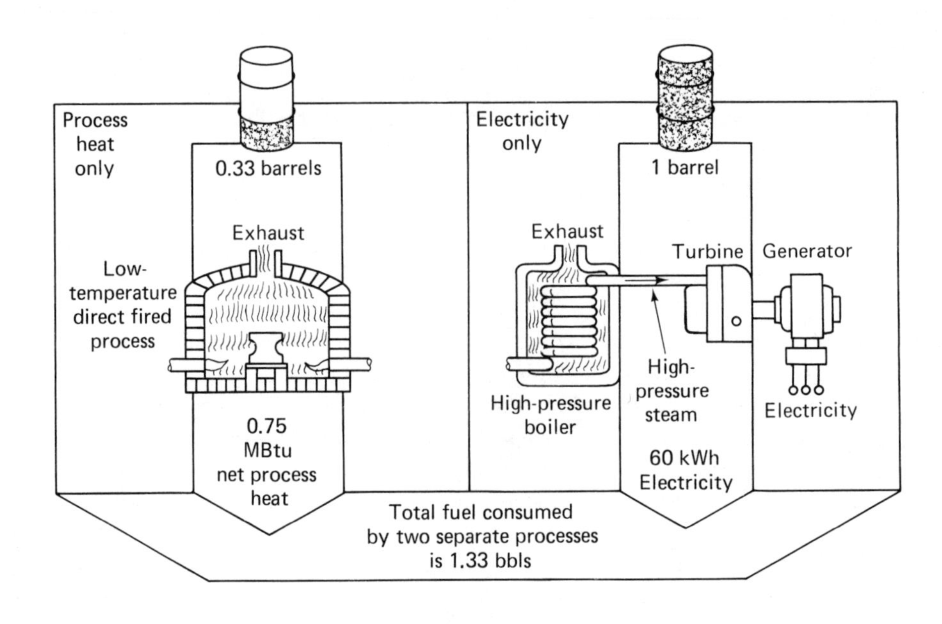

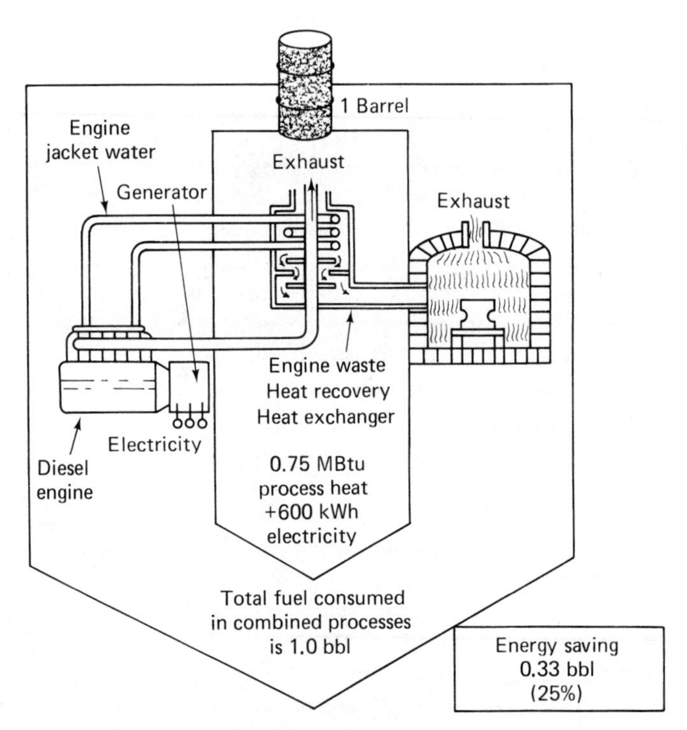

Figure 6.7. Diesel cogeneration of low-temperature process heat. (11) Heating processes such as baking, drying, curing, etc., which have relatively low-temperature requirements (200 to 300°F [95 to 150°C]) can be accomplished more efficiently with a diesel topping engine cogenerating electricity. Ebullient jacket cooling and heat recovery from the diesel exhaust can provide heat in this temperature range, save fuel, and provide valuable electricity also.

recovery of this heat is possible at a rate of approximately 2,000 Btu/kWh generated, and at pressures up to 15 psi. Today, single-package units are available which recover both jacket water and exhaust heat in a single unit. The typical savings possible are illustrated in Figure 6.7, where an application of the diesel system to a process requiring very low temperatures is illustrated. (11) For intermediate temperatures, opportunities exist for using the waste heat derived from the exhaust either directly or with a high-temperature boiler. Diesel electric generators are available in sizes from 50 kW to over 2,000 kW and can operate on a variety of fuels including natural gas, residual oils, and some industrial wastes. Because of their high efficiency, low cost, and moderately low maintenance costs, diesel engine systems offer significant potential for cogeneration in both retrofit and new installations.

Gas and Steam Turbine Systems

Two examples of the use of the gas turbine will be discussed. In one of these examples a very high overall efficiency is possible because the exhaust heat from the gas turbine is utilized for its high quality to make high-pressure steam for a steam turbine. This may be compared with the example in Table 6.4, where low-temperature steam is produced for an absorption chiller. (13)

Two different systems were considered for this building; 1) a conventional design with an oil/gas fired boiler and chillers, and 2) a total energy system. The following criteria were used in the selection process: diversity of energy sources for supply security, energy saving, and economy with respect to expected future fuel price increases.

Based on these criteria, the decision was made to install a total energy unit using two

Table 6.4. Total Energy System for a Commercial Building (13)

Demand	*Power kW*[a]	*Energy, MWh/year*[a] *(MBtu/yr)*
Electricity[b]	3,500	17,900
Cooling	4,000	8,400 (28,700)
Heating	2,600	3,800 (13,000)
Emergency power	3,000	

Note: The air-conditioned building has a net office area of 150,000 ft^2 (14,000 m^3) including a 56,000 ft^2 (5,200 m^3) computer center.

[a]With heat recovery.

[b]Without refrigeration units.

gas turbines (oil/gas-fired), each with a capacity of 700 kW_e. One of them is used for emergency power generation and the second is held on reserve as a standby unit. A gas turbine was selected for the system instead of a steam turbine or diesel engine for the following reasons:

A high pressure boiler for a steam turbine would result in higher installation costs.

The heat recovery system for a diesel engine would be somewhat more expensive to design, because recovery is from several temperature levels: cooling water, exhaust gas, and oil. With the gas turbine, heat is only recovered at one temperature from the exhaust gases.

The reliability for gas turbines is as high as for diesel engines.

No water cooling system is required for a gas turbine; only air cooling is used.

Because of the relatively light weight of the gas turbine, it could be installed on the roof of the building.

The gas turbine consists of a compressor, a fuel chamber, and a turbine which drives the generator through a gearbox. The outside air is compressed to 90 psi (6 atmospheres). In the combustion chamber, oil or gas is burned and the gas is heated to 1,560°F (850°C), after which it passes through the turbine. The exhaust at the exit from the turbine is about atmospheric pressure and the temperature is about 810°F (430°C). The energy of the exhaust gas is recovered in a waste heat boiler. The overall efficiency of this cycle is shown in the following comparison with the production of electricity alone.

1. Total energy system

Net electrical power	700 kW	
Net heat load	1,600 kW	(5.5 MBtuh)
Total output	2,300 kW	(7.85 MBtuh)
Fuel used	3,300 kW	(11.3 MBtuh)

Efficiency (peak) with cogeneration $\dfrac{2{,}300 \text{ kW}}{3{,}300 \text{ kW}} = 69$ percent

2. For the production of electricity alone the efficiency is

$$\frac{700 \text{ kW}}{3{,}300 \text{ kW}} = 21 \text{ percent}$$

An absorbtion refrigeration unit and the heating system are supplied with hot water from the heat recovery boiler.

			Load	Energy (annual)
1.	Energy from the total energy system			
	Electricity		700 kW	4,800 MWh
	Energy for cooling	4.1 MBtuh	(1,200 kW)	22,200 MBtuh (6,500 MWh)
	Energy for heating	5.5 MBtuh	(1,600 kW)	6,500 MBtuh (1,900 MWh)
2.	Energy supplied by the conventional system			
	Electricity supplied by the utility		2,800 kW	13,100 MWh
	Energy for cooling (compressor cooling units)	9.56 MBtuh	(2,800 kW)	6,500 MBtuh (1,900 MWh)
	Heating fuel burned in boilers	3.4 MBtuh	(1,000 kW)	6,500 MBtuh (1,900 MWh)

Investment for the total energy system

Gas turbines	$ 544,000.00
Waste heat boiler	120,000.00
Gas supply	168,800.00
Electrical equipment	56,000.00
Absorption refrigeration unit	163,200.00
Miscellaneous (radiator, noise silencer, heat protection)	54,400.00
Total	$1,106,400.00

With a saving of $664,400 on emergency power generation, boilers, and compressor refrigeration units, the additional investment is therefore only $442,000.

Yearly energy use	Total energy unit	Conventional systems
Oil (No. 2)	8,200 MBtu (2,400 MWh)	6,400 MBtu (1,885 MWh)
Gas (interruptible rate)	71,700 MBtu (21,000 MWh)	12,100 MBtu (3,540 MWh)
Electricity supplied by utility	13,100 MWh	19,600 MWh

If electrical energy is converted to its fossil fuel equivalent (using a factor of 3), the primary energy saving from the total energy system is 3,500 MBtu (1,000 MWh). (The substitution of more efficient compressor chillers for the absorption cooling units would change this comparison.)

In conclusion, the annual energy cost comparison of the total energy system with a conventional system:

Oil	$ 27,840	$ 21,840
Gas	$ 294,000	$ 49,600
Electricity	$ 812,000	$1,216,000
	$1,133,840.00	$1,287,440

The initial annual saving of $153,600, may be compared with the additional investment of $442,000 (which is 3 percent of the cost of building utility systems).

A Combined Cycle System

Cascading a second stage of cogeneration is a more efficient use for high temperature gas turbine exhaust than making low temperature steam discussed in the previous example. Figure 6.8 shows an example of cascading processes requiring decreasing levels of energy quality. The exhaust heat of a simple cycle gas turbine-generator makes steam to generate more electricity using a steam cycle turbine. (This combined cycle is the most efficient method of electricity generation with overall efficiencies greater than 40%, which is better than central station power plants.) The Table on page 347 shows the performance of this system for Dow Chemical of Canada, Ltd. (14)

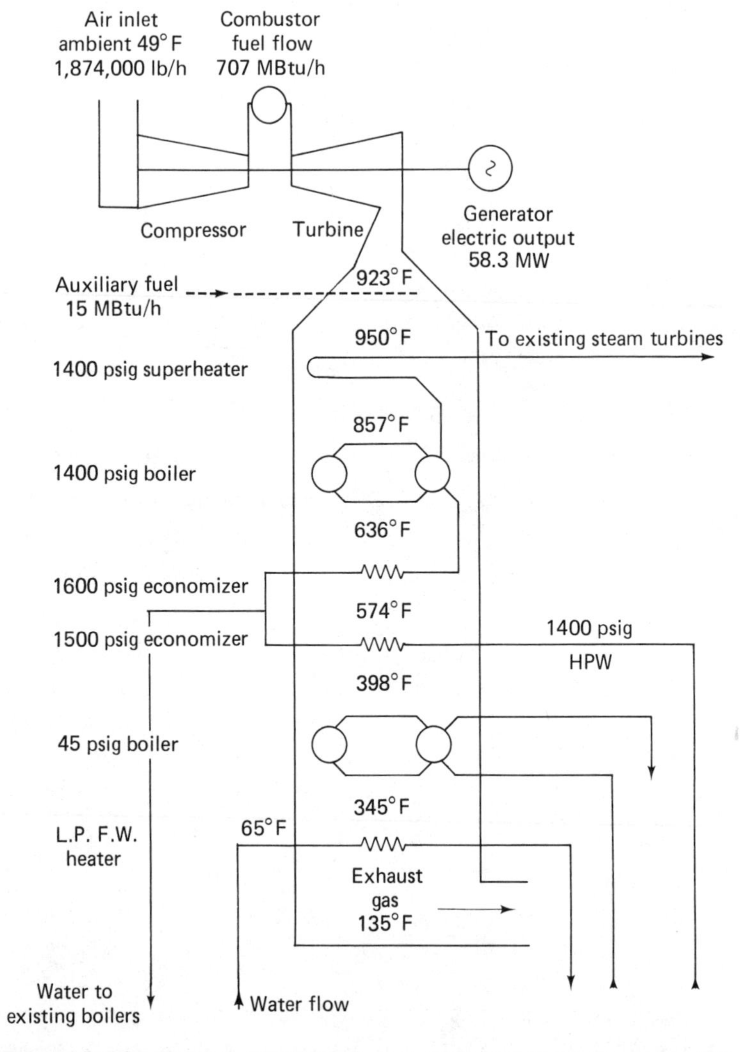

Figure 6.8. Dow Chemical of Canada Ltd. gas turbine–waste heat boiler system. (14)

Data for Figure 6.8:

Expected Performance for Turbine Operating with an Exhaust Flow of 1,860,000 lb/h at 940°F

	High pressure superhtr.	*High pressure boiler*	*Final H.P. economizer*	*Final I.P. economizer*	*Low pressure boiler*	*Storage water coil sect 1 & 2*
Turbine exhaust gas flow (lbs./h.)	1,860,000	1,860,000	1,860,000	1,860,000	1,860,000	1,860,000
Steam or feed flow (lb/h)	195,000	195,000	195,000	500,000	27,700	500,000
Inlet temp. (°F)	594	584	455	292	293	65
Outlet temp. (°F)	880	596	582	455	296	270
Gas inlet temp. (°F)	950	857	636	574	398	345
Gas outlet temp. (°F)	857	636	574	398	345	132
Heating surface (sq ft)	61,800	131,000	76,000	127,000	75,500	216,000
Heat pick-up (MM Btu/h)	48.4	111.2	30.9	86.0	25.4	101.0
					TOTAL	403.9

The combined cycle system is a good example of cascading energy through a variety of different tasks requiring progressively lower energy quality. At each point, from the high combustion temperature in the gas turbine, through to low temperature process steam, the task for which energy is used closely matches the quality of the remaining energy rejected from the previous process. The gas turbine, which is relatively inefficient when used for power alone, is an efficient element in the combined cycle system. It can extract available work at temperatures between the fuel combustion temperature and the input for conventional, high temperature steam turbines. [This available work is unused (lost) in conventional power plants.] Cascading is continued by extracting process steam from the back pressure steam turbine after it has been used to generate power.

In the scheme shown in Figure 6.8, additional process steam is made after the turbine exhaust has been cooled below temperatures suitable for efficient power generation. (14) The Dow Chemical combined cycle system uses a 60 MW turbine generator with an exhaust temperature of 925°F (500°C) to generate:

1. 195,000 lb. (88.5 metric ton) of steam/hour at 880°F (470°C) to supply a steam turbine driving a second generator,
2. 500,000 lb/h (225 metric ton/h) at 455°F (235°C) for high-pressure process heat, and
3. 27,700 lb/h (12.6 ton/h) at 296°F (147°C) for low-pressure process heat. (15)

Since a back-pressure turbine-generator essentially skims its energy off the hot exhaust from the gas turbine, less waste heat is rejected to the environment. The recovered heat from these two turbine generators is used either to make high-pressure steam for generating electricity, or for process heat. The power purchased, if considered as a byproduct of the process heat, should be charged with the fuel consumption over and above that required when process steam is produced directly without intervening power-producing machinery. On that basis, the Dow system produces electricity at a little more than 1 kWh of fuel input per 1 kWh of electricity.

By itself, the gas turbine has a low efficiency in converting fuel to shaft horsepower. This conventional disadvantage is of less consequence in the Dow system because its high quality exhaust heat is recovered to produce high-pressure steam. Consequently, if used alone, or if its exhaust heat is used inefficiently (e.g., with an absorption chiller), the overall efficiency of cogeneration using gas turbines will fall short of the efficiency which could be achieved with a more efficient prime mover. A total energy system using gas turbines with an absorption chiller for an office building and computer center, as discussed above and in reference 13, provides scant energy-saving compared to the conventional equipment it replaces. For medium- and large-size installations, the combined-cycle steam and gas turbine-generators represent a total energy configuration that compares more favorably with the high efficiency possible with diesel-electric cogeneration.

6.3 HEAT RECOVERY AND BOTTOMING CYCLES

The preceding discussion refers to cogeneration of electricity with process steam or comfort heat. There are additional large opportunities for cogeneration of electricity from direct-fired process heaters (e.g., kilns, furnaces, drying ovens, etc.). The engines employed to recover heat from industrial processes are known as bottoming engines. Depending on the temperature of the rejected heat, steam or organic Rankine-cycle engines can be used in bottoming plants using established technology.

More than 30 percent of the energy input into fossil-fuel direct-fired heaters eventually leaves through the stack gases. In the United States' chemical, petroleum, and paper industries, over 80 percent of the energy released into the atmosphere is at a temperature between 30°F and 1,000°F (150–540°C). (10) With bottoming cycles recovering this waste heat, about 6.7 percent of the United States' electricity could be potentially generated fuel-free.

The increase in overall efficiency depends on what fraction of the waste heat can be put to useful purpose. Some manufacturing processes are exothermic, that is, they have an energy-rich waste product which can be burned as fuel, or used directly for its heating (or cooling) capacity. As the energy content (temperature) of the waste heat steam decreases, it becomes more economical to recycle the energy directly in a subsequent process such as heating or pre-heating drying air, heating (or cooling) another product stream, etc., than to use a bottoming plant. For example, the heat of condensation from one distillation column could be cycled to another column requiring a lower temperature level. Other examples of cascading are described below.

Recover Heat from Hot Air Exhaust Streams

A heat recovery system in the exhaust stream of a tenter frame of a textile plant cost $22,750 installed and resulted in an energy savings of 14,145 million Btu per year and a cost savings of $10,500 per year. A major source of waste heat discharge in textile fabric finishing plants is the exhaust from tenter frames which are used for either finishing or heat setting (preshrinking) fabric. Finishing operations typically run at 275°F while heat setting requires about 400°F. Actual tests were performed in a finishing plant on tenter frames used for heat setting to determine the feasibility of recovering heat from the exhaust air stream.

EXAMPLE

Natural gas was used to heat outside air to a temperature of 400°F for heat setting in a tenter frame. The exhaust air which was typically 30°F to 60°F lower than the inlet temperature was exhausted to the atmosphere. A waste heat recovery system was installed to preheat outside air using the tenter frame exhaust.

Tests were conducted using a heat wheel equipped with a pre-filter. Equation 1 approximates the possible energy savings from a heat recovery system as a function of the standard volumetric flow rate and the temperature entering the recovery system (or the process exhaust temperature).

$$\text{Equation 1: Heat recovery} = (\text{volume flow}) \times (T_{exh} - T_{out})\ \eta \left(\frac{1.08\ \text{Btuh}}{\text{scfm °F}}\right)$$

Consider an example for an outdoor air temperature T_{out} of 60°F and a heat recovery efficiency η of 65%. With an air flow rate of 11,000 scfm and an exhaust air temperature T_{exh} of 355°F, from equation 1,

$$\text{Heat recovery} = 2.3\ \text{MBtuh}$$

Equipment costs were obtained from a manufacturer and installation costs were estimated.

Cost of heat wheel = $13,750

Filter modification and estimated installation cost = $ 9,000

Total equipment and installation cost = $22,750

If the energy cost is $0.74 MBtu and the system is used 6,150 hours per year,

$$\text{Annual cost savings} = 2.3\ \text{MBtuh} \times \$0.74/\text{MBtu} \times 6150\ \text{h/yr}$$

$$= \$10{,}500\ \text{per year}$$

$$\text{Annual energy savings} = 2.3\ \text{MBtuh} \times 6150\ \text{h/yr}$$

$$= 14{,}145\ \text{MBtu per year}$$

SUGGESTED ACTION

Heat from drying operations can be utilized for heating makeup water for boilers, for other wet process operations, or for preheating air. Consider recovery of waste heat from all exhaust air streams.

SOURCE: National Bureau of Standards (15).

A second example of cascading energy is in the heat treatment of alloys, a common manufacturing process. There are several methods for cutting energy costs below those for conventional practice. The current practice for high-temperature hardening and tempering of alloy steel parts such as gears, shafts, and bearings, is illustrated in Figure 6.9. (11) This process consists of three operations:

1. Heating steel parts to 1,650°F (1,000°C).
2. Quenching parts in oil at 350°F (175°C).
3. Tempering parts at 1,400°F (750°C).

Current practice in the United States uses about 1,400 Btu's of fuel (usually gas) per pound of heat treated material, with an efficiency of about 19 percent. (16) A list of measures to improve overall efficiency is given in Table 6.5. This includes measures such as improved insulation, which could cut gas use by 18 percent and increase overall efficiency to 22.5 percent. However, it is important to note that even with perfect insulation (shown as item three in Table 6.5), the limiting efficiency is only about 31 percent. There

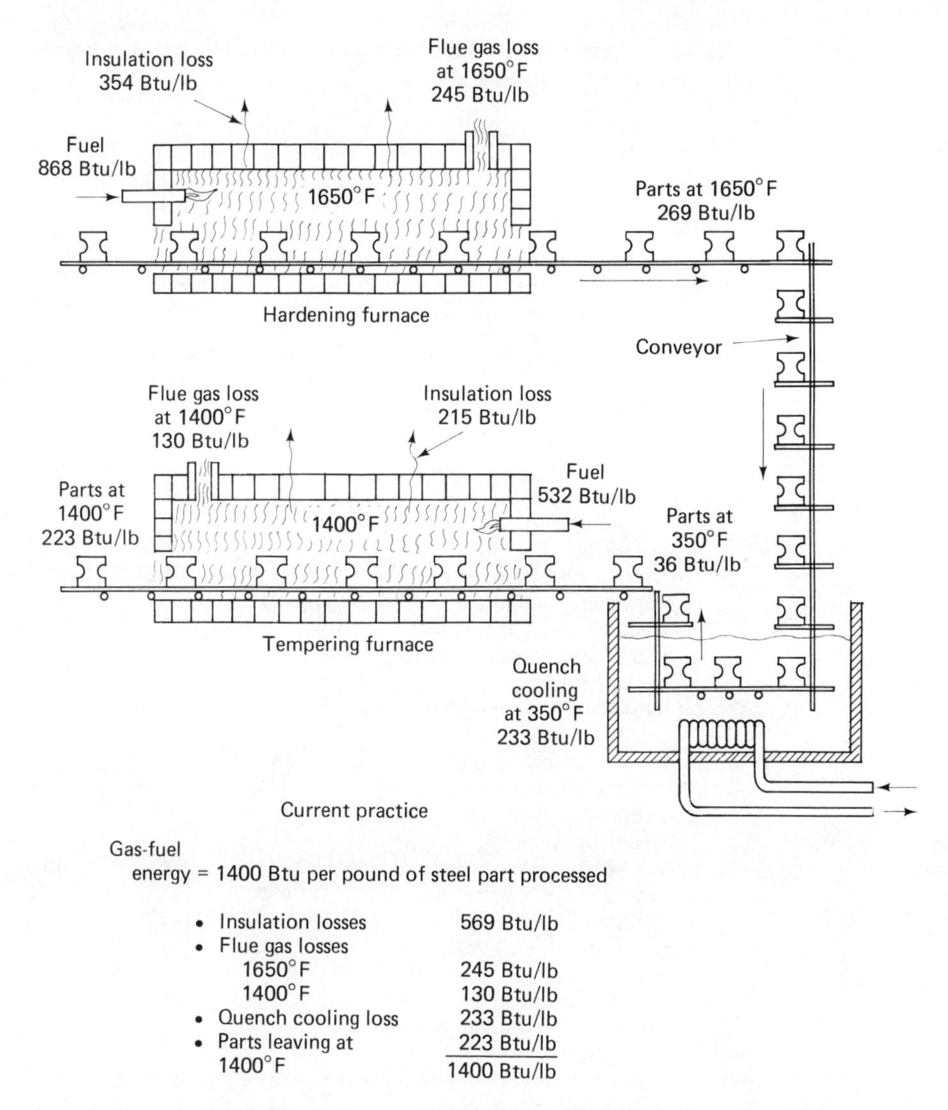

Figure 6.9. The conventional process for hardening and tempering of alloy steel parts. (11) The cost for fuel could be reduced by cascading heat from the hardening furnace exhaust to the tempering furnace.

are seven other measures shown in this table which could improve efficiency beyond that resulting from perfect insulation; one of them yields an efficiency of 51 percent. All these measures rely on recovering waste heat, internal process regeneration, or a combination of both to overcome losses in the conventional heat treatment process. Typically, the recovery of waste heat for direct use will be possible at a lower cost than bottoming engines. However, if demand or time-of-use rates are high, generation of electricity with

Table 6.5. A Comparison of Different Methods for Improving the Efficiency of the Heat Treating of Alloy Metals (11)

	Conservation measures	*Gas fuel used per pound of parts heat treated (Btu/lb)*	*Energy saved (%)*	*Efficiency (%)*
(1)	Present practice	1,400	Baseline	19.0
(2)	Improved insulation	1,153	17.7	22.5
(3)	Perfect insulation	826	41.0	31.3
(4)	Improved insulation plus recuperators on the hardening and tempering furnaces	986	29.5	26.2
(5)	Improved insulation plus bottoming engine for electricity	1,153 (−496)[a]	53.1	37.0
(6)	Improved insulation plus bottoming engine for electricity plus heat recovery from quench tank for steam raising	1,153 (−642)[b]	63.5	40.5
(7)	Improved insulation plus total process regeneration	728	48.0	35.5
(8)	Improved insulation plus total process regeneration plus bottoming engine for electricity	728 (−127)[a]	57.1	41.4
(9)	Improved insulation plus total process regeneration plus bottoming engine for electricity plus quench tank recovery for steam raising	728 (−273)[b]	67.5	46.9
(10)	Improved insulation plus total process regeneration plus recuperation	582	58.5	44.5
(11)	Improved insulation plus total process regeneration plus recuperation plus quench tank recovery for steam raising	582 (−146)[c]	69.0	51.3

Note: It is noteworthy that there are at least seven different measures which would save more fuel than if the furnaces could be equipped with perfect insulation (item 3 above).

[a]Credit for fuel saved because the electricity generated by the bottoming engine need not be produced by a utility.

[b]Credit for fuel saved because of the electricity generated by the bottoming engine and the steam raised by the heat recovery system.

[c]Credit for fuel saved because of the steam raised by the heat recovery system.

Table 6.6. Typical Temperatures for Consideration in the Multiple Use of Energy

Source or use	Typical temperatures: Primary Source		Secondary Use	
Basic oxygen furnace	3,000–3,500°F	(1,650–1,930°C)		
Copper reverberatory furnace	2,000–2,500°F	(1,095–1,370°C)		
Zinc-fuming furnace	1,800–2,000°F	(980–1,095°C)		
Forge & billet-heating furnaces	1,700–2,200°F	(925–1,205°C)		
Sulfur ore processing	1,600–1,900°F	(870–1,040°C)		
Garbage incinerator	1,550–2,000°F	(845–1,095°C)		
Ammonia oxidation process	1,350–2,100°F	(730–1,150°C)		
Open-hearth steel furnace (oxygen)	1,300–2,100°F	(705–1,150°C)		
Cement kiln (dry process)	1,150–1,500°F	(620–815°C)		
Open-hearth steel furnace (air)	1,000–1,300°F	(540–705°C)		
Diesel engine exhaust	1,000–800°F	(540–150°C)		
Petroleum refinery	1,000–1,100°F	(540–595°C)		
Gas turbine exhaust	850–1,025°F	(455–550°C)		
Back pressure steam turbine	530–230°F	(275–110°C)		
Aluminum annealing			575–1,025°F	(300–550°C)
Heat treating steel & alloys			440–1,200°F	(225–650°C)
Rankine engine			400–1,100°F	(200–600°C)
Process heat	430–160°F	(220–70°C)	200–450°F	(100–230°C)
Low pressure steam turbine			300–750°F	(150–400°C)
Absorption chiller (2 stage)			300–750°F	(150–400°C)
Refrigeration heat of compression	250–195°F	(120–90°C)		
Diesel engine water jacket	250–160°F	(120–70°C)		
Process cooling water	230–68°F	(110–20°C)		
Absorption chiller (single stage)			160–250°F	(70–120°C)
Combustion air preheat			200–1,200°F	(100–650°C)
Waste water sewers	212–60°F	(100–15°C)		
Condensate from absorption chillers	195°F	(90°C)		
Water from chiller condenser	150–80°F	(65–25°C)		
Drying			125–250°F	(50–120°C)
Space heat			105–195°F	(40–90°C)
Domestic hot water			105–185°F	(40–85°C)
Exhaust air	85–65°F	(30–18°C)		
Fuel oil viscosity preheat			85–260°F	(30–125°C)
Washing water			70–185°F	(20–85°C)
Water from cooling towers	35–85°F	(2–30°C)		

Table 6.7. Examples of Costs and Savings from Improvements to Process Heat Systems

Energy-saving measure	Cost ($)	Savings ($/year)	Energy saving MBtu (kWh)	SCP, $/BPD ($/kW)	Comment	Reference
Steam generating muffler	20,000	13,900	6,800 (2×10^6)	3,000 (43)	An exhaust boiler replaces a conventional muffler on a 500 hp gas engine used to drive a compressor. The engine discharges 4,250 cfm (7,200m^3/h) of gas 1,100°F (600°C) and 6 in. (15 cm) of water back pressure. With a temperature drop of 750°F (420°C) across the heat recovery muffler, about 70% of the exhaust heat is recovered. Recovered heat generates process steam.	(16)
Upgrade steam pressure control	13,300	6,270	6,300 (1.84×10^6)	4,400 (62)	Mechanical control between steam pressure and the fuel and air supply was replaced with electric regulation. Gas rate was determined by steam pressure on 25 MBtuh (6.8 MW) boiler. Average excess air was reduced from 40% to 10% during 8,000 operating hours per year.	(17)
Steam output increased with automatic flue gas analysis	6,500	140,000	3,400 (76×10^6)	70 (1)	A small chemical firm could not get more than 95 MBtuh (28 MW) from their 120 MBtuh (35 MW) boiler. Condensed moisture in the combustion air flow controller was causing 4% combustible gas to be wasted up the flue. A flue gas analysis was installed to monitor the composition of the flue gas.	(18)
Blowdown heat recovery	30,000	30,000	15,000 (4.4×10^6)	4,250 (60)	Monsanto Fulton plant	(19)
Cascading process heat	22,750	10,500	14,000 (4.14×10^6)	3,400 (48)	Air used for pre-shrinking textiles at 400°F (200°C) is recycled via a heat wheel to heat a second process air stream at 275°F (135°C).	(14)
Combustion air preheaters	1,915,000	1,330,000	670,000 (200×10^6)	6,060 (86)	Three furnaces retrofitted with air preheaters to increase thermal efficiency. The heat exchangers installed in the stacks are expected to have a lifetime of 15 years before major overhaul is required. Savings based on oil at $11.50/bbl.	(20)

a bottoming engine may be attractive. Turnkey systems are available which can turn a plant's waste heat into useful power.

Table 6.6 gives a list of input and output temperatures that may suggest possible schemes to utilize otherwise wasted heat. Also described below is an example of the recovery of heat from the waste product incinerator of an asphalt roofing manufacturer. (15)

Waste Heat Recovery from Incinerator Flue Gas[a]

EXAMPLE

A potential saving of $45,000 annually was identified in the recovery of waste heat from an incinerator of an asphalt roofing plant to produce 15,000 lb/h of steam at 150 psig.

An asphalt roofing plant disposes of 16,000 scfm of asphalt-saturated air by burning the mixture in an incinerator at 1400° F. The normal procedure was to discharge the incinerator flue gas to the atmosphere. At the same time, the plant purchases 15,000 lb/h of process steam at 150 psig from a neighboring steam plant. A brief study showed that a waste heat boiler installed in the incinerator stack could produce the required steam, and more if needed.

The plant has 200° F feedwater available. To produce 1 lb of 150 psig saturated steam (366° F) from 1 lb of 200° F water requires 1036 Btu. Heat (q) to produce 15,000 lb/h of steam.

$$q = 1036 \text{ Btu/lb} \times 15{,}000 \text{ lb/h}$$

$$= 15.54 \text{ MBtuh}$$

Flue gas flow rate (m) at a gas density of 0.0763 lb/cu ft at standard conditions,

$$m = 16.000 \text{ scfm} \times .0763 \text{ lb/cu ft} \times 60 \text{ min/h}$$

$$= 73{,}248 \text{ lb/h}$$

With a 2% heat loss and with C = 0.28 Btu/lb° F, the required temperature drop in the flue gas is

$$\Delta T = \frac{15.54 \text{ MBtuh}}{0.28 \text{ Btu/lb}^\circ \text{ F} \times 73{,}248 \text{ lb/h} \times (1 - 0.02)}$$

$$= 773^\circ \text{ F}$$

Therefore, the flue gas need only be cooled to 1400 − 773 = 627° F to produce the required 15000 lb/h of 150 psig steam. At a cost of $1.00 per 1000 lbs of steam and operating 3000 hours per year, the cost savings if waste heat is recovered is

$$\text{Cost savings} = 15000 \text{ lb/h} \times \$1.00/1000 \text{ lb} \times 3000 \text{ h/yr}$$

$$= \$45{,}000 \text{ per year}$$

[a]This example illustrates savings at previously low energy prices when there is a use for the recovered heat. The need to avoid condensation (asphalt on heat exchanger, or acids, etc.) sets a lower limit to the temperature of the heat recovered.

If the flue gas were cooled to a lower temperature, say around 400° F, instead of 627° F, a total of 19,400 lb/h of 150 psig steam could be generated with a cost savings of $58,200/yr.

At an installed cost of approximately $37,000, the first year savings paid for the boiler, even at the lower steam output.

SUGGESTED ACTION

Evaluate plant exhaust streams with temperatures higher than 300° F as potential sources of heat for steam generation. Consider selling unneeded steam to a neighboring plant. Consult waste heat boiler equipment manufacturers for recommendations.

CAUTION

Flue gas containing condensible components must be kept above the dew point temperature to minimize corrosion problems. Usually the flue gas temperature should not be cooled below 300° F.

SOURCE: National Bureau of Standards (16).

We should be seeing many more examples in the future of the cascading of two processes requiring successively lower temperatures. In Table 6.7 there is a summary of some investments for industrial heat recovery and other improvements for process heat systems showing the savings and comparing the specific capital productivity (SCP).

The following example from the liquefied methane facility of the Boston Gas Company illustrates the recovery of energy from several sources, which in the former period of cheap energy were uneconomical to utilize. Natural gas from the interstate pipeline enters their plant at a pressure 20 times that needed for distribution to their customers. The excess energy represented by this higher pressure is converted to work in turbines and used to operate the refrigeration compressors of their methane liquefier. About 13 percent of the gas is liquefied using the energy content of the gas at its original pipeline pressure. Another recovery scheme uses the gas boiled off due to heat leakage into the storage tanks. This gas is used to run gas-engine-driven alternators for the plant's total energy system. And waste heat recovered from these gas engines provides the plant's process heat, building heat, and air conditioning. As the cost of energy increases, many more innovative ways will be found to make multiple use of energy formerly rejected to the environment at high temperatures.

REFERENCES

(1) DOE, *Managing the Energy Dilemma: Increasing Energy Efficiency,* Technical Reference Manual for Industrial Workshops (Washington, D.C.: United States Department of Energy, 1977).

(2) Kennard L. Bowlen, "Energy Recovery from Exhaust Air," *ASHRAE Journal* (April 1974), pp. 49–56.

(3) DOE, *Guidelines for Saving Energy in Existing Buildings, Engineers, Architects and Operators Manual ECM 2,* Conservation Paper number 21 prepared by Dublin, Mindel and Bloome (Washington, D.C.: Department of Energy, 1975), pp. 229–43.
(4) Fred S. Dubin, "More Efficient Utilization of Energy in Buildings," in T. Nejat Veziroğlu, ed., *Energy Conservation: A National Forum,* proceedings of the Forum sponsored by ERDA, School of Engineering and Environmental Design (Coral Gables, Fla.: Clean Energy Research Institute, University of Miami, 1–3 December 1975), p. 508.
(5) M. Kiss, H. Mahon, and H. Leimer, *Energiesparen Jetzt!* (Berlin, Germany: Bauverlag, 1978), pp. 201–202.
(6) Ibid., pp. 291–297.
(7) Lewis Allen Felton, *Energy Conservation; A Case Study for a Large Manufacturing Plant* (thesis for the Master of Science degree in the Department of Ocean Engineering, Massachusetts Institute of Technology, June 1974).
(8) NBS, *Waste Heat Management Guidebook,* NBS Handbook 121, National Bureau of Standards and the Federal Energy Administration, (Washington, D.C.: U.S. Government Printing Office, February 1977).
(9) James Connelly, Nancy K. Juren, Pentti Aalto, et al., *Cogeneration: Its Benefits to New England,* Final Report of the Governor's Commission on Cogeneration, (Boston, Mass.: The Commonwealth of Massachusetts, October 1978).
(10) *A Study of Inplant Electric Power Generation in the Chemical, Petroleum Refining and Paper and Pulp Industries,* a report for the United States Department of Energy, (Waltham, Mass.: Thermo Electron Corporation, 1976).
(11) Thomas F. Widmer and George N. Hatsopoulos, "Summary Assessment of Electricity Cogeneration in Industry," (Waltham, Mass.: Thermo Electron Corporation, 1977).
(12) Carter B. Horsley, "Building in Midtown to Supply its Power," *New York Times,* (June 20, 1978), p. 1.
(13) "Planung und Installation einer Totalenergieanlage fur ein Gewerbehaus." (Winterthur, Switzerland: Gebruder Sulzer, Aktiengesellschaft, 1975).
(14) R. S. Spencer and G. L. Decker, "Energy and Capital Conservation Through Exploitation of the Industrial Stream Base," Dow Chemical Co. Report for the NSF "Energy Industrial Center Study," in T. Nejat Vezirŏglu, ed., *Energy Conservation: A National Forum,* proceedings of the Forum sponsored by ERDA, School of Engineering and Environmental Design (Coral Gables, Fla.: Clean Energy Research Institute, University of Miami, 1–3 December 1975, pp. 343–362.
(15) Robert R. Gatts, Robert G. Massey, and John C. Robertson, *Energy Conservation Program Guide for Industry and Commerce,* National Bureau of Standards Handbook 115 (Washington, D.C.: U.S. Department of Commerce, September, 1974; Supplement, December 1975), pp. 3–42G.
(16) Ibid., pp. 3–42.
(17) Ibid., pp. 3–39.
(18) Ibid., pp. 3–47.
(19) Ibid., pp. 3–50.
(20) R. E. Doeer, "Six Ways to Keep Score on Energy Savings," *Oil and Gas Journal,* (May 17, 1976), pp. 130–45.
(21) Carroll Willson, "The National Investment Tradeoff," from briefing notes from the Workshop on Alternative Energy Strategies (Cambridge, Mass.: October 1, 1974), p. 2.

7

Examples of Energy-Saving Programs

In Chapter 5, many examples of possible energy-saving methods were given in the check lists. In this chapter, examples show the application of these energy-saving methods to actual situations which illustrate:

Some methods that have been used in practice

Some of the actual costs of saving energy

The results in practical applications

In Chapter 5, costs for saving energy could usually only be handled in a general manner. In this chapter it will be possible to give more detailed economic data for some of the examples. The reader can develop her or his own rule of thumb for the investment costs. Examples have been chosen from both the United States and Europe and were selected to give a cross-section of different kinds of residential, commercial, and industrial installations.

7.1. EXAMPLE: SCHOOLS AND OTHER PUBLIC BUILDINGS (1)

The Massachusetts Department of Community Affairs, in cooperation with six cities and towns, evaluated the potential for saving energy in municipal buildings. Opportunities to reduce fuel and electricity consumption were examined in 112 buildings including schools, libraries, police stations, and garages. Yearly energy costs of the 6 municipalities in 1975 were $2.2 million, or 3 percent of their total annual budgets.

Table 7.1. Audit Checklist

TOWN: Wattsville *BUILDING: High School*	*Status*				
Measures	*Not applicable*	*Already imple-mented*	*Recom-mended[a]*	*Requires further study*	*No cost measure*
1. Set back indoor temperatures during unoccupied periods to a recommended level of 55°F (13°C).			H		X
2. Shut down ventilation system during unoccupied periods.		X			
3. Reduce ventilation rates during occupied periods.			H	X	
4. Reduce conductive heat loss transmission through the building envelope by adding wall insulation, roof insulation and storm windows.			L	X	
5. Measure the burner-boiler/furnace efficiency to ascertain that boiler is operating with a combustion efficiency of 75 to 80%.			H		
6. Reduce consumption of hot water through low flow shower heads and automatic shut off lavatory faucets.			L		
7. Add timeclock to recirculating system.		X			
8. Turn off cooling system during unoccupied periods.		X			
9. Use switching and timers on school lights by installing recommended devices in these locations.		X			
10. Reduce power for lighting by disconnecting ballasts when delamping.			H		X
11. (a) Maintain steam traps every 3 months.		X			
(b) Check filters on central air handling units and replace every month.		X			
(c) Insulate distribution in the following areas [None].		X			
(d) Eliminate reheat.	X				
12. Check window units and chillers.		X			
13. Use outdoor air for cooling.		X			

Table 7.1. Audit Checklist (*cont.*)

TOWN: Wattsville
BUILDING: High School

	Status				
Measures	*Not applicable*	*Already implemented*	*Recommended*[a]	*Requires further study*	*No cost measure*
14. Reduce winter indoor temperatures during occupied periods to a recommended level of 68°F (20°C).			H		X
15. Increase summer indoor temperature and relative humidity levels during occupied hours up to a recommended maximum of 78°F/60% (25°C/60%).			H		X
16. Reduce hot water temperatures to a recommended temperature of 110°F (43°C).			H		X
17. Reduce solar heat gains through addition of blinds, curtains, etc.			H		X
18. Reduce illumination levels by replacing existing lamps with ______ or by removing about $^1/_3$ of lamps from existing fixtures in Administrative area.			H	X	
19. Turn off lights in unused areas.			H		X
20. Use task lighting in the areas of Administration.			H	X	
21. Utilize daylight for natural illumination in the following locations: Library, Shops			H		X
22. Reduce energy consumption for equipment and machines by adjusting the following equipment: ______.	X				
23. Reduce electric demand by turning off A/C units.	X				
24. Install separate domestic hot water heater.		X			
25. Check controls calibration.		X			

[a]In the RECOMMENDED column, "H" indicates high priority; "L" indicates low.
SOURCE: Massachusetts Department of Community Affairs.

Table 7.2. Typical Applicability of No-Cost Measures According to Building Type[a]

	Municipal building types					
Recommended no-cost measures	*Pre-1945 schools*	*Post-1945 schools*	*Town offices*	*Fire/ police stations*	*Libraries*	*DPW garages*
1. Set back thermostats to 55°F (13°C) during unoccupied periods.	√	√	√		√	
2. Shutdown ventilation system during unoccupied periods.		√			√	
3. Shutdown cooling system during unoccupied periods.			√		√	
4. Reduce unnecessary lighting by delamping.	√	√	√	√	√	
5. Reduce domestic hot water temperature to 110°F (43°C).	√	√	√	√		√
6. Reduce ventilation rates during occupied periods.	√	√				
7. Measure and adjust burner/boiler efficiency.	√	√	√	√	√	√
8. Calibrate thermostats and other controls.		√		√		
9. Eliminate reheat in HVAC system (where applicable).		√	√		√	
10. Disconnect ballasts when delamping.		√	√			
11. Reduce winter indoor temperature to 68°F (20°C).	√	√	√	√	√	√
12. Increase summer indoor temperature to 78°F (25°C).			√	√	√	√
13. Turn off unused lights.	√	√	√	√	√	
14. Use outdoor air for summer cooling.			√	√		√
15. Use blinds/curtains to reduce solar heat gain in summer.			√	√		
16. Use natural lighting when available.	√	√	√		√	

[a]Measures checked would be expected to be applicable; however all measures should be examined for their suitability with respect to the specific conditions at each location.

SOURCE: Massachusetts Department of Community Affairs.

To evaluate the potential for saving, an energy audit was made for each building. Actual energy use was compared with the theoretical demand related to weather and method of operation. Opportunities for saving were identified in the course of the energy audit with the help of a checklist shown in Table 7.1. This accompanied a letter to the person responsible for maintenance of the building explaining the results of the audit and the most important energy-saving opportunities.

Results of the 112-building analysis showed that savings of 30 percent or more were possible in many buildings and that energy costs could be reduced by $500,000 by implementing cost-effective measures. About half of the savings came from measures that could be implemented at no cost.

As a conclusion to this study, a comprehensive energy-saving program was prepared by the Massachusetts Department of Community Affairs. This program is described in the publication, "Energy Management in Municipal Buildings" (1), which has been made available for use by other municipal governments interested in controlling energy costs.

No-cost measures which would be particularly applicable to certain types of buildings are organized by the function of the building, as shown in Table 7.2. A similar organization

Table 7.3. Typical Applicability of Capital Investment Measures According to Building Type

	Municipal building types					
Capital investment measures	*Pre-1945 schools*	*Post-1945 schools*	*Town offices*	*Fire/police stations*	*Libraries*	*DPW garages*
Install separate room thermostats.	√		√	√		
Heat recovery from boiler stacks.	√	√				
Heat recovery (thermal wheels).		√				
Multiple boiler controls.	√	√	√			√
Reduce over-ventilation.		√				
Automatic door closers.				√		√
Roof insulation.	√	√	√	√	√	√
Wall insulation.	√		√	√	√	
Storm windows.	√		√	√	√	
Window insulation/replacement.	√		√	√		√
More efficient lighting system.	√		√		√	
Install task lighting.		√	√		√	

SOURCE: Massachusetts Department of Community Affairs

of measures requiring capital investment is shown in Table 7.3. Considerable emphasis was given to cost-benefit analysis of the proposed measures, taking into consideration the increase in the future price of energy.

A summary is shown in Table 7.4 of 29 different energy saving measures (ESM) from the Massachusetts State Program to improve municipal energy use. Some of these

Table 7.4. Summary of 29 Energy-Saving Examples in Municipal Buildings

		Cost ($)	Initial annual savings	Energy savings		SCP	
				MBtu	kWh	$/kW	$/BPD
1.	Separate room thermostats	400	269	114.3	33,500	105	7,400
2.	Thermostatically controlled valves	1,000	457	157.7	55,000	106	7,500
3.	Separate domestic hot water heater	240	96	87.7	25,700	82	5,800
4.	Stack heat recovery	5,000	735	332.2	97,350	450	31,800
5.	Stack heat recovery	3,500	1,000	465.0	136,270	225	15,900
6.	Air-to-air heat recovery	20,000	2,600	1,177.2	345,000	510	36,000
7.	Lead/lag boiler controls	150	115	49.8	14,600	90	6,400
8.	Reduction of ventilation air	400	480	199.3	58,400	60	4,250
9.	Reduction of ventilation air	800	780	341.6	100,100	70	4,960
10.	Reduction of ventilation air	500	625	266.1	78,000	56	3,970
11.	Automatic fire station door-closure device	800	585	208.1	61,000	115	8,100
12.	Garage door control of heating system	750	465	160.4	47,000	140	9,900
13.	Roof insulation	22,000	2,900	1,126.6	330,000	585	41,400
14.	Sprayed-on roof insulation	1,900	715	237.1	69,500	240	17,000
15.	Sprayed-on roof insulation	8,500	650	212.6	62,300	1,195	84,600
16.	Roof insulation	750	65	25.6	7,500	850	60,200
17.	Sprayed-on roof insulation	16,700	1,360	580.1	170,000	870	61,600
18.	Roof insulation	12,900	1,400	597.1	175,000	650	46,000
19.	Wall insulation	13,350	2,400	887.2	260,000	450	31,900
20.	Wall insulation	9,400	1,750	580.1	170,000	485	34,300
21.	Storm windows	2,900	425	238.8	70,000	365	25,800
22.	Storm windows	1,800	310	71.7	21,000	765	54,200
23.	Storm windows	950	155	51.2	15,000	555	39,300
24.	Plastic bubble window insulation	605	170	62.3	18,250	290	20,500
25.	Elimination of window areas	1,200	460	151.5	44,400	235	16,600
26.	Replacing incandescent with fluorescent lighting	12,700	850	58.0		1,150	81,400
27.	Task lighting	2,400	230	17.6	5,150	840	59,500
28.	Task lighting	600	70	5.3	1,550	940	66,600
29.	Task lighting	1,000	110	8.6	2,520	715	50,600
	Total	143,220	22,227	8,500.2	2,491,000		

Note: The specific capital productivity (SCP) is the capital investment for a given measure that results in decreasing the average rate of fuel use by one barrel of oil per day (BPD) or the average rate of electric energy use by one kilowatt.

SOURCE: Massachusetts Department of Community Affairs (1).

measures have been discussed earlier in Chapter 5. The sum of the annual energy saved with these measures is 8500 MBtu or 2,500,000 kWh.* The average specific capital productivity represented by these measures saved fuel at the rate of $35,800/BPDE or $500/kW. Evaluated with the lower costs indicated at the time this program was initiated, the dollar saving from these measures yielded better than 20 percent return on the average cost of putting them in place. (1) Another way of looking at the attractiveness of these investments is that by the time energy prices had reached the equivalent of $1.00 per gallon of oil, these ESM were yielding an average of 43 percent simple annual return on their initial cost.

7.2. EXAMPLE: PUBLIC BUILDINGS (2)

The management of the Federal Public Buildings of Switzerland has long given high priority to the energy-efficient operation and maintenance of its heating systems. As an example, a contract for yearly burner service has been in effect since the late 1960s which provides for a regular annual checkout of the burners.

A provisional Energy Plan for Federal buildings was released in 1975. Even earlier, in 1973, the Swiss Federal Government (Bundesrat) has instructed the General Administration, the Postal Service, and the Federal Institute of Technology (ETH) to follow the guidelines for room temperatures as given below.

Maximum Room Temperature (2)

Offices	Day to 68°F (20°C) Night and Weekend 59°F (15°C)
Apartments	Day 68°F (20°C) Night 63–65°F (17–18°C)
Workshops with manual work	61–65°F (16–18°C)
Empty apartments and offices (in cases of vacation or longer absences)	To 54°F (12°C)
Unused rooms	Reduced temperatures

*Most of these measures were directed at saving heating fuel. However, this summary combines both the energy savings in heating fuel and the smaller energy savings in electricity counted at its value at the building boundary (secondary level). Thus the attractiveness of these average data are understated somewhat. The saving of 8500 MBtu, or 2,500,000 kWh, per year (i.e. per 8760 hours) represents a reduction in the annual rate of energy use in these municipal buildings of 168 gallons of No. 2 heating oil per day (i.e., 4 forty-two gallon barrels of oil per day) or 284 kW.

The main features of the energy plan can be summarized as follows:

Reduction of comfort criteria (restrictions on room temperature and on the operating time of heating systems, and more stringent criteria for the installation of air conditioning systems for new buildings)

Improved insulation of buildings

Improved operation and maintenance of building mechanical systems

Diversification of energy sources

Environment protection receives greater emphasis.

Some of the recommendations are summarized below.

Some Selected Criteria from the Energy Plan for Swiss Federal Buildings (2)

1.1 Reduction of room temperatures according to the levels described above.

1.2 Restricted use of air conditioning systems (need must be demonstrated). The cost-benefit analysis for air conditioning must include operating cost. Air conditioning in Switzerland is actually required only in exceptional cases, e.g., high-rise buildings, or for selected rooms or areas.

2. Improved insulation of the building, improved design criteria for planning

   Outside walls U 0.08 to 1.2 Btu/ft^{2}h°F (.45 to .7W/m^2°C)

   Windows U smaller than 0.54 Btu/ft^{2}h°F (3.1 W/m^2 °C)

   Window frame U smaller than 0.51 Btu/ft^{2}h°F (2.9 W/m^2 °C)

   For the ratio of windows to wall area no general value for the heat transmission coefficient U is given; the optimum value must be selected for each individual case.

   Windows must be shaded to protect them from 70 percent or more of the sun's direct radiation.

   Roof U-value 0.062 to 0.12 Btu/ft^{2}h°F (.35 to .7 W/m^2 °C)

   Heat storage capacity of the wall: insulation and head storage must be chosen to limit temperature variations on the inner surface of the wall to 1/15th or less of the amplitude of the temperature excursion at the exterior surface.

   Windows must be sealed to restrict volume flow to less than 0.2 cfm per linear foot of window crack (1 m^3/hr per meter of crack length) at a pressure difference of 0.04 inches of water (1 mm).

3. Measures for existing buildings: The design values as given for new buildings must be applied according to the actual conditions in existing buildings.

4. Control and maintenance of mechanical systems: Periodic and regular checking of heating systems, together with continuous monitoring of energy use. Scheduled replacement of uneconomical components in HVAC systems.
5. The diversification of energy sources: Connection to district heating systems. Reduction of the use of heating oil through increased use of other fuels such as gas, coal, or wood. Limitation of electrical spare heating to a few cases which must be individually examined. Special attention is given to extended storage of fuels.
6. Environmental protection: If buildings are renovated or if the boiler, burner, chimney, or oil tanks are replaced new, more stringent environmental guidelines are applicable. Existing buildings, furnaces, and mechanical systems will be upgraded to conform with the new regulations. Oil and coal should be used only in relatively small furnaces and in areas with a low population density because of the difficulty of achieving acceptable emission standards in small units with these fuels.

7.3. EXAMPLE: CENTRAL HEATING OF A SMALL COMMUNITY OF ONE FAMILY HOUSES (3)

Forty-five single-family houses are supplied from one central heating center with an installed power of 5 million Btu (1,455 kW) per hour. The distribution line has a length of 5000 ft (1,500 m) not counting the smaller connection lines to the individual houses. The heat supplied is used for heating of the houses and for the heating of domestic hot water. This district heating was installed between the years 1970–72 and was justified on the basis of the improvement in air quality compared to the emissions from 45 individual burners, and on the basis of lower installation costs.

The heating costs during the 1974–75 heating season were extraordinarily high in comparison with other buildings. The heating costs for a house of medium size was $800 corresponding to $.60 per ft^2 ($6.40 per m^2) of floor area.

Because of these high costs, the entire heating system has been subjected to detailed analysis in which the original concept was reviewed and the economics of its operation were studied in detail. Comprehensive measurements showed the main losses as given in Figure 7.1 and also indicated possible improvements. The economic analysis demonstrated a good rate of return for several of the proposed improvements.

No-cost measures:

1. Reduction of the temperature at which heat is distributed by readjustment of the existing control thermostats.
2. The system is shut down at night during the summer.
3. Improved instruction for the individual users as to how to achieve energy-saving operation of their connected heating systems, for example, by regulation of the heat exchanger thermostat to 130°F (55°C). This would be accomplished through (improved) communication with the

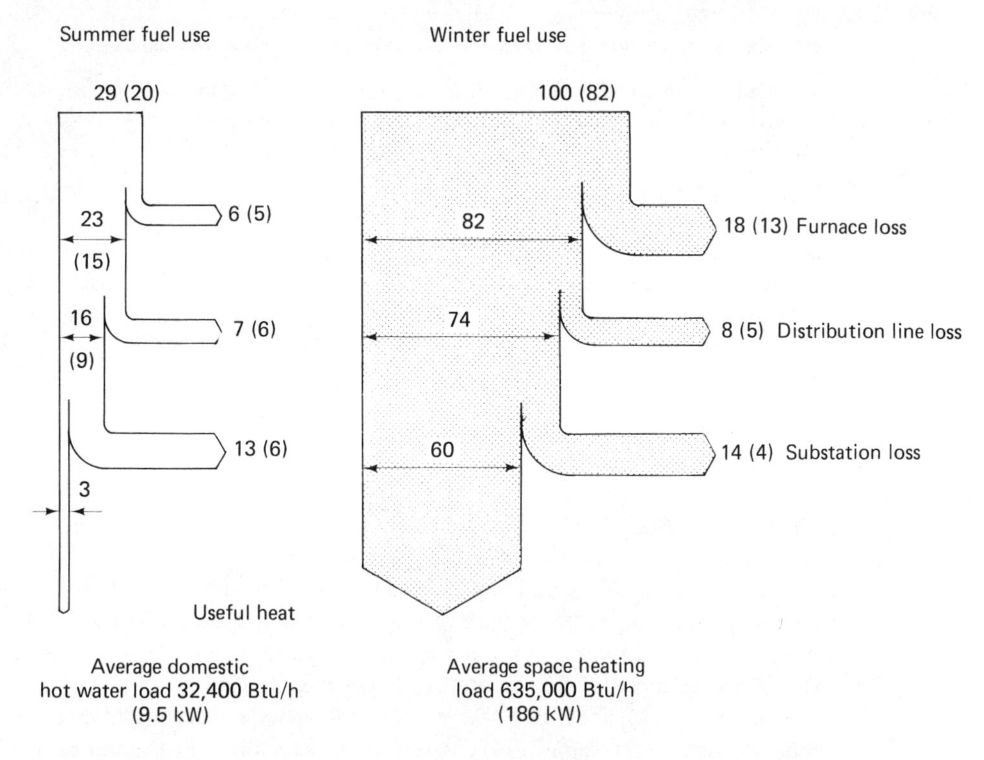

Figure 7.1. Measured losses for a small district heating system serving 43 one-family houses from a central boiler. (5) Numbers in parentheses indicate percent losses after completion of the energy-saving program.

operator of the district heating station, rather than by the individual's altering the setting of his own thermostat.

4. Better information for the operator of the district heating unit on the main aspects of energy-saving operation, e.g., how to more closely control the combustion air regulation under changing load conditions or improved summer operation (for which period the smaller furnace is in operation), and more economical operation in winter using the two furnaces in cascade instead of parallel.

Measures with investment:

5. Insulation of the uninsulated pipes in the substations.
6. Central automatic control and regulation of the distribution temperature depending on the outside weather conditions.

Further improvements could be realized in the existing unit only with a relatively high investment. For new installations, however, the following measures would be employed:

Limitation of water flow rate to each house to moderate peak power demand

Increased size for domestic hot water storage, sufficient to meet the needs for one day, so that charging could be limited to only twice a day. This would reduce losses during light load operation.

Heat metering for each house.

Measures 1 through 6 were carried out at a total cost of $12,000 (including the analysis). During two subsequent heating periods, a 25 percent reduction of oil use—12,600 gal (40,000 ℓ)—was achieved, after correction for the different weather conditions of the two heating seasons. The investment was returned from annual savings in about two years.

7.4. EXAMPLE: OFFICE BUILDING (4)

This corporate headquarters building is occupied by 800 employees (1–16 persons per office) and consists of offices, service rooms such as corporate archives, restaurants for personnel, conference rooms, etc. For air conditioning of the offices, a dual duct system is used. The annual energy use before EIP is given in Table 7.5.

The mechanical systems of a new office building were being completed just at the time that the price of heating oil increased by a factor of 3, in the fall of 1973. The

Table 7.5. The Yearly Energy Costs in an Office Building before an EIP. Energy Costs Are for 87,725 ft^2 (8150 m^2) Floor Area

Oil consumption (winter 1974–75)	Heating	40,000 gal	(150,000 ℓ)
	Air conditioning	53,000 gal	(200,000 ℓ)
	Domestic hot water	26,500 gal	(100,000 ℓ)
	Total	119,500 gal	(450,000 ℓ)
HVAC system operating costs	Heating oil	$ 62,000	
	Cooling (electricity)	8,000	
	Ventilation (electricity)	40,000	
	Maintenance (supplies)	17,600	
	Total	$127,600	

Notes:

Oil price: $.50/gal ($0.134/ℓ)

Price of electricity: $.032 per kWh day; $.016 per kWh night

Energy cost per unit area: $1.45/ft^2 ($15.60/m^2)

owner asked for an analysis of energy-saving measures. An earlier analysis had shown that the rate of return from energy saved, using, for example, heat recovery with a heat pump, was too long due to the low energy prices then.

The EIP was organized according to the priorities given in Chapter 4. In this energy-saving program however, measures requiring investment came under consideration as early as the first phase. On the other hand, it was the policy that the only improvements to the HVAC system would be those which could be carried out without any effect on the occupants. The energy-saving improvements investigated were:

1. Heat recovery between exhaust air and outside air. The air inlet and exhaust were not in the same location; therefore, only a heat recovery system with two separate heat exchangers—one in the outside air intake and one in the exhaust air connected by a water circulation system—could be considered.
2. In addition to this installation, the ventilation air from the archives was ducted to the garages (instead of exhausting it directly outside as earlier).
3. An additional measure consisted of using an enthalpy controller (instead of simple temperature regulation) to achieve minimum energy mixing of outside and recirculated air.
4. For humidity control of the files and personnel restaurant, a control system was used with hygrostat sensors instead of dew point regulation. With this system, it is no longer necessary to cool the air to the dew point and reheat it again to the desired level.
5. The mechanical rooms for the air conditioning, heating, and refrigeration systems are ventilated using intermittent operation.

An investment of $90,000 was required. In addition, engineering costs of $29,000 were incurred. The resulting savings, according to Table 7.6, were $18,000 per year. This corresponds to a write-off time of 7 to 8 years at an interest rate of 6 percent. The annual energy savings for heating and air conditioning were 15 percent. The saving in oil was 20 percent.

Additional energy-saving methods which were considered during the analysis:

Table 7.6. Energy Saving in an Office Building after an EIP

Measure	Cost[a] ($)	Saving[b] ($)
Heat recovery	67,200	9,200
Humidity control	14,200	2,800
Intermittent ventilation of machine rooms and air ducting from archives to garage	8,800	6,000
Total	90,200	18,000
Oil use was reduced by 26,000 gallons per year		

[a]In addition to engineering costs of $29,000.
[b]The operating costs before the EIP were $127,600.

Periodic shut-off of the humidification (except in winter). Periodic shut-off of the dehumidification (except in the summer) and shut-down of the air conditioning (except during the summer).

Reduction of illumination levels and of air flow rate.

Periodic checks on the system regulation.

Adjustment of the temperature and humidity levels (for example, to lower temperatures in winter and to higher temperatures in summer).

A heat pump using a large nearby lake as the heat source was analyzed as well. It was concluded that this would become economically viable only after a further increase of the price of oil relative to the cost of the heat pump on the order of 30 percent.

Automatic shut-off of office lighting every 3 hours if level of illumination from outside light is sufficient. (The lighting could be switched on again if individuals actually require it.)

7.5. EXAMPLE: CONVERSION OF ROOFTOP AIR CONDITIONERS TO A HEAT PUMP OPERATION (5)

The capital cost of heat pumps is relatively high. However, for the conversion of an existing refrigeration system this is less of a problem. After electric rates had increased by a factor of three, and winter heating costs reached $400,000 per season, engineers at the Raytheon Corporation investigated the conversion of their rooftop air conditioners to heat pump operation.

The Raytheon plant in West Andover, Massachusetts, was built in 1970 when electric rates were low enough to foster all-electric design. There are 7,000 employees who fabricate components and assemble high-reliability electronic systems. The plant has an area of 700,000 ft^2 (65,000 m^2) on one level and it is completely air conditioned, with 24 zones regulated according to heat produced by local manufacturing processes to provide close temperature control. Because of fumes produced by plating and other manufacturing processes, large quantities of fresh air are circulated and this adds to the winter heating requirements. Heating from all sources such as lighting, manufacturing, and resistance heaters accounted for 51 percent of the plant's energy use before the heat pump conversion.

There are 24 rooftop units, each of which has a nominal capacity of 130 tons (457 kW heat transfer capacity for cooling) and multiple stage resistance heaters which supply up to 310 kW for air heating. These heaters supply 990,000 Btu (290 kW) per hour for a typical zone when the outside temperature is −8°F (−22°C). This corresponds to 12,420 Btu/hr for each Fahrenheit degree difference between indoor (70°F or 21°C) and outdoor temperatures (6.55 kW per centigrade degree difference). The annual heating requirements can be determined by multiplying this quantity by the indoor-to-outdoor temperature differences (in 5-degree increments) and by the number of hours during the heating season when such temperatures prevail.

The heat that can be supplied by the heat pump is obtained from calculated performance curves supplied by the manufacturer after accounting for the duration of defrost periods, the season energy requirements to operate the heat pump, and the supplemental resistance heat required during cold spells when the heat pump output would be inadequate to hold the building at 70°F. The internal heat gain in each zone must be taken into consideration, of course. Figure 7.2 shows the performance curve for a typical heat pump system. This performance curve is not identical to the units converted at the West Andover plant.

The key steps in the conversion are outlined below and the modifications to the

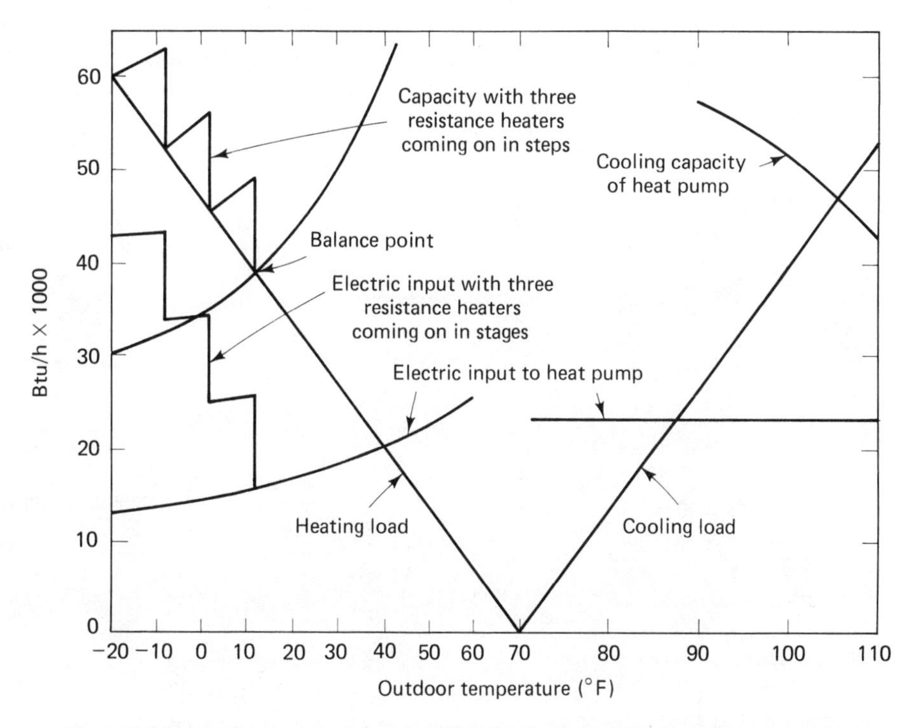

Figure 7.2. Performance curves for a typical heat pump. The building heating and cooling loads shown in this figure are assumed to be linear and directly proportional to the difference between 70°F (21°C) and outdoor temperature. In practice, internal heat from equipment and lighting would alter the shape of these curves. In the cooling mode (right-hand portion of the chart), the power input to the compressor is considered constant, with the cooling capacity dropping from about 57,000 Btuh (16.7kW) heat removal capacity) at 90°F (32°C) to 43,000 Btuh (12.6kW) at 110°F (43°C). At the design temperature of 105°F (41°C), such a heat pump would transfer approximately 48,000 Btuh (14kW).

In the heating mode, electrical input to the compressor drops as the temperature drops, but the compressor heating capacity and its coefficient of performance drop even faster. Electrical resistance heaters are added in stages to supplement the heat pump when temperature drops to about 12°F (−11°C). Even at −20°F (−29°C) the heat pump delivers 30,000 Btu of heat for every 13,000 Btu of electrical input energy (i.e., 8.8 kW per 3.8 kW electrical input power), for a coefficient of performance of 2.3. The application of heat pumps to recovering energy from waste heat flows, such as from ventilation exhaust air, could make their economics even more attractive.

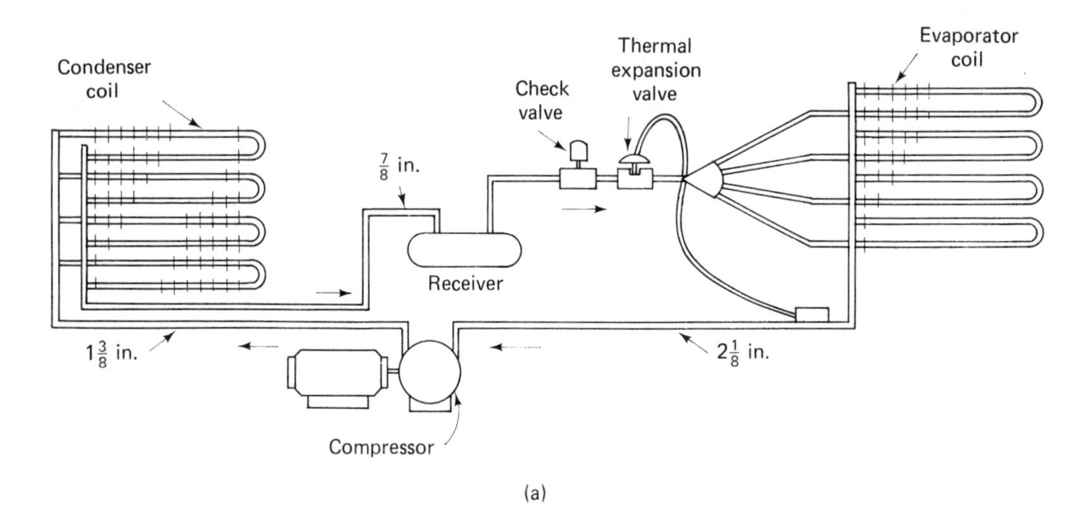

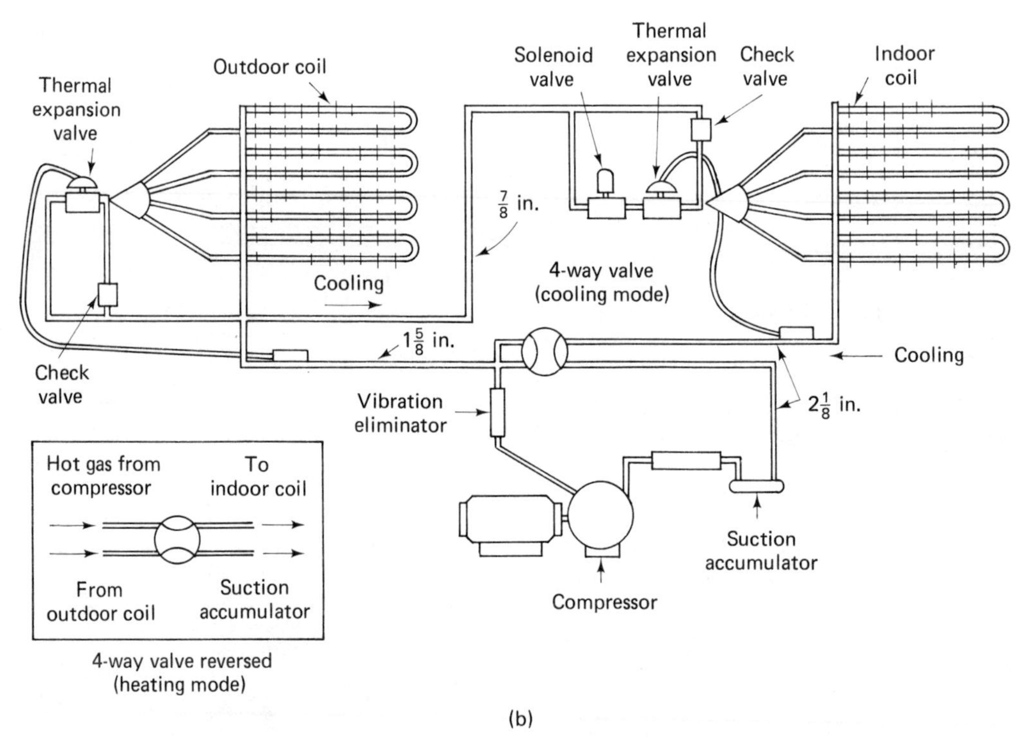

Figure 7.3. Original standard refrigeration system and the converted system for heat pump operation. Some manufacturers offer package conversion kits. (a) The original refrigerant system of the rooftop air conditioner. Each rooftop unit used 4–40 hp shaft-driven compressors for a total of 130 tons of nominal cooling capacity. (b) The converted heat pump system required almost all new larger-diameter refrigerant lines, a large 4-way valve for converting flow from cooling to heating, replacement of each receiver with a much smaller suction-accumulator, and replacement of the condensors with higher capacity coils that will pick up heat from or discharge heat to the outside air.

refrigerant system are shown in Figure 7.3. Retrofitting a refrigeration system to operate as a heat pump involves a substantial amount of engineering and analysis. Considerable assistance was provided for this conversion by the manufacturer of the rooftop unit. A more elaborate control system was required, and this was assembled from time delay relays and other standard components. The rewiring necessary is outlined below.

Key Steps in the Conversion of the Rooftop Air Conditioners to Heat Pump Operation[a]

MECHANICAL

1. Remove and save refrigerant charge from four refrigerant circuits.
2. Remove eight existing condenser (outdoor, air-cooled) coils.
3. Remove intermediate condenser coil supports.
4. Disconnect existing four receivers from liquid and discharge lines. Cap connections to prevent water or dirt accumulation.
5. Prepare and seal entire condenser section floor with mastic sealer. Fabricate and install sheet metal condensate deflectors along periphery of condenser section.
6. Install eight new outdoor coils in place of condenser coils with distributors and heaters attached on condensate deflectors and bolt to unit frame.
7. Mount four 4-way valves in condenser section.
8. Mount four horizontal suction line accumulators in condenser section.
9. Provide new refrigerant piping, check valve, thermal expansion valve, and thermostat, as shown in Fig. 7.3, for each circuit.
10. Add four defrost-initiate control capillaries to respective suction lines between compressor and accumulator.
11. Add four head-pressure, fan-switch capillaries to respective discharge lines between compressor and four-way valves.

ELECTRICAL

1. Remove existing heating sequencer and cooling sequencer.
2. Mount new central solid-state logic panel.
3. Mount time-delay relays, one for each compressor.
4. Mount new sequencer for heating mode.
5. Mount new sequencer for cooling mode.
6. Mount defrost cycling timer.
7. Mount two changeover (cooling to heating) relays and require ambient changeover control.
8. Mount eight defrost relays on four outdoor and four indoor coils.
9. Mount four pressure-switch, initiate-defrost controls (set at 50 psig).
10. Mount four temperature stat defrost-terminate controls (set at 80F), with bulbs strapped on lines to outdoor coil discharge head.
11. Mount four head-pressure, fan-switch controls.
12. After removing unused wiring, run all new wiring between controls, valves, relays, and control power sources as outlined in master wiring diagram supplied. All wiring is to meet local and NE Code requirements.

[a]Diverting the warm exhaust air from the plant through the heat pump evaporator coil would improve the annual coefficient of performance.

Because of the increased annual operating hours on the compressor and electrical components of the system, more frequent and thorough maintenance checks are required. Operating hour meters on the compressors are a help in scheduling maintenance.

The cost of modifying each rooftop unit was approximately $16,000. The total energy required for the converted operation was found to be 52 percent of those units not yet converted. This represents a saving of approximately 320,000 kWh, which at 2 1/2¢/kWh amounts to $8,000 per unit over the initial test period. Conversion of the entire 24 units will save 10 million kWh per year and return the cost of conversion in under two years.

7.6. EXAMPLE: UNIVERSITIES

Massachusetts Institute of Technology Cambridge, Mass. (7)

The physical plant at MIT includes about 100 buildings with an area of approximately 6,000,000 ft^2 (600,000 m^2) and a variety of environmental control systems and operating requirements.

Earlier buildings, built before 1916, showed a specific energy use of only about 112,000 Btu/ft^2 (350 kWh/m^2), leaky windows and poorly controlled heating systems.* Newer buildings have higher illumination and ventilation levels and higher electrical consumption for the mechanical equipment, so that these used up to five times more electricity per unit area. They have an average specific energy use of 210,000 Btu/ft^2 (650 kWh/m^2).* The total energy costs in 1975 were $2.5 million for electricity and $2.5 million for fuels. At that time, electricity cost $.026 per kWh, and heat (steam) cost $2.05 per 10^6 Btu ($7 per 1000 kWh).

Phase I: Operational changes. Reduction of illumination level to 50–70 fc (700 Lux) for work areas and to 5 fc (50 Lux) in corridors; decorative lighting and unneeded lighting was shut off. The setting and regulation of the air-conditioning systems was checked. Unneeded air conditioning was shut off and closer control was maintained in matching use to hours of occupancy. Larger variations were tolerated in room temperatures 68–79°F (20–26°C). The refrigeration unit was shut off in winter for some areas, and the heating was turned off in summer for the dual duct systems. Outside air flow rates were reduced.

Phase II: Measures with short rates of return. Installation of timers, improvements in the control and regulation of HVAC systems to minimize the need for manual readjustment for different weather conditions, automatic reduction of room temperatures during the night, reduction of air flow rate in corridors and in rooms with low occupancy. Improvements in the distribution of chilled water.

Phase III: Longer-return investments. Installation of a central building monitoring and control system to serve the major energy-using buildings, and programmed to closely control energy

*Combined fuel and the secondary energy value of electricity.

use as a function of the occupancy and outside weather conditions. The central control system also has the capability to optimize the operation of the central steam and chilled water plant and to schedule maintenance activities. Conversion of incandescent lighting to fluorescent or other high efficiency lighting was accelerated.

The energy reductions achieved as a result of Phase I were 20 percent for electricity and 25 percent for steam. This represented a saving in operating costs of $1,138,000. These results were part of an initial EIP that had begun in 1971. The cumulative saving in electricity is shown in Figure 7.4. Similar results have been achieved in reduced steam use. The energy intensity of electricity use per unit area at MIT has taken a sharp break downward from its trend during the period between 1960 and 1975.

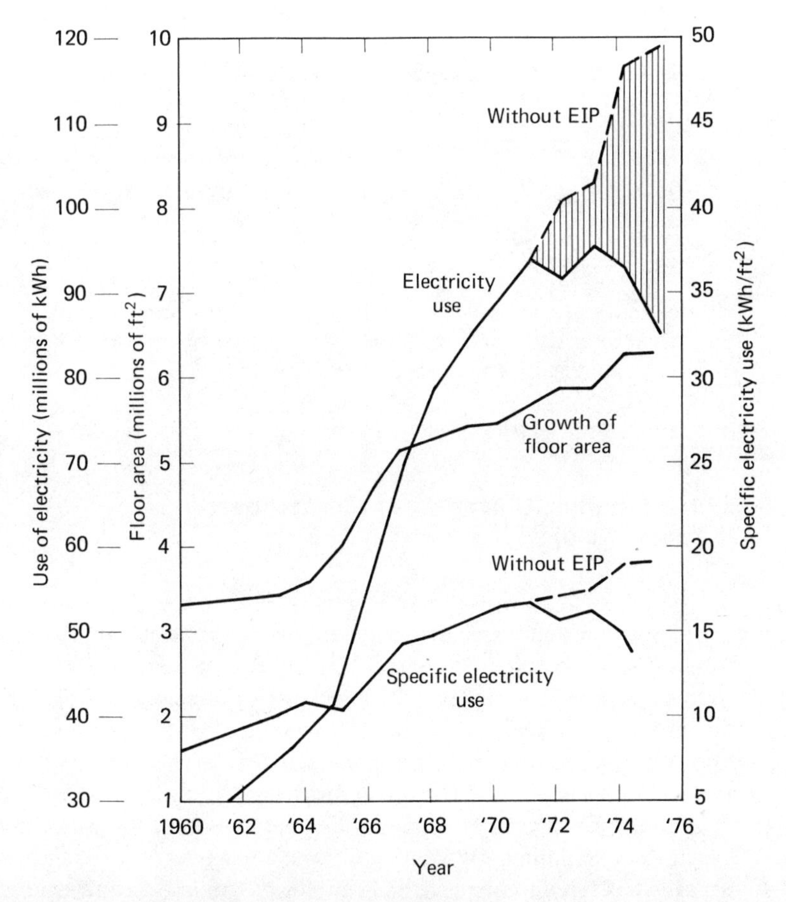

Figure 7.4. Annual use of electricity compared with growth of floor area and the intensity of electricity use per unit floor area. The shaded area represents savings in electricity. The savings in steam follow a similar pattern. The energy-saving program at MIT had been instituted two years before the energy price shock of 1973-74. (7)

Example of Improvement to the Allied Medical Facility Complex at Ohio State University, Columbus, Ohio (8)

The annual savings from 4 significant measures:

Removal of 22 percent of fluorescent bulbs to reduce energy cost by 2 percent.

savings $1,581
cost $1,000

Reduction of air flow rates to spaces where lighting was removed, reducing warm air supply to interior spaces, and providing variable volume cooling supply to areas with changing loads.

savings $13,340
cost $8,000

Shut-off of air conditioning to unoccupied rooms.

savings $21,640
cost $14,000

Use of two-speed motors for air-conditioning fans for television studio.

savings $4,470
cost $7,000

Total measured energy savings of over 50 percent have been achieved. The entire investment was returned in nine months. In addition, improved control over the environment had been obtained, resulting in fewer complaints.

Harbor Campus, University of Massachusetts at Boston (8, 9)

The university consists of six electrically heated buildings, with 1,350,000 ft^2 (125,000 m^2) of space which were first occupied during the winter of 1973–74. A view of the Harbor Campus is shown in Figure 7.5. The cost of $2 million for electricity during the first year in the new complex was an increase by a factor of 13 over the energy costs for the university's former campus in older buildings. One reason for the unexpectedly high costs was the escalation in the price of electricity, as is also shown for several other large institutions in Figure 7.6. The initial distribution of energy use is shown in Figure 7.7.

A comprehensive EIP was mounted to reduce energy use which initially was running at the rate of 90 million kWh during the first few weeks of occupancy. A full-time energy manager was hired to organize and run this program. Some selected items from a 1977 status report are shown in Table 7.7.

By 1977 energy use had been reduced 40 percent from an annual rate of 75 million kWh to 44 million kWh, saving $2.5 million (see Figure 7.8). The investment to achieve this saving was only $128,633 including the salary of the EIP manager.

Figure 7.5. Harbor Campus of the University of Massachusetts in Boston. The walls are brick veneer and hollow concrete block, separated by an air space that has between 0 and 2 in. (5 cm) of insulation. The percentage of glass in the six different buildings varies between 11% and 25%; the average window area is 15%.

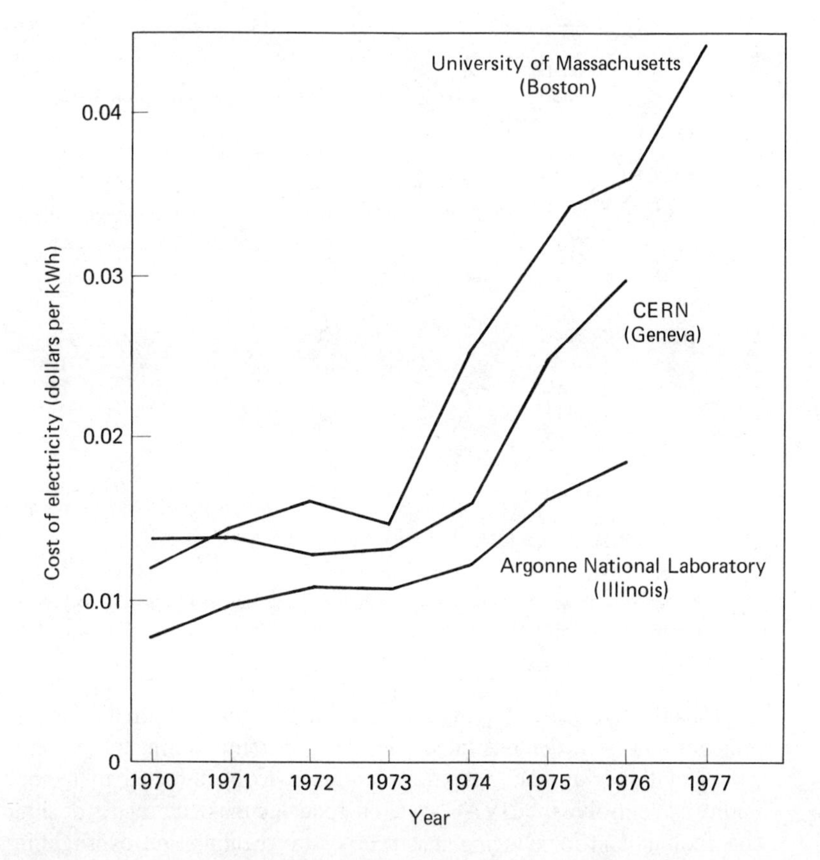

Figure 7.6. The increase in electricity prices for some public institutions. The average yearly increase in cost since 1970 has been 23.2%, 12.7%, and 15.3% for the University of Massachusetts (Boston), CERN, and the Argonne National Laboratory, respectively.

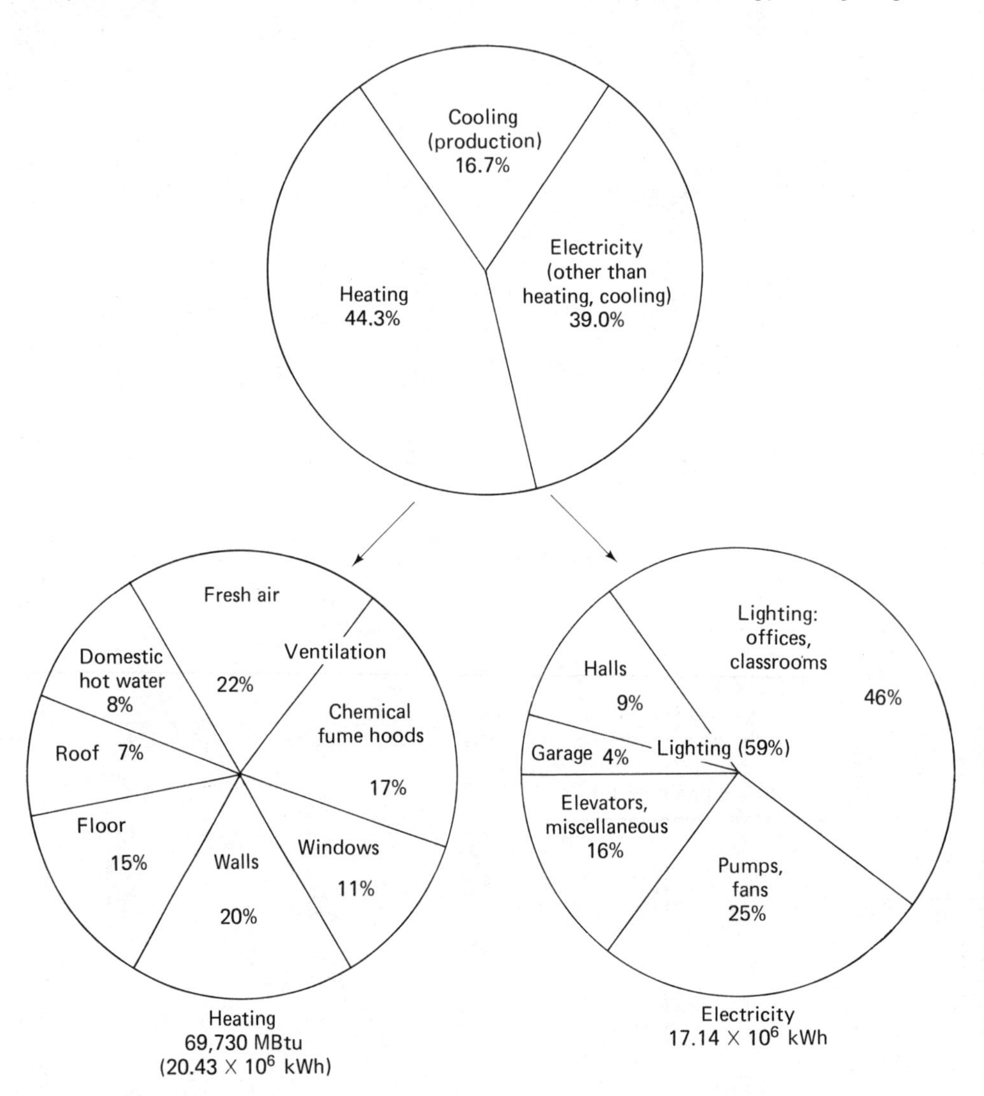

Figure 7.7. Initial distribution of energy use at the Harbor Campus of the University of Massachusetts in Boston.

The effectiveness of having a competent, full-time person directing the energy saving program had been demonstrated and the university administration authorized a continuation of this program. Attention continued to be directed to methods for obtaining improved control of the HVAC systems–reducing the occurrences of simultaneous heating and cooling–and to reducing instances of overheating and overlighting. By 1980, this program had reduced the University's energy use to 93,500 Btu/ft^2 (295 kWh/m^2). This was already 33 percent less than the national average for universities in similar climates.

Table 7.7 Energy-Saving Actions at the University of Massachusetts Harbor Campus, Boston (Selected Examples from 1977 Status Summaries)

	Cost ($)	*Date started*	*Potential annual kWh savings*	*Status (%)*
Delamping	N/C	Sep. 74	1,655,000	100
Removal of 16% of bulbs in classrooms, offices, and corridors				
Reduction of heating temperature				
72 to 68°F (22 to 20°C)	N/C	Oct. 75	1,990,000	33
68 to 65°F (20 to 18°C)	N/C	Aug. 77	1,550,000	N/A
Improved control of fan coil units in Building 110	2,000	Feb. 76	297,000	100
After-working-hours shut-off of perimeter heat in Building 080	5,000	no action	674,000	0
Reduced hours for chiller	N/C	Dec. 76	400,000	100
Fan control for lab exhaust hoods	N/A	Apr. 75	578,000	100
Switches and relays for after-hours shut-off of public area lighting	20,000	Dec. 76	1,606,000	100
Install phantom tubes	7,000	Apr. 77	630,000	50
HVAC computer control system				
Reprogramming (rescheduling)	N/C	Mar. 74	3,900,000	100
Upgrading	N/A	in design	1,500,000	0

Note:
N/C = no cost
N/A = data not available

Its near-term goal was to reduce this level to 87,100 Btu/ft^2 (275 kWh/m^2). The progress of this improvement is shown in Figure 7.8.

During this program, the manager kept the university community informed of the progress that had been achieved, and the measures that were to be addressed during the next phase of the program. This was quite helpful in soliciting their participation, although in fact the activities of the occupants had very little effect on the efficiency of the energy using systems.

As an indication of the effectiveness of the University's efficiency improvement program, it had saved $5 million by 1980 at a cost of $350,000. This amounted to an annual saving of $100 per student, which was returned to the university community in

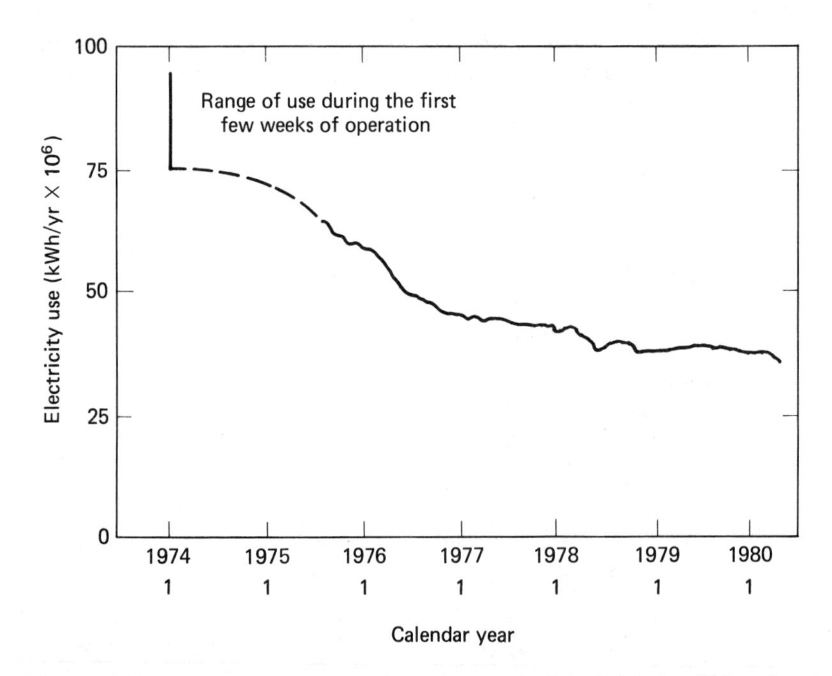

Figure 7.8. Decrease in the annual energy use as the result of the University of Massachusetts Harbor Campus energy-saving program. (9)

the form of increased services and programs. Figure 7.9 shows a comparison of the annual expenditure for electricity, and what that cost would have been without the EIP.

Summary of Some Investments for Saving Energy from References 8 and 9

In Table 7.8 are listed, for comparison, some specific energy-saving investments, together with their savings in energy and dollars. In only one of the examples listed was the ratio of cost to annual savings greater than unity.

7.7. EXAMPLE: SWIMMING POOL (10)

An analysis of the exceptionally high operating cost of this indoor swimming pool was prompted at the time its owners wanted to sell the entire complex of which the indoor pool was a part. Records for the yearly operating costs showed $72,000 was spent for heating ventilation air and $24,000 for water consumption. Next to salaries, expenses of $96,000 per year for energy and water were the largest components of the total yearly

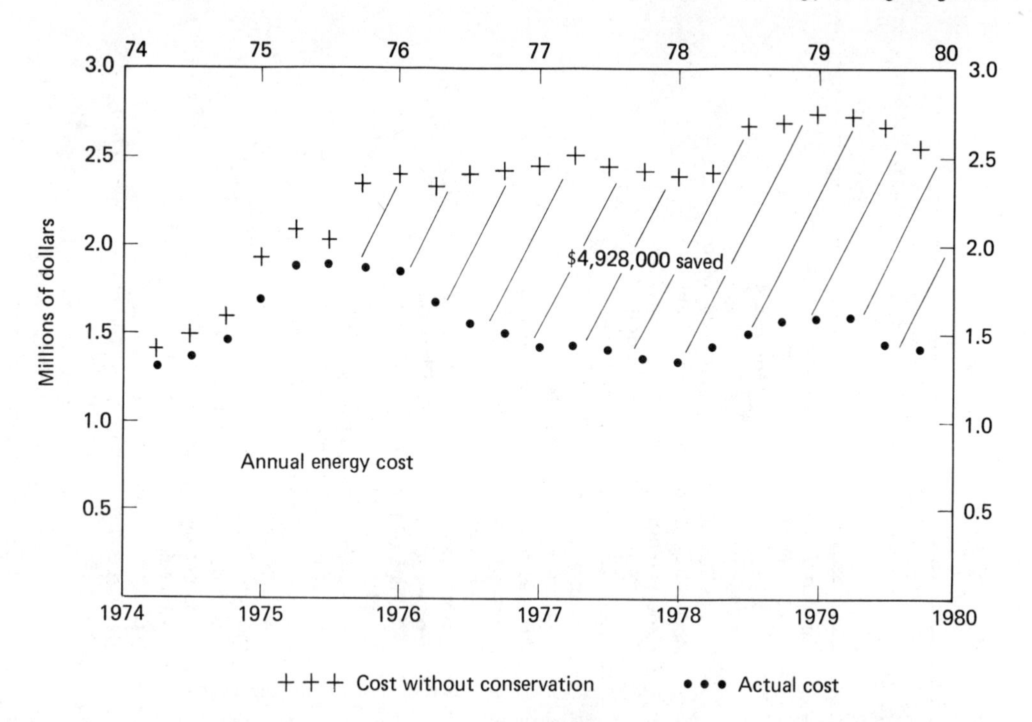

Figure 7.9. Comparison of the actual cost to the University for electricity, with what that cost would have been without the efficiency improvement program. Such comparisons are useful in obtaining support for an EIP. (The data shown have been computed on an annual average basis).

operating costs. Although operating expenses generally tend to run higher for indoor pools than for other buildings of the same area, the expenses for this case were unusually high. This was traced in part to poor management of the heating and ventilation.

The dimensioning, function, and control of the HVAC systems for the pool and locker rooms were checked. The economic analysis of the energy-saving measures which were proposed showed them to have a good rate of return. Some of the measures were immediately implemented after the analysis. The measures proposed were:

1. Reduction of the operating time of the pool and dressing room air-conditioning systems to periods of use (previously these systems ran continuously day and night). Use of the existing capability for two-speed operation (previously only the full-speed setting was used).
2. Reduction in the amount of fresh water consumption with a concomitant reduction in the energy used for heating it.
3. Reduction in the temperature of the ventilation air for the pool from 88 to 84°F (31 to 29°C).
4. Reduction of pool water temperature from 84 to 80°F (29 to 27°C).
5. Proposal of a heat recovery system to exchange energy between exhaust air and outside air.

Table 7.8. Comparison of Investments Made by Several Universities for Saving Energy, with the Dollar and Energy Amount of the Savings

Energy-saving measure	*Cost ($)*	*Savings ($/year)*	*Energy savings, MBtu (kWh)*	*SCP, $/BPD ($/kW)*	*Comments*	*References*
Time switches	15,000	250,000	——	——	Installed time clocks to shut down night and weekend ventilation and air conditioning in buildings totaling 4 million ft^2 (400,000 m^2) of floor space. Ohio State University, Columbus, Ohio.	(7)
Electric energy-saving program	16,000	180,000	6,000,000	24	More efficient lighting installed. Interior and exterior lighting levels reduced. A 70,000 gal. (265 m^3) water storage tank used as a cold sink for building cooling. Clarkson College of Technology, Potsdam, N.Y.	(7)
SEP for fuel	36,000	57,500	11 (6,800,000)	3,700 (46)	Installed roof insulation, run-around heat recovery system, new insulation on steam and condensate lines; reduced steam pressure by 44 percent plus other no-cost measures. Clarkson College of Technology, Potsdam, N.Y.	(7)
Infrared survey	11,800	21,800	—	—	25 miles (40 km) of steam lines on the Cornell University campus (Ithaca, N.Y.) checked for leaks and poor insulation by using aerial survey with infrared techniques.	(7)
Unoccupied shut-down of motorized equipment	14,000	21,640	833,000	(86)	Motors with a total capacity of 193 hp shut down for 14 hours/day during non-occupied hours at the Allied Medical Facility at Ohio State University, Columbus.	(7)
Controls for unoccupied fan coil units	2,000	13,000	1,000 (297,000)	4,250 (60)	Heating and cooling distribution units underneath windows in the administration building were fitted with relays which allowed them to be shut down during non-occupied periods,	(9)

Energy-saving measure	Cost ($)	Savings ($/year)	Energy savings, MBtu (kWh)	SCP, $/BPD ($/kW)	Comments	References
					reducing their use of electricity for heating by 50% at the University of Massachusetts Boston Harbor Campus.	
Central control for public area lighting	20,000	70,660	1,606,000	110	Lighting in public areas was connected to central panel-mounted contactors for shut-down during unoccupied periods for a 34% saving in the electricity used for this lighting purpose. University of Massachusetts Boston Harbor Campus.	(9)
Incandescent lighting replaced with mercury	7,000	5,390	123,000	500	Two hundred 150-watt reflector floods were replaced with 80-watt mercury lamps for a 40% saving. University of Massachusetts Boston Harbor Campus.	(9)
Central control of perimeter heaters	5,000	29,650	2,900 (674,000)	4,600 (65)	Perimeter heaters with 803 kW were connected with relays to allow central shut-off after hours to achieve temperature set-back. University of Massachusetts Harbor Campus.	(9)
Control consolidation	16,500	29,000	660,000	220	Controls on certain mechanical equipment to allow selective shut-down during operating hours. University of Massachusetts Boston Harbor Campus.	(9)
Demand controls on fume exhaust hoods	35,200	68,750	2,750,000	112	Laboratory fume hoods, used to exhaust toxic gases released by teaching and research experiments, were fitted with individual controls. These 310 hoods now can be operated according to actual need. University of Massachusetts Boston Harbor Campus	(9)
Phantom tubes	7,000	30,000	630,000	97	Phantom tubes were substituted for one 40 W fluorescent lamp in 2-lamp fixtures. Much of this program was coordinated with regular maintenance and replacement programs. University of Massachusetts Boston Harbor Campus.	(9)

Table 7.9. Summary of Energy Savings for an Indoor Pool

ESM[a]	*Energy saved* 1,000 *kWh/yr*	*MBtu/yr*	*Saving ($)*
Indoor pool			
Reduction of air supply temperature to 84°F (29°C)	119	406	1,275
Reduced operating period	372	270	4,000
Heat recovered to preheat fresh air	127	433	1,365
Heat recovered to preheat pool water	617	100	6,635
Locker rooms			
Reduced operating period	499	1,700	5,360
Heat recovered to preheat fresh air	112	382	1,200
Other savings			
Reduced water consumption[b]	31,500m^3	(1,112,000 ft^3)	6,300
Reduced electricity used for equipment (mostly fans)[c]	130,000kWh		4,160
			30,300

[a]Savings in oil heat at \$3.15/10^6Btu (\$10.75/1000 kWh)
[b]Savings in water at \$0.75/1000 gal (\$0.20/1000 ℓ)
[c]Savings in electricity at \$0.032/kWh

These proposals were capable of reducing the annual operating costs by \$30,000. The required investment, including the cost of the energy analysis, was \$60,000 for a payback period of about 2 years. Additional details on savings and costs are given below and in Table 7.9.

Costs and Savings for Improvements in Energy Use for an Indoor Pool (10)

ESTIMATED COSTS

1. Improvements to the controls for better operation of the heating and ventilation systems.	\$ 2,000
2. Heat recovery system for preheating the ventilation for the pool and dressing rooms.	\$42,000
3. Analysis, design of the improvements, coordination of contractors and supervision during installation, and training of personnel in energy-conscious operation.	\$16,000
Total cost for improvements	\$60,000

7.8. EXAMPLE: DISTRIBUTION CENTER FOR A SUPERMARKET CHAIN (11)

The distribution center in Dierikon, Switzerland, is used for processing, packaging, storage, and distribution of the products sold by the Migros marketing chain. The Dierikon distribution center services a large area of central Switzerland. The Migros chain has a yearly sales volume of $2.3 billion. The major part of its volume consists of food products; however, there is also a large non-food assortment of products. The distribution center has a floor area of 235,000 ft^2 (22,000 m). The capacity of the boilers is 35 million Btuh (10 MW).

For the purposes of the energy-saving program, the equipment has been divided according to the following classifications: boiler house, air-conditioning systems, domestic

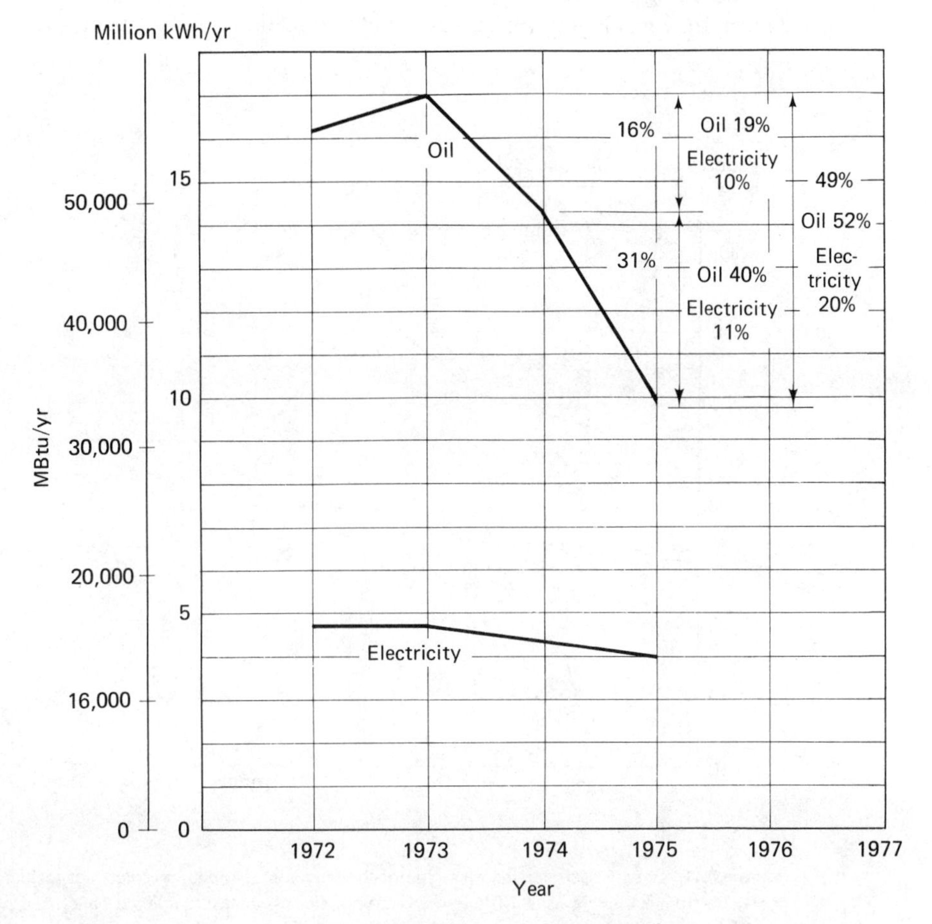

Figure 7.10. Energy use by the Migros supermarket distribution center near Dierikon, Switzerland. Process heat recovered by burning waste is not included. (11)

hot water, laundry and utensil-washing systems, cool and deep freeze storage rooms, pressurized air system, battery system, and transformers.

In addition to these service systems, each production department has been treated separately, for example, meat, fruit and vegetables, groceries, dairy, staff restaurant, etc. The consumption of oil and electricity before the EIP is given in Figure 7.10 and the energy costs in Figure 7.11.

General description of the Migros energy-saving program:

1. Energy analysis.
2. Determination of the main energy using departments.
3. Production factors determining each department's use of energy.
4. List of energy-saving methods.

The energy analysis of the entire distribution center was completed after the introduction of the energy-saving methods had already begun. Despite the fact that the energy flow diagram shown in Figure 7.12 gives useful information about the potential for energy

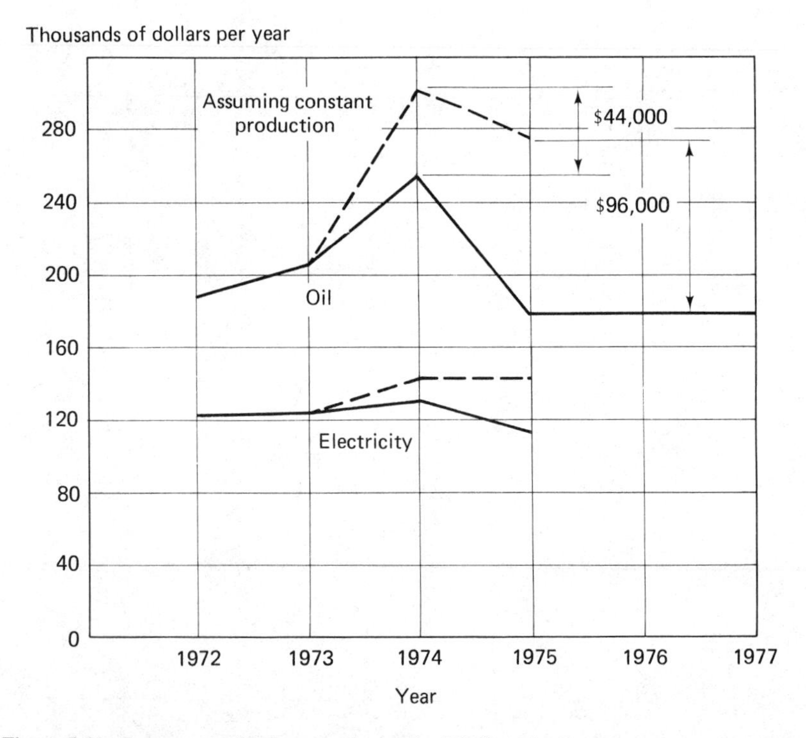

Figure 7.11. Energy costs for the operation of the Dierikon distribution center, not including the value from burning wastes. The dashed curve indicates what costs would have been had production not declined due to economic recession. (11)

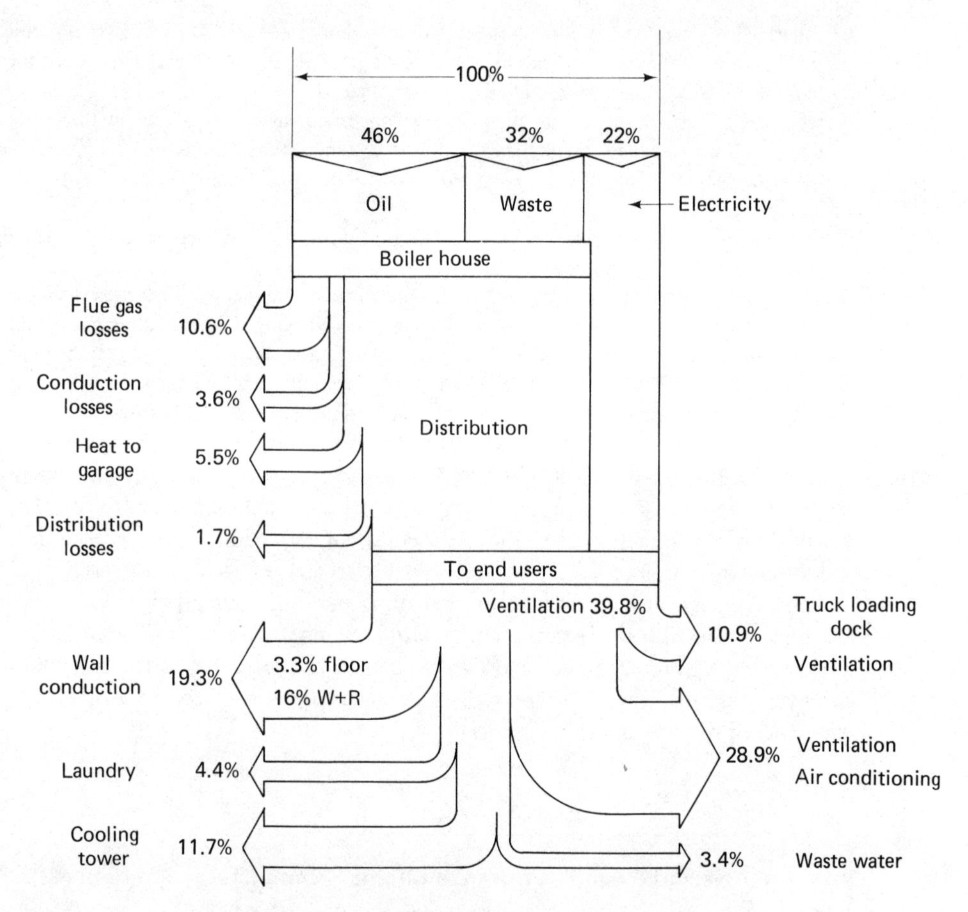

Figure 7.12. Energy flow diagram for the Migros supermarket distribution center in Dierikon, Switzerand. (11)

Assumptions:	Exterior temperature	31.6°F (−0.25°C)
	Wind	Weak
	Date	End of November 1975

saving, it would be hoped that such useful information could be obtained more quickly in order to provide better planning for subsequent EIP.

Examples of ESM used by Migros:

1. Improved boiler management. Formerly, oil was burned in both the heat recovery waste burning boiler and in the more efficient boiler fired only by oil. By burning the waste (much of it wood) in larger batches, costly oil could be reserved for use in the more efficient boiler.

2. Improved management of the storage of process hot water at 350°F (175°C). Previously, the boilers were run during low production periods at night and on weekends to keep the storage tank nearly fully charged. Improved operation charged the storage completely at the end of the work day. The control setting was modified to permit using almost all of the energy stored during a long weekend. At the start of the next work period, the large accumulation of trash could be burned efficiently for a sustained period with efficient heat transfer to the depleted storage tank.
3. Domestic hot water circulation was shut down during non-working hours for a saving of 40 percent of this use (5 percent of the total).
4. The laundry machine was reconstructed to make more efficient use of heated drying air.
5. Timers were used to turn off equipment during non-working hours. Heating temperatures were lowered, and the outside air volume reduced; operating schedules for air conditioned areas have been improved. These measures resulted in a 50 percent reduction in oil for space heating and a 20 percent saving of electricity used for ventilation fans and air conditioning.

No investment was required for the first phase of the program. The total energy saved as shown in Figure 7.10 was 20 percent for electricity and 40 percent for oil. During the period shown, production also declined because of economic recession. The influence of this is shown in Figure 7.11.

The program at this distribution center is part of a comprehensive energy-saving program of the Migros Society. The goal of this energy-saving program is a 6 percent to 8 percent saving each year and the use of part of the money saved for investment to achieve additional energy savings. Some selected details of the Migros efficiency improvement program are given below.

Selected Examples from the Migros Saving-Energy Program (12)

1. *Heating and Ventilation*

1.1 *Reduced room temperatures*

Office room temperature in winter: during working time 68°F (20°C), nights and weekends 59°F (15°C).

Storage room temperature: heat to the lowest possible level; separate storage areas according the minimum allowable room temperature.

Staff restaurant: reduce temperature after use.

Reduce temperature to the lowest possible level in production areas with forced ventilation.

For processes requiring high very high ventilation rates, try to introduce unheated outside air as close as possible to the area requiring ventilation, as for example chemical hoods, cooking processes, etc.

Block thermostats so that unauthorized personnel cannot tamper with their setting.

Do not heat traffic ramps to loading docks and garage; use salt to melt snow.

1.2 *Minimize heat loss through doors, openings, and exhaust fans*

All doors connecting to the outside should be self-closing.

Replace air curtain entries with doors.

Hermetically seal all windows and doors. Check the sealing of all of the doors and windows before each winter heating season.

Open warehouse doors for loading purposes only. Do not open large doors simply for personnel traffic.

Shut off exhaust fans when the associated equipment is not being operated.

1.3 *Conserve heat with improved insulation*

Check insulation of all piping.

Insulate the liquid surfaces of warm storage tanks using plastic domes or other means.

1.4 *Reduce heat radiation losses from warm surface*

Insulate all equipment with a surface temperature over 86°F (30°C), such as ovens, steam vessels, heat exchangers, reaction tanks, compressors, sterilization autoclaves, etc.

Reduce radiation of equipment which is hard to insulate, for example by painting pipe valves, unions, pressure relief valves, etc., with silver paint, or cover with aluminium foil, and by keeping bright surfaces clean and polished.

1.5 *Check heating and ventilation and carry out necessary improvement*

Check for leaking pipes, regulating valves, taps, steam traps, etc.; if necessary, use ultrasonic leak detection equipment.

Recycle clean condensate to save this expensive, chemically prepared water and to reclaim its remaining heat content.

1.6 *Improve efficiencies of furnaces, boiler, steam generators*

Adjust the oil burners periodically to achieve highest efficiency. Repair or replace if necessary to achieve expected performance.

When the burner is shut off, close the exhaust gas damper at the same time to avoid rapid cooling down of the furnace.

2. *Refrigeration and air conditioning units*

2.1 *Reduce cooling losses*

Reduce lighting in cold storage rooms to a minimum. Use fluorescent lighting.

Install door-closing equipment on cold storage room entrances. If possible, use air locks with one door closed all of the time.

Avoid allowing warm air to enter cold storage rooms, and vice-versa. This double loss costs the energy to heat the air and it requires the expenditure of 5 kWh just to remove one kWh from the cold storage room.

2.2 *Improve the efficiency of the refrigeration system*

Optimize defrosting. Defrost periodically to minimize the ice layer thickness. Defrosting should be short to minimize warming up the cold storage room and its contents.

The refrigeration units should be operated during weekends for short periods at full power and then be shut off completely.

3. *Electricity*

3.1 *Reduce lighting to the necessary level*

Improve individual task lighting where required to improve production. Reduce general illumination levels, but maintain sufficient lighting for safety.

Remove the tubes in every second lighting fixture in the corridors. This will help to remind personnel to save energy.

Use "Save Energy, Shut off Light" stickers at every light switch.

List the operating hours on every switch which has to be left on at night. Provide adequate instruction for night watchmen.

Replace incandescent lights with fluorescent lights. A 40-watt incandescent bulb gives 400 lumens. A 40-watt fluorescent lamp gives 2,000 lumens.

3.2 *Use electrical energy sparingly*

Reduce peak power demand with a plan for when electrical equipment needs to be running.

Install a warning signal which will sound if peak demand lasting longer than one minute occurs, and shut off equipment not related to the production. Priorities have to be planned in advance.

Use "Save energy, Shut off Equipment" stickers on equipment with electric motors.

Replace oversized motors.

Check processes, seek alternative processes which allow a reduction in the electric energy used.

4. *General recommendations*

Do maintenance and cleaning during normal working hours, to reduce heating and lighting needed after regular working hours.

Train safety personnel to perform energy measurement, check switches left on, and to report energy waste to the responsible departments.

5. *Implementation of the EIP*

In each department, someone must be designated as an energy manager.

The goals for energy saving should be well publicized. Specific responsibilities should be defined and methods for controlling energy waste should be explained.

A status display of the progress toward the goals for energy and water use should be in a prominent location.

Make in-house films on energy saving showing how to do it, how not to do it.

Set up a training course for energy saving. This should be done periodically and should be a continuous effort rather than a task force effort. The goal is to develop energy consciousness on the part of the users.

7.9. EXAMPLE: AN INDUSTRIAL FACILITY (13)

This industrial facility employs about 1,000 workers. The products manufactured include electric insulation, capacitors, high voltage electrical equipment, and vacuum systems for the electric power industry.

The motivation for the organization of an energy-saving program originated with one of the members of the board of directors. The success of this program was due in part to the strong leadership exerted by the management to develop enduring, energy-conscious work habits on the part of every employee. The individual steps of the EIP were:

Analysis of the existing situation of how and where energy is used

Estimation of the trend for future energy prices

Attention first to the larger energy users

Setting of attainable goals

Examination of all processes

Checking on the interrelationships among major factors such as transfer of heat, operating time of equipment, possible changes in products, etc.

Analysis of specific energy-saving methods

Formulation of detailed steps to achieve more efficient energy use

Some examples of measures used:

Reduction of room temperatures from 75 to 68°F (24 to 20°C)

Insulation of heated storage tanks for residual oil

Addition of thermostatic valves to radiators

Weatherstripping and insulation of windows, doors, and walls

Improvement of the waste-burning capability of one of the process heat boilers

During a period of one-and-a-half years, a 30 percent saving was achieved in oil used for space heating. Between 1974 and 1975 an investment of $300,000 was made. Of this investment, about $200,000 was used for a cooling water recirculation facility, which resulted in a reduction in the use of water by roughly a factor of 3. Because of the savings, existing equipment was able to meet new demands; for example, a $60,000 outlay for a new oil tank was avoided.

Previous yearly energy costs were $400,000. In 1975 a saving of $120,000 was achieved. The largest portion of the saving resulted from the reduced use of fresh water. The investments have had very short pay-back periods of 2 to 3 years.

Additional improvements being implemented or planned include improvement of the waste-burning capability by adding more storage; a shredder and a burner which will fire liquid wastes. The insulation of the process heating lines will be improved, more thermostatic valves will be installed on the radiators, and a new workshop will be outfitted with a heating system in the floor. The program of replacing faulty windows and improving the heat-tightness of walls, windows, and doors will continue. The budget for these improvements is $175,000 which, together with the measures already carried out, will reduce the oil use by 55 percent over the 1973 consumption, despite increased sales growth.

7.10 EXAMPLE: A HEAT PUMP FOR USE OF WASTE HEAT (3)

In a sub-station of an electric utility, a power factor improvement unit (with a capacity of 60 MVar) required continuous cooling with water. The cooling power requirement of 650 kW corresponds to a heat rate sufficient for the heating needs of 60 to 80 apartments. Because the cooling water could have an outlet temperature no higher than 77 to 86°F

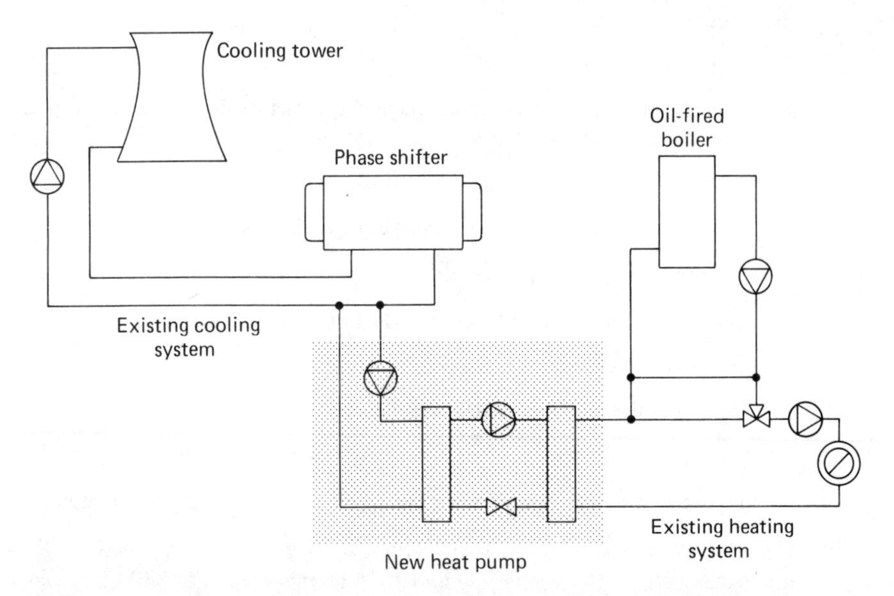

Figure 7.13. Elements of the heat pump system for the use of the waste heat from cooling the power factor correction unit (phase shifter) in the public utility's electric distribution system. The addition of the heat pump enabled the replacement of heating oil using 45% less raw source energy to the power plant supplying electricity for the heat pump. (3)

(25 to 30°C) direct use of the cooling water for heating is not possible. For this reason the cooling water was sent originally to a cooling tower, which resulted in wasted energy. An analysis of the use of a heat pump showed that the cooling water could be used as an economical source of energy for a heat pump (Figure 7.13). Next to the sub-station, there was an existing office building with a maximum heat demand of 0.29 MBtuh (85 kW), and it was possible to supply 0.18 MBtu/hr (52 kW) of this with the heat pump.

The condenser heat of the heat pump has a temperature of about 130°F (55°C) allowing an easy integration into the existing oil heating. During subsequent operation the calculations were shown to have been correct in predicting that for ambient weather temperatures above 28°F (−2°C), a water temperature of 130°F (55°C) to the radiators was adequate, and the heat pump could handle the full heat demand. At lower outside temperatures, the oil furnace is used in series with the heat pump.

The costs of operation:

Maintenance	$ 320/yr
Energy costs (at $.024/kWh)	$1,525/yr
Total operating costs for the heat pump	$1,845/yr
This is to be compared with savings of fuel (with oil at $0.45/gal)	$4,600/yr

The difference between cost saving in fuel and operating cost of the heat pump is $2,755/yr. With an interest rate of 6 percent, the investment of $24,000 has a payback period of about 12 years. The heat pump has a yearly average COP (including motor efficiency) of 4.2.

The heat pump has substituted for oil with an energy content of 130 MBtu (380,000 kWh) per year with 63,000 kWh of electricity. With the assumption that the electrical energy is produced in a nuclear power plant, 63,000 kWh corresponds to 720 MBtu (210,000 kWh) in the heat content of the nuclear fuel. Losses in the electrical distribution network are included. This example shows that a saving of raw source energy of 580 MBtu (170,000 kWh), or 45 percent more than the energy content of the oil not burned, has been achieved.

7.11 EXAMPLE: MANUFACTURING COMPLEX AND HIGH-RISE OFFICE BUILDING (3)

The headquarters for a large, multinational industrial concern is shown in Figure 7.14. About 10,000 people are employed in this administrative and manufacturing complex. The oil price crisis of 1973 stimulated a detailed analysis of the energy used for heating, air conditioning, and industrial processes. The result was surprising. As shown in Figure 7.15, it was possible within a matter of one year to reduce energy use by 20 percent! This meant a $400,000 reduction in fuel costs. At the same time the consumption of water was reduced by 50 percent. The reductions were achieved despite the fact that

Figure 7.14. The headquarters of the Sulzer Brothers industrial complex in Winterthur, Switzerland. The energy facilities include a total energy system. Heat recovered from the furnaces in the casting foundry is supplied to the city's district heating system. (3)

overall production increased by 20 percent. We summarize below some of the energy saving methods used.

EIP for the Administration Building

This high rise building shown in Figure 7.14 has 26 floors and an air-conditioned office area of 194,000 ft^2 (18,000 m^2). It houses engineering disciplines and specialists for HVAC systems. Accordingly, the administration building was designed to use the most advanced technology available for the mechanical systems of the building at the time it was built. Despite this, an economic analysis after the oil price shock indicated cost-effective measures for reducing energy use by 20 percent at a saving of $60,000 yearly. This example shows again the tremendous potential for saving that exists in yesterday's energy-intensive designs. Some of the specifications and features at the time of its construction:

Facade: Aluminum skin on a reinforced concrete framework.

Windows: Insulating double glazing with outside venetian blinds. The ratio of windows to the exterior wall area is 56 percent.

Illumination: 3.35W/ft^2 (36W/m^2).

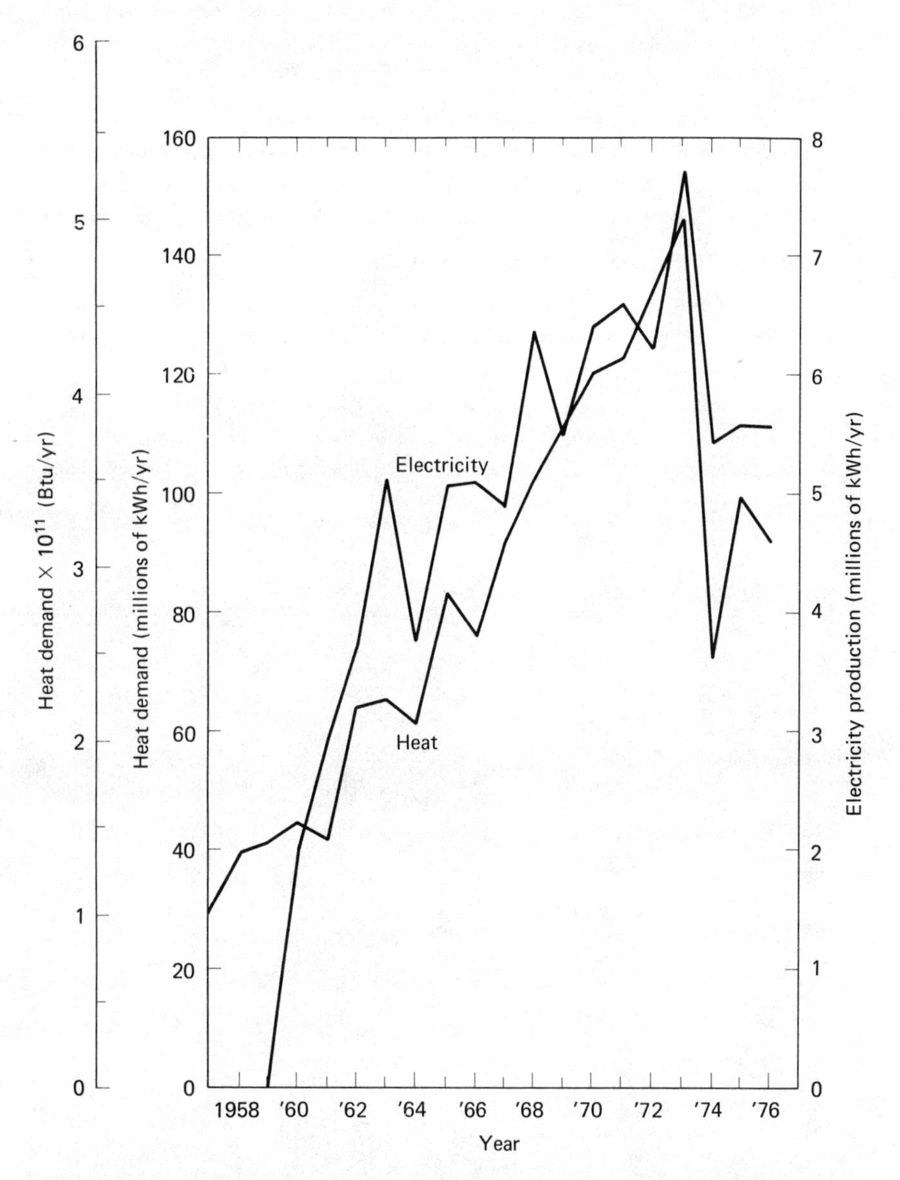

Figure 7.15. Heat and electricity demand for the Sulzer administrative and manufacturing complex. Electricity is generated in a total energy facility using a back-pressure turbine to supply the process steam requirements. (3)

HVAC systems: One dual duct unit serves the north and west facades and a second system serves the south and east facades. There is separate heating and cooling for each of the four facades of the building. The total volume flow of air in both systems is 120,000 cfm (206,000 m^3/h). In addition, the window areas are served by four other dual duct HVAC systems, with an air handling capacity of 40,600 cfm (69,000 m^3/h).

The ratio of return air to outside air is a maximum of 50 percent at 5°F (−15°C). Heat demand at 5°F (−15°C) is 14 MBtuh (4.1 MW) (conduction heat losses of the building and heating of ouside air). Maximum cooling demand at 86°F (30°C) is 400 tons (1.4 MW). The following energy-saving actions were implemented between 1973 and 1975:

Oct. 1973: Request made to personnel to adjust room thermostats to a maximum of 68°F (20°C).

Nov. 1973: The outside air flow rate was reduced and the recirculated air flow rate increased. The resulting saving was 25 percent of the heating energy per year. The outside airflow rate of 35 cfm (60 m^3/h) per person was found to be sufficient.

Nov. 1974: The operating time of the dual duct system was reduced by one hour daily. The window air-conditioning system was run in the recirculation mode during the night with no outside air mixing. In the summer, the air conditioning is operated between 4 and 6 a.m. without any mechanical cooling, using only outside air.

Table 7.10. Energy Used in the Administration Building

			1972	*1973*	*1974*	*1975*
Heating energy	MBtu		24,870	23,270	17,130	18,080
	(MWh)		(7,290)	(6,820)	(5,020)	(5,300)
Cooling energy	MBtu		2,560	3,530	2,130	2,200
	(MWh)		(750)	(1,035)	(624)	(645)
Electrical energy[a]	(MWh)		1,350	1,403	1,180	1,045
Lighting	(MWh)		(1,152)	(1,059)	(1,152)	(1,149)
Energy factor	1,000 Btu/ft^2		274	268	273	220
	(kWh/m^2)		(864)	(845)	(705)	(695)
Water	gal		993,000	1,011,800	837,685	795,170
	(m^3)		(3,759)	(3,830)	(3,152)	(3,010)
Comparative ambient weather data:						
Winter months (Dec., Jan., Feb.)						
Mean outside temperature		°F	31.5	30.2	34.5	33.3
		(°C)	−0.3	−1.0	+1.4	+0.7
Summer months (June, July, Aug.)						
Mean outside temperature		°F	59.4	62.4	60.6	60.6
		(°C)	+15.2	+16.9	+15.9	+15.9
Hours of sunshine			492	565	629	587

[a]For the operation of the HVAC system

1975: Due to the colder winter, slightly more heating energy was used than in 1974. There were some additional minor changes in the operation of the HVAC; for example, a change was made to a one-duct system for the window air supply, thus avoiding mixing losses. Controls were installed to reduce the difference of the two hot and cold deck temperatures. Results of the energy-saving program are shown in Table 7.10.

Other Buildings and Workshops

Following the energy price shock, management introduced the following energy-saving measures:

1. Requests were put up on the bulletin boards asking all employees to limit winter room temperatures to 68°F (20°C) in offices and to 61°F (16°C) in workshops.
2. Changes were made in about 20 room heating systems to convert them from manual operation to automatic operation and temperature control. The cost was $24,000.
3. In factory entrances, the heating ventilator is normally shut off. If the workshop temperature drops lower than a given value, a smaller heater will operate under thermostat control. The large heater will come on only if the door is closed. The cost was $4,000 for 10 entrances. The energy saving is 785 MBtu/yr (230,000 kWh/yr) corresponding to a cost saving of $5,600 per year.
4. In workshops, the water supply temperature for heat is regulated as a function of the outside air temperature. The temperature is regulated by area thermostats which control the valves of the radiators (cost $12,000). This scheme utilizes the heat from equipment, lighting, and people and avoids overheating. The result is that heating in the winter has only been necessary on Monday morning.(!) During the rest of the week, the internal heat sources are sufficient for room heating.

Heat Recovery from the Compressed Air System

The cooling water for the compressor is heated to 122–132°F (50–55°C) and can be used for domestic hot water. Because the pressurized air system uses an oil-free compressor, fresh water used for cooling can be subsequently used for domestic hot water. With an investment of $20,000 for a storage tank and pump, monthly savings of 4 million gallons (15,000 m^3) domestic hot water were achieved. This resulted in a yearly fuel saving corresponding to 34,000 MBtu (10,000 MWh) worth $140,000 at that time.

Heat Recovery in the Foundry

An economizer was installed on the foundry furnace to use the waste heat of the exhaust gases. The system is shown in Figure 7.16. To avoid corrosion in the economizer, a special feature of the control system is needed to guarantee a minimum exhaust gas temperature. The recovered heat is transferred via a heat exchanger to the existing district heating system. To allow operation of the furnace during periods of insufficient heat demand in the district heating network, a standby cooling system using city water was

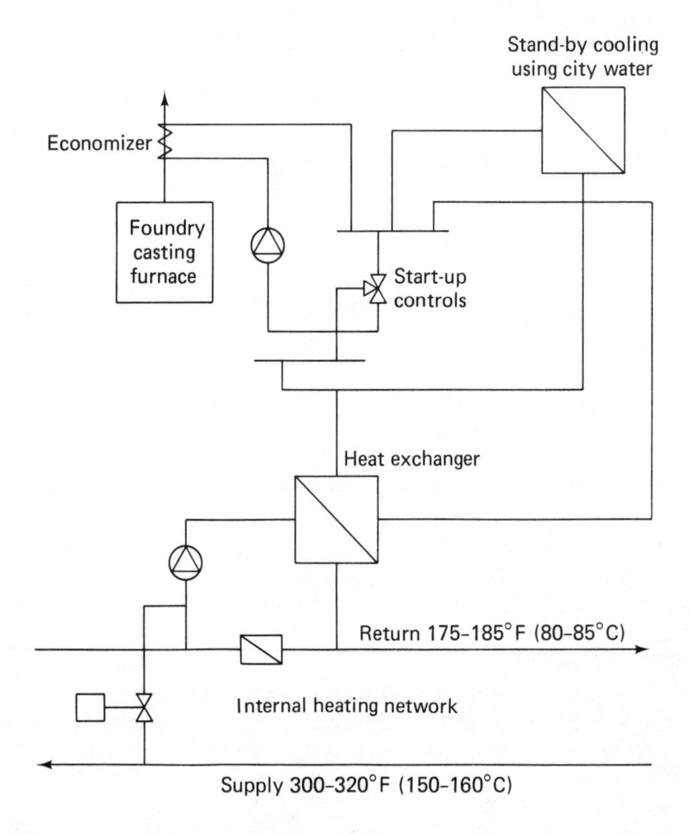

Figure 7.16. Scheme for using the high temperature heat recovered from the foundry casting furnace exhaust gases to supply the internal heating system. Special controls in the economizer keep the exhaust temperature above the condensation temperature of corrosive flue-gas acids. Standby cooling is used when the internal heating users cannot utilize the rejected heat from the casting furnace. (3)

installed. The entire system for heat recovery required an investment of $160,000. The corresponding yearly energy savings amounted to 3,500 MWh worth $84,000. In addition, there was a significant reduction in the use of water. This installation resulted in a yearly saving of 16 million gallons (60,000 m^3) of potable water, corresponding to a cost saving of $20,000.

Another example is shown in Figure 7.17. The cooling water from another casting furnace supplies heat at a temperature of 117°F (47°C), suitable for space heating and as preheated boiler feed water. To cover the requirements of the space heating system when the casting furnace is not in operation, the system draws energy from the district heating system. The additional cost for this heat recovery system amounted to $240,000. With this investment it was possible to save 3.2×10^{10} Btu (9.3×10^6 kWh) of heat energy, thereby avoiding an annual expenditure of $220,000. In addition, a further, substantial saving in cooling water was also achieved.

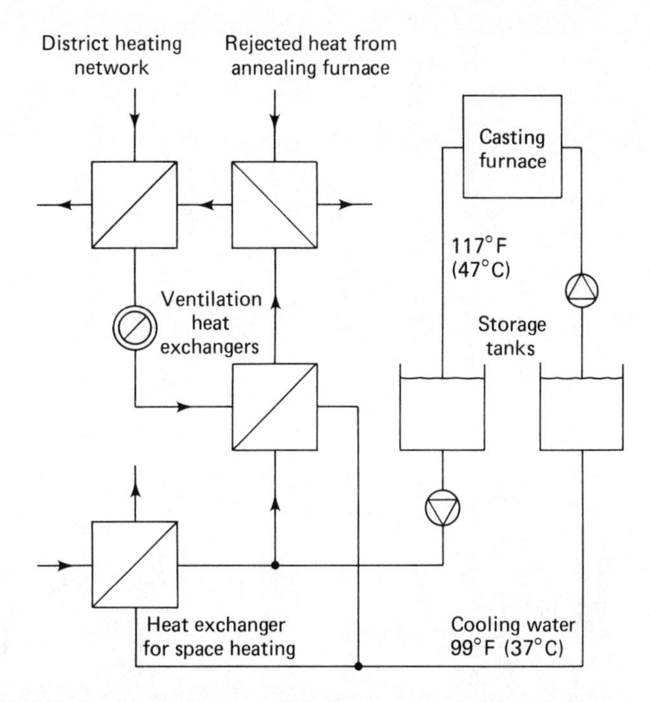

Figure 7.17. Heat recovery scheme to use the low temperature heat from the cooling water of the casting furnace and the annealing oven to heat ventilation air and the building. This heat is obtained from the district heating network when the other sources of rejected heat are not in operation. (3)

Cogeneration of Heat and Electricity

To save energy some of the heat energy needed in industrial factories should be obtained from cogeneration, for example, from a back-pressure steam turbine. The simultaneous production of heat and electricity can obtain large overall energy savings. In 1973, at the time of the sudden increase in oil prices, the price for electricity generated from hydropower plants remained stable, thus changing the ratio of oil and electricity prices. The result of this situation was that cogeneration of heat and electricity became more costly. Production was reduced and electricity was purchased from the city network. The consequence of this undesirable situation is shown in Figure 7.15 as the sharp drop in the amount of cogenerated electricity. Cogeneration becomes economically more favorable the higher the ratio of electricity price to the price of fuel burned to make steam.

Summary

The examples show that a yearly saving of 1.4×10^{11} Btu (40,000 MWh) of heat energy could be achieved with operational measures and with investments of $480,000. If the entire energy saving is evaluated using the cost of fuel oil, the cost avoidance during the

Table 7.11. Comparison of the Costs and Savings of Some of the Examples in this Chapter

Example	*Cost ($)*	*Savings ($/year)*	*Energy savings MBtu (kWh)*	*SCP, $/BPD ($/kW)*	*Comments*	*Reference*
Community heating system	12,000	5,600	6,300 (1.85×10^6)	4,600 (57)	Improvements to a 43-house central heating system reduced annual average rate of energy use by 0.72 MBtuh (210 kW) (see Section 7.3).	(3)
Apartment building	131,000	6,600	1880 (5.5×10^5)	53,300 (660)	Multi-family dwelling located in Germany. Savings averaged over heating season.	(3)
Office building	119,000	18,000	4,440 (1.3×10^6)	65,400 (810)	Savings of fuel and electricity at its raw source value averaged together. EIP reduced annual energy costs by 15% (see Section 7.4).	(4)
Indoor swimming pool	60,000	30,300	7,650 (2.24×10^6)	19,000 (235)	Savings of both fuel and electricity (at primary value) are combined and averaged over the year (see Section 7.7).	(10)
Computer control	64,000		3,750 (1.1×10^6)	40,400 (500)	Intermittent operation of HVAC and other equipment limits demand. Energy saving shown is estimate of total raw source energy.	(3)
Industrial EIP	100,000	46,000	20,000 (6×10^6)	(145)	The first phase program at Micafil (see Section 7.9).	(13)
Industrial EIP	170,000	26,200	11,600 (3.4×10^6)	34,700 (430)	The potential future saving program at Micafil (see Section 7.9).	(13)
Heat pump	24,000	2,750	580 (1.7×10^5)	100,000 (1,240)	Because of heat pump's high COP (4.2), it saves 580 MBtu (170,000 kWh) in oil not burned, over the raw source energy in the fuel consumed at the power station to supply electricity for the heat pump (see Section 7.10).	(3)
EIP for heavy industry	480,000	600,000	136,500 (4×10^7)	8,500 (105)	Energy savings are in addition to fresh water savings worth an additional $136,000 (see Section 7.11).	(3)
EIP for a university	128,638	1.3 million	100,000 (3.1×10^7)	2,900 (36)	Savings achieved at the all-electric Harbor Campus of University of Massachusetts, Boston (see Section 7.6).	(9)
Municipal EIP	143,220	22,227	8500 (2.5×10^6)	40,000 (495)	Energy saved includes electrical savings at their approximate raw source energy value. Data is the summary of the 29 examples from Section 7.1.	(1)
Rooftop conversion to heat pump	384,000	250,000	3.4×10^4 (10^7)	12,300 (244)	Average demand during heating season is reduced by 1572 kW by converting 24 rooftop air conditioners with resistance heaters to heat pump operation at the Raytheon manufacturing plant, West Andover, Mass.	(5)

first year was more than $600,000. This means an extremely short payback period for the investments. An additional result was that previously planned expansion of the energy production facilities for this factory could be put off. As a result of the energy-saving program, there is spare capacity to meet future growth. Furthermore, a saving of 110 million gallons (420 million liters) of fresh water was achieved, thus avoiding an expenditure for this water of $136,000. The bottom line is that the examples described saved close to $750,000 with an investment that was recovered in less than eight months.

A summary of some selected energy-saving programs is given in Table 7.11. These data show the return from investments in saving energy and the productivity of capital spent for more efficient use of energy.

REFERENCES

(1) Daniel A. Harkins, John R. Haynes, et al., "Energy Management in Municipal Buildings" (Boston: Energy Conservation Project, Massachusetts Department of Community Affairs, 1977).
(2) E. Stocker, "Probleme und Erfolge von Energiesparmassnahmen bei öffentlichen Bauten," Proceedings of the Seminar Bauliche Massnahmen zum Energiesparen in der Gemeinde (Rüschlikon, Switzerland: Gottlieb Duttweiler-Institut, June 1976), pp. 161–172.
(3) M. Kiss, H. Mahon, and H. Leimer, *Energiesparen Jetzt!* (Berlin, Germany: Bauverlag, 1978), pp. 198–204.
(4) Ibid., pp. 201–202.
(5) Thola Theilhaber, Internal Report (Lexington, Mass.: Raytheon Company, 1975).
(6) Department of Physical Plant, "Energy Conservation at MIT," (Cambridge: Massachusetts Institute of Technology, April 1975).
(7) DOE, "Energy Conservation on Campus," vols. I and II (Office of Energy Conservation and Environment, U.S. Department of Energy, Washington, D.C., December 1976).
(8) Frank O'Brien, "Energy Conservation at the Harbor Campus of U Mass/Boston in Fiscal Year 1976," unpublished report to the University Physical Plant, Dorchester, MA. (August 1976).
(9) Tom McNeil, Internal Reports of the Physical Plant of the University of Massachusetts at Boston, dated 1977 and 1980.
(10) E. Wiedmer, "Vorschläge zur Energieeinsparung" unpublished engineering report from Proctecta-sol AG, Stäfa, Switzerland, 2 February 1977.
(11) H. Renfer, "Energiebilanz der Migros: Erste Ergebnisse eines Versuches," in the proceedings of the Seminar "Wie Spare Ich Energie im Betrieb," 9 to 10 January 1976 (Gottlieb Duttweiler-Institut, Rüschlikon, Switzerland, 1976), pp. 28–41.
(12) R. W. Peter, "Ein nationaler Energiesparplan" (Gottlieb Duttweiler-Institut, Rüschlikon, Switzerland, 1977).
(13) A. Fischer and H. Steinmann, "Das Energiesparprogramm am Beispiel der Micafil Spartechnik in der Industrie," in the proceedings of the Seminar "Wie Spare Ich Energie im Betrieb," 9 to 10 January 1976 (Gottlieb Duttweiler-Institut, Rüschlikon, Switzerland, 1976), pp. 47–53.

8

Calculation Examples, Studies

Productive energy use, as compared with less energy-conscious operation, requires a certain amount of additional analysis. For example, operating costs must be calculated at an early point in the EIP, and the economics of a large number of variants must be compared.

Calculation methods for some typical examples are illustrated in the following sections. The examples selected have been actual projects: a heat pump using ground water as the heat source, a solar energy unit for heating and cooling, air-conditioning systems for computer centers, heat recovery for several different systems, and the analysis of energy use in a university.

8.1 A HEAT PUMP USING AN AIR CONDITIONER CHILLER: AN ANALYSIS OF DIFFERENT SYSTEMS

To illustrate the possibilities for the use of heat pumps, an interesting example has been selected. (1, 2, 3, 4) The owner became interested in the possibilities of using heat pumps after the project was already under way. The design work for the administration building of a regional government in Austria was already well advanced and the contract for the heating system using an oil-heated furnace had already been given to a supplier. The building includes 96,850 ft^2 (9000 m^2) of office area, rooms for display, conference rooms, and auditoriums for meetings. Despite the initial impression that the cost for the heat pump would be very high—probably precluding its use (except perhaps for the desire to reduce the adverse environmental problem from soot and SO_x—a study was commissioned to evaluate the following systems. Results are given in Table 8.1.

Table 8.2. Investment, Energy, and Operation Costs of the Different Heating Systems (System 3 Was Used for the Final Design.)

System			*Additional investment ($)*	*(A) Annual capital retirement ($)*	*(B) Energy costs ($)*	*Total yearly costs (total = A + B) ($)*
System 1	Oil heating	3 × 3.4 MBtuh	— (3 × 1000 kW)	—	73,200	73,200
System 2	Gas/oil heating	3 × 3.4 MBtuh (3 × 1000 kW)	+ 23,600	+ 2,700	77,600	80,300[a]
System 3	Gas/oil heat with 1 heat pump	3 × 2.56 MBtuh (3 × 750 kW) 1 × 2.56 MBtuh (1 × 750 kW)	+102,400	+12,000	63,600[c]	75,600
System 4	Oil heating only with 1 heat pump	3 × 2.56 MBtuh (3 × 750 kW) 1 × 2.56 MBtuh (1 × 750 kW)	+ 82,800	+ 9,600	55,600	65,200
System 5	Gas/oil heat with 2 heat pumps	2 × 2.56 MBtuh (2 × 750 kW) 2 × 2.56 MBtuh (2 × 750 kW)	+143,200	+17,200	65,200	82,400
System 6	Oil heating only with 2 heat pumps	2 × 2.56 MBtuh (2 × 750 kW) 2 × 2.56 MBtuh (2 × 750 kW)	+126,400	+15,600	61,200	76,800
System 7	Gas heat only with 1 heat pump without oil back-up heating	3 × 2.56 MBtuh (3 × 750 kW) 3 × 2.56 MBtuh (1 × 750 kW)	− 60,400	−19,200	63,600	44,400[b]
System 8	Gas heat only with 2 heat pumps without oil back-up heating	2 × 2.56 MBtuh (2 × 750 kW) 2 × 2.56 MBtuh (2 × 750 kW)	+ 48,400	+ 8,400	65,200	73,600

Note: For comparison of costs, the basic case is System 1 (System 7 requires a lower investment than System 1; all others are larger). Interest rate is 8%; write-off period for electromechanical systems is 15 years; for buildings, 30 years.

[a] Annual cost using a conventional system for comparison with System 3.

[b] Most economical solution.

[c] Details in Table 8.2.

Data from Table 8.1 show the following results:

1. The investment for the gas heating in combination with the heat pump is lower than for an oil-fired furnace (space for the oil tank also reduced.)
2. The total annual costs (energy costs plus capital costs for the additional investment) are minimum for this combination.
3. If, for reasons based upon the need to obtain a more diversified energy supply, both oil and gas should be used as well (this diversification was a requirement of the owner), then the heating with a mixed-fuel, gas/oil furnace in combination with the heat pump still results in lower annual costs than the same system without heat pump. Compare systems 2 and 3.

Heating Systems Considered in the Analysis

System 1: Oil Heating Only. The heat is produced by three oil-fired furnaces each of 3.4 million Btuh (1 MW) capacity.

System 2: Natural Gas and Oil Heating. The heat is produced in three gas-fired furnaces, each of 3.4 million Btuh (1 MW) capacity. To reduce peak demand of natural gas and to obtain a lower gas price, the burners are dual fuel equipped for oil and gas.

System 3: Gas and Oil Heating in Combination with a Heat Pump. As shown in Figure 8.1, the centrifugal chiller used for air conditioning is also operated for the production of heat. The heating capacity of the heat pump is 2.56 million Btuh (750 kW). In addition to this, for the peak heating demand, three furnaces with gas/oil firing are foreseen, each 2.56 million Btuh (750 kW). Annual heat production by the heat pump is 8,940 million Btu (2,620 MWh) per year. With outside temperatures over 52°F (11°C), the heat pump is adequate without additional gas or oil heating. With a lower outside temperature, one or more of the furnaces will be used according to heat demand. For daytime operation, the heat demand as a function of the outside temperature is shown in Figure 8.2.

System 4: Oil Heating Combined with a Heat Pump. The heating system is operated with oil only. Other than this, the system corresponds to System 3.

System 5: Gas/Oil Heating Combined with Two Heat Pumps. The system is expanded by incorporating a second chiller from the air-conditioning system, to use as a heat pump, both of 2.56 million Btuh (750 kW) heat output. The heat demand is covered with two furnaces using gas/oil firing: each 2.56 million Btuh (750 kW). The two pumps produce 50 percent of the power, and cover a large portion of the yearly energy use. During the year, 11,800 million Btu (3,460 MWh) are produced with the heat pumps and 1,300 million Btu (380 MWh) with the gas/oil heating. The two heat pumps produce the necessary amount of heat when outside temperatures are above 34°F (1°C). For lower outside temperatures the two furnaces are switched on automatically as required.

System 6: Oil Heating Combined with Two Heat Pumps. The two furnaces use oil heating only. All other aspects of the system correspond to System 5.

System 7: Gas Heating Combined with One Heat Pump. The furnaces are operated with gas only; otherwise, the system corresponds to System 3.

System 8: Gas Heating with Two Heat Pumps. The furnaces are fired with gas only; otherwise the system corresponds to System 5.

The project was modified as a result of this analysis and it was decided to install system 3: a gas- and oil-fired boiler combined with heat pump. Ground water is used as the heat source. The refrigeration unit of the air-conditioning system is used as the heat pump. For other systems using two heat pumps (systems 5, 6, and 8), the additional investment for the second heat pump and a second ground water heat source is too high. It was intended that gas should be the main supplementary fuel, with oil used as a back-up to obtain security through diversity of energy sources. Economic and energy data for System 3 are given in Table 8.2. Figures 8.1 through 8.4 show details of System 3.

Table 8.2. Economic and Energy Data for System 3

Economic analysis	*Additional cost for equipment*	*Additional cost for building*
Gas, oil heating + heat pump 3 furnaces, each 2.56 MBtuh (750 KW) 1 Heat pump 2.56 MBtuh (750 kW) Dual fuel burner for gas, oil Gas supply lines, gas metering Centrifugal heat pump compressors Heat exchangers Storage tank, pump Controls for heat pump Increased radiator surfaces	$118,000	
Space for storage tanks		+ $12,800
Smaller oil tanks and furnace (saving)	− $ 12,000	
Less space requirement for oil tank (saving)		− $16,400
Total	+ $106,000	− $ 3,600

Calculated energy requirements

1.	Yearly heat demand		13,100 MBtu	(3,840 MWh)
1.1	Heat supplied by gas heating:			
	Monday to Friday,	daytime	3,340 MBtu	(980 MWh)
		nighttime	480 MBtu	(140 MWh)
	Weekends	daytime	140 MBtu	(40 MWh)
		nighttime	200 MBtu	(60 MWh)
	Total		4,160 MBtu	(1,220 MWh)

Table 8.2. Economic and Energy Data *(cont.)*

1.2 Heat supplied by the heat pump:

(a) Domestic hot water:
The demand for domestic hot water is 4,000 gal/day (15 m^3/day) at a temperature of 113°F (45°C) from a source at 50°F (10°C).

The annual heat demand is:

$$Q = 4{,}000 \text{ gal/day} \times 260 \text{ day} \times 8.35 \text{ Btu/gal°F} \times (113 - 50°\text{F}) = 547 \text{ MBtu}$$

$$\left[Q = 15 \times 10^3 \frac{\ell}{\text{day}} (260 \text{ day}) \frac{(1 \text{ kcal})}{\ell°\text{C}} (45°\text{C} - 10°\text{C}) \frac{(\text{MWh})}{860 \times 10^3 \text{ kcal}} = 160 \text{ MWh}\right]$$

(b) For the air-conditioning system and radiators: 8,390 MBtu (2,460 MWh)

(c) Total supply by heat pump: 8,937 MBtu (2,620 MWh)

2. Electrical energy required for the operation of the heat pump with a COP of 3.2:

$$E = \frac{8{,}937 \text{ MBtu/yr}}{3.2} \times 0.293 \frac{\text{MWh}}{\text{MBtu}} = 820 \text{ MWh/yr}$$

3. Energy cost for the production of heat:

3.1 Energy used for gas heating:

Gas price: \$3.18/10^6 BTU (\$10.85/1000 kWh) referred to the upper heat value.

Fee for connection to gas system:	\$6,400/year.
Upper heat value of natural gas:	H_u = 1,024 Btu/ft^3 (10.6 kWh/m^3)
Lower heat value of natural gas:	H_l = 980 Btu/ft^3 (9.4 kWh/m^3)
Yearly average efficiency of system:	80% of H_l

Furnace power 3 × 2.56 MBtuh (3 × 0.75 MW)

Costs for natural gas for 4,160 MBtu/yr (1220 MWh/yr):

$$K_1 = 4{,}160 \text{ MBtu} \cdot \$3.18/\text{MBtu} \cdot \frac{1{,}024}{908} \frac{1}{80\%}$$

$$\left(K_1 = 1220 \text{ MWh/yr} \frac{\$10.85}{\text{MWh}} \frac{10.6}{9.4} \frac{1}{80\%}\right) = \$18{,}650$$

K_2 Annual connection fee	6,400
Energy costs of gas	25,050

3.2 Cost for heat supplied by the heat pump:

Electricity price:	*Day* 6.00 - 22.00	*Night* 22.00 - 06.00
Summer (April 1 to Sept. 30):	\$0.0228/kWh	\$0.0168
Winter (Nov. 1 to March 31):	0.0316/kWh	0.0232
Tariff on maximum demand:	40.80/kW per year	

Electrical energy needed for heat pump

Summer:		Winter:	
Day	210 MWh/yr	Day	440 MWH/yr
Night	60 MWh/yr	Night	110 MWh/yr

Table 8.2. Economic and Energy Data *(cont.)*

	Costs for electricity:		
	Summer:		
	Day	210 MWh/yr × $22.80/MWh	$ 4,790
	Night	60 MWh/yr × $16.80/MWh	1,010
	Winter:		
	Day	440 MWh/yr × $31.60/MWh	13,900
	Night	110 MWh/yr × $23.20/MWh	2,550
	K_3 =		$22,250/yr
	Costs for peak demand:		
	Peak demand at full load		
	P = 240 kW		
	K_4 = 240 kW × 40.80/kW		$ 9,790/yr
4.	Total energy costs for heat production		
	$K_5 = K_1 + K_2 + K_3 + K_4$		$57,090/yr
	Energy costs including taxes	$63,600/yr	

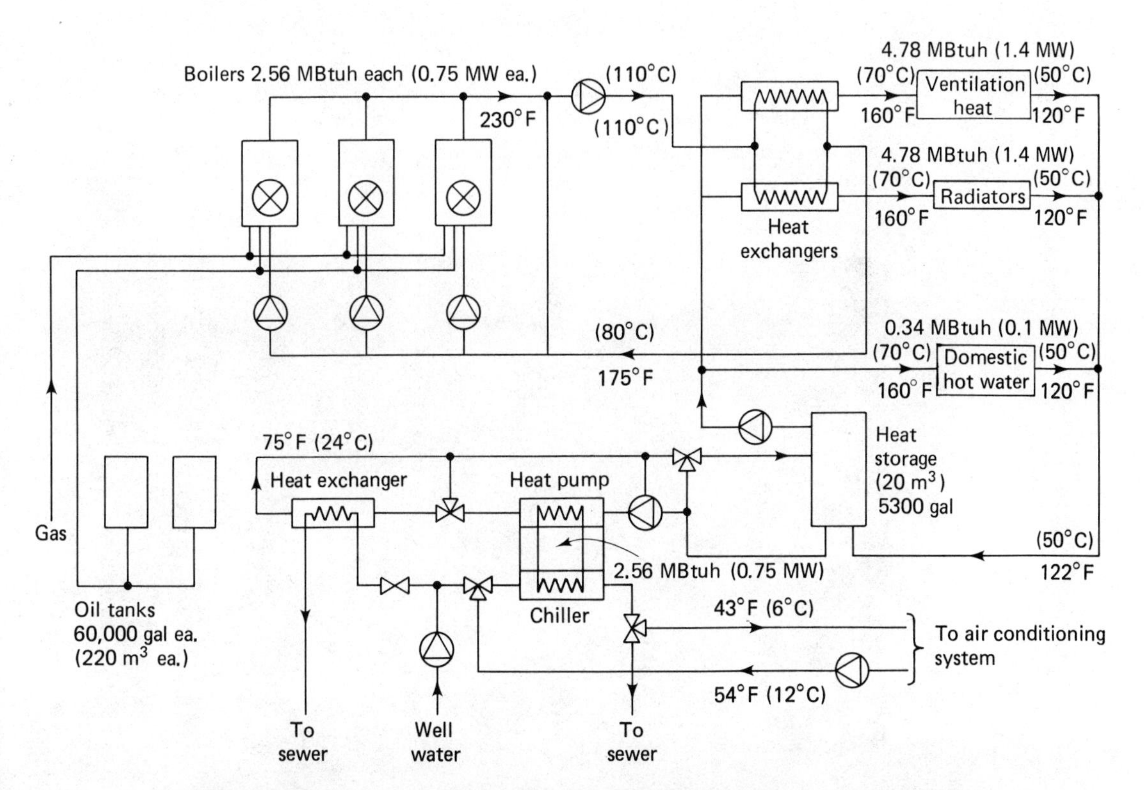

Figure 8.1. System 3 with a single heat pump supplemented by three dual fuel, 2.56 MBtuh (0.75 MW) boilers. The heat pump, which is used as a chiller for cooling during the summer, uses well water at 350 gal/min (22ℓ/sec) and at 50°F (10°C) as a heat source, returns water at 39°F (4°C) to the sewer. In the summer well water is used to cool the chiller condensers, returning 64°F (18°C) water to the sewer.

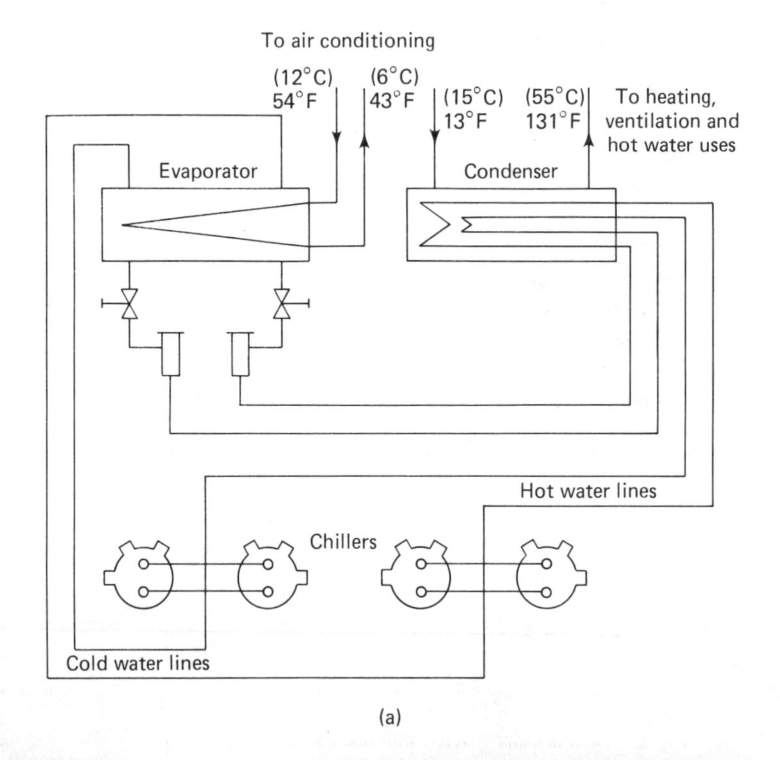

(a)

(b)

Figure 8.2. (a) Scheme for a heat pump using well water as a source for space and domestic hot water heat. (b) A photograph of the installation.

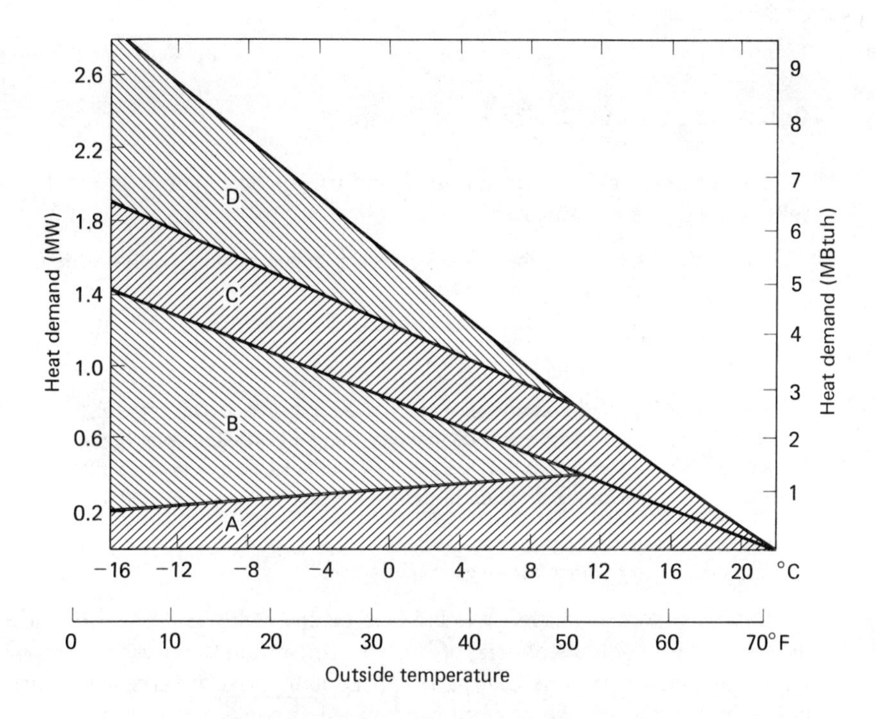

Figure 8.3. Heat demand as a function of the outside temperature during the day.

A + B: Heat demand of the radiators
C + D: Heat demand of the air-conditioning system
A + C: Heat demand covered with one heat pump
B + D: Heat demand covered with oil/gas firing

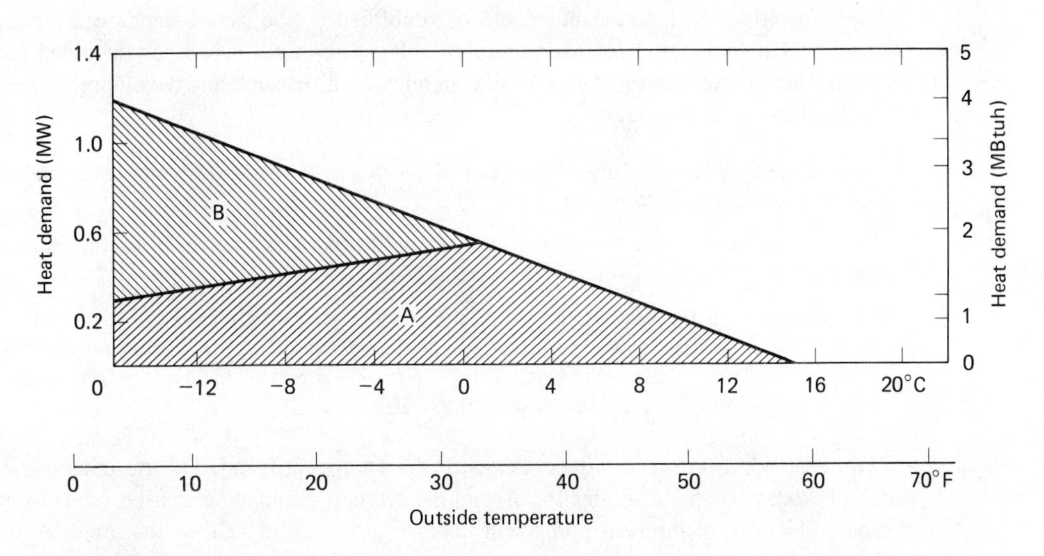

Figure 8.4. Heat demand as a function of the outside temperature during nights and weekends, using System 3.

A + B: Heat requirements for the radiators
A: Heat demand met by one 2.56 MBtuh (750 kW) heat pump
B: Heat demand covered by the use of supplementary oil/gas firing

Comments on the Use of Heat Pumps

This example shows that an economic use of heat pumps is possible today. The following criteria are important:

To avoid peak loads on the electrical system, supplementary heating with another source is required or heat storage should be used.

Whenever possible the refrigeration unit which is used for the air-conditioning system should be used as a heat pump.

The heat source, in this case well water, can be transported over some distance and still remain economical. In another case, a community using a heat pump was 1½ miles (2 to 3 km) from a river. With oil priced at $.45/gal. ($.12/ℓ) and electricity costing $.024/kWh, heating with a heat pump, including bringing in the river water, was economical (see Reference 2).

Industrial waste heat can be used if its quality is improved with heat pumps. The heat can be distributed also to other buildings in the region.

It is even more avantageous to use heat pumps with waste heat. Heat pumps can use the waste energy of transformers, of waste water (sewage), of wash water, or of industrial plants. In addition to these heat pumps, heat pumps that use renewable heat (e.g., river or lake water, the air, the ground) can be used as well.

The investment costs for a heat pump that makes use of a given heat source are in many cases considerable. The operating costs for medium- and large-size heat pumps, and hence the relative costs for the heat, will be lower than for small pumps. The economics for smaller heat pumps may not be competitive. Another problem with small heat pumps is that the control system is often inadequate because of cost reasons. (The cost of using heat pumps would be considerably lower if they could be mass produced in larger quantities, if maintenance could be simplified by the use of standardized components, and if they could be made more durable.) Therefore, it can be concluded that for small heat pumps there is still a significant amount of research and development work yet to be done by manufacturers.

Heat pumps may be especially advantageous in sports centers, particularly those that combine an ice rink and indoor swimming pools.

The internal heat sources present in larger buildings make ideal sources of energy for heat pumps.

The economics of the heat pump system also involves the distribution and end use of the energy. The lower the temperature produced, the larger must be the distribution and heat exchanger surfaces. A temperature of 120°F (50°C) can be used with increased radiator surface area or with ceiling or preferably floor heating systems.

The heat pump is not a wonder cure for the energy problem. On the other hand, especially today when substitutes for oil and gas are being sought, increased use is to be expected. The use of the heat pump will become more economical as the prices of oil and gas increase in relation to electrical energy prices. To optimize economical operation of a heat pump, however, it should be included in the design from the beginning.

8.2
THE ECONOMICS OF A SOLAR ENERGY SYSTEM FOR HEATING AND COOLING

The use of solar energy for heating, and if necessary cooling, of buildings today is technically possible. There are solar collectors available which are mass produced and which have a sufficiently high efficiency. The economic side of the operation will depend on the following factors:

First cost, required write-off period, interest rates, maintenance costs.

Amount of usable energy collected by the system.

Price inflation of the energy sources it replaces.

To improve efficiency, the investment and maintenance costs must be kept as low as possible and the lifespan of the system must be as long as possible. At the same time, the output of the system in terms of energy must be as high as possible. A reduction in the cost of solar equipment can be expected as sales increase and the economies achievable in large-scale mass production can be realized. The development of commercially viable solar energy will occur in three stages:

1. First Phase: Demonstration systems, financed by the government or by large corporations. Main objective is not the economics of the demonstration plant, but publicity, development of experience, and the support of development systems to obtain performance data. One example of such a demonstration is described below.
2. Second Phase: Commercial systems in areas where very advantageous climatic conditions favor the economics of solar energy. These areas will be enlarged as the prices for conventional energy sources increase. Solar energy will also be preferred to relieve certain environmental problems, and where supplies of conventional energy (electricity, gas, or oil) are relatively expensive or not sufficiently reliable.
3. Third Phase: General application of solar energy systems in areas with favorable climatic conditions. Solar energy can be found today in all three phases of development in different parts of the world with the corresponding conditions. For most of the industrial countries, however, widespread use of solar energy will require a significant reduction of investment costs for active solar energy systems or a significant increase in energy prices.

The maximum output of the system depends on the climatic conditions of the site, and also on the optimum use of the available solar radiation. One problem here is energy storage. Storage of large quantities of low-quality heat over long periods of time with today's technology increases the investment. In the example below, for economic reasons, the collected solar energy must be used within a relatively short period.

Thus for maximum use of the energy from the solar collectors, it should be used both for heating in winter and for refrigeration in the summer (the energy collected is used immediately). In addition, some amount of storage for the energy collected over the weekend is usually necessary to achieve economic operation. The collector area should be moderately sized so that the peak demand is covered with conventional systems using a fuel that can be stored on the premises.

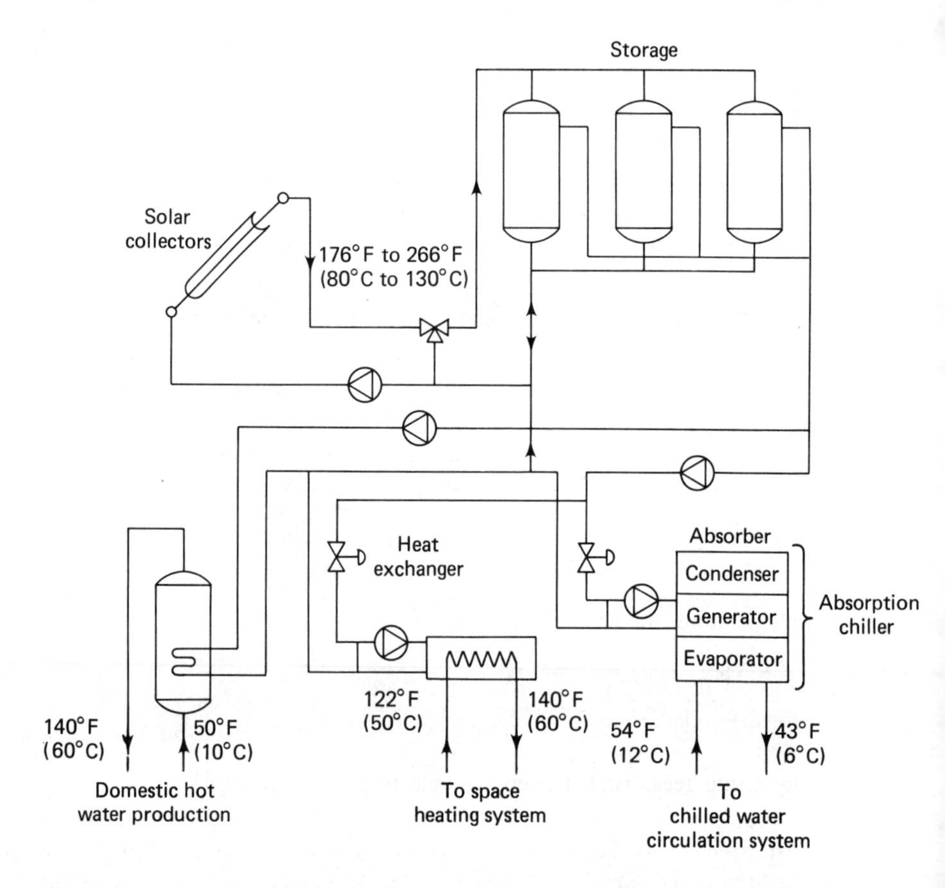

Figure 8.5. Schematic presentation of the solar energy system. There are 21,000 ft² (1,960 m²) of cylindrical-parabolic focusing collectors and three storage tanks, each with a volume of 12,000 gal (45 m³).

The following example shows a typical example for industry: a solar heating and cooling system for a manufacturing facility. This large-scale demonstration unit has been designed to supply part of the heat and cooling energy needed by a factory and office building in Rome. (5) The essential components of this industrial solar energy system are shown in Figure 8.5

The total collector area has an active surface of 21,100 ft² (1,960 m²) of which 4,950 ft² (460 m²) are installed on the roof of the factory and the rest are located on the roof of the covered parking lot.

The collectors have been developed especially for this project. They are cylindrical-parabolic collectors with sun-following controls. A special feature of these new collectors is their comparatively low price. They allow the production of heat at a high temperature level and with a reasonably good efficiency. This makes it possible to use a conventional design, absorption refrigeration unit, and therefore no new development work is necessary. Compared with flat collectors, the parabolic collectors have a higher-energy production and a more even distribution of the output water temperature during the course of the day. In addition, their higher output temperatures allow better use of storage.

Operation of the System

During the heating season, the water is heated to 175°F (80°C) and used in the radiators. During the cooling period, heated water at 240°F (115°C) is used in an absorption cooling machine. The absorption machine has a cooling capacity of 200 tons or 2.4 million Btuh (700 kW). The total cooling demand of the entire factory is 17.1 MBtuh (5,000 kW). The solar energy cooling produced by this machine is used during the spring to fall cooling season to compensate for the large internal heat sources in the laboratories and test facilities.

During the entire year, domestic hot water is heated with solar energy. Over the weekends, solar energy is collected and stored in the storage system. The microprocessor-based control system operates the pumps and valves, based on the availability of solar radiation and on the cooling, heating and domestic hot water demands, the stored water temperature, the day of the week, etc.

The investment for the solar energy system is $760,000. This value corresponds to $36 per ft^2 ($385 per m^2) of collector area, which breaks down to $18/ft^2 ($190 per m^2) for the collectors themselves, $15/ft^2 ($165/m^2) for the absorption refrigeration system, storage, and heat distribution, and $3/ft^2 ($30/m^2) for controls, building costs, and engineering fees. Had it been possible to plan this system at the time the building was

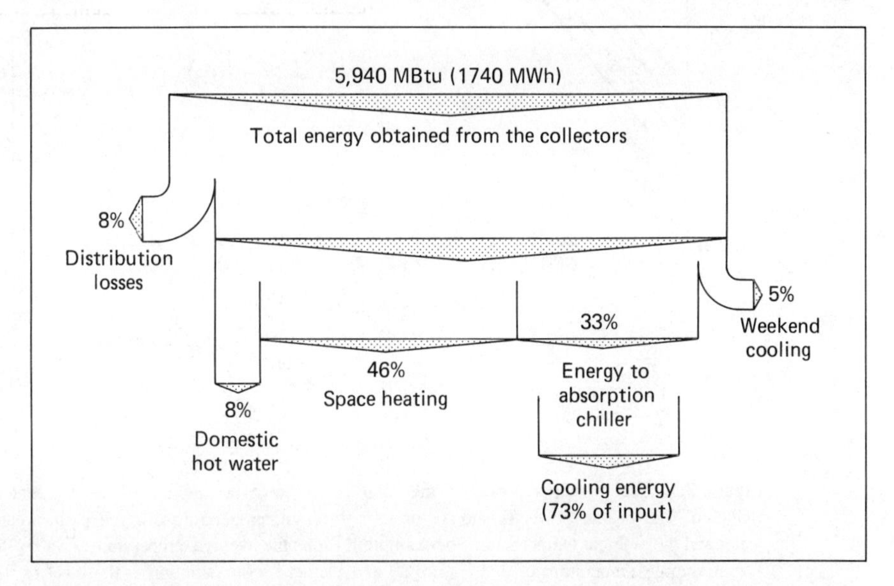

Figure 8.6. Distribution of the energy from the solar collectors during the course of the year. Ninety-two percent of the collected energy is utilized.

planned, the cost would have been $200,000 lower. The yearly energy flow is shown in Figure 8.6. The output of the solar energy system is:

Energy for cooling	1,430 MBtu/yr	(420 MWh/yr)
Energy for heating domestic hot water	480 MBtu/yr	(140 MWh/yr)
Energy for space heating	2,730 MBtu/yr	(800 MWh/yr)

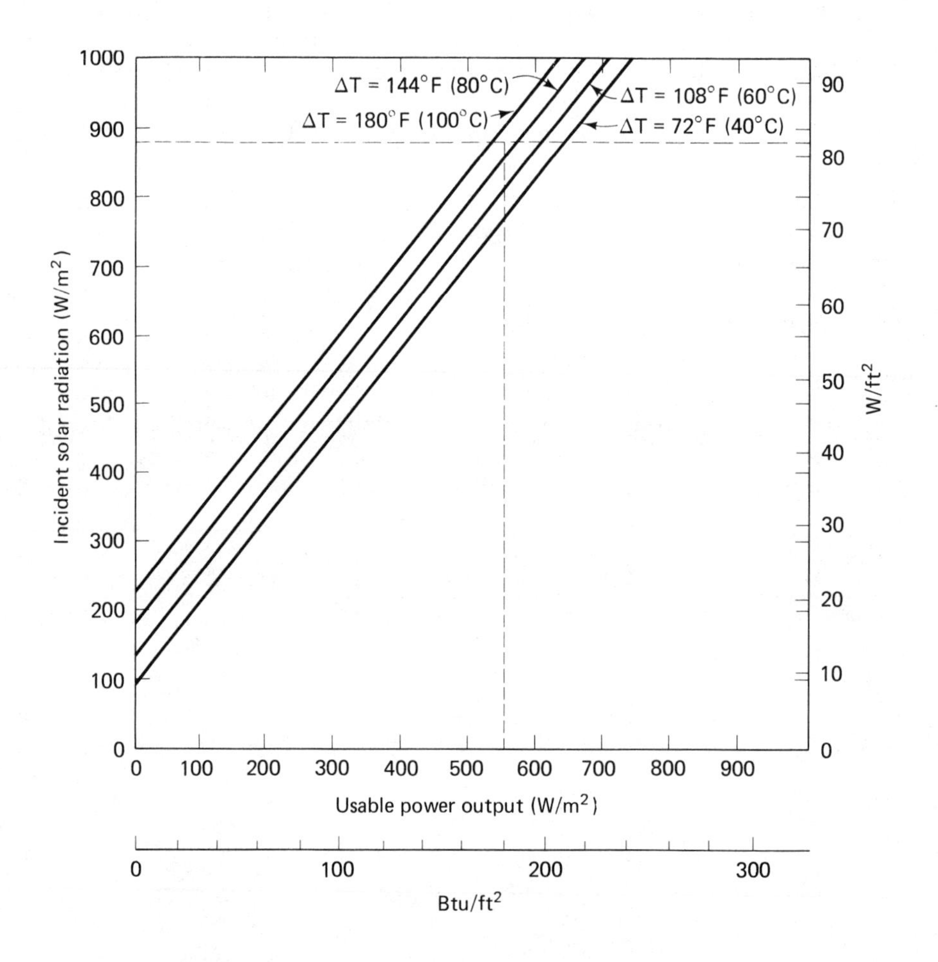

Figure 8.7. Usable output power of the solar collectors as a function of the incident solar radiation. The usable power is shown for four different temperature differences between the water and the ambient temperature. For example, in June the average temperature is 86°F (30°C). For an output temperature of 248°F (120°C) and incident solar radiation at the level of 82 W/ft^2(882 W/m^2), 175 Btu/ft^2 (550 W/m^2) of usable output can be obtained.

Table 8.3. Data Used for the Analysis of Available Solar Energy: Solar Radiation and Duration of Solar Radiation (5)

Month	ϕ (degrees)	p_E (Btuh ft^2)	p_E (W/m^2)	N (h/day)	n_1 (h/day)	n_2 (h/day)
January	33	240	756	9.4	4.3	4.1
February	25	260	819	10.4	4.7	4.1
March	14.5	277	873	11.8	6.6	5.1
April	2	285	900	13.2	7.0	4.8
May	−6	283	891	14.4	8.6	5.4
June	−11	280	882	15.0	9.4	5.7
July	−10	280	882	14.7	10.8	6.6
August	2.5	285	900	13.8	9.9	6.5
September	9	283	891	12.3	8.1	5.9
October	20	237	846	10.9	0.4	5.3
November	30	219	783	9.7	4.1	3.8
December	35	206	738	9.0	3.3	3.3

ϕ: The angle between the sun's radiation and a line perpendicular to the collector. The collectors are installed with an angle of 30° to the horizontal. The latitude of Rome is $47\frac{1}{2}°$.

p_E: Radiation incident on the collectors.

N: Theoretical maximum possible duration of solar radiation (in hours per day).

n_1: Measured duration of daily solar radiation including cloud cover (in hours per day).

n_2: Effective duration of solar radiation corrected for the shading of one element of the collectors by another.

Useful data for evaluating the performance of the system is given in the following figures and tables. Performance is given in Figure 8.7. Data showing the availability of solar radiation are given in Table 8.3. The collector system power conversion and the usable energy is shown in Figure 8.8.

The energy cost saving (see p. 418) in the first 10 years is \$615,000, using an initial price of \$0.064/kWh for electricity and \$0.45/gal (\$0.118/ℓ) for oil. It was assumed that there will be a 10 percent yearly increase in energy prices. Together with the investment costs of \$760,000, the write-off period of the demonstration unit is 18 years (see Figure 8.9). The write-off period of future systems with reduced investment costs will be shorter.

Summary

This demonstration example allows economical use of solar energy for heating and cooling. A conventional system for heating and cooling is necessary because the solar energy system supplies only 20–25 percent of the yearly energy use for heating and cooling. For future systems, the collector element might be integrated into the design of the roof of the factory, so that an additional cost reduction could be obtained.

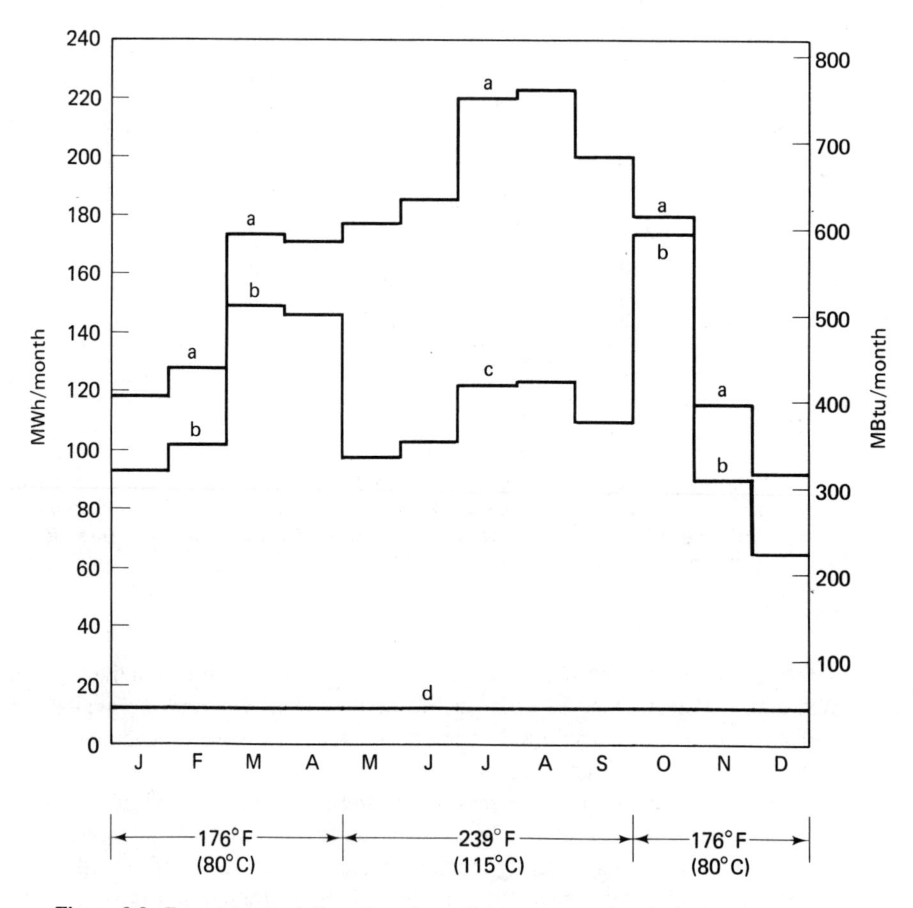

Figure 8.8. Energy received from the solar collectors compared with the energy used. The operating temperatures required 176°F (80°C) for heating, 239°F (115°C) in the summer for cooling, and 266°F (130°C) for cooling on the weekends, are shown for the corresponding months.

a: Collect energy, including the losses from the 36,000 gal (135m^3) storage.
b: Energy used for space heating. System losses are the difference between a and b.
c: Energy used for cooling. System losses are the difference between a and c.
d: Energy used for domestic hot water.

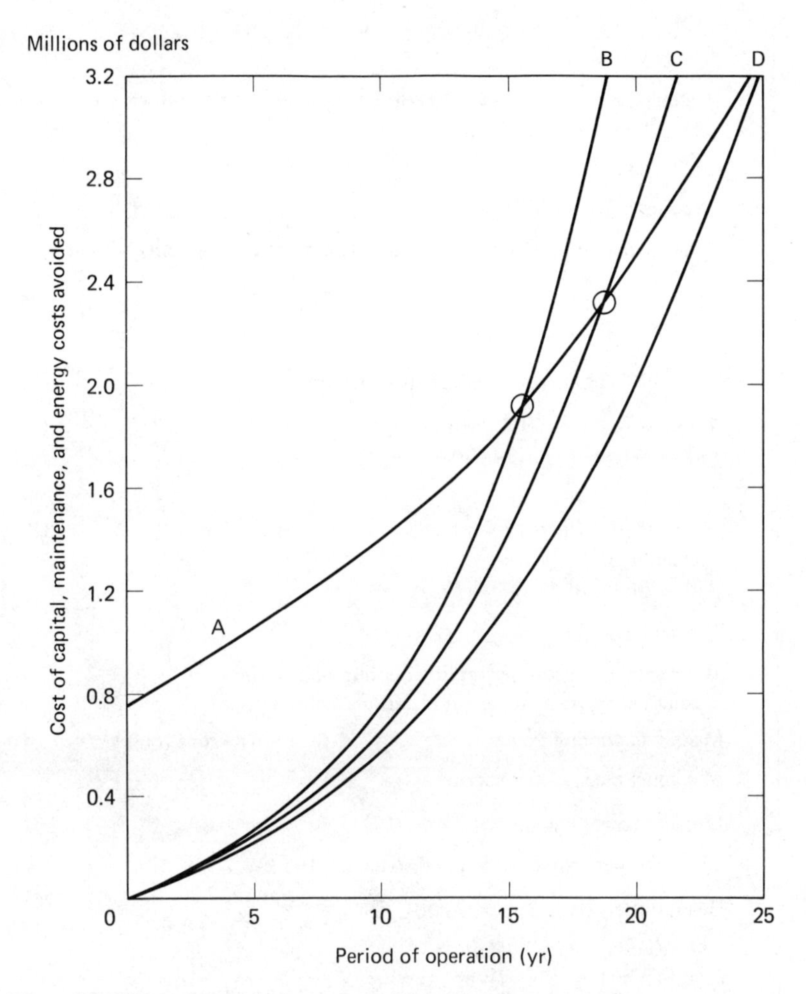

Figure 8.9. Years to return the investment of the solar energy system.

A: Annual expense for maintenance and capital retirement, including interest at 6% and amortization of the investment.

B: Value of the annual energy saved with interest at 6%, assuming an energy inflation rate of 13% yearly.

C: As in B, but with 10% annual energy inflation.

D: As in B, but with 7% annual energy inflation.

Analysis of Energy Cost Savings

1. *Energy prices*

   Heating: $4.70/$10^6$ Btu ($16/MWh) corresponding to an oil price of $0.45/gal ($0.118/ℓ) and a system efficiency of 75%

   Cooling: $19/MWh

   Electricity: $6.40/MWh

   Demand charge per month for electricity for the cooling unit, $2.40/kW.

2. *Energy cost saving*

   Value of solar energy used for space heating

   2,736 MBtu/yr × $4.70/MBtu
   (800 MWh/yr × $16/MWh) = $12,860/yr

   Value of solar energy used for air conditioning = $ 8,930/yr
   470 MWh/yr × $19/MWh

   Demand charge:

   (Cost saving because absorption cooling unit of the solar energy system has no electricity consumption)

Maximum cooling power	Q =	788 tons (660 kW)
Maximum electric motor power	P =	160 kW

   Demand power charge:

   $2.40/kW per month × 6 months/year × 160 kW = $ 2,305/yr

   Heating energy for domestic hot water

   480 MBtu/yr × $4.70/MBtu = $ 2,255/yr
   (140 MWh/yr × $16/MWh)

   Total energy costs = $26,350/yr

3. *Yearly yield of the energy cost saving*

   If a yearly increase of energy prices and the investment of energy cost savings are assumed, the yearly yield is calculated as follows:

   Year 1: Yield $K_1 = \$26{,}350$
   Year 2: Yield $K_2 = K_1 + a \times K_1 + b \times K_1$
   Year 3: Yield $K_3 = K_1 + 2a \times K_1 + b \times K_2$

Year 4: Yield $K_4 = K_1 + 3a \times K_1 + b \times K_3$

Year n: Yield $K_n = K_1 + (n - 1) \times a \times K_1 + b \times K_{(n-1)}$

Where a = rate of energy price inflation per year

b = interest rate on invested energy savings

The yearly yield is shown in Figure 8.9.

4. Capital cost

—Collector, installed	\$378,000
—Storage, absorption, cooling unit, domestic hot water boiler, heat exchanger for room heating, piping, control and regulation installed	\$328,000
—Measurements	\$ 6,000
—Additional building costs	\$ 8,000
—Engineering fee	\$ 40,000
Total capital cost	\$760,000

The return on the investment at an interest rate of 6 percent is shown in Figure 8.9. As shown, the return on the investment and of the energy cost savings are equal (with an assumed energy price increase of 13 percent yearly) after 15 years. This means the write-off period of the system is 15 years.

8.3 COMPUTER CENTER: ANALYSIS OF ENERGY USE AND METHODS FOR IMPROVEMENT

The air conditioning for present-day computer centers requires considerable energy, which in some cases is comparable to the energy used by the computers they cool. Computer centers have high internal heat loads of up to 160 Btu/ft^2 (500 W/m^2), compared to typical lighting loads of 2.3 to 4.6 W/ft^2 (25 to 50 W/m^2), and are often in operation 24 hours per day. In this section, an analysis of the energy use in two computer centers is described. First, the analysis of energy as used in the computer center of CERN (The European Center for Nuclear Research) (6) is discussed. A different method of analysis will then be described, which was used for the computer center of the Swiss Federal Technical Institute (ETH), in Zurich. Following this, energy-saving possibilities for newly designed air-conditioning systems in computer centers are discussed. If these possibilities are used, significant reductions in energy use may be expected. Some possibilities are: reduction of air flow rates, direct cooling with outside air, and the transferal of energy from heat sources to places where it is needed.

Energy Use for Air Conditioning the CERN Computer Center

CERN is located outside of Geneva, Switzerland, along the French border. The computing center, Building 513, consists of a 20-ft (6 m) high hall with an area of 22,400 ft^2 (2,080 m^2) where the computers are located, an operating center, and a storage room for tapes. Programming services are housed in an adjacent building.

The air-conditioning system was designed to satisfy extremely strict temperature, humidity, and air cleanliness conditions: dew-point temperature 50 to 54°F (10 to 12°C), dry bulb temperature 70.7 ± 1.8°F (21.5 ± 1°C), and 40 to 60 percent relative humidity. Heat can be removed from the computer hall at rates between 0 and 1,460 Btu/ft^2hr (0 and 4.6 kW/m^2). The cooling system was designed to remove an average load of 147 Btu/ft^2h (465 W/m^2).

Treated air is delivered and exhausted through ceiling ducts by ten air handling units each with a total name plate fan motor capacity of 44 kW. Each of the ten units can supply 50,000 cfm (85,000 m^3/h) of air. Two of the units supply outside air at the rate of 15,000 cfm (25,000 m^3/h). Between one and eight of these units is operated at any one time, according to the need for heat removal. Three Westinghouse centrifugal chillers, each of 370 tons (1.3 MW) cooling capacity, provide chilled water circulated at 760 gallons per minute (172 m^3/h). Their condenser heat is rejected through three roof-mounted cooling towers. A computer-controlled building automation system provides for 250 remote alarm points and 20 remotely metered functions at the main CERN control center.

The energy flows into and out of the building were determined by a team of people making nearly simultaneous measurements over a warm 24-hour period in July. The analysis included electrical energy used by the computers, cooling system, lighting, air and water flow rates and temperatures, room conditions, and the measurement of outside air conditions and the surface temperatures of this windowless, metal-clad building, to account for the solar loading.

The energy distribution of the CERN cooling system is shown in Figure 8.10. One of the results determined by the energy analysis is that the coefficient of performance (COP) of the entire air-conditioning system is only 1.2. This means that only 1.2 units of heat can be removed with one unit of electricity. The reason for this cannot be found in a faulty chiller; the COP of the compressor alone is above 4.7. It is due to the considerable energy used for pumps and (primarily) the fans. Consequently, even with free cooling (using cold outside air with the chiller off), the COP for heat removal is only 1.5.

The distribution of annual energy use for the CERN Computer Center is given in Figure 8.11. Some of the results from the energy analysis are presented in Table 8.4.

Energy Use for the ETH Computer Center

Next we describe a different method used to analyze energy use in the Swiss Federal Technical Institute (ETH) computer building. The computer center for the ETH has a CDC 6600 system housed in a modern, glass-faced building with 33,000 net square feet

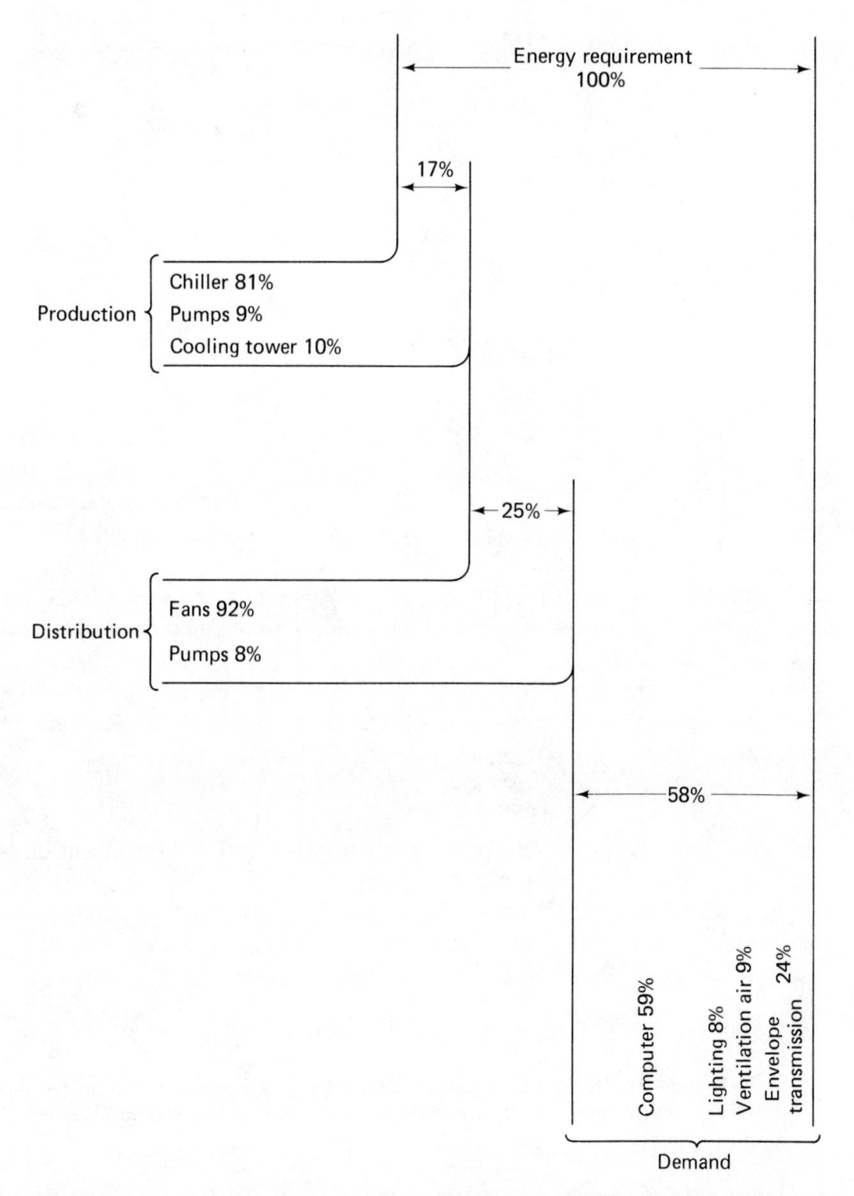

Figure 8.10. Energy flow in cooling system for Building 513 during measurement period of July 5.

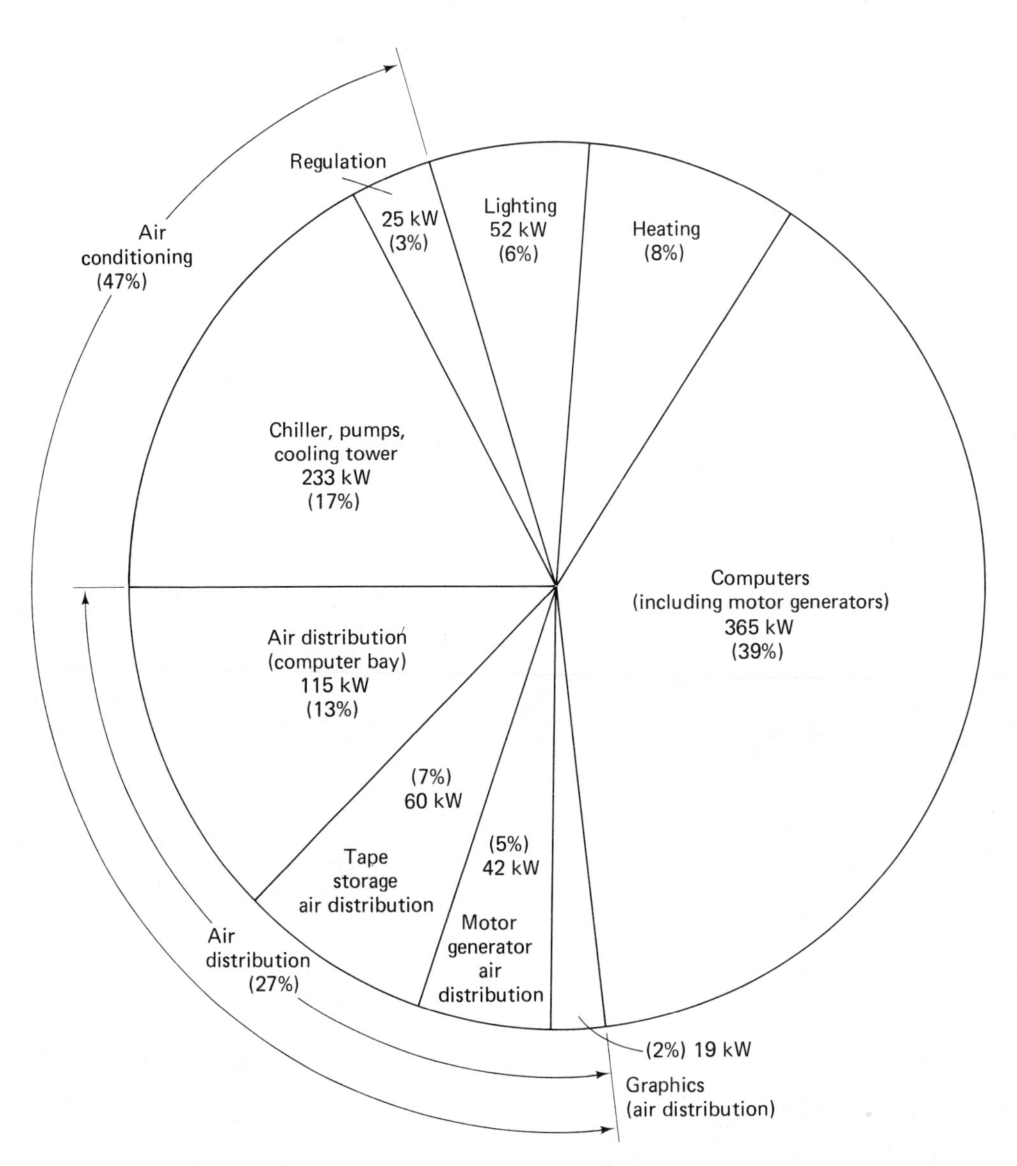

Figure 8.11. Distribution of annual energy use and average electric demand for Building 513 (Computer Center), CERN, European Organization for Nuclear Research, Geneva.

(3,065 m^2), most of which is used for office space. A recording power meter was connected to the input power lines. The electrical load due to different components within the building could be determined from the change in the meter reading during a period in which weather conditions varied (and while internal operating conditions remained constant). Then, during an interval of constant outside conditions, different pieces of equip-

Table 8.4. Coefficient-of-Performance Calculation, Building 513

Heat load to be removed

Computers	1.25 MBtuh	(365 kW)
Lighting	0.18 MBtuh	(52.2 kW)
Solar heat transmitted through building envelope	0.54 MBtuh	(158.6 kW)
Ventilation air (net)	0.15 MBtuh	(45.3 kW)
Total	2.12 MBtuh	(621.1 kW)

Power required by the air-conditioning system (most of which must itself be removed)

Chiller compressor	192 kW
Condenser pump	20.7 kW
Cooling tower fans	16.2 kW
Evaporator pump (chilled water circulation)	22.4 kW
Regulation	24.5 kW
Air distribution fans	235.9 kW
Total	511.7 kW

Operation with mechanical cooling

$$\text{COP} = \frac{\text{Computers} + \text{lighting} + \text{solar radiation and transmission}}{\text{Compressor} + \text{air distribution} + \text{pumps} + \text{cooling tower} + \text{regulation}}$$

$$= \frac{621.1\ \text{kW}}{511.7\ \text{kW}} = 1.21$$

Operation using cool outside air ("free" cooling)

$$\underset{\text{(free cooling)}}{\text{COP}} = \frac{\text{Computers} + \text{lighting} + \text{solar radiation and transmission}}{\text{Air distribution} + \text{pumps} + \text{regulation}}$$

$$= \frac{417.2\ \text{kW}}{282.8\ \text{kW}} = 1.47$$

ment were deactivated to determine their loading from changes on the chart recording. In this way the COP of the chiller could be determined.

For example, the chiller power required to remove heat from the mid-afternoon sun could be determined as the difference in load during the cool, very early morning load after correction for the lighting and other loads not connected at that time. As another example, the characteristic of the chiller as a function of its heat load could be determined after the chiller had been shut down long enough for the chilled water temperature to rise

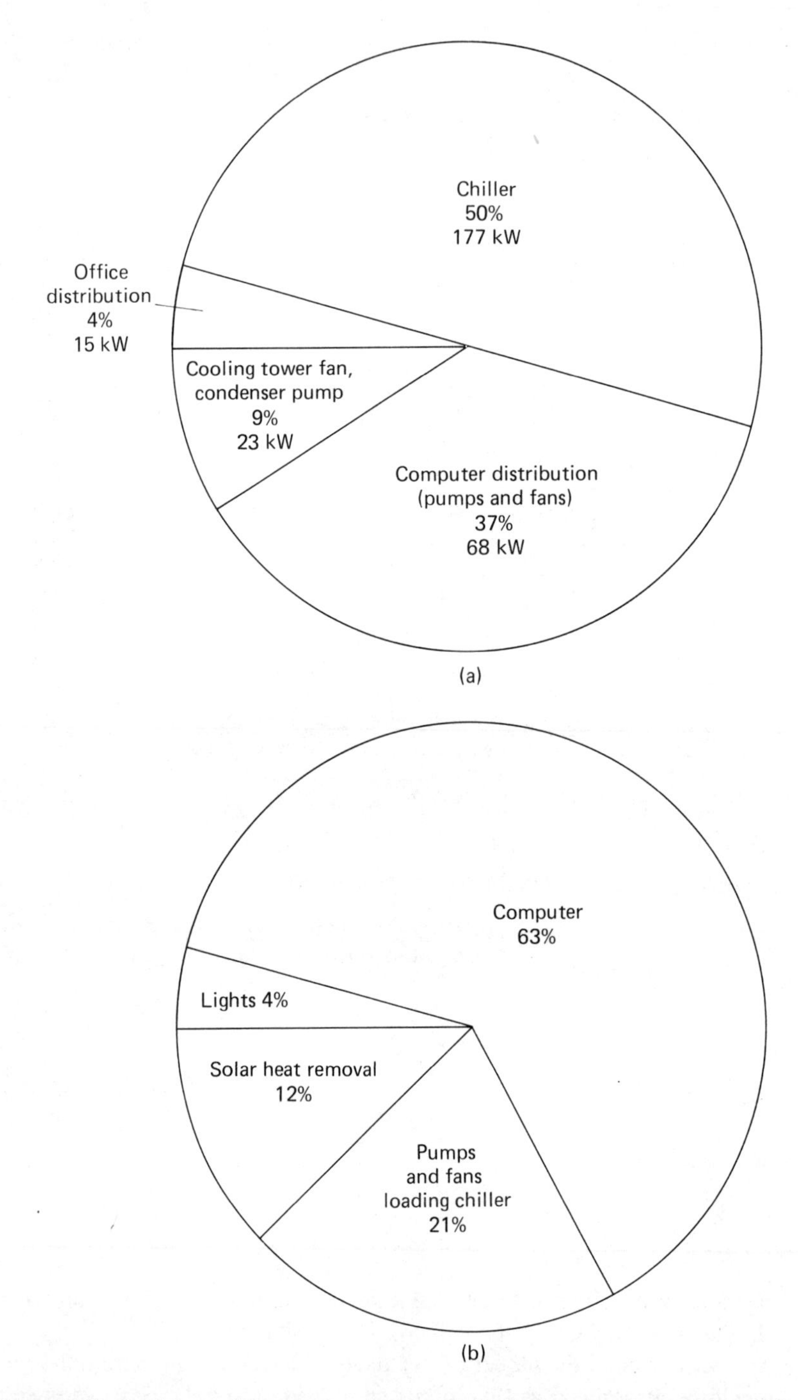

Figure 8.12. Breakdown of the annual production and distribution of cooling, and demand for cooling, for the ETH Computer Center. (a) Energy for cooling production and distribution, 1.4 $\times$ 10^6 kWh per year and average kW demand. (b) Cooling (heat removed) 8,360 MBtu (2.45 $\times$ 10^6 kWh) per year. The annual coefficient of performance of the chiller alone is typically 3.5. The coefficient of performance of the complete central air conditioning system is much lower: COP = (2.45 $\times$ 10^6 kWh)/(1.4 $\times$ 10^6 kWh = 1.75).

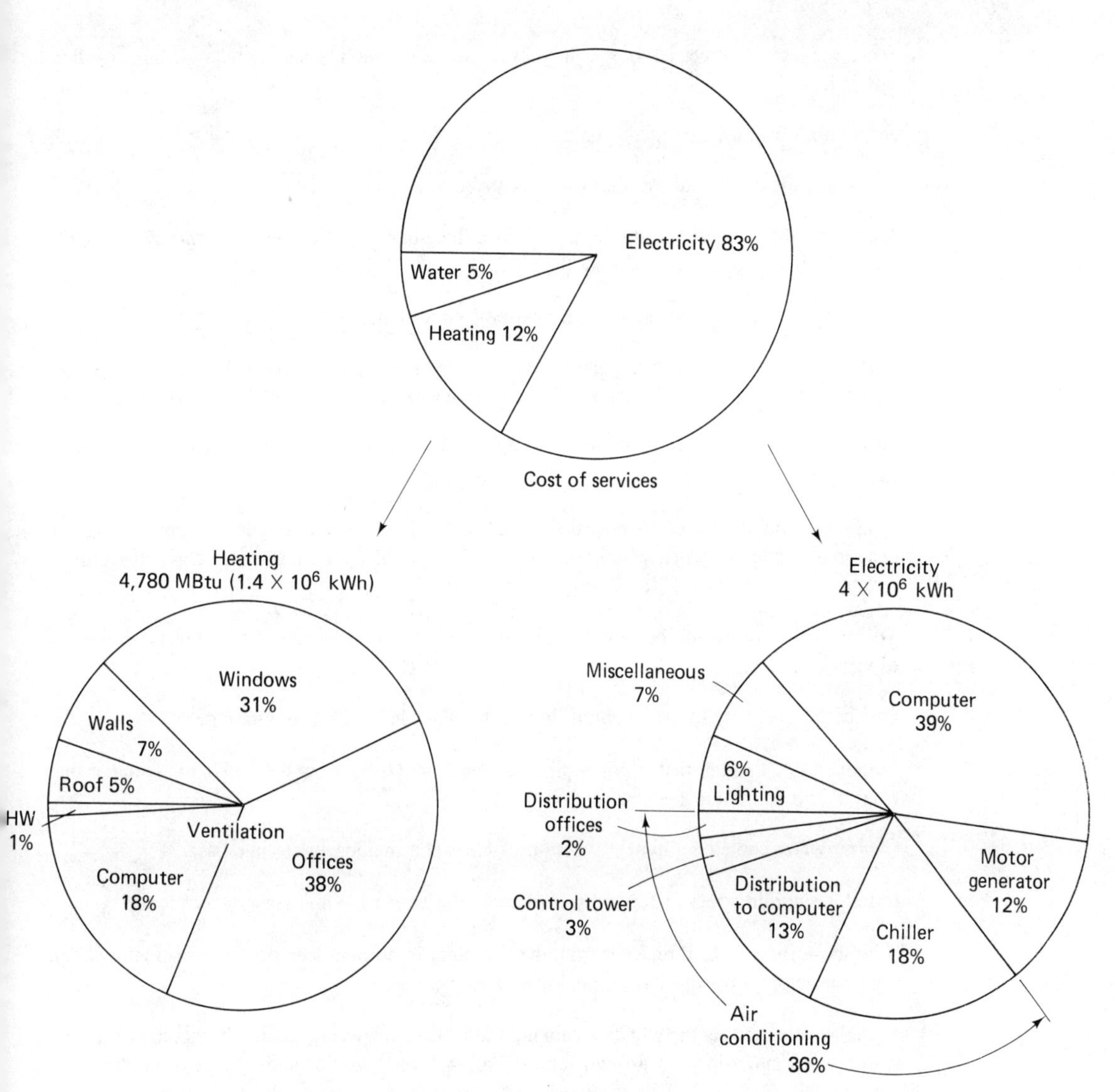

Figure 8.13. Initial distribution of annual energy use for ETH Computer Center.

to its upper limit. By recording the decrease in the water temperature after the chiller was again started, its COP under load could be determined. The power-factor-corrected load of other components could be determined from similar tests.

The components of the production and distribution of cooling, and of demand, are summarized in Figure 8.12. Annual energy use for heating and electricity is summarized in Figure 8.13.

Some of the possible energy-saving actions suggested as the result of these two studies:

Increased use of direct cooling with outside air, e.g., nights, automatic switching to direct cooling.

Switching off unused computer units.

Simplification of the duct system for air distribution.

Increasing the temperature difference between the supply and exhaust air from the computer hall, which would permit reducing the air flow rate.

Reducing the rate of outside air under extreme conditions.

Using exhaust air from the computer hall for storage ventilation and to cool the motor generators. (The CERN 50 Hz to 60 Hz motor generators have a capacity of 1.1 MW and run continuously.)

Using outside air at night and during the day over a greater period of the year for cooling the motor generators.

Using the cooling tower to provide chilled water for the water-cooled computer units by evaporative cooling when conditions permit. This would permit shutting down the chillers during the winter during "free" cooling period.

The power factor should be raised to above 90 percent with locally installed phase correcting capacitors.

Reduce the intensity of area lighting in the computer hall with task lighting.

Operate all equipment that is not required to run from 60 Hz from the 50 Hz lines, rather than from the 60 Hz motor generators.

Choose water-cooled equipment where possible when making future purchases.

Install a recording heat meter and record computer center kWh at least every month.

With the surplus of heat rejected from the computer in normal operation, in principal it should not be necessary to supply additional heat during the winter.

Humidification during the winter: steam humidification in the cool air supply should be replaced with a water spray in the warm air return. This will also provide some evaporative cooling.

Conclusions

The relatively low COP of the central air-conditioning system for the computer center is not unusual. Similarly low COP's may be found for the central air-conditioning systems of many office buildings. It is the consequence of using large volumes of air to transport heat. In fact, a window air conditioner, despite its faults, may have a higher COP as the result of the smaller air transport losses.

The COP of 1.2 would correspond to burning 7 units of oil in order to supply 1 unit for the service required (i.e., the operation of a computer) and then to remove the heat generated. This low efficiency carries with it a high cost for the service provided. If cooling is used to remove heat for a service that is itself inefficient or unnecessary, such as unused lighting, that is indeed throwing good money after bad.

Possibilities for Energy Saving in New Air-Conditioning Systems for Computer Systems

In the design of new air-conditioning systems for computers, there are many possibilities for the reduction of the energy used. Important factors for a computer center are related to the high dollar value of the equipment and its high energy dissipation per unit floor area, and the need for dependability, security, and a controlled environment. The requirements for computer HVAC systems differ from those designed for comfort air conditioning; the low occupancy results in there being little moisture released and very low outside air needs, and the concentration of equipment to be cooled in a small area results in much greater cooling loads per unit floor area. For efficient dust and humidity control the space must have a good vapor barrier and should be pressurized. Heat must be removed so that there are no hot spots; effective fire and smoke control should be planned. The reliability of the HVAC system should be cost-effectively balanced with that required of the computer service. Some energy saving methods are:

Use less stringent requirements for temperature and humidity tolerances of the room. The earlier requirements set by computer manufacturers were unnecessarily strict. Today, reduced requirements, e.g., air supply temperatures of 64°F (18°C) and even lower, room temperatures (in the lower part of the room) of 72 to 79°F (22 to 26°C), and 50 percent relative humidity (± 20 percent) may be acceptable in some cases.

The air flow rate and pattern should be arranged so that the exhaust air temperature can be raised to 82–86°F (28–30°C) (see discussion with Figures 8.14 and 8.15). This allows more opportunities to use the energy directly for heating elsewhere, and for the heating of the outside air in other air handling systems. With this method of design, the air flow to the occupied portion of the room must be separated from air flows for equipment cooling. With this, the air flow rate can be reduced by 40 percent, or more.

The ventilation system should be made up of multiple, smaller units. This allows the combination of individual units to supply the actual requirements. For example, only four of the ten CERN air handling units were in use during the above analysis because a large part of the computer system had been shut down.

The outside air should be used for direct cooling, whenever favorable, if the total cost, including that for air transport, would be less. It should be possible to use a maximum of up to almost 100 percent of outside air, depending on outside air conditions. The possibilities of using heat recovery and transporting the surplus heat to where it is needed should be included in the decision concerning the mode of operation. With occupancy reduced to a functional minimum, fresh air during hot conditions can be reduced to that required for building pressurization. (For simplicity of control, this air can be supplied by a small fan and filter so that the main fresh air dampers may be closed as tightly as they permit during hot weather.)

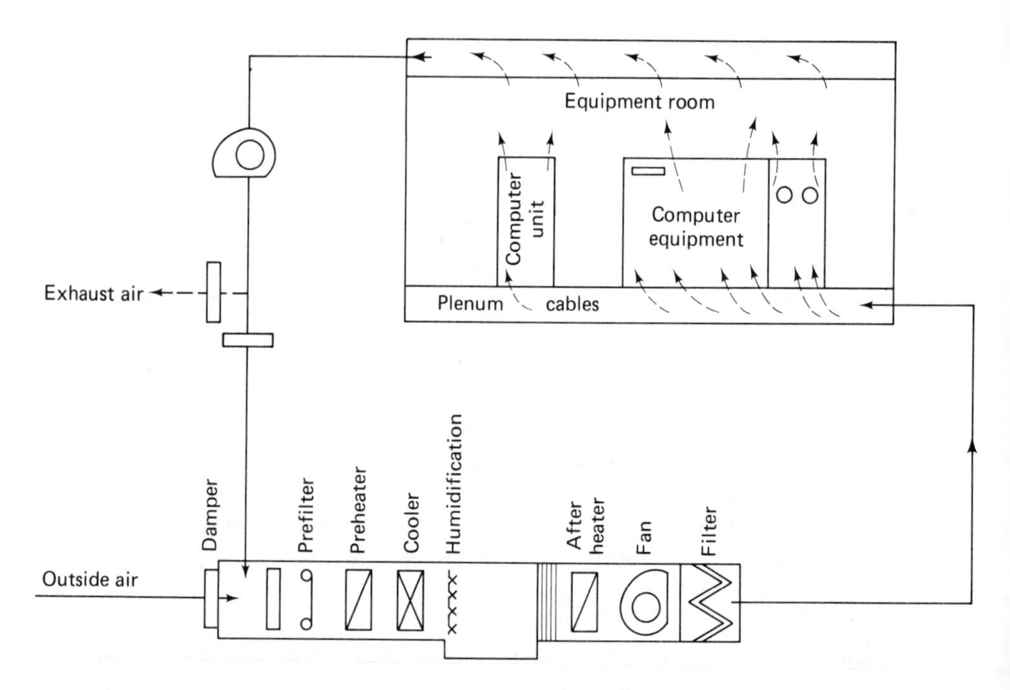

Figure 8.14. Air handling system for computer cooling.

Equipment that does not require cooling or close proximity to the computer area (power supplies, hard copy peripherals, motor generators, etc.) should be located outside of the space that requires a high quality environment. Repair work programming and all other activity that does not require hands-on operation of the computer should also be located outside of the computer area.

Windows are unnecessary for illumination, increase the cost of climate control, and are a weak element in the security of the building.

Other important factors to keep in mind when setting up an energy-saving program for a computer center stem from the need for security and reliability. Data must be protected in storage, and from loss or misuse. Particular attention should be given to minimizing fire hazard as well as possible water damage either from water entering from a source above the ceiling, or due to a break in one of the HVAC water lines. Cleanliness, freedom from dust, and security will be improved by better control over the entry of authorized operators into the computer areas. Energy-saving measures should be part of a comprehensive program of improved maintenance and operation that results in more reliable operation and better user satisfaction, as well as more efficient energy use. On the other hand, improvements should not be rejected for fear of upsetting a delicate system. Energy waste does not buy reliable operation.

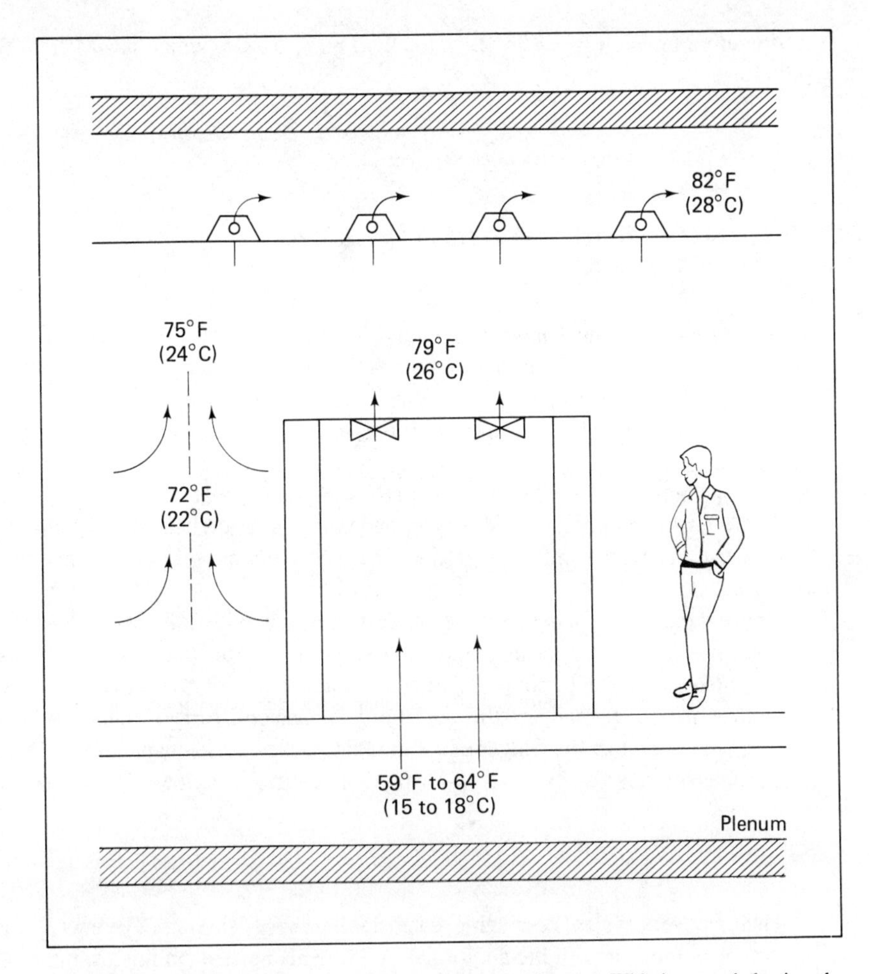

Figure 8.15. Air distribution for cooling computer equipment. With improved ducting the computer units serve as room air heaters. Cooling air can be introduced at the lowest temperature permitted by equipment humidity requirements. With the increased temperature difference, a smaller air flow is required with lower fan energy requirements.

Reuse of Computer Heat

Of course, every effort should be made to reduce the waste heat of computers. However, the waste heat from a computer center, similar to the heat rejected from other energy-intensive industrial processes, should be considered as a resource to be further exploited—as a heat source for other buildings in the neighborhood. By using the refrigeration unit in the heat pump mode, it is possible to raise the temperature of the rejected heat to a level sufficient for space heating. The condenser temperature of the refrigeration unit is 130°F (55°C) in the heat pump mode, which is sufficient for heating purposes.

8.4 A HOSPITAL: CALCULATION OF A REGENERATIVE HEAT RECOVERY SYSTEM

As an example for the analysis of energy recovery for heating and cooling using the exhaust air of an air-conditioning system as the energy source, a regenerative rotary heat exchanger has been selected. The regenerative system with a hygroscopic wheel is discussed in Chapter 6. The advantage of the hygroscopic wheel for the exchange of energy between exhaust air and fresh air is that it is efficient in the recovery of the enthalpy content of the air, or sensible heat as well as the latent heat.

In winter operation, the cold, dry outside air can be preheated and humidified using the exhaust air. With summer operation, the warm and humid outside air will be cooled and dried.

The example (7) is based on a modern hospital with 500 beds. The choice of a regenerative heat exchanger in a hospital was questioned during the design phase because of the potential for contamination. Because some mixing between exhaust and inlet air is possible in this exchanger, the possibility of bacteria carryover had to be analyzed. Experts concluded that the danger of contamination is nonexistent and that the regenerative heat exchanger can also be used in hospitals without any hygienic consequences.

Results

Heat recovery systems can achieve substantial energy savings. The energy cost savings have to be compared with the additional investments needed. In this example, 45 percent of the yearly heating and cooling energy could be saved with the recovery system (see Table 8.5).

The use of electricity by the HVAC system is increased by 17 percent. The total energy costs saving is $46,500 per year. The additional investment, according to Table 8.6, has been only $53,200 because the heat recovery had already been foreseen in the design phase. The additional investment has a short amortization period.

Calculation Method

The calculation performed for the heat recovery system has two purposes: a comparison of the economics of a system without and with heat recovery, and a comparison of the advantages of several different methods of heat recovery, for example a heat exchanger with heat pipes, plates, or a runaround heat exchanger, and optimization of the selected system.

Table 8.5. Energy Saving for a Hospital

Energy use per year	*Without heat recovery*[a]		*With heat recovery*	
Heating of air	10,663 MBtu	(3,125 MWh)	4,777 MBtu	(1,400 MWh)
Humidification	17,060 MBtu	(5,000 MWh)	9,997 MBtu	(2,930 MWh)
Cooling of air	1,706 MBtu	(500 MWh)	1,620 MBtu	(475 MWh)
Water	2.27×10^6 gal	(8,600 m^3)	1.37×10^6 gal	(5,000 m^3)
Electrical energy		960,000 kWh		1,130,000 kWh
Yearly Energy Costs[a]				
Heating of air	$42,000		$18,560	
Humidification	66,800		39,400	
Cooling of air	16,000		15,180	
Water	345		2,020	
Electrical energy	38,400		45,200	
Total	$163,545		$120,360	

[a]Energy prices: Heating energy $3.96/MBtu ($13.50/MWh)
Cooling energy $10.11/MBtu ($34.50/MWh)
Electrical energy $0.04 kWh
City water $1.50/1,000 gal ($0.40/m^3)

The Steps of Calculation

1. For the given heat recovery system, the total costs including capital costs, maintenance, and operating energy costs should be analyzed for at least three cases if possible.
2. The total costs should be evaluated as well for different types of heat recovery systems.
3. The decision on the best heat recovery system should be based on the total yearly costs and on its compatibility with the HVAC system and operating requirements.

For a regenerative heat exchanger, the calculation of the total yearly cost is shown in the following data for the hospital. An example of the cost saving with heat recovery

Table 8.6. Cost of the Heat Recovery System and Savings from the Reduced Capacity Required in the Air-Conditioning System with Heat Recovery

	Additional costs
Heat exchanger	$110,720
Air distribution[a]	18,000
Reduction in the size of the heating system	−44,560
Reduction in the size of the cooling system	−30,800
Additional investment for the heat recovery	$53,360

[a]The exhaust air duct and outside air duct have to be routed next to each other to allow the installation of the heat wheel.

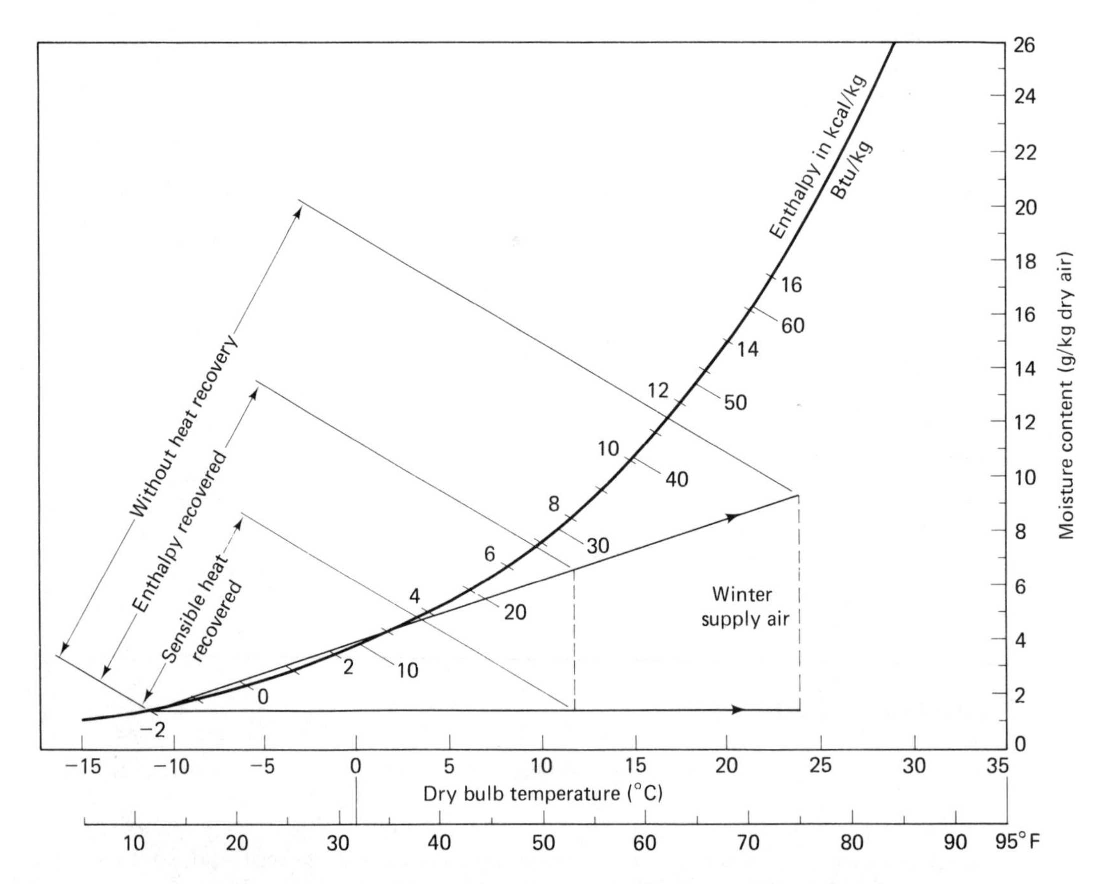

Figure 8.16. Example for the warming and humidification of outside air with and without heat recovery as represented on a psychrometric diagram (Molier diagram). At an outside air rate of 53,000 cfm (90,000 m³/h) the energy requirement of 6.14 MBtuh (1.8 MW) is reduced to 2.05 MBtuh (0.6 MW) with heat recovery.

is shown. This requires 53,000 cfm (90,000 m³/h) for 12 hours each day. The remaining annual operation is at one-half this ventilation rate.

The first step involves calculating the enthalpy* for conditioning the ventilation air—the energy for heating, cooling, and changing the water content of the air. On a psychrometric diagram such as shown in Figure 8.16, points representing the outside air and the supply air are located. For this example, we show 12°F (−11°C) at 90 percent relative humidity (RH), and 75°F (24°C) at 50 percent RH, respectively. The enthalpy representing the weather conditions at the site is plotted to obtain the distribution function such as shown in Figure 8.17.

*Enthalpy includes the energy required to change the temperature of the air plus the energy required to change the water content of the air. Enthalpy, therefore, is measured in the same units as energy.

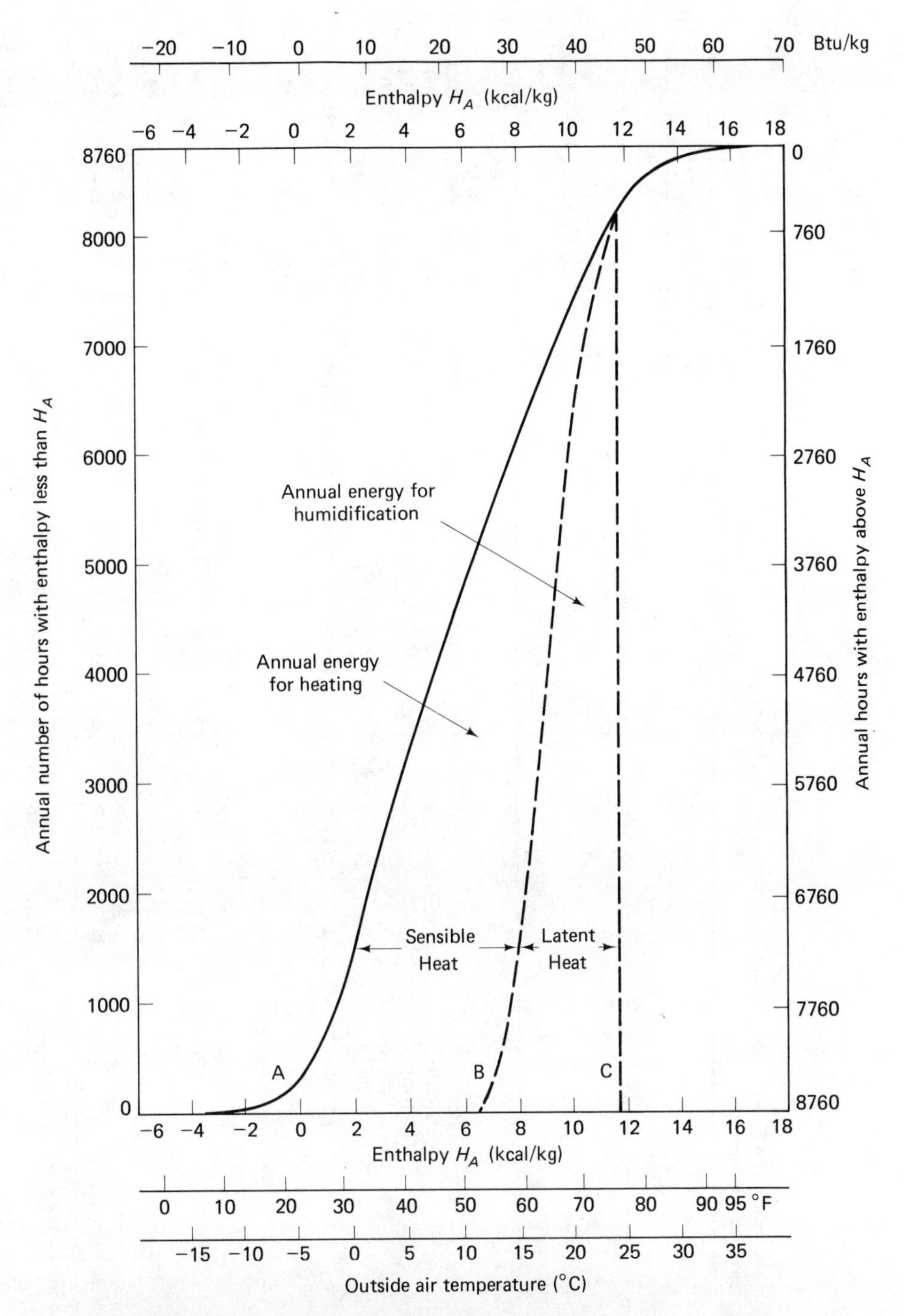

Figure 8.17. Distribution of the annual air enthalpy values H_A for Zurich. Energy for heating and humidification can be recovered for 8,000 hours each year.

A: Outside air conditions

B: Air conditions after heating

C: Room conditions after heating and humidification which correspond to the exhaust air conditions for this example

Table 8.7. Comparison of the Economics and Design Parameters for Several Heat Recovery Systems for a Hospital

		Runaround coils			Plate type	Heat pipe			Rotary wheel		
		I	*II*	*III*	*I*	*I*	*II*	*III*	*I*	*II*	*III*
	°F	50	55	64.4	64	50	63.7	68			
Δt with t_a = 14°F (−10°C)	(°C)	(10)	(13)	(18)	(17.8)	(10)	(17.6)	(20)			
η with t_a = 14°F (−10°C)	%	31	40	56	66	31	55	62	70	73	76
η with t_a = 50°F (+10°C)	%	31	40	56	60	28	50	56	70	73	76
Energy saved with heat recovery											
heating	MBtu/yr	498	648	908	972	460	802	921	168	1,774	1,853
	(MWh)	(146)	(190)	(266)	(285)	(135)	(235)	(270)	(492)	(520)	(543)
cooling	MBtu/yr	10	13	19	19	9	15	17	62	64	67
	(MWh)	(2.9)	(3.7)	(5.5)	(5.5)	(2.6)	(4.4)	(5.0)	(18.1)	(18.9)	(19.7)
water	gal/yr							64	460	67,100	70,000
	(m^3/yr)								(244)	(254)	(265)
(a) heating value	($23/MWh)	$3,350	$4,365	$6,115	$6,560	$3,110	$5,410	$6,200	$11,300	$11,970	$12,480
(a) cooling value	($34.50/MWh)	$100	$130	$170	$190	$90	$150	$170	$620	$650	$675
(a) water value	($0.30/$m^3$)								$70	$75	$75
Additional electricity use											
supply air	kW	1.2	1.7	2.75	3.75	4.1	4.1	4.1	4.75	3.8	3.35
exhaust air	kW	1.1	1.6	3.15	3.75	3.5	3.5	3.5	4.75	3.8	3.35
pumps	kW	0.6	0.8	1.0							
rotary heat exchange	kW								0.25	0.25	0.4
Total additional electricity	kW	2.9	4.1	6.9	7.5	7.6	7.6	7.6	9.75	7.85	7.1
(b) Total yearly electrical costs	$/yr	$275	$375	$630	$690	$695	$695	$695	$895	$720	$650
Credit for additional electricity use											
Annual MWH		2.8	4.0	6.4	8.9	9.7	9.7	9.7	11.1	8.9	8.0
(c) heat energy value	$/yr	$65	$90	$150	$205	$220	$220	$220	$255	$205	$185
Additional energy cost for cooling											
MBtu/yr	(MWh/yr)	2 (0.5)	2 (0.6)	3 (1.0)	5 (1.4)	5 (1.5)	5 (1.5)	5 (1.5)	5 (1.6)	4 (1.2)	4 (1.2)
(d)	$/yr	$15	$20	$35	$50	$50	$50	$50	$55	$40	$40

		Runaround coils			Plate type	Heat pipe			Rotary wheel		
		I	II	III	I	I	II	III	I	II	III
	°F	50	55	64.4	64	50	63.7	68			
Δt with t_a = 14°F (−10°C)	(°C)	(10)	(13)	(18)	(17.8)	(10)	(17.6)	(20)			
η with t_a = 14°F (−10°C)	%	31	40	56	66	31	55	62	70	73	76
η with t_a = 50°F (+10°C)	%	31	40	56	60	28	50	56	70	73	76
Capital costs											
Cost for heat recovery	$ × 10³	12	14.2	17.6	21.6	16.8	24	29.6	35.2	39.2	47.2
(e) annuity	$/yr	$1,945	$2,300	$2,850	$3,495	$2,720	$3,890	$4,800	$5,700	$6,350	$7,650
(n = 8 year P = 6%)											
a = 16.2%											
(f) Maintenance	$/yr						600	740	1,320	1,470	1,770
1%/yr (1.5% for heat wheel)	$/yr	$120	$140	$175	$215	$170	$240	$295	$530	$590	$710
(A) *Energy cost saving a − b + c − d*		$3,225	$4,190	$5,770	$6,215	$2,675	$5,035	$5,845	$11,295	$12,140	$12,725
(B) *Fixed costs e + f*		$2,065	$2,440	$3,025	$3,710	$2,890	$4,130	$5,095	$6,230	$6,940	$8,360
(C) *Annual dollar saving A − B*		$1,160	$1,750	$2,745	$2,505	−$215	$905	$750	$5,065	$5,200[a]	$4,365[b]

Note:

Energy demand without heat recovery			*Energy cost without heat recovery*		
Heating (net)	760 MBtu/yr	(810 MWh/yr)	Heating	$6.74/MBtu	($23/MWh) $18,650
Cooling	213 MBtu/yr	(64 MWh/yr)	Cooling	$10.11/MBtu	($34.50/MWh) $ 2,200
Ventilation	33.5 MWh/yr		Ventilation motors		($0.032/kWh) $ 1,135

[a]Heat recovery system with maximum return on investment. Net raw source energy saved 1,760 MBtu/yr (515 MWh/yr)

[b]Heat recovery system with maximum annual raw source energy savings of 1,860 MBtu/yr (545 MWh/yr)

The enthalpy for conditioning the outside air is the enthalpy difference between the fresh air and supply air. This is calculated using the air density, which is also given on the psychrometric diagram,* along with the moisture content per kg of air.

With heat recovery that recovers 66 percent of the enthalpy in the exhaust air, the fresh air is prewarmed to 54°F (12°C). Its moisture content is also increased from 1.3 grams of water per kg of outside air to 6.7 grams of water per kg of air. For a heat recovery device that responds only to a difference in temperature, i.e., just warming the outside air to 54°F (12°C), 35 percent less energy would be recovered than if part of the moisture in the exhaust air were also transferred. The enthalpy recoverable under different weather conditions is plotted in Figure 8.17 for the yearly distribution of operating conditions.

Additional electrical energy is required to operate the heat recovery system (for the pump or wheel, etc.) and to supply additional fan energy to overcome the increased resistance caused by the added heat exchanger surface in the ventilation ducts. Much of this electricity is converted into usable heat. Therefore a reduction can be made in the costs for heating. In summer, however, additional energy use for cooling is required as a consequence of the additional electrical energy used.

Maintenance

The yearly costs for maintenance, for example cleaning, depend on the type of heat recovery systems. Normally, 1 percent of the investment is assumed as yearly maintenance costs.

Filter Costs

With seriously contaminated or dirty exhaust air, for example from a foundry, a prefilter may be necessary before the heat exchanger. The costs of filter replacement have to be included in the maintenance costs. Some systems, for example the rotary heat exchanger, require filters, but the runaround coil heat exchanger is less critical in this respect. Also, cleaning equipment with sprays can sometimes be installed for the cleaning of plate exchanger.

In Table 8.7, the results of the calculation for several heat exchanger systems are presented.

8.5 AN ANALYSIS OF ENERGY USE IN A UNIVERSITY

Analysis of energy use in large buildings can be facilitated with the use of computer-based modeling programs such as BLAST-3 and DOE-2. (8,9) These programs are capable of approximating thermal characteristics of multiple path energy flows, heat storage, and complex HVAC control systems to obtain an approximate annual energy use based on

*For clarity, the density information has been omitted from Figure 8.16.

Table 8.8. Data for Heat Conduction Through the Building Envelope[a]

Building	*Administration (110)*		*Science (080)*	
Gross area	9,000 m^2	(97,000 ft^2)	27,600 m^2	(297,000 ft^2)
Net area	5,615 m^2	(60,400 ft^2)	13,200 m^2	(141,900 ft^2)
Wall area (opaque)	3,185 m^2	(34,250 ft^2)	10,300 m^2	(110,750 ft^2)
Average U-value	0.98 W/m^2°C	(0.173 Btuh°F ft^2)	2.4 W/m^2°C	(0.423 Btuh°F ft^2)
Wall heat conduction	3.12 kW/°C	(5,925 Btuh°F)	24.7 kW/°C	(46,850 Btuh°F)
Window	1,070 m^2	(11,500 ft^2)	1,275 m^2	(13,700 ft^2)
Single glass	400 m^2	(4,260 ft^2)	860 m^2	(9,210 ft^2)
Double glass	670 m^2	(7,215 ft^2)	415 m^2	(4,480 ft^2)
Window heat conduction	4.2 kW/°C	(7,935 Btuh°F)	7.7 kW°C	(14,570 Btuh°F)
Roof area	3,220 m^2	(34,650 ft^2)	6,975 m^2	(75,050 ft^2)
U-value	0.82 W/m^2°C	(0.144 Btuh°F ft^2)	0.68 W/m^2°C	(0.12 Btuh°F ft^2)
Roof heat conduction	2.6 kW/°C	(5,000 Btuh°F)	4.75 kW/°C	(9,000 Btuh°F)
Exposed floor area	2,700 m^2	(29,120 ft^2)	7,100 m^2	(76,320 ft^2)
U-value	0.96 W/m^2°C	(0.169 Btuh°F ft^2)	0.744 W/m^2°C	(0.131 Btuh°F ft^2)
Floor heat conduction	2.6 kW/°C	(4,925 Btuh°F)	5.3 kW/°C	(10,000 Btuh°F)
Total envelope heat conduction	12.6 kW/°C	(23,800 Btuh°F)	42.4 kW/°C	(80,450 Btuh°F)
Heat loss per degree day	300 kWh/°C-day	(5.7×10^5 Btu/°F-day)	1,000 kWh/°C-day	(1.9×10^6 Btu/°F-day)

[a]Details do not add due to rounding.

hourly data for solar radiation and other weather data, and occupant activities. Furthermore, building energy performance may be compared for many different configurations to enable the designer to select an optimum benefit/cost configuration.

For an existing building with a given design, calculation by hand may enable the energy analyst to obtain a better intuitive feeling for the thermal characteristics of the building. A hand calculation should also be part of any computer analysis in order to verify the methodology.

The analysis of energy use described below was done with the help of a desk calculator. Data are discussed for two buildings at the Harbor Campus of the University of Massachusetts. (10)

Together with weekly readings of the actual energy used, as measured by 29 kilowatt-hour meters, the analysis provided information on how energy was being used and how well the control systems were responding to the variables of weather and operating conditions. Although the buildings represent a large area, hand calculation was practical because of the relatively small amount of energy used for air conditioning, and the small variation in the internal loads during occupied hours.

Hand calculation is satisfactory when the variety of internal conditions to be analyzed

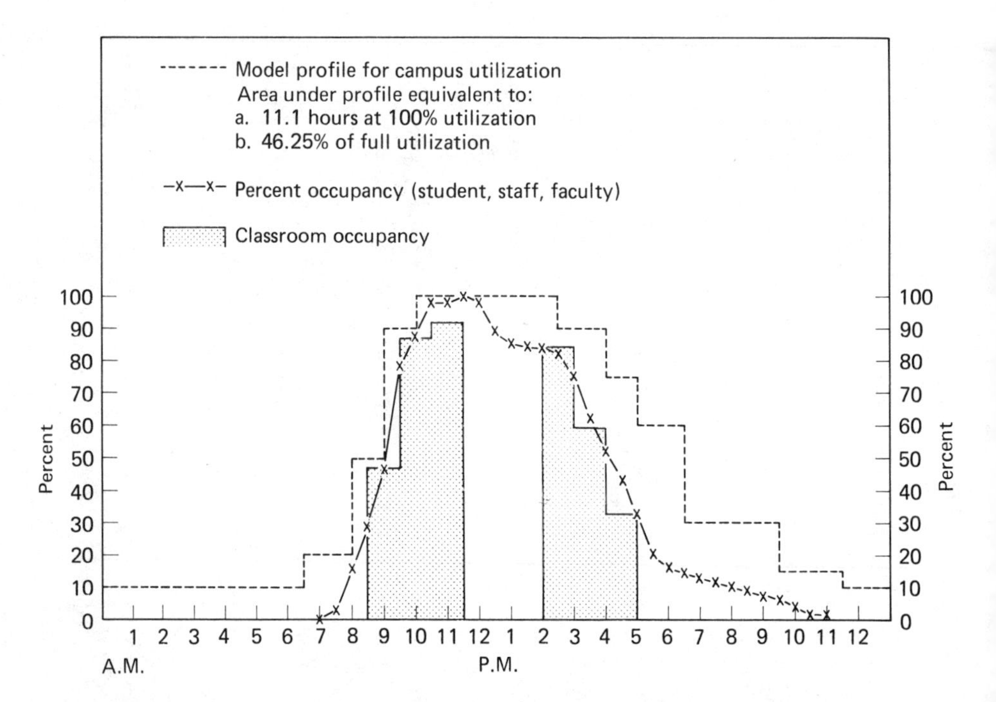

Figure 8.18. Classroom occupancy, population, and model profile for campus utilization and lights on, shortly after initial occupancy on the buildings. There are 11.1 hours of operation for a normal class day and 60.3 hours per week. Because of reading periods and exams, holidays and vacations, the weekly hours vary as below.

are small and when the effects of weather can be simplified. The quality of the final results depends more on the degree to which the data approximates the actual conditions and on the methods used than on the quantity of conditions analyzed or the device used for the calculation. In fact, the practical simplicity demanded by a manual analysis may be an important factor in tailoring the analysis to represent the actual condition.

Data used included the construction details of the envelope (different for each of the six buildings), together with the respective window and opaque area; the lighting wattage per unit area in lecture halls, classrooms, offices, and laboratories; data for the installed equipment; and traffic counts for building occupancy together with annual occupancy as modified by examination periods and vacations. The analysis covered the period of one year, starting five months after the initial occupancy. The results give the coefficients for demand as a function of occupancy and for heating demand as a function of the exterior temperature.

A summary of data for the heat conduction of the envelope is contained in Table 8.8. The model used for the occupancy profile is shown in Figure 8.18. The "lights-on" profile represents 11.1 hours per day at 100 percent equivalent operation. The number of week-to-week operating hours, shown in Figure 8.19, varies because of reading periods and examinations, holidays and vacations, and totals 2,772 hours for the year. Using the rates of energy use for items listed in Table 8.9, along with the weekly operating hours and weather data, the yearly energy use shown in Figure 8.20 is obtained.

The results show that adjustment of the control systems could be improved to regulate energy use according to changing conditions. Control of air flows and energy use are typical problems for new buildings. However, if a comparison of actual and expected energy use is not made, excessive energy use due to an inadequately regulated control system may continue indefinitely. Without an energy analysis, the administration would have no way of knowing that their operating costs were needlessly excessive. At the end of the analysis period shown in Figure 8.20, the control system of the administration

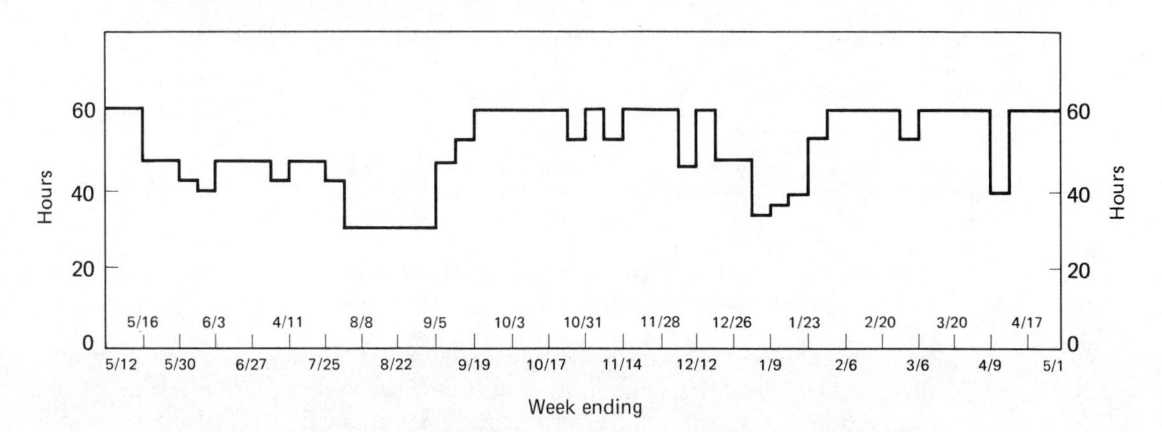

Figure 8.19. Weekly hours of University operation. There are 2325 hours during the heating season and 2772 operating hours during the year.

Table 8.9. Summary of the Calculated Components of Building Energy Compared with the Metered Kilowatt Hours[a]

Building	*Administration*		*Science*	
Number	*110*		*080*	
Gross area	*(97,023 ft²) 9,020 m²*		*(296,716 ft²) 27,575 m²*	
Net area	*(60,407 ft²) 5,615 m²*		*(141,887 ft²) 13,185 m²*	
Conductive heat loss	23,880 Btuh	(12.6 kW/°C)	79,620 Btuh°F	(42 kW/°C)
Ventilation: Fresh air	8,910 Btuh°F	(4.7 kW/°C)	26,540 Btuh°F	(14 kW/°C)
Fume hoods			54,970 Btuh°F	(29 kW/°C)
Season heat loss	4,435 MBtu	(1.3×10^6 kWh)	22,180 MBtu	(6.5×10^6 kWh)
Heat recovered from lights, etc.	−375 MBtu	(−(1.1×10^5 kWh))	−2,150 MBtu	(−(6.3×10^5 kWh
Heat trace cables	15 MBtu	(4.5×10^3 kWh)	330 MBtu	(9.7×10^4 kWh)
Domestic hot water	1,433 MBtu	(4.2×10^5 kWh)	989 MBtu	(2.9×10^5 kWh)
Total heat energy	5,493 MBtu	(1.61×10^6 kWh)	21,360 MBtu	(6.26×10^6 kWh)
Interior lights:				
Offices, classrooms, labs	3.3×10^5 kWh		1.6×10^6 kWh	
Halls	3×10^4 kWh		4.3×10^5 kWh	
Miscellaneous electric	9×10^4 kWh		5×10^5 kWh	
Total recoverable heat	(164 kW)		(903 kW)	
Exterior lighting				
Garage lighting	1.1×10^5 kWh		1.7×10^5 kWh	
Elevators	4×10^4 kWh		7×10^4 kWh	
Fume hood motors			5.1×10^5 kWh	
Fans, other motors	1.7×10^5 kWh		7.5×10^5 kWh	
Total lighting and mechanical cooling	7.7×10^5 kWh		4×10^6 kWh	
Total calculated energy	2.38×10^6 kWh		10.3×10^6 kWh	
Metered energy	2.22×10^6 kWh		14.1×10^6 kWh	
Percent of metered energy	107%		73%	

Notes:

(1) Fresh air rate is 0.25 cfm per gross ft² (4.6 m³/hr per gross m²) modulated according to the occupancy profile.

(2) During the analysis period there were 5,696°F-days (3,164°C-days)

(3) 2,325 hours during heating season.

[a]The calculation does not include cooling (it is separately metered for the total campus) nor is reheat used with cooling.

building (Building 110) was regulating energy use very close to the demand variables. However, the controls for the more complex energy systems in the science building (Building 080) were still not regulating energy use according to operating and weather variables. The introduction of energy-saving measures that resulted in significant reductions have been described in Chapter 7.

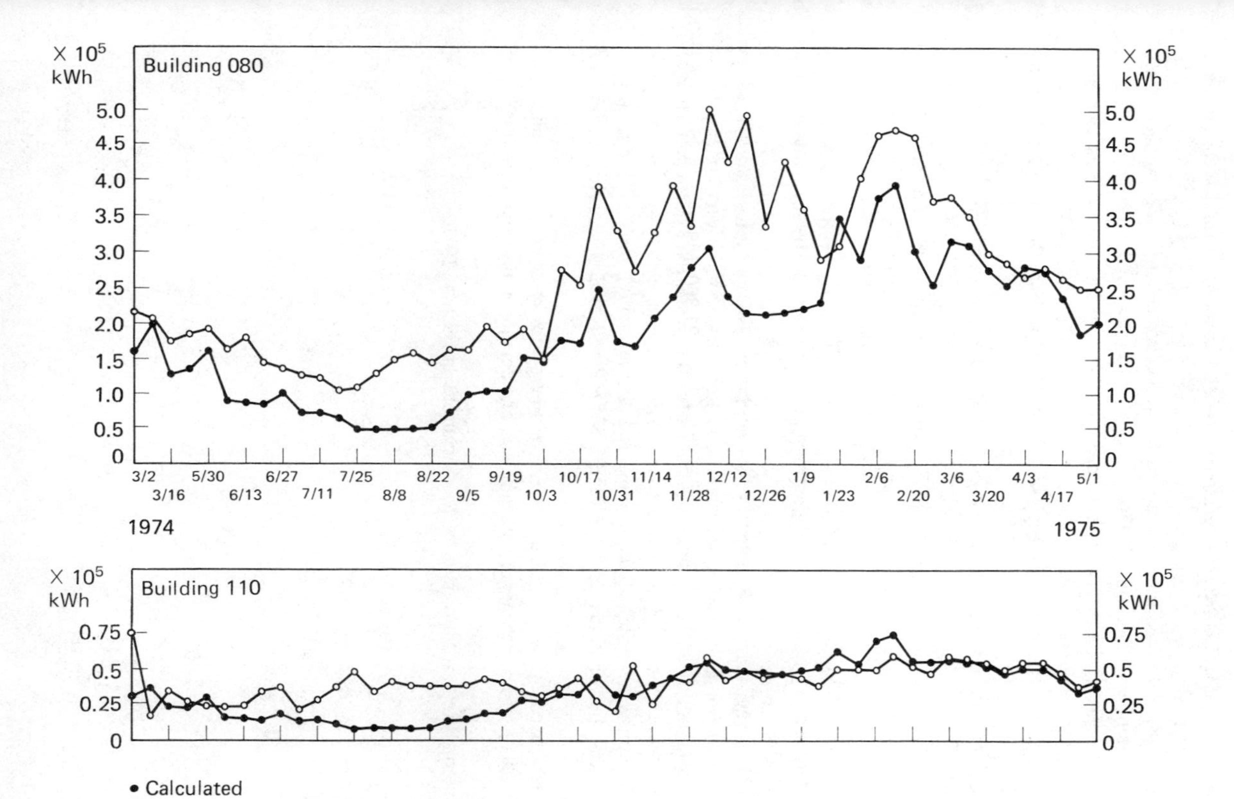

Figure 8.20. Energy use by the Science (080) and Administration (110) buildings between May 2, 1974, and May 1, 1975. The agreement between calculated and metered energy use improved during the last six months of operation shown for Building 110. The control system for Building 080 functioned more erratically. Control systems for new buildings typically operate erratically following the start-up of a new building. Unless their energy performance is examined, they may continue using excessive energy for their entire operation.

REFERENCES

(1) M. Kiss, H. Mahon, and H. Leimer, *Energiesparen Jetzt!* (Bauverlag, Berlin, West Germany, 1978), pp 255–264.

(2) Studienabteilung des Eidgenössischen Institutes für Reaktorforschung, Würenlingen, "Mechanische Wärmepumpen," December 1976.

(3) M. Kiss, H. Mahon, and H. Leimer, *Einergiesparen Jetzt!* (Bauverlag, Berlin, West Germany, 1980), and reference 3 therein, p. 300.

(4) M. Kiss, H. Mahon, and H. Leimer, *Einergiesparen Jetzt!* (Bauverlag, Berlin, West Germany, 1980), p. 219.

(5) M. Kiss, H. Mahon, and H. Leimer, *Einergiesparen Jetzt!* (Bauverlag, Berlin, West Germany, 1980), pp. 270–278.

(6) The authors are indebted to the CERN SB Facilities Group under the direction of Pierre LaPorte and for the assistance of P. Ciriani and V. Dany in obtaining the data presented in the technical report "Balance Sheet of all Operating Energies in Building 513 (Computer Center)." (5 October 1976).

(7) M. Kiss, H. Mahon, and H. Leimer, *Einergiesparen Jetzt!* (Bauverlag, Berlin, West Germany, 1980), pp. 291–297.

(8) BLAST-3 (1981) is revision 3 of Building Loads and System Thermodynamics developed by U.S. Army Construction Engineering Research Laboratory. BLAST is a very comprehensive code for estimating hourly space heating and cooling requirements, fan system performance, performance of conventional HVAC, and active and passive solar energy systems. Magnetic tapes with program, weather data, sample runs, and documentation are available from U.S. National Technical Information Service (NTIS), 5285 Port Royal Road, Springfield, VA 22150.

(9) DOE-2 was developed by the U.S. Department of Energy for research, design, and code-compliance evaluation. Program tape, weather data, and documentation available from NTIS as in Reference 8.

(10) H. Mahon, Internal Report, University of Massachusetts at Boston, 1975.

Appendix 1: Definitions and Conversion Factors

ENERGY

1 Btu = 2.931×10^{-4}kWh = 0.2520 kcal = 1054 J

1 kWh = 3,412 Btu = 859.8 kcal = 3,600 kJ

1 kcal = 3.968 Btu = 1.163×10^{-3}kWh = 4,186 J

1 MBtu = 1 million Btu

1 MWh = 1 million Wh

1 Therm = 100,000 Btu

POWER

1 kw = 3,412 Btuh = 859.8 kcal/h

1 Btuh = 2.931×10^{-4}kW = 0.252 kcal/h

1 horsepower (hp) = 0.7457 kW = 2,544 Btuh

1 boiler hp = 33,470 Btuh = 9.81 kW

1 ton of refrigeration = 12,000 Btuh* = 3.517 kW*

*Rate of heat transfer.

1 kW = 1,000 watts

1 MW = 1 million watts

1 GW = 1 Gigawatt = 1 million kilowatts

SPECIFIC CAPITAL PRODUCTIVITY

\$1/kW = \$71/BPDE* = \$293/MBtuh

\$1000/BPDE* = \$14/kW = \$0.47/MBtu/yr

\$1000/MBtuh = \$3.40/kW = \$242/BPDE*

HEAT–TRANSFER COEFFICIENT (U-VALUE)

$1\ \text{Btu/ft}^2\text{h°F} = 5.678\ \text{W/°cm}^2 = 4.882\ \text{kcal/m}^2\text{h°C}$

$1\ \text{W/°cm}^2 = 0.1761\ \text{Btu/ft}^2\text{h°F} = 0.8598\ \text{kcal/m}^2\text{h°C}$

THERMAL CONDUCTIVITY

$$1\ \text{Btuh/ft°F} = 0.01730\ \text{W/cm°C} = \frac{1.488\ \text{kcal/h}}{\text{m°C}} = \frac{4.134 \times 10^{-3}\ \text{cal}}{\text{sec cm°C}}$$

$$1\ \frac{\text{W}}{\text{m°C}} = \frac{0.5782\ \text{Btuh}}{\text{ft°F}} = \frac{0.239\ \text{cal}}{\text{sec cm°C}}$$

PRESSURE

1 atm = 1 Atmosphere = 14.7 lb/in.^{2} = 29.92 inches head of mercury = 101.325 kpascal

1 lb/in.^{2} = 2.04 inches head of mercury = 27.72 inches head of water = 6.8948 kpascal

1000 pascal = 1 k pascal = 0.09869 atm = 0.14504 lb/in^2

FLOW

m^3/h = 0.58858 cu ft/min (cfm)

1 cfm = 1.6990 m^3/h

*Barrels of oil per day energy equivalent.

VOLUME

$$1 \text{ cu m } (m^3) = 1000\ \ell = 35.31 \text{ cu ft} = 264 \text{ U.S. gallons} = 6.29 \text{ barrels}$$

$$1 \text{ cu ft.} = 2.83 \times 10^{-2}\ m^3 = 7.48 \text{ U.S. gallons} = 0.178 \text{ barrels}$$

$$1 \text{ U.S. gallon} = 3.79 \times 10^{-3}\ m^3 = 0.134 \text{ cu ft} = 0.0238 \text{ barrels}$$

$$1 \text{ barrel (bbl)} = 0.15898\ m^3 = 5.614 \text{ cu ft} = 42 \text{ U.S. gallons}$$

SURFACE AREA

$$1\ m^2 = 10.764\ ft^2 \text{ (square feet)}$$

$$1\ ft^2 = 0.0929\ m^2$$

LENGTH

$$1 \text{ m} = 3.281 \text{ ft} = 1.0936 \text{ yd}$$

$$1 \text{ ft} = 0.3048 \text{ m}$$

$$1 \text{ yd} = 0.9144 \text{ m}$$

TEMPERATURE

$$\text{Temperature in °C} = \frac{5}{9} \times (\text{temperature in °F} - 32\text{°F})$$

$$\text{Temperature in °F} = \frac{9}{5} \times (\text{temperature in °C}) + 32\text{°F}$$

Example: 20°C = 68°F

ABSOLUTE TEMPERATURE SCALE

Temperature in degrees Kelvin (K) = temperature in °C + 273.15

Example: 100°C = 373.15 K

Temperature in degrees Rankine = temperature in °F + 459.67

Example: 32°F = 491.67 °R

DEGREE DAYS

There is 1 degree-day for each °F the mean daily temperature is below 65°F. Thus the U.S. mean heating season of 212 days duration and with 5,156°F-days (2,864°C-days) corresponds to an average temperature of 40.7°F or 4.8°C.

Appendix 2: Nominal Energy Content of Fuels

NOMINAL ENERGY CONTENT OF FUELS AND OTHER CONVERSION FACTORS ACCORDING TO THE U.S. DEPARTMENT OF ENERGY[a]

Natural gas	1021 Btu/ft^3 (10.57 kWh/m^3)
Distillate fuel oil (No. 2)	5.825 × 10^6 Btu/bbl (1707 kWh/bbl)
Residual fuel oil (No. 6)	6.287 × 10^6 Btu/bbl (1843 kWh/bbl)
Bituminous coal and lignite	25.970 × 10^6 Btu/ton (7610 kWh/1000 kg)
Anthracite	24.69 × 10^6 Btu/ton (7235 kWh/1000 kg)
Gasoline	5.253 × 10^6 Btu/bbl (1540 kWh/bbl)
Kerosene	5.67 × 10^6 Btu/bbl (1662 kWh/bbl)
Crude oil:	
Energy content	5.8 × 10^6 Btu/bbl (1700 kWh/bbl)
1 metric ton (7.33 bbls)	4.251 × 10^7 Btu/1000 kg (12,460 kWh/1000 kg)
1 barrel per day equivalent	70.81 kW or 0.2417 MBtuh
Hydropower	10,435 Btu per kWh of electricity
Nuclear power	10,770 Btu per kWh of electricity

[a]1 barrel (bbl) contains 42 gallons or 159 liters.
1 ton is 2,000 pounds (lb) or 907.18 kg.
1,000 kg is 2,204.62 lb.
SOURCE: U.S. Energy Information Administration, Washington, D.C.

Appendix 3: Tables

Table 1: Present Values as a Function of Discount Rate and Year

Number of years	Discount rate									
	4.500%	6.000%	8.000%	10.000%	12.000%	15.000%	20.000%	25.000%	30.000%	35.000%
1.	.957	.943	.926	.909	.893	.870	.833	.800	.769	.741
2.	.916	.890	.857	.826	.797	.756	.694	.640	.592	.549
3.	.876	.840	.794	.751	.712	.658	.579	.512	.455	.406
4.	.839	.792	.735	.683	.636	.572	.482	.410	.350	.301
5.	.802	.747	.681	.621	.567	.497	.402	.328	.269	.223
6.	.768	.705	.630	.564	.507	.432	.335	.262	.207	.165
7.	.735	.665	.583	.513	.452	.376	.279	.210	.159	.122
8.	.703	.627	.540	.467	.404	.327	.233	.168	.123	.091
9.	.673	.592	.500	.424	.361	.284	.194	.134	.094	.067
10.	.644	.558	.463	.386	.322	.247	.162	.107	.073	.050
11.	.616	.527	.429	.350	.287	.215	.135	.086	.056	.037
12.	.590	.497	.397	.319	.257	.187	.112	.069	.043	.027
13.	.564	.469	.368	.290	.229	.163	.093	.055	.033	.020
14.	.540	.442	.340	.263	.205	.141	.078	.044	.025	.015
15.	.517	.417	.315	.239	.183	.123	.065	.035	.020	.011
16.	.494	.394	.292	.218	.163	.107	.054	.028	.015	.008
17.	.473	.371	.270	.198	.146	.093	.045	.023	.012	.006
18.	.453	.350	.250	.180	.130	.081	.038	.018	.009	.005
19.	.433	.331	.232	.164	.116	.070	.031	.014	.007	.003
20.	.415	.312	.215	.149	.104	.061	.026	.012	.005	.002
21.	.397	.294	.199	.135	.093	.053	.022	.009	.004	.002
22.	.380	.278	.184	.123	.083	.046	.018	.007	.003	.001
23.	.363	.262	.170	.112	.074	.040	.015	.006	.002	.001
24.	.348	.247	.158	.102	.066	.035	.013	.005	.002	.001
25.	.333	.233	.146	.092	.059	.030	.010	.004	.001	.001
26.	.318	.220	.135	.084	.053	.026	.009	.003	.001	.000
27.	.305	.207	.125	.076	.047	.023	.007	.002	.001	.000
28.	.292	.196	.116	.069	.042	.020	.006	.002	.001	.000
29.	.279	.185	.107	.063	.037	.017	.005	.002	.000	.000
30.	.267	.174	.099	.057	.033	.015	.004	.001	.000	.000

Table 2: Annuities as a Function of Discount Rate and Year

Number of years	Discount rate									
	4.500%	*6.000%*	*8.000%*	*10.000%*	*12.000%*	*15.000%*	*20.000%*	*25.000%*	*30.000%*	*35.000%*
1.	.957	.943	.926	.909	.893	.870	.833	.800	.769	.741
2.	1.873	1.833	1.783	1.736	1.690	1.626	1.528	1.440	1.361	1.289
3.	2.749	2.673	2.577	2.487	2.402	2.283	2.106	1.952	1.816	1.696
4.	3.588	3.465	3.312	3.170	3.037	2.855	2.589	2.362	2.166	1.997
5.	4.390	4.212	3.993	3.791	3.605	3.352	2.991	2.689	2.436	2.220
6.	5.158	4.917	4.623	4.355	4.111	3.784	3.326	2.951	2.643	2.385
7.	5.893	5.582	5.206	4.868	4.564	4.160	3.605	3.161	2.802	2.508
8.	6.596	6.210	5.747	5.335	4.968	4.487	3.837	3.329	2.925	2.598
9.	7.269	6.802	6.247	5.759	5.328	4.772	4.031	3.463	3.019	2.665
10.	7.913	7.360	6.710	6.145	5.650	5.019	4.192	3.571	3.092	2.715
11.	8.529	7.887	7.139	6.495	5.938	5.234	4.327	3.656	3.147	2.752
12.	9.119	8.384	7.536	6.814	6.194	5.421	4.439	3.725	3.190	2.779
13.	9.683	8.853	7.904	7.103	6.424	5.583	4.533	3.780	3.223	2.799
14.	10.223	9.295	8.244	7.367	6.628	5.724	4.611	3.824	3.249	2.814
15.	10.740	9.712	8.559	7.606	6.811	5.847	4.675	3.859	3.268	2.825
16.	11.234	10.106	8.851	7.824	6.974	5.954	4.730	3.887	3.283	2.834
17.	11.707	10.477	9.122	8.022	7.120	6.047	4.775	3.910	3.295	2.840
18.	12.160	10.828	9.372	8.201	7.250	6.128	4.812	3.928	3.304	2.844
19.	12.593	11.158	9.604	8.365	7.366	6.198	4.843	3.942	3.311	2.848
20.	13.008	11.470	9.818	8.514	7.469	6.259	4.870	3.954	3.316	2.850
21.	13.405	11.764	10.017	8.649	7.562	6.312	4.891	3.963	3.320	2.852
22.	13.784	12.042	10.201	8.772	7.645	6.359	4.909	3.970	3.323	2.853
23.	14.148	12.303	10.371	8.883	7.718	6.399	4.925	3.976	3.325	2.854
24.	14.495	12.550	10.529	8.985	7.784	6.434	4.937	3.981	3.327	2.855
25.	14.828	12.783	10.675	9.077	7.843	6.464	4.948	3.985	3.329	2.856
26.	15.147	13.003	10.810	9.161	7.896	6.491	4.956	3.988	3.330	2.856
27.	15.451	13.211	10.935	9.237	7.943	6.514	4.964	3.990	3.331	2.856
28.	15.743	13.406	11.051	9.307	7.984	6.534	4.970	3.992	3.331	2.857
29.	16.022	13.591	11.158	9.370	8.022	6.551	4.975	3.994	3.332	2.857
30.	16.289	13.765	11.258	9.427	8.055	6.566	4.979	3.995	3.332	2.857

Table 3a. Present Value of Energy Savings at a 4.50% Fuel Price Inflation Rate

Number of years	Discount rate									
	4.50%	*6.00%*	*8.00%*	*10.00%*	*12.00%*	*15.00%*	*20.00%*	*25.00%*	*30.00%*	*35.00%*
1.	1.000	.986	.968	.950	.933	.909	.871	.836	.804	.774
2.	2.000	1.958	1.904	1.853	1.804	1.734	1.629	1.535	1.450	1.373
3.	3.000	2.916	2.810	2.710	2.616	2.485	2.290	2.119	1.969	1.837
4.	4.000	3.860	3.686	3.524	3.374	3.167	2.865	2.608	2.387	2.196
5.	5.000	4.792	4.534	4.298	4.081	3.786	3.365	3.016	2.723	2.474
6.	6.000	5.710	5.355	5.033	4.741	4.349	3.802	3.357	2.992	2.689
7.	7.000	6.615	6.149	5.732	5.356	4.861	4.181	3.643	3.209	2.856
8.	8.000	7.507	6.917	6.395	5.931	5.326	4.512	3.881	3.384	2.985
9.	9.000	8.387	7.661	7.025	6.466	5.748	4.800	4.081	3.524	3.084
10.	10.000	9.254	8.380	7.624	6.966	6.132	5.051	4.248	3.636	3.162
11.	11.000	10.109	9.076	8.193	7.433	6.481	5.269	4.387	3.727	3.221
12.	12.000	10.952	9.750	8.733	7.868	6.798	5.460	4.503	3.800	3.268
13.	13.000	11.782	10.401	9.247	8.274	7.086	5.625	4.601	3.858	3.303
14.	14.000	12.602	11.032	9.734	8.653	7.348	5.769	4.682	3.905	3.331
15.	15.000	13.409	11.642	10.197	9.007	7.585	5.895	4.750	3.943	3.353
16.	16.000	14.205	12.232	10.638	9.337	7.801	6.004	4.807	3.973	3.369
17.	17.000	14.990	12.803	11.056	9.645	7.998	6.100	4.855	3.998	3.382
18.	18.000	15.764	13.356	11.453	9.932	8.176	6.183	4.895	4.018	3.392
19.	19.000	16.526	13.891	11.830	10.200	8.338	6.255	4.928	4.033	3.400
20.	20.000	17.278	14.408	12.189	10.450	8.486	6.318	4.956	4.046	3.406
21.	21.000	18.020	14.909	12.529	10.683	8.620	6.373	4.979	4.056	3.410
22.	22.000	18.751	15.393	12.853	10.901	8.741	6.420	4.999	4.064	3.414
23.	23.000	19.471	15.862	13.160	11.104	8.852	6.462	5.015	4.071	3.417
24.	24.000	20.181	16.316	13.452	11.293	8.952	6.498	5.028	4.076	3.419
25.	25.000	20.882	16.754	13.730	11.470	9.044	6.530	5.040	4.081	3.421
26.	26.000	21.572	17.179	13.993	11.635	9.127	6.557	5.049	4.084	3.422
27.	27.000	22.253	17.590	14.243	11.789	9.202	6.581	5.057	4.087	3.423
28.	28.000	22.924	17.987	14.481	11.933	9.271	6.602	5.064	4.089	3.424
29.	29.000	23.585	18.372	14.707	12.067	9.333	6.620	5.069	4.091	3.424
30.	30.000	24.237	18.744	14.922	12.192	9.389	6.636	5.074	4.092	3.425

Table 3b. Present Value of Energy Savings at a 6.00% Fuel Price Inflation Rate

Number of years	Discount rate									
	4.50%	6.00%	8.00%	10.00%	12.00%	15.00%	20.00%	25.00%	30.00%	35.00%
1.	1.014	1.000	.981	.964	.946	.922	.883	.848	.815	.785
2.	2.043	2.000	1.945	1.892	1.842	1.771	1.664	1.567	1.480	1.402
3.	3.087	3.000	2.890	2.787	2.690	2.554	2.353	2.177	2.022	1.886
4.	4.146	4.000	3.818	3.649	3.492	3.276	2.962	2.694	2.464	2.266
5.	5.219	5.000	4.729	4.480	4.252	3.942	3.499	3.133	2.825	2.564
6.	6.309	6.000	5.623	5.281	4.970	4.555	3.975	3.504	3.119	2.799
7.	7.414	7.000	6.500	6.053	5.650	5.120	4.394	3.820	3.358	2.983
8.	8.534	8.000	7.361	6.796	6.294	5.641	4.765	4.087	3.554	3.127
9.	9.671	9.000	8.207	7.513	6.903	6.121	5.092	4.314	3.713	3.241
10.	10.824	10.000	9.036	8.203	7.480	6.564	5.382	4.506	3.843	3.330
11.	11.994	11.000	9.850	8.868	8.026	6.972	5.637	4.669	3.949	3.400
12.	13.181	12.000	10.649	9.510	8.542	7.348	5.863	4.808	4.035	3.454
13.	14.384	13.000	11.434	10.127	9.031	7.695	6.062	4.925	4.106	3.498
14.	15.605	14.000	12.203	10.723	9.494	8.014	6.238	5.024	4.163	3.531
15.	16.843	15.000	12.959	11.297	9.931	8.309	6.394	5.109	4.210	3.558
16.	18.100	16.000	13.700	11.849	10.346	8.580	6.531	5.180	4.248	3.579
17.	19.374	17.000	14.428	12.382	10.738	8.831	6.652	5.241	4.279	3.595
18.	20.666	18.000	15.142	12.895	11.109	9.061	6.760	5.292	4.305	3.608
19.	21.977	19.000	15.843	13.390	11.460	9.274	6.854	5.336	4.325	3.618
20.	23.307	20.000	16.531	13.867	11.793	9.470	6.938	5.373	4.342	3.626
21.	24.656	21.000	17.207	14.326	12.108	9.650	7.012	5.404	4.356	3.632
22.	26.024	22.000	17.870	14.769	12.405	9.817	7.077	5.431	4.367	3.637
23.	27.412	23.000	18.520	15.196	12.687	9.970	7.135	5.453	4.376	3.641
24.	28.820	24.000	19.159	15.607	12.954	10.112	7.186	5.472	4.384	3.644
25.	30.248	25.000	19.785	16.003	13.207	10.242	7.231	5.488	4.390	3.647
26.	31.696	26.000	20.401	16.384	13.445	10.362	7.271	5.502	4.395	3.648
27.	33.166	27.000	21.004	16.752	13.672	10.473	7.306	5.514	4.399	3.650
28.	34.656	28.000	21.597	17.107	13.886	10.575	7.337	5.524	4.402	3.651
29.	36.168	29.000	22.178	17.448	14.088	10.669	7.364	5.532	4.405	3.652
30.	37.702	30.000	22.749	17.777	14.280	10.756	7.388	5.539	4.407	3.653

Table 3c. Present Value of Energy Savings at an 8.00% Fuel Price Inflation Rate

Number of years	Discount rate									
	4.50%	6.00%	8.00%	10.00%	12.00%	15.00%	20.00%	25.00%	30.00%	35.00%
1.	1.033	1.019	1.000	.982	.964	.939	.900	.864	.831	.800
2.	2.102	2.057	2.000	1.946	1.894	1.821	1.710	1.610	1.521	1.440
3.	3.205	3.115	3.000	2.892	2.791	2.649	2.439	2.255	2.094	1.952
4.	4.346	4.192	4.000	3.821	3.655	3.427	3.095	2.813	2.571	2.362
5.	5.525	5.290	5.000	4.734	4.489	4.158	3.686	3.294	2.966	2.689
6.	6.744	6.409	6.000	5.630	5.293	4.844	4.217	3.710	3.295	2.951
7.	8.003	7.549	7.000	6.509	6.068	5.488	4.695	4.070	3.568	3.161
8.	9.305	8.710	8.000	7.372	6.816	6.093	5.126	4.380	3.795	3.329
9.	10.650	9.893	9.000	8.220	7.537	6.661	5.513	4.648	3.984	3.463
10.	12.040	11.099	10.000	9.053	8.232	7.195	5.862	4.880	4.140	3.571
11.	13.477	12.327	11.000	9.870	8.902	7.696	6.176	5.081	4.270	3.656
12.	14.962	13.578	12.000	10.672	9.549	8.167	6.458	5.254	4.378	3.725
13.	16.496	14.854	13.000	11.460	10.172	8.609	6.712	5.403	4.468	3.780
14.	18.082	16.153	14.000	12.233	10.773	9.024	6.941	5.532	4.543	3.824
15.	19.722	17.476	15.000	12.993	11.352	9.414	7.147	5.644	4.605	3.859
16.	21.416	18.825	16.000	13.738	11.911	9.780	7.332	5.740	4.656	3.887
17.	23.166	20.199	17.000	14.470	12.450	10.124	7.499	5.824	4.699	3.910
18.	24.976	21.599	18.000	15.189	12.970	10.447	7.649	5.896	4.735	3.928
19.	26.846	23.025	19.000	15.895	13.471	10.750	7.784	5.958	4.764	3.942
20.	28.778	24.479	20.000	16.588	13.954	11.035	7.906	6.012	4.789	3.954
21.	30.776	25.959	21.000	17.268	14.420	11.302	8.015	6.058	4.809	3.963
22.	32.840	27.468	22.000	17.936	14.869	11.553	8.114	6.098	4.826	3.970
23.	34.973	29.005	23.000	18.591	15.302	11.789	8.202	6.133	4.840	3.976
24.	37.178	30.571	24.000	19.235	15.720	12.011	8.282	6.163	4.852	3.981
25.	39.457	32.167	25.000	19.867	16.123	12.219	8.354	6.189	4.861	3.985
26.	41.812	33.793	26.000	20.488	16.512	12.414	8.419	6.211	4.870	3.988
27.	44.246	35.449	27.000	21.097	16.886	12.598	8.477	6.230	4.876	3.990
28.	46.761	37.137	28.000	21.695	17.247	12.770	8.529	6.247	4.882	3.992
29.	49.361	38.857	29.000	22.283	17.596	12.932	8.576	6.261	4.886	3.994
30.	52.048	40.609	30.000	22.859	17.931	13.084	8.618	6.274	4.890	3.995

Table 3d. Present Value of Energy Savings at a 10.00% Fuel Price Inflation Rate

Number of years	Discount rate									
	4.50%	*6.00%*	*8.00%*	*10.00%*	*12.00%*	*15.00%*	*20.00%*	*25.00%*	*30.00%*	*35.00%*
1.	1.053	1.038	1.019	1.000	.982	.957	.917	.880	.846	.815
2.	2.161	2.115	2.056	2.000	1.947	1.871	1.757	1.654	1.562	1.479
3.	3.327	3.232	3.112	3.000	2.894	2.747	2.527	2.336	2.168	2.020
4.	4.555	4.392	4.189	4.000	3.825	3.584	3.233	2.936	2.681	2.461
5.	5.847	5.595	5.285	5.000	4.738	4.384	3.880	3.463	3.114	2.820
6.	7.207	6.844	6.401	6.000	5.636	5.150	4.474	3.928	3.481	3.112
7.	8.639	8.140	7.538	7.000	6.517	5.883	5.018	4.336	3.792	3.351
8.	10.147	9.485	8.696	8.000	7.383	6.584	5.516	4.696	4.055	3.545
9.	11.733	10.881	9.876	9.000	8.234	7.254	5.973	5.012	4.277	3.703
10.	13.404	12.329	11.077	10.000	9.069	7.895	6.392	5.291	4.465	3.832
11.	15.162	13.832	12.301	11.000	9.889	8.508	6.776	5.536	4.624	3.938
12.	17.012	15.392	13.547	12.000	10.694	9.095	7.128	5.752	4.759	4.023
13.	18.960	17.010	14.817	13.000	11.486	9.656	7.451	5.942	4.873	4.093
14.	21.011	18.690	16.110	14.000	12.263	10.193	7.746	6.109	4.970	4.150
15.	23.169	20.433	17.426	15.000	13.026	10.706	8.018	6.256	5.051	4.196
16.	25.441	22.242	18.768	16.000	13.775	11.197	8.266	6.385	5.120	4.234
17.	27.833	24.119	20.134	17.000	14.511	11.667	8.494	6.499	5.179	4.265
18.	30.351	26.067	21.525	18.000	15.234	12.116	8.703	6.599	5.228	4.290
19.	33.001	28.088	22.942	19.000	15.945	12.546	8.894	6.687	5.270	4.310
20.	35.790	30.186	24.386	20.000	16.642	12.957	9.070	6.765	5.305	4.327
21.	38.727	32.363	25.856	21.000	17.327	13.350	9.231	6.833	5.335	4.340
22.	41.817	34.622	27.353	22.000	18.000	13.726	9.378	6.893	5.361	4.351
23.	45.071	36.966	28.878	23.000	18.660	14.086	9.513	6.946	5.382	4.360
24.	48.496	39.398	30.431	24.000	19.309	14.430	9.637	6.992	5.400	4.368
25.	52.101	41.923	32.013	25.000	19.947	14.759	9.751	7.033	5.416	4.374
26.	55.896	44.543	33.625	26.000	20.573	15.074	9.855	7.069	5.429	4.379
27.	59.890	47.261	35.266	27.000	21.187	15.375	9.950	7.101	5.440	4.383
28.	64.095	50.082	36.938	28.000	21.791	15.663	10.038	7.129	5.449	4.386
29.	68.521	53.010	38.640	29.000	22.384	15.939	10.118	7.153	5.457	4.388
30.	73.180	56.048	40.374	30.000	22.967	16.202	10.191	7.175	5.463	4.391

Table 3e. Present Value of Energy Savings at a 12.00% Fuel Price Inflation Rate

Number of years	Discount rate									
	4.50%	6.00%	8.00%	10.00%	12.00%	15.00%	20.00%	25.00%	30.00%	35.00%
1.	1.072	1.057	1.037	1.018	1.000	.974	.933	.896	.862	.830
2.	2.220	2.173	2.112	2.055	2.000	1.922	1.804	1.699	1.604	1.518
3.	3.452	3.353	3.228	3.110	3.000	2.846	2.617	2.418	2.243	2.089
4.	4.771	4.599	4.384	4.185	4.000	3.746	3.376	3.063	2.794	2.563
5.	6.185	5.916	5.584	5.279	5.000	4.622	4.085	3.640	3.269	2.956
6.	7.701	7.307	6.828	6.394	6.000	5.475	4.746	4.158	3.678	3.282
7.	9.325	8.778	8.118	7.528	7.000	6.306	5.363	4.621	4.030	3.552
8.	11.067	10.331	9.455	8.683	8.000	7.116	5.938	5.037	4.334	3.777
9.	12.933	11.972	10.842	9.859	9.000	7.904	6.476	5.409	4.595	3.963
10.	14.932	13.707	12.281	11.057	10.000	8.672	6.977	5.742	4.820	4.117
11.	17.076	15.539	13.773	12.276	11.000	9.420	7.446	6.041	5.015	4.246
12.	19.373	17.475	15.320	13.517	12.000	10.148	7.883	6.309	5.182	4.352
13.	21.835	19.521	16.925	14.781	13.000	10.857	8.290	6.549	5.326	4.440
14.	24.474	21.683	18.588	16.068	14.000	11.548	8.671	6.764	5.450	4.513
15.	27.303	23.967	20.314	17.378	15.000	12.220	9.026	6.956	5.557	4.574
16.	30.344	26.380	22.103	18.713	16.000	12.875	9.358	7.129	5.649	4.624
17.	33.583	28.930	23.959	20.071	17.000	13.513	9.667	7.283	5.728	4.666
18.	37.065	31.624	25.883	21.454	18.000	14.135	9.956	7.422	5.797	4.701
19.	40.797	34.470	27.879	22.862	19.000	14.740	10.226	7.546	5.856	4.730
20.	44.797	37.478	29.949	24.296	20.000	15.329	10.477	7.657	5.906	4.753
21.	49.083	40.656	32.095	25.756	21.000	15.903	10.712	7.757	5.950	4.773
22.	53.678	44.014	34.321	27.243	22.000	16.462	10.931	7.846	5.988	4.790
23.	58.602	47.562	36.629	28.756	23.000	17.007	11.136	7.926	6.020	4.803
24.	63.880	51.311	39.022	30.297	24.000	17.537	11.327	7.998	6.048	4.815
25.	69.536	55.272	41.505	31.866	25.000	18.054	11.505	8.062	6.072	4.824
26.	75.599	59.457	44.079	33.464	26.000	18.557	11.671	8.120	6.093	4.832
27.	82.096	63.879	46.749	35.090	27.000	19.046	11.827	8.171	6.111	4.838
28.	89.060	68.552	49.517	36.746	28.000	19.523	11.972	8.217	6.126	4.843
29.	96.524	73.488	52.388	38.433	29.000	19.988	12.107	8.259	6.140	4.848
30.	104.523	78.705	55.365	40.150	30.000	20.441	12.233	8.296	6.151	4.852

Table 3f. Present Value of Energy Savings at a 16.00% Fuel Price Inflation Rate

Number of years	Discount rate									
	4.50%	*6.00%*	*8.00%*	*10.00%*	*12.00%*	*15.00%*	*20.00%*	*25.00%*	*30.00%*	*35.00%*
1.	1.110	1.094	1.074	1.055	1.036	1.009	.967	.928	.892	.859
2.	2.342	2.292	2.228	2.167	2.108	2.026	1.901	1.789	1.689	1.598
3.	3.710	3.602	3.467	3.339	3.219	3.052	2.804	2.588	2.399	2.232
4.	5.228	5.037	4.798	4.576	4.370	4.088	3.678	3.330	3.033	2.777
5.	6.914	6.606	6.227	5.880	5.562	5.132	4.522	4.018	3.599	3.246
6.	8.785	8.324	7.762	7.255	6.796	6.185	5.338	4.657	4.103	3.648
7.	10.862	10.203	9.412	8.706	8.075	7.248	6.126	5.250	4.554	3.994
8.	13.167	12.260	11.183	10.235	9.399	8.319	6.889	5.800	4.956	4.291
9.	15.726	14.511	13.085	11.848	10.770	9.401	7.626	6.310	5.314	4.546
10.	18.567	16.975	15.129	13.549	12.191	10.491	8.338	6.784	5.634	4.766
11.	21.720	19.670	17.323	15.342	13.662	11.591	9.027	7.223	5.920	4.954
12.	25.220	22.620	19.681	17.234	15.185	12.700	9.693	7.631	6.175	5.116
13.	29.105	25.849	22.212	19.228	16.763	13.819	10.336	8.010	6.402	5.255
14.	33.419	29.381	24.932	21.332	18.398	14.948	10.959	8.361	6.605	5.375
15.	38.206	33.248	27.853	23.550	20.091	16.087	11.560	8.687	6.786	5.478
16.	43.521	37.479	30.990	25.889	21.844	17.236	12.141	8.990	6.947	5.566
17.	49.420	42.109	34.360	28.356	23.660	18.394	12.703	9.270	7.092	5.642
18.	55.969	47.175	37.979	30.957	25.540	19.563	13.246	9.531	7.220	5.707
19.	63.238	52.720	41.866	33.700	27.488	20.742	13.772	9.773	7.335	5.763
20.	71.307	58.788	46.041	36.593	29.506	21.931	14.279	9.997	7.437	5.811
21.	80.265	65.429	50.526	39.643	31.595	23.130	14.770	10.205	7.529	5.853
22.	90.208	72.696	55.343	42.860	33.759	24.340	15.244	10.398	7.610	5.888
23.	101.245	80.648	60.516	46.252	36.001	25.560	15.703	10.578	7.683	5.919
24.	113.497	89.351	66.073	49.830	38.322	26.791	16.146	10.744	7.748	5.945
25.	127.097	98.874	72.041	53.602	40.726	28.033	16.574	10.899	7.806	5.968
26.	142.194	109.296	78.452	57.581	43.217	29.285	16.989	11.042	7.857	5.987
27.	158.952	120.702	85.337	61.776	45.796	30.549	17.389	11.175	7.904	6.004
28.	177.554	133.183	92.733	66.200	48.467	31.823	17.776	11.298	7.945	6.018
29.	198.204	146.842	100.676	70.865	51.234	33.108	18.150	11.413	7.981	6.030
30.	221.125	161.789	109.207	75.785	54.099	34.405	18.512	11.519	8.014	6.041

Table 3g. Present Value of Energy Savings at a 20.00% Fuel Price Inflation Rate

Number of years	Discount rate									
	4.50%	*6.00%*	*8.00%*	*10.00%*	*12.00%*	*15.00%*	*20.00%*	*25.00%*	*30.00%*	*35.00%*
1.	1.148	1.132	1.111	1.091	1.071	1.043	1.000	.960	.923	.889
2.	2.467	2.414	2.346	2.281	2.219	2.132	2.000	1.882	1.775	1.679
3.	3.981	3.865	3.717	3.579	3.449	3.269	3.000	2.766	2.562	2.381
4.	5.720	5.507	5.242	4.996	4.767	4.454	4.000	3.616	3.288	3.006
5.	7.717	7.366	6.935	6.541	6.179	5.691	5.000	4.431	3.958	3.561
6.	10.010	9.471	8.817	8.226	7.692	6.982	6.000	5.214	4.577	4.054
7.	12.643	11.854	10.908	10.065	9.313	8.329	7.000	5.965	5.148	4.492
8.	15.666	14.552	13.231	12.071	11.049	9.735	8.000	6.687	5.675	4.882
9.	19.138	17.606	15.812	14.259	12.910	11.202	9.000	7.379	6.161	5.228
10.	23.125	21.064	18.680	16.646	14.904	12.732	10.000	8.044	6.610	5.536
11.	27.704	24.978	21.866	19.250	17.040	14.329	11.000	8.682	7.025	5.810
12.	32.961	29.409	25.407	22.091	19.328	15.996	12.000	9.295	7.408	6.053
13.	38.999	34.425	29.341	25.191	21.780	17.735	13.000	9.883	7.761	6.270
14.	45.931	40.104	33.712	28.572	24.407	19.549	14.000	10.448	8.087	6.462
15.	53.893	46.533	38.569	32.260	27.222	21.443	15.000	10.990	8.388	6.633
16.	63.035	53.811	43.966	36.283	30.238	23.418	16.000	11.510	8.666	6.785
17.	73.532	62.050	49.962	40.673	33.469	25.480	17.000	12.010	8.922	6.920
18.	85.588	71.377	56.625	45.461	36.931	27.631	18.000	12.490	9.159	7.040
19.	99.431	81.936	64.027	50.685	40.641	29.876	19.000	12.950	9.378	7.147
20.	115.327	93.890	72.253	56.384	44.615	32.219	20.000	13.392	9.579	7.241
21.	133.581	107.423	81.392	62.600	48.873	34.663	21.000	13.816	9.766	7.326
22.	154.543	122.743	91.546	69.382	53.436	37.213	22.000	14.224	9.937	7.401
23.	178.614	140.086	102.829	76.781	58.324	39.875	23.000	14.615	10.096	7.467
24.	206.255	159.720	115.366	84.852	63.561	42.652	24.000	14.990	10.243	7.526
25.	237.997	181.947	129.296	93.656	69.173	45.550	25.000	15.350	10.378	7.579
26.	274.446	207.110	144.773	103.261	75.185	48.574	26.000	15.696	10.503	7.626
27.	316.302	235.596	161.970	113.740	81.627	51.729	27.000	16.029	10.618	7.667
28.	364.365	267.845	181.078	125.171	88.529	55.022	28.000	16.347	10.724	7.704
29.	419.558	304.353	202.308	137.641	95.924	58.457	29.000	16.654	10.822	7.737
30.	482.938	345.682	225.898	151.244	103.847	62.043	30.000	16.947	10.913	7.766

Table 3h. Present Value of Energy Savings at a 25.00% Fuel Price Inflation Rate

Number of years	Discount rate									
	4.50%	*6.00%*	*8.00%*	*10.00%*	*12.00%*	*15.00%*	*20.00%*	*25.00%*	*30.00%*	*35.00%*
1.	1.196	1.179	1.157	1.136	1.116	1.087	1.042	1.000	.962	.926
2.	2.627	2.570	2.497	2.428	2.362	2.268	2.127	2.000	1.886	1.783
3.	4.339	4.210	4.047	3.895	3.752	3.553	3.257	3.000	2.775	2.577
4.	6.386	6.144	5.842	5.563	5.303	4.949	4.434	4.000	3.630	3.312
5.	8.835	8.424	7.919	7.458	7.035	6.466	5.661	5.000	4.452	3.993
6.	11.764	11.113	10.323	9.611	8.968	8.115	6.938	6.000	5.242	4.623
7.	15.268	14.284	13.105	12.058	11.125	9.908	8.269	7.000	6.002	5.206
8.	19.459	18.024	16.325	14.838	13.532	11.856	9.655	8.000	6.733	5.747
9.	24.473	22.434	20.053	17.998	16.219	13.974	11.099	9.000	7.435	6.247
10.	30.470	27.635	24.366	21.589	19.217	16.276	12.603	10.000	8.111	6.710
11.	37.643	33.767	29.359	25.669	22.564	18.778	14.170	11.000	8.760	7.139
12.	46.224	40.999	35.138	30.306	26.299	21.498	15.802	12.000	9.385	7.536
13.	56.488	49.527	41.826	35.575	30.468	24.455	17.502	13.000	9.986	7.904
14.	68.766	59.584	49.568	41.562	35.120	27.668	19.273	14.000	10.563	8.244
15.	83.452	71.443	58.527	48.366	40.313	31.161	21.118	15.000	11.118	8.559
16.	101.019	85.428	68.897	56.098	46.108	34.957	23.040	16.000	11.652	8.851
17.	122.032	101.920	80.900	64.884	52.576	39.084	25.041	17.000	12.166	9.122
18.	147.167	121.368	94.791	74.868	59.795	43.570	27.126	18.000	12.659	9.372
19.	177.234	144.302	110.870	86.214	67.851	48.445	29.298	19.000	13.134	9.604
20.	213.198	171.347	129.479	99.106	76.843	53.745	31.561	20.000	13.590	9.818
21.	256.218	203.239	151.017	113.757	86.878	59.505	33.917	21.000	14.029	10.017
22.	307.677	240.848	175.945	130.406	98.078	65.767	36.372	22.000	14.451	10.201
23.	369.231	285.199	204.798	149.325	110.579	72.573	38.930	23.000	14.857	10.371
24.	442.860	337.498	238.192	170.824	124.530	79.970	41.593	24.000	15.247	10.529
25.	530.933	399.173	276.843	195.255	140.100	88.011	44.368	25.000	15.622	10.675
26.	636.284	471.902	321.577	223.017	157.478	96.751	47.258	26.000	15.983	10.810
27.	762.301	557.667	373.353	254.564	176.872	106.251	50.269	27.000	16.330	10.935
28.	913.040	658.805	433.279	290.414	198.518	116.578	53.405	28.000	16.663	11.051
29.	1093.349	778.072	502.638	331.152	222.677	127.802	56.672	29.000	16.984	11.158
30.	1309.030	918.717	582.914	377.446	249.639	140.002	60.075	30.000	17.292	11.258

Table 3i. Present Value of Energy Savings at a 30.00% Fuel Price Inflation Rate

Number of years	Discount rate									
	4.50%	*6.00%*	*8.00%*	*10.00%*	*12.00%*	*15.00%*	*20.00%*	*25.00%*	*30.00%*	*35.00%*
1.	1.244	1.226	1.204	1.182	1.161	1.130	1.083	1.040	1.000	.963
2.	2.792	2.731	2.653	2.579	2.508	2.408	2.257	2.122	2.000	1.890
3.	4.717	4.575	4.397	4.229	4.072	3.853	3.528	3.246	3.000	2.783
4.	7.112	6.837	6.496	6.180	5.887	5.486	4.906	4.416	4.000	3.643
5.	10.091	9.612	9.023	8.485	7.994	7.332	6.398	5.633	5.000	4.471
6.	13.798	13.015	12.065	11.210	10.439	9.419	8.014	6.898	6.000	5.268
7.	18.409	17.188	15.726	14.430	13.278	11.778	9.766	8.214	7.000	6.036
8.	24.145	22.306	20.133	18.235	16.572	14.444	11.663	9.583	8.000	6.776
9.	31.281	28.583	25.438	22.733	20.396	17.459	13.718	11.006	9.000	7.488
10.	40.158	36.281	31.824	28.048	24.835	20.866	15.944	12.486	10.000	8.173
11.	51.201	45.721	39.510	34.329	29.987	24.718	18.356	14.026	11.000	8.834
12.	64.939	57.300	48.762	41.753	35.967	29.073	20.969	15.627	12.000	9.469
13.	82.029	71.500	59.898	50.526	42.908	33.996	23.800	17.292	13.000	10.082
14.	103.290	88.915	73.304	60.894	50.965	39.560	26.867	19.024	14.000	10.671
15.	129.739	110.273	89.440	73.148	60.316	45.851	30.189	20.825	15.000	11.239
16.	162.642	136.467	108.862	87.629	71.171	52.962	33.788	22.698	16.000	11.786
17.	203.574	168.591	132.242	104.744	83.769	61.000	37.687	24.645	17.000	12.312
18.	254.493	207.989	160.384	124.970	98.393	70.087	41.911	26.671	18.000	12.819
19.	317.839	256.308	194.258	148.873	115.367	80.359	46.487	28.778	19.000	13.307
20.	396.641	315.566	235.033	177.123	135.069	91.971	51.444	30.969	20.000	13.777
21.	494.673	388.242	284.114	210.509	157.937	105.098	56.815	33.248	21.000	14.230
22.	616.627	477.372	343.192	249.965	184.480	119.937	62.633	35.618	22.000	14.666
23.	768.340	586.682	414.306	296.595	215.290	136.711	68.935	38.083	23.000	15.086
24.	957.074	720.742	499.905	351.703	251.051	155.674	75.763	40.646	24.000	15.490
25.	1191.862	885.156	602.941	416.831	292.559	177.109	83.160	43.312	25.000	15.879
26.	1483.943	1086.795	726.966	493.800	340.738	201.341	91.174	46.084	26.000	16.254
27.	1847.298	1334.088	876.256	584.764	396.660	228.733	99.855	48.968	27.000	16.615
28.	2299.318	1637.372	1055.956	692.266	461.570	259.699	109.259	51.966	28.000	16.963
29.	2861.640	2009.324	1272.262	819.315	536.911	294.703	119.448	55.085	29.000	17.297
30.	3561.179	2465.492	1532.630	969.463	624.361	334.273	130.485	58.328	30.000	17.620

Table 3j. Present Value of Energy Savings at a 35.00% Fuel Price Inflation Rate

Number of years	Discount rate									
	4.50%	*6.00%*	*8.00%*	*10.00%*	*12.00%*	*15.00%*	*20.00%*	*25.00%*	*30.00%*	*35.00%*
1.	1.292	1.274	1.250	1.227	1.205	1.174	1.125	1.080	1.038	1.000
2.	2.961	2.896	2.813	2.733	2.658	2.552	2.391	2.246	2.117	2.000
3.	5.117	4.961	4.766	4.582	4.409	4.170	3.814	3.506	3.237	3.000
4.	7.902	7.592	7.207	6.851	6.520	6.069	5.416	4.867	4.400	4.000
5.	11.500	10.943	10.259	9.635	9.065	8.298	7.218	6.336	5.607	5.000
6.	16.149	15.210	14.073	13.052	12.132	10.915	9.246	7.923	6.862	6.000
7.	22.154	20.645	18.842	17.245	15.828	13.987	11.526	9.637	8.164	7.000
8.	29.912	27.567	24.802	22.392	20.284	17.594	14.092	11.488	9.516	8.000
9.	39.934	36.383	32.253	28.709	25.655	21.828	16.979	13.487	10.291	9.000
10.	52.881	47.610	41.566	36.461	32.129	26.798	20.226	15.645	12.379	10.000
11.	69.607	61.909	53.208	45.974	39.932	32.632	23.879	17.977	13.894	11.000
12.	91.215	80.120	67.760	57.650	49.338	39.481	27.989	20.495	15.467	12.000
13.	119.129	103.314	85.949	71.980	60.675	47.521	32.613	23.215	17.100	13.000
14.	155.191	132.852	108.687	89.566	74.340	56.960	37.814	26.152	18.796	14.000
15.	201.777	170.472	137.109	111.149	90.812	68.040	43.666	29.324	20.558	15.000
16.	261.961	218.385	172.636	137.638	110.666	81.047	50.249	32.750	22.387	16.000
17.	339.711	279.405	217.045	170.147	134.597	96.316	57.655	36.450	24.286	17.000
18.	440.152	357.119	272.556	210.043	163.443	114.240	65.987	40.446	26.259	18.000
19.	569.910	456.095	341.945	259.008	198.213	135.282	75.361	44.762	28.307	19.000
20.	737.539	582.150	428.681	319.101	240.122	159.983	85.906	49.423	30.434	20.000
21.	954.093	742.691	537.101	392.851	290.639	188.980	97.769	54.457	32.643	21.000
22.	1233.853	947.153	672.626	483.362	351.529	223.020	111.115	59.893	34.937	22.000
23.	1595.264	1207.554	842.033	594.444	424.923	262.981	126.130	65.765	37.320	23.000
24.	2062.159	1539.196	1053.791	730.773	513.389	309.890	143.021	72.106	39.793	24.000
25.	2665.326	1961.570	1318.489	898.085	620.023	364.958	162.023	78.954	42.362	25.000
26.	3444.535	2499.500	1649.361	1103.422	748.554	429.603	183.401	86.351	45.030	26.000
27.	4451.170	3184.599	2062.952	1355.427	903.480	505.490	207.452	94.339	47.801	27.000
28.	5751.607	4057.131	2579.939	1664.706	1090.222	594.576	234.508	102.966	50.678	28.000
29.	7431.598	5168.374	3226.174	2044.276	1315.312	699.154	264.946	112.283	53.665	29.000
30.	9601.921	6583.637	4033.968	2510.111	1586.626	821.920	299.190	122.346	56.768	30.000

Table 3k. Present Value of Energy Savings at a 50.00% Fuel Price Inflation Rate

Number of years	Discount rate									
	4.50%	6.00%	8.00%	10.00%	12.00%	15.00%	20.00%	25.00%	30.00%	35.00%
1.	1.435	1.415	1.389	1.364	1.339	1.304	1.250	1.200	1.154	1.111
2.	3.496	3.418	3.318	3.223	3.133	3.006	2.813	2.640	2.485	2.346
3.	6.453	6.251	5.997	5.759	5.535	5.225	4.766	4.368	4.021	3.717
4.	10.699	10.261	9.718	9.217	8.753	8.119	7.207	6.442	5.794	5.242
5.	16.792	15.936	14.886	13.932	13.061	11.895	10.259	8.930	7.839	6.935
6.	25.539	23.966	22.064	20.361	18.832	16.819	14.073	11.916	10.199	8.817
7.	38.094	35.329	32.034	29.129	26.561	23.242	18.842	15.499	12.922	10.908
8.	56.116	51.409	45.880	41.085	36.912	31.621	24.802	19.799	16.064	13.231
9.	81.985	74.163	65.112	57.389	50.775	42.549	32.253	24.959	19.689	15.812
10.	119.117	106.363	91.822	79.621	69.342	56.802	41.566	31.150	23.872	18.680
11.	172.417	151.929	128.919	109.938	94.208	75.395	53.208	38.581	28.698	21.866
12.	248.923	216.409	180.443	151.279	127.510	99.645	67.760	47.497	34.267	25.407
13.	358.742	307.654	252.004	207.654	172.112	131.276	85.949	58.196	40.693	29.341
14.	516.375	436.775	351.395	284.528	231.847	172.534	108.687	71.035	48.107	33.712
15.	742.644	619.493	489.437	389.356	311.848	226.349	137.109	86.442	56.662	38.569
16.	1067.432	878.056	681.163	532.304	418.993	296.542	172.636	104.931	66.533	43.966
17.	1533.634	1243.946	947.448	727.233	562.491	388.098	217.045	127.117	77.923	49.962
18.	2202.824	1761.717	1317.289	993.004	754.675	507.520	272.556	153.740	91.065	56.625
19.	3163.384	2494.410	1830.957	1355.515	1012.065	663.286	341.945	185.688	106.229	64.027
20.	4542.178	3531.241	2544.385	1849.793	1356.784	866.460	428.681	224.026	123.726	72.253
21.	6521.308	4998.455	3535.257	2523.809	1818.461	1131.470	537.101	270.031	143.914	81.392
22.	9362.165	7074.700	4911.468	3442.922	2436.778	1477.135	672.626	325.237	167.209	91.546
23.	13439.949	10012.783	6822.873	4696.257	3264.881	1928.002	842.033	391.484	194.087	102.829
24.	19293.228	14170.447	9477.601	6405.350	4373.948	2516.090	1053.791	470.981	225.101	115.366
25.	27695.065	20053.935	13164.724	8735.932	5859.305	3283.161	1318.489	566.377	260.885	129.296
26.	39755.117	28379.625	18285.728	11913.998	7848.622	4283.688	1649.361	680.853	302.175	144.773
27.	57066.196	40161.262	25398.233	16247.724	10512.887	5588.723	2062.952	818.223	349.818	161.970
28.	81914.636	56833.390	35276.712	22157.351	14081.099	7290.943	2579.939	983.068	404.790	181.078
29.	*********	80426.023	48996.823	30215.933	18859.954	9511.231	3226.174	1180.882	468.219	202.308
30.	*********	*********	68052.531	41204.909	25260.206	12407.257	4033.968	1418.258	541.407	225.898

Index